MSFC-MAN-206

SKYLAB
SATURN IB
FLIGHT MANUAL

NATIONAL AERONAUTICS AND SPACE ADMINISTRATION

SEPTEMBER 30, 1972

LIST OF EFFECTIVE PAGES

TOTAL NUMBER OF PAGES IN THIS PUBLICATION IS 272, CONSISTING OF THE FOLLOWING:

Page No.	Issue
Title	Original
A	Original
i thru iv	Original
1-1 thru 1-13	Original
2-1 thru 2-18	Original
3-1 thru 3-18	Original
4-1 thru 4-64	Original
5-1 thru 5-72	Original
6-1 thru 6-33	Original
7-1 thru 7-24	Original
8-1 thru 8-13	Original
9-1 thru 9-6	Original
A-1 thru A-5	Original

September 30, 1972

TABLE OF CONTENTS

			Page
SECTION	I	General Description	1-1
SECTION	II	Performance	2-1
SECTION	III	Emergency Detection and Procedures	3-1
SECTION	IV	S-IB Stage	4-1
SECTION	V	S-IVB Stage	5-1
SECTION	VI	Instrument Unit	6-1
SECTION	VII	Ground Support Interface	7-1
SECTION	VIII	Mission Control Monitoring	8-1
SECTION	IX	Mission Variables and Constraints	9-1
APPENDIX	A	Abbreviations, Signs, and Symbols	A-1

FOREWORD

This Saturn IB Flight Manual provides launch vehicle systems descriptions and predicted performance data for the Skylab missions. Vehicle SL-2 (SA-206) is the baseline for this manual; but, as a result of the great similarity, the material is representative of SL-3 and SL-4 launch vehicles, also.

The Flight Manual is not a control document but is intended primarily as an aid to astronauts who are training for Skylab missions. In order to provide a comprehensive reference for that purpose, the manual also contains descriptions of the ground support interfaces, prelaunch operations, and emergency procedures. Mission variables and constraints are summarized, and mission control monitoring and data flow during launch preparation and flight are discussed.

This manual was prepared under the direction of the Saturn Program Engineering Office, PM-SAT-E, Marshall Space Flight Center, Alabama 35812.

SATURN IB

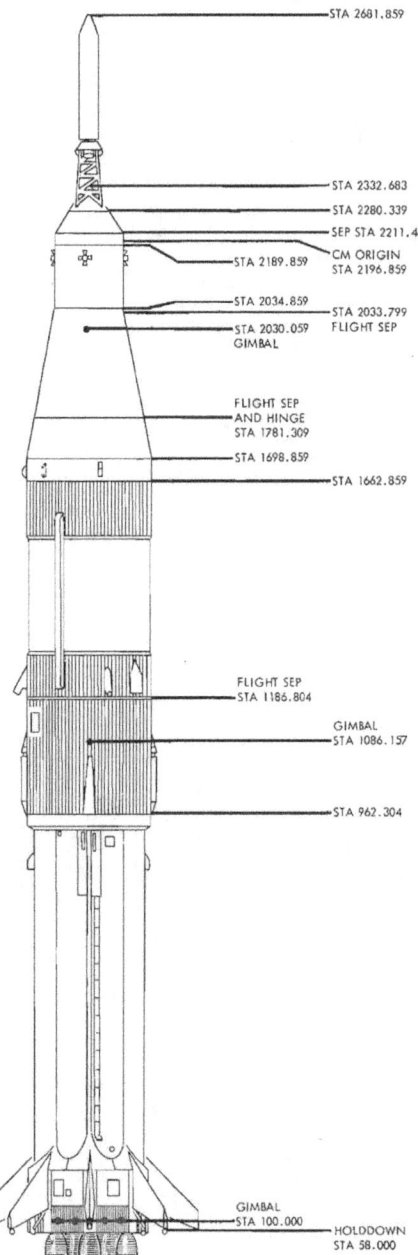
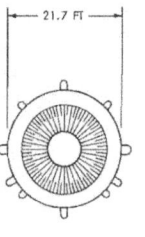

S-IVB STAGE

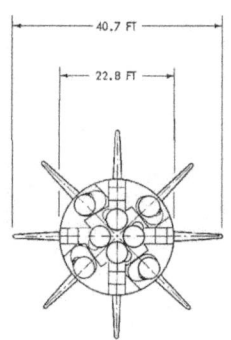

S-IB STAGE

Figure 1-1

SECTION I
GENERAL DESCRIPTION

TABLE OF CONTENTS

Skylab Program	1-1
Vehicle Profile	1-1
Saturn History	1-4
Design Ground Rules	1-6
Saturn IB Production	1-8
Range Safety	1-9

SKYLAB PROGRAM.

The two-stage, liquid-propellant Saturn IB launch vehicle (figure 1-1) is utilized in the Skylab program to transport the three-man crews to the Saturn Workshop (SWS) in earth orbit.

Objectives of the Skylab program are to establish an experimental laboratory in earth orbit and to conduct medical, scientific, and solar astronomy experiments. The laboratory, SWS, includes an orbital workshop (OWS), which is a modified S-IVB stage; a multiple docking adapter (MDA); an airlock module (AM); an apollo telescope mount (ATM); and a payload shroud, which is jettisoned during launch.

The SWS will be launched and inserted into orbit (figure 1-2) from LC-39A at KSC as part of vehicle SL-1, which utilizes a two-stage Saturn V launch vehicle. Approximately 24 hours later, a manned command and service module (CSM) will be boosted into a rendezvous phasing orbit from LC-39B aboard vehicle SL-2, which utilizes two-stage Saturn IB launch vehicle SA-206. After rendezvousing and docking with the SWS the crew will activate and inhabit the SWS for a period of up to 28 days and then return to earth via the CSM. On Skylab missions SL-3 and SL-4, launch vehicles SA-207 and SA-208 will place CSM's into orbit for revisitation of the SWS for up to 56 days on each mission.

After a pitch and roll maneuver, initiated 10 sec after liftoff, the launch vehicle will fly a time-tilt program that provides near zero angle of attack through the high dynamic pressure region of flight until about 2 min 10 sec after liftoff. This attitude will be maintained through S-IB/S-IVB separation at 2 min 22.9 sec after liftoff, until iterative guidance is initiated at 2 min 50 sec after liftoff. During this time period, the expended ullage rockets are jettisoned at 2 min 34.9 sec. The launch escape tower will be jettisoned by astronaut command about 23.5 sec after S-IB/S-IVB separation. In the iterative guidance mode the S-IVB pitch and yaw attitude commands are issued to obtain an optimal path (that which requires the least propellant consumption) to achieve the desired end conditions of flight. Guidance cutoff signal is predicted to occur at 9 min 50.1 sec after liftoff.

The S-IVB/IU/CSM will be inserted into an 81 by 120 NM elliptical orbit inclined 50 deg to the equator, at 9 min 60.1 sec after liftoff. The CSM will separate from the S-IVB at six min after orbit insertion. At 15 min after insertion, S-IVB tank venting will commence to passivate the S-IVB. Pressure sphere safing will be initiated at approximately 1 hr 23 min after insertion.

At approximately 36 min after orbital insertion for SA-206, the S-IVB/IU will begin maneuvers to maintain a sun-vehicle orientation in support of Skylab experiment M415 for the remainder of the mission. The SA-208 S-IVB/IU will perform attitude maneuvers during orbit in support of Skylab experiment S150.

See figure 1-3 for a detailed sequence of events.

VEHICLE PROFILE.

Figure 1-4 shows a cutaway profile of the Saturn IB vehicle and identifies the first powered stage (S-IB), the second powered stage (S-IVB), the instrument unit (IU), and the major features of these stages.

S-IB STAGE.

The S-IB stage is an uprated Saturn I series booster manufactured by Chrysler Corporation Space Division (CCSD). The basic design concept incorporates Jupiter and Redstone components because of their high reliability and qualification status. The S-IB stage is analogous to the R&D S-I stage but has lightened structure, uprated engines, a simplified propulsion system, and reduced instrumentation.

The main stage body is a cluster of nine propellant tanks. The cluster consists of four fuel tanks and four lox tanks arranged alternately around a larger center lox tank. Each tank has anti-slosh baffles to minimize propellant turbulence in flight. Stage electrical and instrumentation equipment is located in the forward and aft skirts of the fuel tanks.

A tail unit assembly supports the aft tank cluster and provides a mounting surface for the engines. Eight fin assemblies support the vehicle on the launcher and improve the aerodynamic characteristics of the vehicle. A stainless steel honeycomb heat shield encloses the aft tail unit to protect against the engine exhausts. A firewall above the engines separates the propellant tanks from the engine compartment. Eight H-1 Rocketdyne engines boost the vehicle during the first phase of powered flight. The four inboard engines are stationary and the four outboard engines gimbal for flight control. Two hydraulic actuators position each outboard engine on signal from the inertial guidance system.

A spider beam unit secures the forward tank cluster and attaches the S-IB stage to the S-IVB aft interstage. Seal plates cover the spider beam to provide an aft closure for the S-IVB stage engine compartment.

S-IVB STAGE.

The S-IVB stage is manufactured by McDonnell Douglas Astronautics Co. (MDAC). It uses a single propellant tank with common-bulkhead design and is powered by one J-2 engine. The aft interstage connects the S-IVB skirt to the S-IB spider beam unit. The aft skirt/aft interstage junction is the separation plane.

A closed loop hydraulic system gimbals the J-2 engine for pitch and yaw control during flight. An auxiliary propulsion system (APS), using two APS modules on the exterior aft skirt, provides vehicle roll control during flight and three-axis control during the coast mode. The exact propellant mass load needed for orbital insertion with minimum residuals at cutoff is determined before launch. A propellant utilization (PU) system helps load this accurate mass.

Section I General Description

INSTRUMENT UNIT.

The instrument unit (IU) is a three-segment ring structure manufactured by International Business Machines (IBM) Corporation. It is sandwiched between the S-IVB stage and the spacecraft LM adapter (SLA). The S-IVB tank dome actually extends into the IU ring. The unit is an unpressurized compartment with honeycomb-panel cold plates mounted around the inside periphery to accommodated the stage equipment. These panels are thermally conditioned by a stage-oriented system, which also conditions

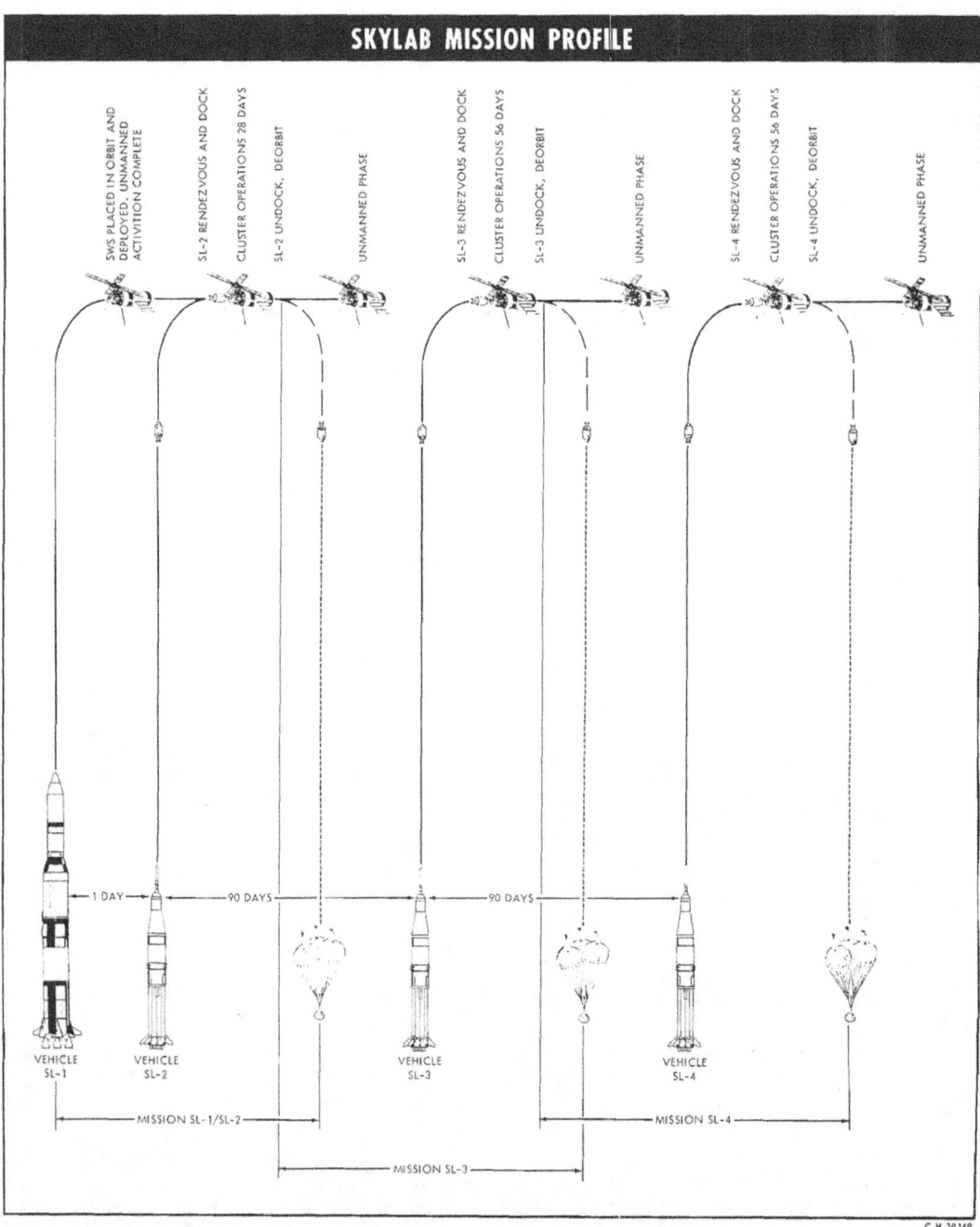

Figure 1-2

Section I General Description

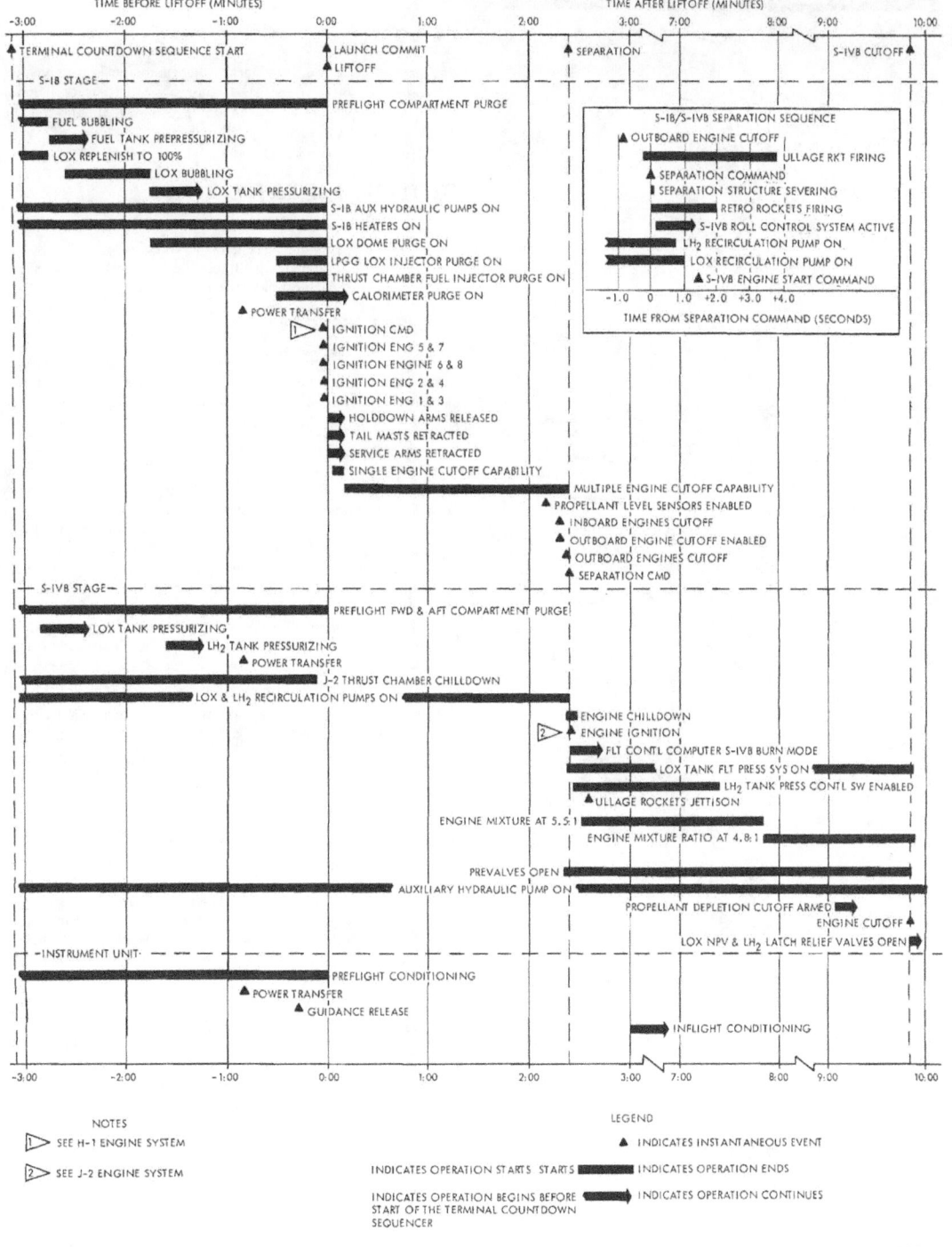

Figure 1-3

Section I General Description

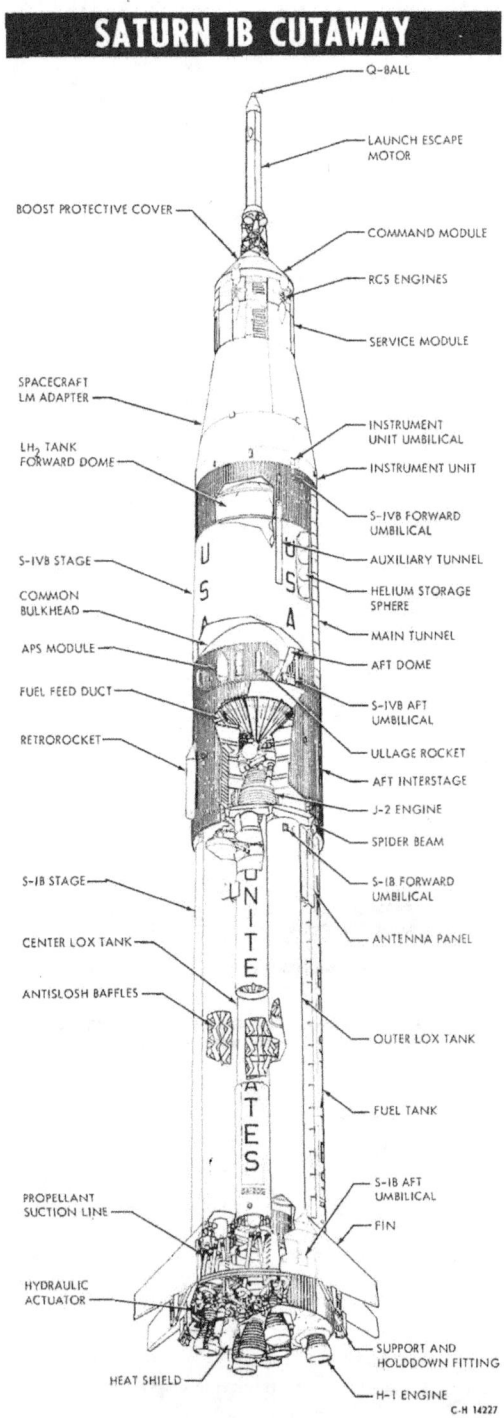

Figure 1-4

equipment in the S-IVB stage forward skirt because of the proximity. The IU houses equipment that guides, controls, and monitors vehicle performance from prelaunch operation to the end of active lifetime in orbit.

PAYLOAD.

The payload is a modular stack attached to the launch vehicle above the IU stage. It consists structurally of a spacecraft lunar module adapter (SLA), a service module (SM), a command module (CM), and a launch escape system (LES).

SATURN HISTORY.

Figure 1-5 compares the three launch vehicles now in the Saturn family. The following narrative traces the historical events leading to development of the Saturn IB launch vehicle.

PROJECT HISTORY.

In April 1957, members of the Army Ballistic Missile Agency (ABMA) initiated studies to establish possible vehicle configurations to launch a payload of 20,000 to 40,000 lbm for orbital missions, and 6,000 to 12,000 lbm for escape missions.

By July 1958, representatives of the Advanced Research Projects Agency (ARPA) showed interest in a clustered booster that would achieve 1.5 Mlbf thrust with available engine hardware. ARPA formally initiated the development program by issuing ARPA Order 14-59 on August 15, 1958. The immediate goal was to demonstrate the feasibility of the engine clustering concept with a full scale captive test firing using Rocketdyne H-1 engines and available propellant containers. In September 1958, ARPA extended the program to include four flight tests of the booster. ARPA Order 47-59, dated December 11, 1958, requested that the Army Ordnance Missile Command (AOMC) design, construct, and modify the ABMA captive test tower and associated facilities for booster development, and determine design criteria for suitable launch facilities.

In November 1958, ARPA approved the development of a clustered booster to serve as the first stage of a multi-stage carrier vehicle capable of performing advanced space missions. The project was unofficially known as Juno V until, on February 3, 1959, an ARPA memorandum made the name "Saturn" official.

Because of the presidential order proposing transfer of the Development Operations Division of ABMA to the National Aeronautics and Space Administration (NASA), an interim agreement was reached between ARPA, NASA, and the Department of Defense on November 25, 1959. The agreement provided for transfer of technical direction of the Saturn program to NASA and for retention of administrative direction by ARPA. ABMA officially transferred responsibility for the Saturn program to NASA on July 1, 1960.

On July 28, 1960, the Douglas Aircraft Company was awarded a contract to develop and fabricate the second stage (S-IV) of the recommended configuration. The original design concept specified four Pratt & Whitney 17.5 klbf liquid hydrogen/liquid oxygen engines (LR-119). The design was later modified to utilize six Pratt & Whitney 15 klbf liquid hydrogen/liquid oxygen engines (RL10A-3).

This S-IV stage flew in the final six Saturn I flight tests. Enlarging and refining of this stage design and replacing the six engines with a single Rocketdyne J-2 engine produced the S-IVB stage used in the Saturn IB launch vehicle. The basic S-IVB stage described here also serves as the third stage of the Saturn V launch vehicle, but for Saturn V it has engine restart capability, larger APS modules

Section I General Description

SATURN CONFIGURATIONS

WEIGHT	SATURN IB	SKYLAB SATURN V	APOLLO SATURN V
DRY	159,000 LB	585,000 LB	553,000 LB
LIFTOFF	1,296,000 LB	6,221,000 LB	6,495,000 LB

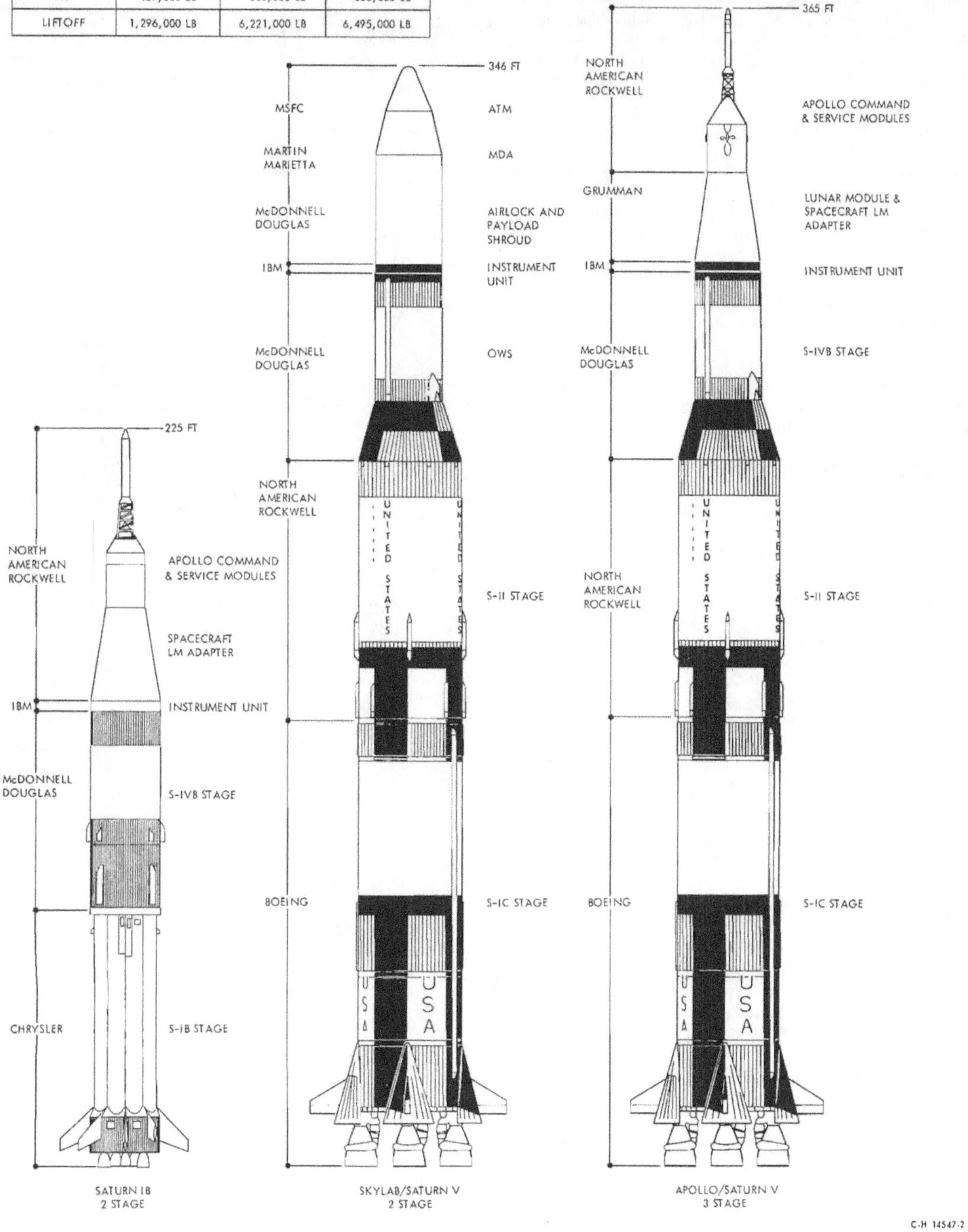

Saturn IB — 2 Stage
- NORTH AMERICAN ROCKWELL — APOLLO COMMAND & SERVICE MODULES
- SPACECRAFT LM ADAPTER
- IBM — INSTRUMENT UNIT
- McDONNELL DOUGLAS — S-IVB STAGE
- CHRYSLER — S-IB STAGE
- 225 FT

Skylab/Saturn V — 2 Stage
- 346 FT
- MSFC — ATM
- MARTIN MARIETTA — MDA
- McDONNELL DOUGLAS — AIRLOCK AND PAYLOAD SHROUD
- IBM — INSTRUMENT UNIT
- McDONNELL DOUGLAS — OWS
- NORTH AMERICAN ROCKWELL — S-II STAGE
- BOEING — S-IC STAGE

Apollo/Saturn V — 3 Stage
- 365 FT
- NORTH AMERICAN ROCKWELL — APOLLO COMMAND & SERVICE MODULES
- GRUMMAN — LUNAR MODULE & SPACECRAFT LM ADAPTER
- IBM — INSTRUMENT UNIT
- McDONNELL DOUGLAS — S-IVB STAGE
- NORTH AMERICAN ROCKWELL — S-II STAGE
- BOEING — S-IC STAGE

Figure 1-5

Section I General Description

with ullage thrust capability, and a flared aft interstage to mate with the S-II stage.

The first eight booster stages (S-I) were initially designed, developed, manufactured, and tested by MSFC personnel. Responsibility for subsequent stages was transferred to Chrysler Corporation Space Division (CCSD). On August 20, 1963 CCSD was awarded a contract for 14 booster stages, which would be built at the Michoud plant, New Orleans, Louisiana. Two S-I stages were produced for the last two R&D vehicles of the Saturn I program, the remaining 12 stages were to be fabricated to the Saturn IB configuration.

The first four Saturn I flights (SA-1 through SA-4) had no instrument unit; instead, these vehicles had instrument cannisters. The next three flights (SA-5 through SA-7) had a pressurized compartment that provided a conditioned environment for navigation, guidance, and control equipment. The unpressurized prototype used on the Saturn IB vehicles first appeared on the SA-8 through SA-10 Saturn I flights. New equipment packaging techniques permitted each assembly to be pressurized individually as necessary and eliminated the need for pressurizing the entire instrument unit.

The instrument unit was designed and developed by MSFC personnel. In the Saturn IB program, responsibility for the IU was gradually transferred to IBM. Work on the first four flight models (S-IU-201 through S-IU-204) was the responsibility of MSFC, with IBM doing the actual assembly and testing at its Huntsville, Alabama facility. The S-IU-205 instrument unit was the first produced entirely under IBM responsibility.

CONFIGURATION HISTORY.

Ten R&D vehicles (SA-1 through SA-10) constitute the Saturn I project. Three R&D vehicles (SA-201 through SA-203) and nine operational vehicles (SA-204 through SA-212) constitute the Saturn IB program.

The Saturn I launch vehicles were of two basic configurations. Block I (SA-1 through SA-4) consisted of a live booster stage (S-I) and dummy upper stages (S-IVD and S-VD); and, Block II (SA-5 through SA-10) consisted of an S-I stage, an S-IV stage, and an instrument unit. Each Saturn IB launch vehicle consists of an S-IB stage, an S-IVB stage, and an instrument unit. In the Apollo program when the launch vehicle combined with the Apollo payload, its configuration was designated Apollo-Saturn (AS); hence. AS-204 and AS-205.

Vehicles SA-206, -207, -208, and -209 have been assigned to the Skylab program. As a result they are designated SL-2, -3, and -4, respectively (SA-209 is the backup vehicle designated SL-R).

DESIGN GROUND RULES.

Saturn IB design has one philosophical ground rule: mission achievement with a safe crew—even under the most adverse flight conditions. This concept reflects in vehicle subsystem design, quality control, structural safety factors, and performance reserves.

STANDARDIZED NOTATION.

When designing a complex structure like the Saturn IB launch vehicle, accurate location of vertical levels is made possible by establishing a datum to which all vertical dimensions are referenced. The measurements in inches from this datum are called stations (figure 1-1). Likewise, structural areas are given a reference designation as shown in figure 1-6. This designator, when used on equipment, indicates its location in the vehicle.

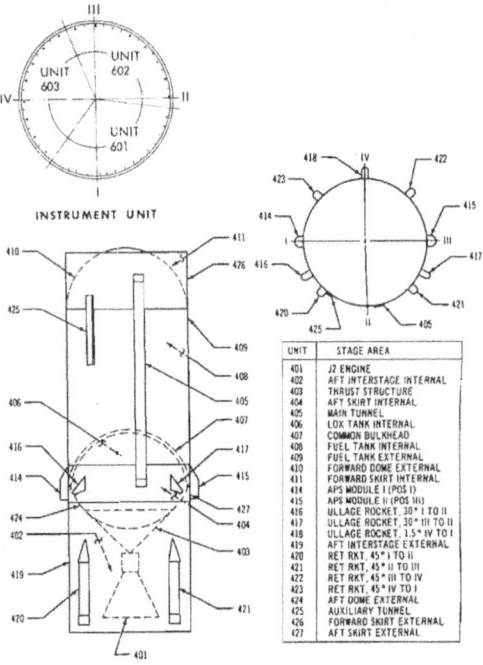

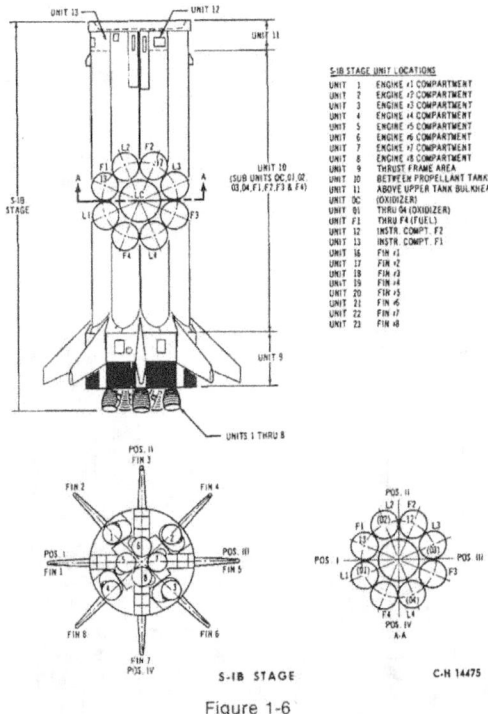

Figure 1-6

OVERALL DESIGN.

Specific ground rules establish the baselines for each particular vehicle mission. However, certain overall design requirements have been upheld throughout the Saturn IB design program. These requirements were levied to assure the success of all launch missions assigned to the vehicle, and are summarized in the following paragraphs.

Reliability.

The launch vehicle must be reliable through the complete flight phase. Reliability performance must be greater than or equal to a 0.88 factor.

Emergency Detection System.

An emergency detection system (EDS) network must be provided that can instantly detect vehicle failure conditions, and either initiate automatic abort or warn the crew that a manual abort is necessary. Manual abort must be initiated on at least two separate and distinct indications. In the event of conflicting information between onboard crew displays and telemetry data relayed to the ground, the onboard information shall always take precedence. Abort decisions shall be made by the flight crew whenever time permits; abort shall be manual rather than automatic when possible.

Triple Redundancy. Triple-redundant circuits with 2-of-3 voting logic must be used for all automatic-abort signals. Redundant circuitry must be used for manual-abort indications from the launch vehicle to the spacecraft.

Single-Point Failure. A single-point electrical failure in the onboard crew safety system cannot result in abort, neither can a single-point failure in the EDS circuitry cause a true or false failure from being detected. All electrical failure possibilities that jeopardize crew safety must be designed out. Keep the circuitry simple, and use a minimum number of sensors. The object is to sense effect rather than cause of failure when possible.

Reliability Goals. As a design goal, the probability of detecting a failure is 0.9973; whereas, the probability of not detecting a false failure is 0.9997.

Separation Systems.

Physical separation of the S-IVB stage from the S-IB stage must be accomplished with retromotors located on the S-IVB aft interstage, using the short coast mode in a single plane. Successful separation shall occur even should the following conditions combine (in a reasonable statistical combination):

a. Engine thrust decay variation

b. Engine dynamic thrust vector deviations

c. Retro and ullage motor static and dynamic misalignment

d. Aerodynamic disturbances

e. Single retro and ullage "motor out" condition

Launch vehicle control will not be lost during stage separation when influenced by maximum limits on angle-of-attack, pitch rate, attitude angle, and dynamic pressure.

Winds.

a. Structural design will assure a free standing capability in 99.9 percent probability non-directional ground winds and associated gusts.

b. The vehicle will be capable of launch in 95 percent probability non-directional ground winds and associated gusts.

c. The vehicle will be capable of flight in 95 percent probability non-directional winds, plus 99 percent associated wind shears and gusts.

Structure.

The general vehicle structure will have a yield safety factor of 1.10 and an ultimate safety factor of 1.40. The erected vehicle will be supported by eight holddown-and-support arms secured on the launcher pad during all prelaunch operations. The vehicle structure must arrest and discharge lightning without damage to the vehicle system. All exterior protrusions will be minimized.

Propulsion.

Propellants and combustibles will not vent into closed compartments. During prelaunch all hydrogen will duct to adequate disposal systems. The launch vehicle will be held down following ignition signal to provide sufficient time (approximately 3.0 sec) for the engines to reach mainstage thrust. The propellant system design will permit the running engines to consume all propellant in the tanks should one booster engine fail during flight. Ullage motors on the S-IVB stage will settle propellants prior to J-2 engine ignition.

Guidance and Control.

Guidance system equipment will accommodate the preset time-tilt program during S-IB powered flight, and will accommodate path-adaptive guidance during S-IVB powered flight. Refer to the coordinate system illustrated in figure 1-7; this system is described more fully in Section VI (Navigation, Guidance, and Control). The launch vehicle flight control computer in the IU will provide control signals to each stage.

Electrical Systems.

Each stage will have an independent electrical power system. Elec-

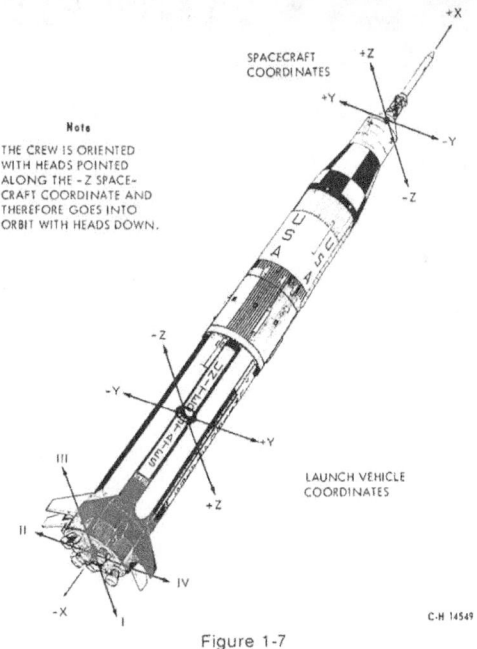

Figure 1-7

Section I General Description

trical distributors will centrally distribute signals and power to minimize cable complexity. Simplicity of design (minimum components), operation, maintenance, and construction shall be prime considerations. Modular components will be used whenever possible. All astrionics systems components will be isolated from their chassis or case, which will bond to a unipotential structure for electro-magnetic interference elimination. Total checkout of all components must be accomplished while the vehicle is on the pad. This checkout procedure shall test pad components also, without interrupting vehicle circuits.

Instrumentation.

Each stage must have integral measuring, signal conditioning, telemetering, and RF subsystems. The entire system must be independent of other electrical systems. No low-level signals (mV range) will be brought through the umbilical to ground electrical support equipment. All measuring signals transmitted to the ground must first be preconditioned to the 0- to 5- Vdc range. Instrumentation will provide sufficient flexibility to accommodate vehicle-to-vehicle changes. Measurements shall be limited to the minimum number required to operate the launch vehicle and to monitor and analyze success or failure.

Environmental Control.

All vehicle interstages and compartments requiring preflight environmental control will be conditioned from a source external to the vehicle. The conditioning medium will be changed to gaseous nitrogen 20 min before loading lox on the S-IVB stage. This procedure reduces the oxygen content to 4 percent (by volume) or less in the conditioned compartments. All inflight conditioning requirements will be stage oriented; however, the IU stage ECS shall also control environment in the S-IVB forward skirt.

Ordnance Systems.

Every vehicle ordnance system will have a dual ignition power source. The vehicle destruct ordnance shall be activated by exploding bridgewire firing units (see Range Safety contained in this section). Dual linear shaped charges will sever each fuel and oxidizer tank required to effectively disperse propellants.

Orbital Coast Period.

The combined IU and S-IVB stages will provide attitude stabilization while attached to the spacecraft. The stage combination shall be capable of sustaining attitude control up to 7 hr and 30 min.

SATURN IB PRODUCTION.

The S-IB, S-IVB, and instrument unit stages of the Saturn IB vehicle are manufactured in Louisiana, California, and Alabama, respectively. The S-IB is static fired at MSFC, and the S-IVB is static fired at the Sacramento Test Center in California. The individual stages are transported to KSC for launch as an integrated vehicle. Figure 1-8 presents the various sites at which production of the Saturn IB launch vehicle is conducted.

The S-IB stage is manufactured and assembled at the Michoud Assembly Facility near New Orleans, Louisiana. The S-IB is transferred by river barge from the Michoud facility to MSFC at Huntsville for static testing and is then returned to Michoud for post static checkout. From there the stage is transferred by river-ocean barge to KSC.

The S-IVB stage is manufactured and assembled at Huntington Beach, California. It is transferred to the Sacramento Test Center for static firing and checkout; then, to KSC by air.

The instrument unit is fabricated and assembled at Huntsville, Alabama. After checkout at the Huntsville facility it is packaged in special environmental containers and transferred by air or water to KSC.

STAGE PRODUCTION AND TRANSPORTATION

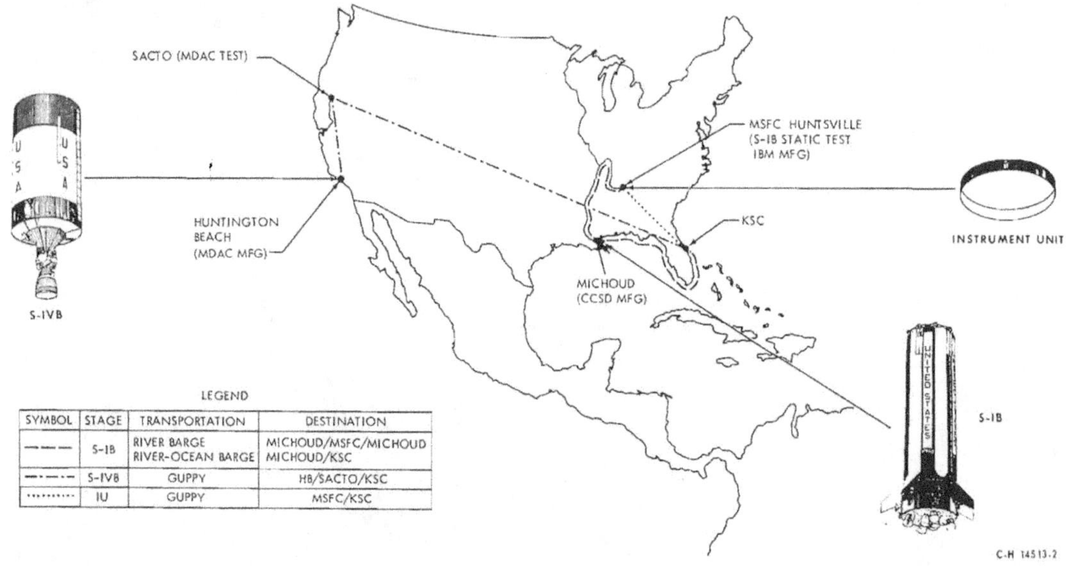

SYMBOL	STAGE	TRANSPORTATION	DESTINATION
— — —	S-IB	RIVER BARGE RIVER-OCEAN BARGE	MICHOUD/MSFC/MICHOUD MICHOUD/KSC
—·—·—	S-IVB	GUPPY	HB/SACTO/KSC
··········	IU	GUPPY	MSFC/KSC

Figure 1-8

RANGE SAFETY.

The Saturn IB launch vehicle range safety system (figure 1-9) enables the range safety officer at Air Force Eastern Test Range (AFETR) to intentionally destroy the vehicle if it should deviate beyond the acceptable limits of the intended trajectory, or if an explosion were imminent during the boost phase of powered flight. The S-IB stage and S-IVB stage each contain an independent range safety system. Each system consists of redundant secure range safety command systems and a propellant dispersion system (PDS). The secure range safety command systems consist of receiving, decoding, and control equipment. Upon receipt of command signals from range safety, these systems provide electrical outputs to shut down the engines and to detonate the PDS ordnance. A built-in time delay of 4 sec between range commanded engine shutdown and propellant dispersion provides crew escape time with the launch escape system (LES) during first stage flight. During second stage flight, after LES jettison, the range has agreed not to send the launch vehicle dispersion command after an abort providing the launch vehicle engines have terminated thrust. This allows adequate time for crew escape during abort using the Service Module propulsion system. The PDS shaped charges rupture the propellant tanks, allowing the propellants to disperse and burn, rather than to explode. The burning propellant results in only a fractional amount of the theoretical yield if the vehicle should explode.

SECURE RANGE SAFETY COMMAND SYSTEMS.

The secure range safety command systems used on the S-IB and S-IVB stages (figure 1-10) consist basically of the same type of equipment. Each stage system uses redundant receiving antennas, power dividers, command receivers, digital decoders, and controllers—all connected in parallel. The command systems receive power from separate batteries in each stage to increase the overall reliability of the systems. The receiving antennas are located on opposite sides of each stage to insure the reception of range safety commands, regardless of vehicle orientation to the transmitting station. All S-IB stage command-system components are installed in instrument compartment no. 1, except the four receiving antennas which are panel-mounted in pairs on opposite sides of the stage at positions I and III. Only two receiving antennas are used on the S-IVB stage. These antennas are mounted on opposite sides of the forward skirt assembly between positions I and II and positions III and IV. The remainder of the S-IVB stage command system is mounted on thermo-conditioning panels in the forward skirt assembly. The panel at position 14 contains the power dividers, command receivers, and the controllers. The panel at position 16 contains the digital decoders. The thermo-conditioning panels provide a heat sink to keep the electrical components cool during flight. The secure range safety command systems in each stage perform the same basic operation. The receiving antennas couple command signals from the ground-based range safety transmitter through the power dividers to the command receivers. The S-IB stage uses a power divider with each pair of antennas. The output from each power divider is directed through a directional coupler to a command receiver. The S-IVB stage uses a hybrid power divider to which both antennas are connected. This power divider splits the inputs into equal strength signals and applies them through a directional power divider to the command receivers.

Command Receiver.

Each command receiver operates at 450 MHz and demodulates and amplifies the range safety command signals for application to the digital decoder. The receiver consists of a preselector assembly, first IF amplifier, IF bandpass filter, second IF amplifier, limiter-discriminator, audio amplifier, isolated outputs, high-level telemetry output assembly, and RFI filter-voltage regulator. The receiver has two isolated audio outputs, but only one is coupled to the decoder. Two telemetry outputs from the receiver (high-level signal strength and low-level signal strength) provide measurement of the RF input signal strength. Only the low-level telemetry signals from command receivers are fed into the stage telemetry systems. During flight an RF carrier is continuously transmitted to the command system. The low-level output from each receiver is then telemetered back to ground receiving stations where the signal strength is monitored on auxiliary pad safety panel. Signal strength measurements received from previous Saturn IB flights indicated adequate signal strength throughout flight and that the command system would have performed satisfactorily if needed. A 28-Vdc output from the command receiver provides a power on signal to the measuring system as soon as power is turned on. The receivers are turned on during countdown and remain on until the safe command after J-2 engine cutoff.

Digital Decoder.

The digital decoder provides timing and gating functions, which determine the validity of received range safety commands. The decoder rejects command messages containing erroneous signal characters or erroneous character sequencing, thereby providing security against enemy intervention and unintentional interrogation of the secure range safety command system during flight. A 21-character high alphabet derived from seven basic audio-frequency tone symbols provides a wide range of message formats that can be transmitted as command messages to the secure range safety command system. Each basic symbol is an audio-frequency tone in the 7.35 kHz to 13.65 kHz range. Simultaneous transmission of two of the seven basic symbols creates a high-alphabet character. Eleven high-alphabet characters contained in two words constitute a command message. An address word uses nine characters to condition the decoder for receipt of a function word. The function word contains two characters that produce a decoder output. Because only nine of the available 21 characters are used in a command message, the function word uses two of the same characters as the address word; however, a given character is used only once in a given word. Identical code plugs installed on the vehicle digital decoders and on the digital encoder at Range Safety Central Control, Cape Kennedy, Florida, determine the command message format. Since only nine characters are used in the address and function words, the code plug channels the 12 unused characters to decoder circuitry, which will cause rejection of a command message if any of the unused characters appear at the decoder input. Additionally, if any of the correct characters forming the address or function word arrive out of sequence at the decoder input, the decoder will reject the command message. Range Safety Central Control transmits a frame consisting of ten identical command messages to insure reception by the vehicle command system. Each character period including dead time is approximately 8.6 msec except the eleventh character period, which is 25.71 msec. The extended period for the eleventh character insures receipt of the function word. Each command message has a total time period of 111.43 msec.

Controller.

The range safety controller assemblies receive and transfer the decoder outputs to the EBW firing units and also provide all power switching for the command system. All measurement signals except the receiver low-level strength signals are routed through the controller assemblies. The controller assemblies consist of relays, a resistor diode module, and four electrical receptacles, three for electrical cable connection and one for installation of a no-safing plug for S-IB stage controllers or a safing plug for S-IVB stage controllers. During prelaunch operations, ground power supplied through the controller assemblies operates the command system

Section I General Description

RANGE SAFETY SYSTEM

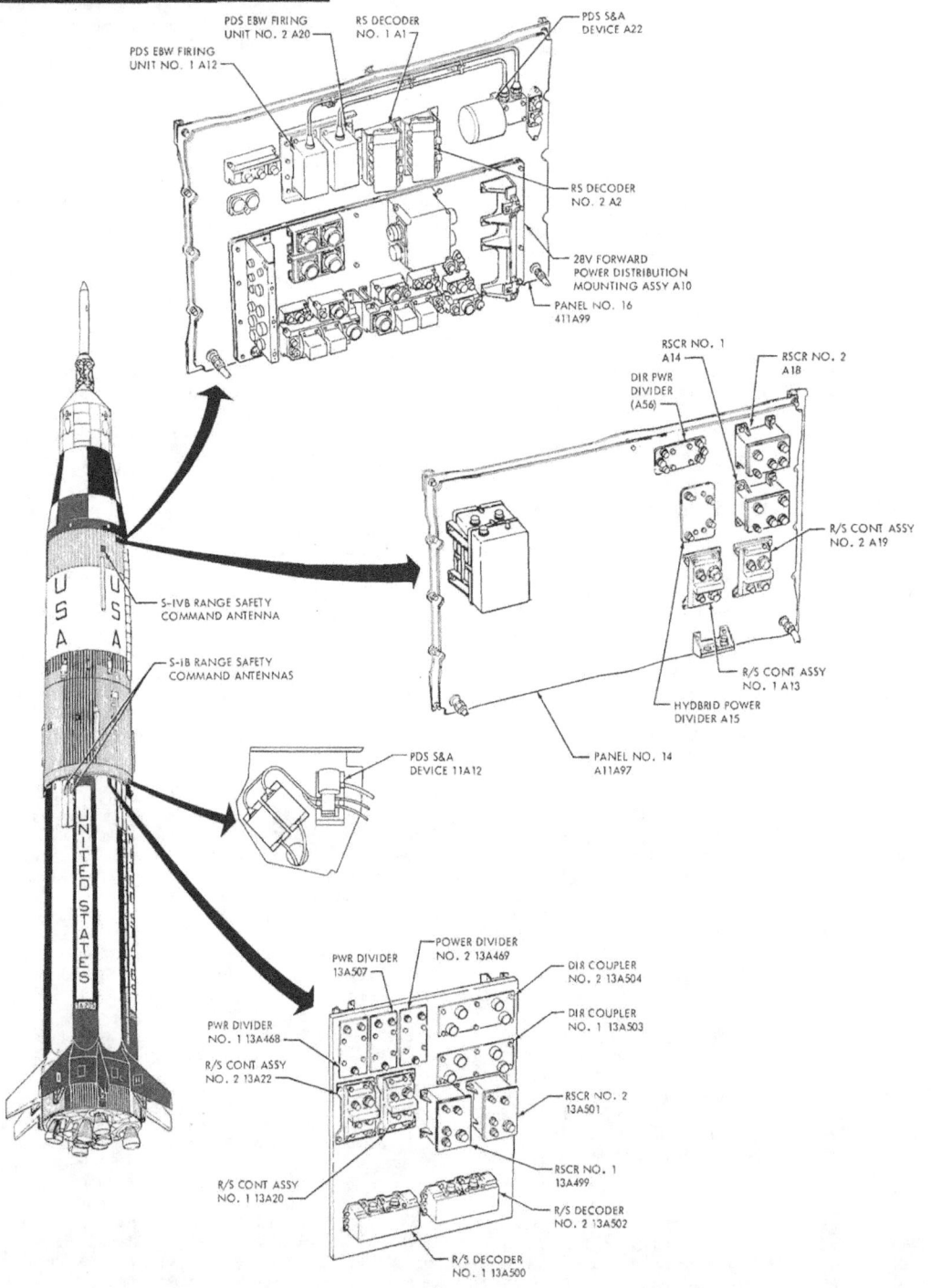

Figure 1-9

Section I General Description

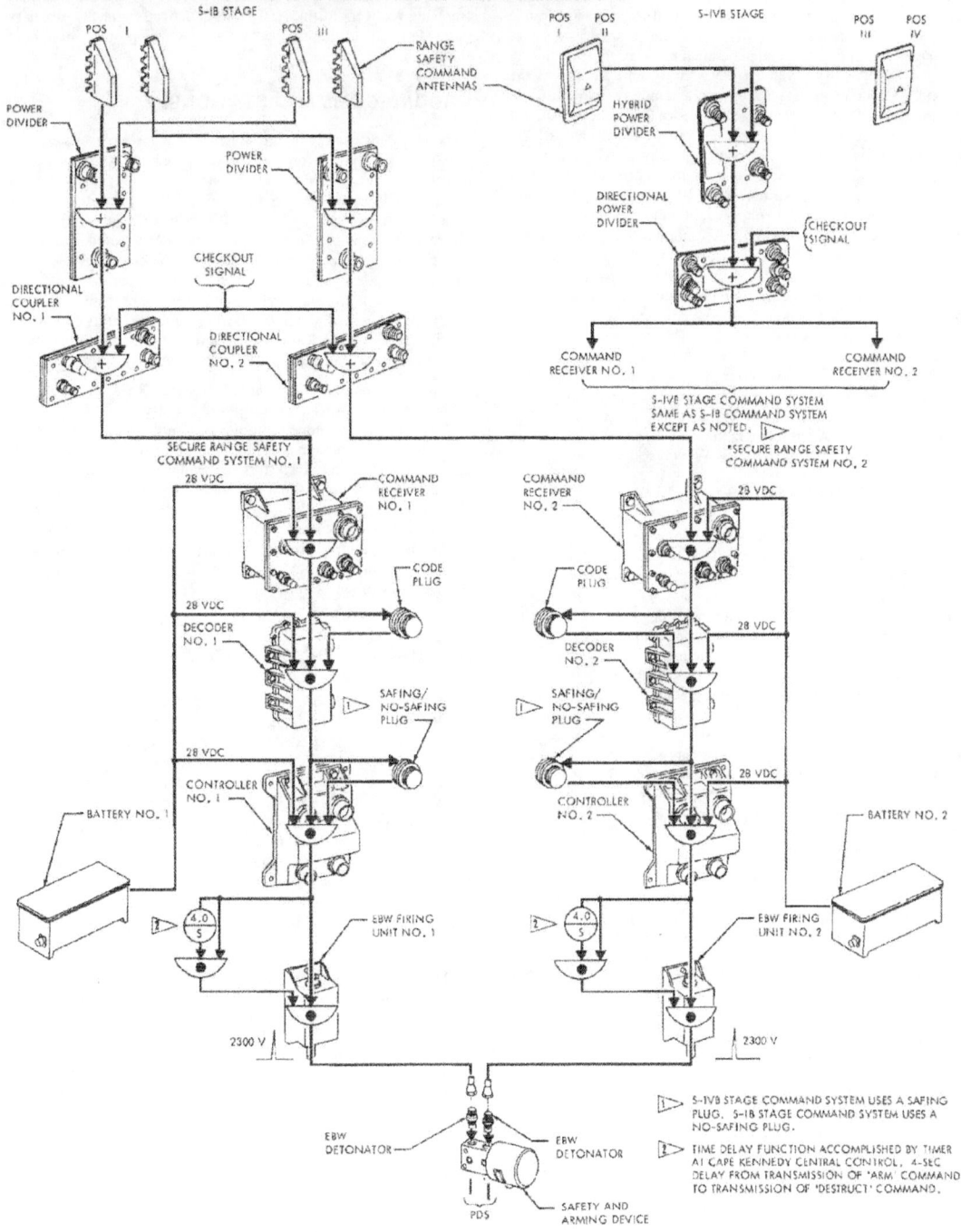

Figure 1-10

Section I General Description

until the power transfer command energizes a magnetic-latching relay that switches the command system to internal power. If the command system must effect propellant dispersion during flight, two command messages must be transmitted to the secure range safety command systems. The first command message will cause the decoder to energize a magnetic-latching relay that applies 28 Vdc to the EBW firing unit charge circuitry and 28 Vdc to the engine cutoff circuitry. The second command message, which arrives approximately 4 sec later, causes the decoder to produce an output that triggers the EBW firing unit discharge circuits. The trigger is coupled to the firing unit through normally closed contacts of a propellant dispersion inhibit relay in the controller. This relay stays energized (with the normally closed contacts open) by ground power until the umbilicals disconnect at liftoff to prevent destruction of the vehicle on the launcher. After normal J-2 engine cutoff, the RSO safes the secure range safety command system since the propellant dispersion capability is no longer necessary. The safe command causes the decoder, through circuitry in the safing plug, to reset the range safety command receiver internal-external power transfer relay to the external position. This removes the 28-Vdc power from the decoder and receiver, rendering the command system inoperable. Once the system has been safed during flight it cannot be reactivated. The S-IB stage secure range safety command system cannot be safed since no-safing plugs are installed on the controllers. After staging, the S-IB propellant dispersion system presents no danger to the S-IVB stage and payload.

PREFLIGHT OPERATIONS.

During preflight operations, test cables connected to the directional power divider in the S-IVB stage and to the power divider between the directional couplers in the S-IB stage permit closed-loop checkout of the secure range safety command systems by using signals transmitted by range safety. This eliminates the necessity for openly transmitting the secure code-of-the-mission command signals, which could seriously compromise mission security. The S-IB and S-IVB propellant dispersion panels control the checkout of the secure range safety command systems. Switches control power application, internal and external, to the command systems. Indicators provide visual monitoring of the command system condition. Receipt of the first range safety system command charges the firing units and issues the engine cutoff signal, which is monitored by the CUTOFF indicators. Receipt of the second command signal, which triggers the EBW firing unit discharge circuits is monitored by the PROPELLANT DISPERSION indicators. The RSCR #1 SIGNAL BLOCKED and RSCR #2 SIGNAL BLOCKED indicators monitor the propellant dispersion inhibit relays to insure that the EBW firing units cannot be triggered during countdown. During checkout operations the inhibit circuit is disabled by relays in the ML S-IB and S-IVB program distributors that are patched into the simulate liftoff circuitry. Placing the simulate liftoff switches on the S-IB and S-IVB propellant dispersion panels in the SIMULATE position energizes the patched relays, which remove power from the inhibit relays in the vehicle controllers. This permits the decoder output to trigger the EBW firing units to check the output pulse. The firing unit output pulse discharges into a pulse sensor that provides a FIRED indication on the propellant dispersion panels. After checkout completion, the pulse sensors are removed and the patched relays are disconnected to energize the propellant dispersion inhibit relays in the controllers. Approximately 60 min before liftoff the secure range safety command systems are switched to internal power. In the S-IB stage, battery D10 supplies power to command system no. 1 and D20 supplies power to command system no. 2. S-IVB stage bus +4D30 supplies power to the S-IVB command system no. 1 and +4D20 supplies power to S-IVB command system no. 2. The internal power command plus safety-and-arming device armed signals (Section IV, Ordnance) provide the S-IB and the S-IVB ORD OK indications on the pad safety supervisor panel. The range safety carrier being transmitted to the command receiver is monitored for signal strength as a result of the low-level outputs from the receivers to the RSCR AGC indicators on the auxiliary pad safety panel. These indicators are also monitored throughout the flight.

GROUND COMMAND STATIONS.

A range safety command transmission may be necessary at any time from liftoff until after S-IVB stage J-2 engine cutoff, just prior to orbital insertion. Tracking stations at Cape Kennedy and at subsequent locations downrange provide vehicle position indications to the range safety officer (RSO). When the vehicle is below the radio horizon of Cape Kennedy, the RSO switches all commands from the Cape Kennedy stations to the downrange stations. The range safety equipment incorporates a priority-interrupt scheme that will interrupt any command being transmitted by the AN/FRW-2A transmitter and will transmit range safety commands selected by either the RSO or the computer. Normal transmission is resumed after transmission of the high-priority command.

To initiate a command from his console, the RSO would actuate a hooded toggle switch. The output of the encoder is then routed in parallel form to a tone remoting transmitter that processes the message for transmission over the 5 mi distance to the transmitter site. A tone-remoting receiver at the transmitter site demodulates the message and feeds it to a modulator that converts the parallel information to the high-alphabet, 11-character format. The 11 dual-tone bursts are then fed to the AN/FRW-2A transmitter system and to the vehicle. For reliability, a completely redundant backup system is provided, with a continuously monitoring error detector that provides automatic transfer to the backup system if the primary chain should fail.

DOWNRANGE REMOTE SITES.

When the vehicle is below the radio horizon of Cape Kennedy, the RF transmission will be made from one of the downrange sites. All downrange sites are connected by cable; when the RSO presses a command switch on his console, the command pulse will be transmitted over the cable via the supervisory control system. (This system will be replaced by a digital remoting system in the near future.) All sites will receive the command, and the transmitter on the air at the moment the command is received will transmit the message to the vehicle. (To keep the onboard receiver captured, one transmitter is always on the air.)

RANGE SAFETY SYSTEM MEASUREMENTS.

Ten measurements (figure 1-11) of the S-IB and S-IVB range safety systems are taken during flight and telemetered to ground receiving stations where they are recorded for postflight evaluation. The four command-receiver, low-level signal-strength measurements are monitored in real time on the auxiliary pad safety panel at KSC. The two S-IVB stage receiver strength signals are flight control measurements that require real-time monitoring by Mission Control Center at Houston, Texas. The S-IB stage telemetry system transmits the S-IB stage range safety measurement data continuously from liftoff until S-IB stage impact. The S-IVB stage telemetry system transmits S-IVB range safety system data from liftoff through orbital coast. All S-IVB signals should be zero-volt signals after the range safety officer safes the S-IVB range safety system. The safe command removes power from the receiver and decoders to deactivate the system after orbit has been attained and the range safety system is no longer necessary.

Section I General Description

RANGE SAFETY SYSTEM MEASURMENTS DATA			
STAGE	MEAS NO.	TITLE	RANGE
S-IB	K65-13	EVENT - R/S 1 CUTOFF/DEST IND	0 TO 5 V
S-IB	K66-13	EVENT - R/S 2 CUTOFF/DEST IND	0 TO 5 V
S-IB	VM505-13 [1]	R/S RCVR 1 LOW-LEVEL SIG STRENGTH	0 TO 5 V
S-IB	VM508-13 [1]	R/S RCVR 2 LOW-LEVEL SIG STRENGTH	0 TO 5 V
S-IVB	K98-411	EVENT - R/S 1 ARM/CUTOFF DEST IND	0 TO 5 V
S-IVB	K99-411	EVENT - R/S 2 ARM/CUTOFF DEST IND	0 TO 5 V
S-IVB	M30-411 [1]	VOLT, FU 1 EBW, RANGE SAFETY	0 TO 5 V
S-IVB	M31-411 [1]	VOLT, FU 2 EBW, RANGE SAFETY	0 TO 5 V
S-IVB	VN57-411 [1][2]	R/S RCVR 1 LOW-LEVEL SIG-STRENGTH	0 TO 5 V
S-IVB	VN62-411 [1][2]	R/S RCVR 2 LOW-LEVEL SIG-STRENGTH	0 TO 5 V

[1] MONITORED IN REAL TIME AT KSC

[2] MONITORED IN REAL TIME AT MCC

Figure 1-11

SECTION II
PERFORMANCE

TABLE OF CONTENTS

Introduction 2-1
Mission Description 2-1
Propulsion Performance 2-7
Separation Dynamics 2-12
Mass Characteristics 2-16
Tracking Coverage 2-18

INTRODUCTION.

This section contains performance data that represent the Saturn IB manned missions for the Skylab program. The launch vehicle data presented is for SA-206 or SA-207, but may be considered typical of the Saturn IB launch vehicles assigned to the Skylab program. The flight sequence illustrated in Figure 2-1 is typical of the launch vehicle mission profile. Figure 2-2 summarizes launch vehicle performance characteristics of SA-206, the first Saturn IB vehicle scheduled for a Skylab flight.

MISSION DESCRIPTION.

The Skylab (SL) program consists of four missions designed to achieve long duration space flights of men and systems and to perform scientific investigations in earth orbit. The first mission, SL-1, will utilize a Saturn V launch vehicle to place the Saturn Workshop (SWS) into a 233.8 NM orbit inclined at 50 degrees. Currently, four Saturn IB launch vehicles are assigned to the Skylab program. SA-206 is assigned for the SL-2 mission, SA-207 the SL-3 mission, and SA-208 the SL-4 mission. SA-209 is assigned to the program as a backup launch vehicle. The SL-2, SL-3, and SL-4 missions are scheduled for launch at 90 day intervals, transporting three man crews to the SWS for operations of 28 to 56 days duration. Mission requirement for the Saturn IB launch vehicle is insertion of the Command and Service Modules (CSM) in an 81 by 120 NM elliptical orbit which is co-planar with the SWS orbit. After separation, the CSM propulsion systems will be utilized to achieve orbit transfer, rendezvous and docking with the SWS.

TRAJECTORY.

The trajectory data presented were extracted from the latest published reports for the SA-206 and SA-207 launch vehicles. It covers all phases of flight for which the launch vehicle has a functional requirement. These tabulations and curves reflect three flight phases. The S-IB stage flight phase starts at guidance reference release, which is assumed to occur 17 sec before first motion. The S-IB stage flight phase ends at physical separation of the S-IB

SL-2 FLIGHT SEQUENCE OF EVENTS

FLIGHT TIME (HR: MIN: SEC)	(SEC)	PROGRAM TIME (SEC)	EVENT
-0:0:17.0	-17.00	-------	GUIDANCE REFERENCE RELEASE (GRR).
-0:0:03.1	-3.10	-------	INITIATE S-IB MAINSTAGE IGNITION SEQUENCE.
0:0:00.0	0.00	-------	FIRST MOTION.
0:0:00.2	0.20	(0.0)1	LIFT-OFF SIGNAL; INITIATE TIME BASE 1.
0:0:10.2	10.20	(10.0)1	INITIATE PITCH AND ROLL MANEUVERS.
0:0:58.9	58.87	-------	MACH ONE.
0:1:13.6	73.61	-------	MAXIMUM DYNAMIC PRESSURE.
0:1:40.0	100.00	(99.8)1	CONTROL GAIN SWITCH POINT.
0:2:00.0	120.00	(119.8)1	CONTROL GAIN SWITCH POINT.
0:2:10.5	130.50	(130.3)1	TILT ARREST.
0:2:11.1	131.15	(130.9)1	ENABLE S-IB PROPELLANT LEVEL SENSORS.
0:2:14.6	134.65	(0.0)2	LEVEL SENSOR ACTIVATION; INITIATE TIME BASE 2.
0:2:17.6	137.65	(3.0)2	INBOARD ENGINE CUTOFF (IECO).
0:2:20.6	140.65	(0.0)3	OUTBOARD ENGINE CUTOFF (OECO). INITIATE TIME BASE 3.
0:2:21.9	141.95	(1.3)3	SEPARATION SIGNAL.
0:2:22.0	142.03	-------	S-IB/S-IVB PHYSICAL SEPARATION;
		(1.4)3	CONTROL GAIN SWITCH POINT.
0:2:23.3	143.35	(2.7)3	J-2 ENGINE START COMMAND.
0:2:25.7	145.70	-------	ULLAGE BURN OUT.
0:2:26.7	146.75	-------	90% J-2 THRUST LEVEL.
0:2:29.3	149.35	(8.7)3	COMMAND 5.5:1 EMR.
0:2:33.9	153.95	(13.3)3	JETTISON ULLAGE ROCKET MOTORS.
0:2:45.2	165.24	-------	DYNAMIC PRESSURE = 1 PSF.
0:2:45.6	165.65	-------	LES JETTISON.
0:2:50.6	170.65	(30.0)3	COMMAND ACTIVE GUIDANCE INITIATION.
0:3:02.6	182.65	(42.0)3	CONTROL GAIN SWITCH POINT.
0:5:46.7	346.75	(206.1)3	CONTROL GAIN SWITCH POINT.
0:7:48.7	468.75	(328.1)3	COMMAND EMR SHIFT TO 4.8:1.
0:9:41.9	581.93	-------	GUIDANCE CUTOFF SIGNAL (GCS).
0:9:42.1	582.13	(0.0)4	INITIATE TIME BASE 4.
0:9:51.9	591.93	-------	ORBIT INSERTION.

Figure 2-1

SL-2 PERFORMANCE CHARACTERISTICS

S-IB STAGE

AVERAGE LONGITUDINAL SEA LEVEL THRUST (LBF)

	H-1 ENGINE	TURBINE	TOTAL
ENGINE #1	207575	663	208238
ENGINE #2	206923	669	207592
ENGINE #3	207031	658	207689
ENGINE #4	207234	706	207940
ENGINE #5	208745	665	209410
ENGINE #6	207955	665	208620
ENGINE #7	207968	667	208635
ENGINE #8	207680	656	208336
TOTAL AVERAGE LONGITUDINAL SEA LEVEL THRUST			1666460

FLIGHT TIME INTERVAL: 0.0 - 137.648 SEC (IECO)

$\dot{W} = [WT. @ (t=0) - WT @ (t = 137.648) - W_{AUX}] / 137.648$

$= 6346.72$ (LBM/SEC)

W_{AUX}: FROST 1100 LBM
SEAL PURGE 6 LBM
FUEL ADDITIVE 27 LBM
TOTAL 1133 LBM

$ISP = F/\dot{W}$

$= 262.57$ (SEC)

S-IVB STAGE

HIGH THRUST LEVEL FLIGHT TIME INTERVAL: 151.00 - 469.50 SEC
LOW THRUST LEVEL FLIGHT TIME INTERVAL: 469.50 - 581.93 SEC

AVERAGE VALUES

	HIGH THRUST LEVEL	LOW THRUST LEVEL
VACUUM THRUST (LBF)	229,714.	198,047.
FLOWRATE (LBM/SEC)	543.87	465.69
SPECIFIC IMPULSE (SEC)	422.37	425.28

Figure 2-2

Section II Performance

SL-2 PREOPERATIONAL FLIGHT TRAJECTORY

EVENT	EARTH FIXED				SPACE FIXED		
	FLIGHT TIME (HR:MIN:SEC)	DECLINATION (DEG)	LONGITUDE (DEG) W.	ALTITUDE (FT)	VELOCITY (FT/SEC)	FLIGHT PATH ANGLE (DEG)	AZIMUTH (DEG)
GUIDANCE REF. RELEASE	-00:00:05.0	28.466	80.621	295.	1,340.44	90.000	90.000
FIRST MOTION	00:00:00.0	28.466	80.621	295.	1,340.44	90.000	90.000
MACH ONE	00:00:58.9	28.478	80.607	24,282.	1,942.46	60.479	79.091
MAX. DYN. PRESSURE	00:01:13.6	28.497	80.586	40,760.	2,416.83	57.840	72.693
TILT ARREST	00:02:10.5	28.767	80.272	158,431.	6,547.02	63.884	54.910
INBOARD ENGINE CUTOFF	00:02:17.6	28.839	80.189	179,884.	7,394.01	65.151	53.816
OUTBOARD ENGINE CUTOFF	00:02:20.6	28.872	80.150	189,284.	7,586.40	65.659	53.603
S-IB/S-IVB PHYSICAL SEP.	00:02:22.0	28.887	80.132	193,605.	7,591.20	65.923	53.593
J-2 ENG. START COMMAND	00:02:23.4	28.902	80.114	196,616.	7,575.35	66.186	53.603
ULLAGE CASE JETTISON	00:02:34.0	29.021	79.974	228,898.	7,621.65	68.197	53.519
LES JETTISON	00:02:45.6	29.157	79.813	260,970.	7,773.19	70.240	53.348
IGM INITIATION	00:02:51.0	29.221	79.738	274,862.	7,852.68	71.134	53.269
EMR SHIFT, 5.5:1 TO 4.8:1	00:07:49.5	34.950	72.168	540,858.	17,955.50	90.671	52.953
GUIDANCE C/O SIGNAL	00:09:41.9	38.727	66.106	519,714.	25,681.12	90.009	38.727
ORBIT INSERTION	00:09:51.9	39.115	65.418	520,175.	25,705.77	89.998	39.303

Figure 2-3

stage from the S-IVB stage, assumed to occur approximately 1.4 sec after outboard engine cutoff signal. The S-IVB stage powered flight phase extends from S-IB/S-IVB separation to orbit insertion. Orbit insertion is defined as 10 sec after the guidance cutoff signal. Guidance cutoff occurs when the space-fixed velocity equals a prespecified value that, accounting for the additional velocity imparted by the J-2 engine thrust decay, provides a velocity of 25,705.8 ft/sec at orbit insertion. The S-IVB/IU orbital flight phase is defined as from orbit insertion to the predicted loss of attitude control capability at approximately 7½ hours of flight. See figures 2-3 through 2-15 for detailed trajectory information.

Trajectory Dispersions.

Since performance predictions and associated calculations are subject to certain tolerances, a dispersion analysis is conducted to establish realistic deviation limits. The error sources considered for the analysis are those associated with predictions of vehicle characteristics, vehicle systems performance, and flight environment. Figure 2-16 summarizes the trajectory dispersion envelopes at S-IB/S-IVB stage separation and orbit insertion. These values reflect the combined S-IB and S-IVB stage three-sigma deviations using the root-sum-square (RSS) technique as follows:

$$+RSS = \sqrt{\Sigma (+\Delta P)^2}$$

$$-RSS = \sqrt{\Sigma (-\Delta P)^2}$$

where
ΔP = Perturbed Parameter − Nominal Parameter

The RSS technique is also utilized to determine the Flight Performance Reserve (FPR) propellant required to ensure a three-sigma probability of achieving the desired S-IVB end conditions of flight.

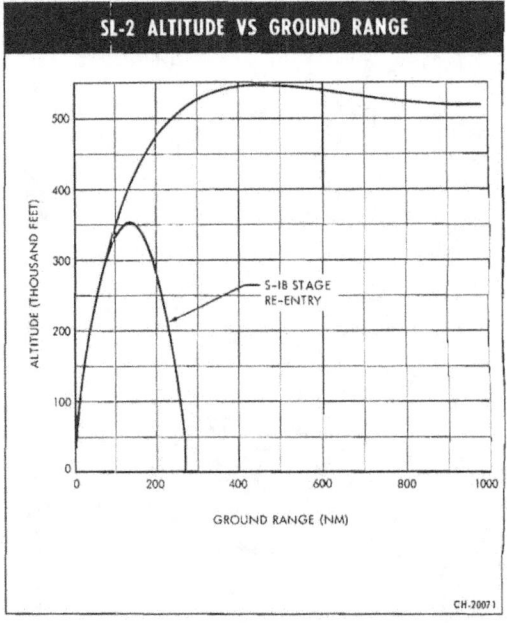

Figure 2-4

Section II Performance

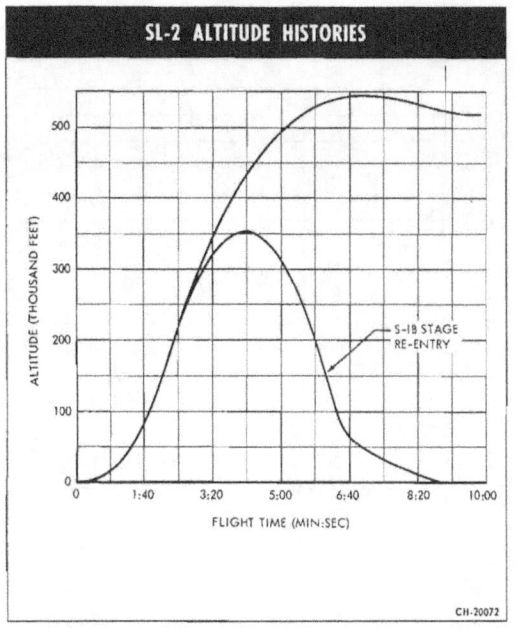

Figure 2-5

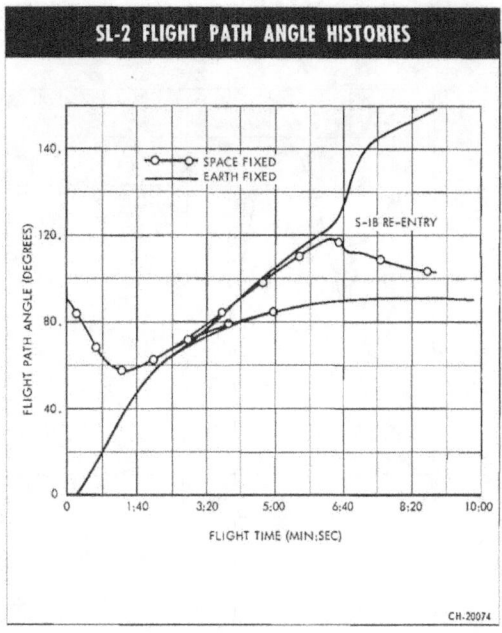

Figure 2-7

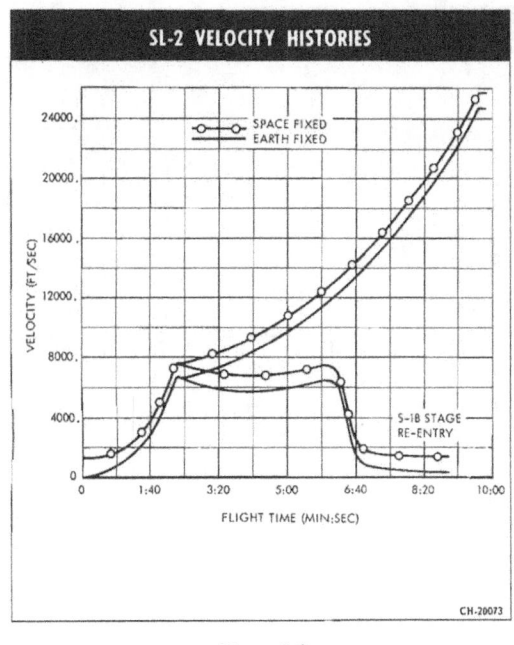

Figure 2-6

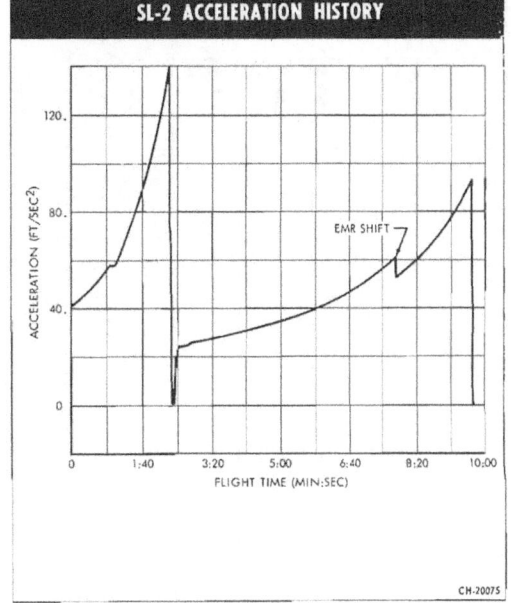

Figure 2-8

2-3

Section II Performance

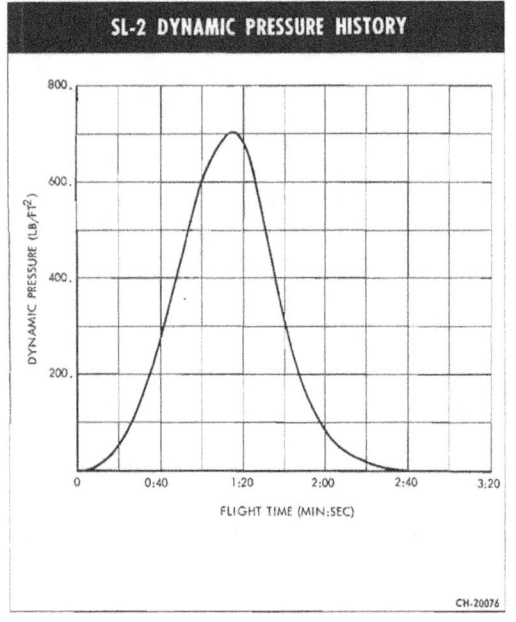

Figure 2-9

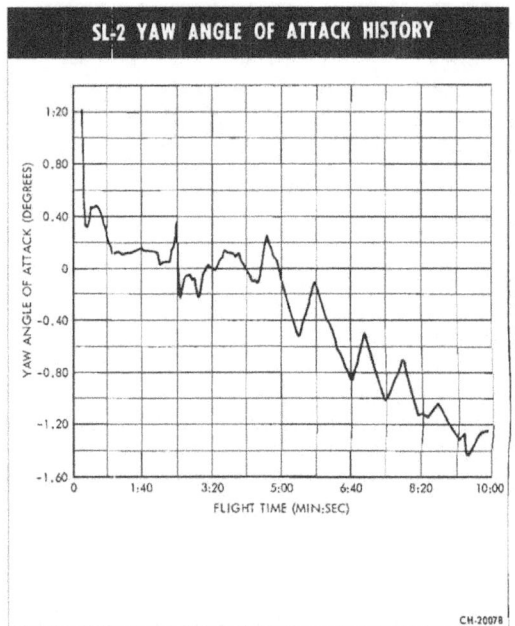

Figure 2-11

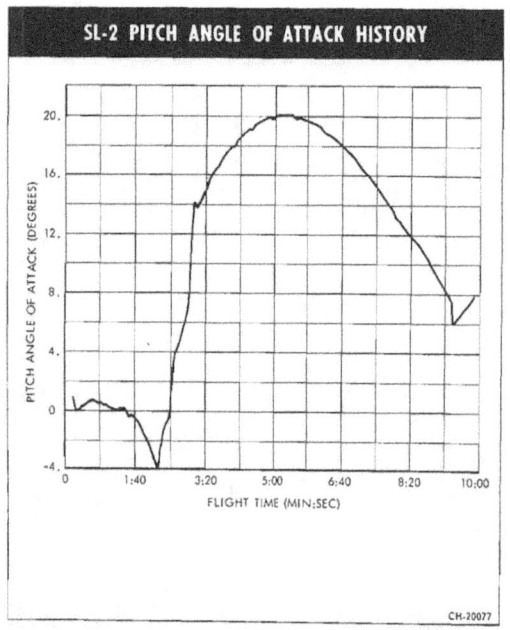

Figure 2-10

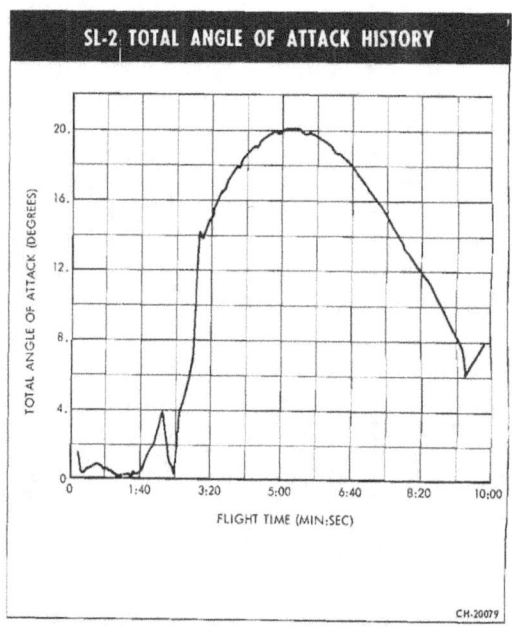

Figure 2-12

SL-2 S-IB STAGE END CONDITIONS OF FLIGHT

FLIGHT TIME (t): OECO + 1.329 SECONDS	142.027	(SEC)
RADIUS (R):	21,102,868.	(FT)
ALTITUDE:	193,605.	(FT)
SPACE FIXED VELOCITY (V):	7,591.19	(FT/S)
SPACE FIXED PATH ANGLE (θ):	65.923	(DEG)
SPACE FIXED FLIGHT AZIMUTH (AZI):	53.593	(DEG)
EARTH FIXED FLIGHT AZIMUTH (AZE):	45.804	(DEG)
GEOCENTRIC DECLINATION (δ):	28.887	(DEG)
GEODETIC LATITUDE (ϕ):	29.050	(DEG)
LONGITUDE (λ): (WEST)	80.132	(DEG)

SPACE FIXED POSITION AND VELOCITY COMPONENTS

X_s	=	21,099,643.	(FT)
Y_s	=	178,280.	(FT)
Z_s	=	322,913.	(FT)
$\dot{X}_s$	=	2,983.85	(FT/S)
$\dot{Y}_s$	=	901.80	(FT/S)
$\dot{Z}_s$	=	6,921.67	(FT/S)

VEHICLE ATTITUDE AND ATTITUDE RATE

PITCH ATTITUDE ANGLE (ϕ_p):	-63.251	(DEG)
YAW ATTITUDE ANGLE (ϕ_y):	-0.073	(DEG)
ROLL ATTITUDE ANGLE (ϕ_r):	0.007	(DEG)
PITCH RATE ($\dot{\phi}_p$):	-0.026	(DEG/S)
YAW RATE ($\dot{\phi}_y$):	0.009	(DEG/S)
ROLL RATE ($\dot{\phi}_r$):	0.007	(DEG/S)

Figure 2-13

SL-2 S-IVB STAGE END CONDITIONS OF FLIGHT (AT GUIDANCE CUTOFF SIGNAL)

FLIGHT TIME (t): GCS	581.926	(SEC)
RADIUS (R):	21,417,871.	(FT)
ALTITUDE:	519,714.	(FT)
SPACE FIXED VELOCITY (V):	25,681.12	(FT/S)
SPACE FIXED FLIGHT PATH ANGLE (θ):	90.009	(DEG)
SPACE FIXED FLIGHT AZIMUTH (AZI):	55.485	(DEG)
EARTH FIXED FLIGHT AZIMUTH (AZE):	53.882	(DEG)
GEOCENTRIC DECLINATION (δ):	38.727	(DEG)
GEODETIC LATITUDE (ϕ):	38.915	(DEG)
LONGITUDE (λ): (WEST)	66.106	(DEG)
INCLINATION (i):	49.999	(DEG)
DESCENDING NODE ARGUMENT (θ):	154.672	(DEG)
INERTIAL RANGE ANGLE:	17.416	(DEG)

SPACE FIXED POSITION AND VELOCITY COMPONENTS

X_s	=	20,447,603.	(FT)
Y_s	=	456,794.	(FT)
Z_s	=	6,357,044.	(FT)
$\dot{X}_s$	=	-7,628.17	(FT/S)
$\dot{Y}_s$	=	2.79	(FT/S)
$\dot{Z}_s$	=	24,522.06	(FT/S)

VEHICLE ATTITUDE ANGLES

PITCH ATTITUDE ANGLE (ϕ_p):	-99.994	(DEG)
YAW ATTITUDE ANGLE (ϕ_y):	-2.899	(DEG)
ROLL ATTITUDE ANGLE (ϕ_r):	0.258	(DEG)

OSCULATING CONIC PARAMETERS

*PERIGEE ALTITUDE	80.97	(NM)
*APOGEE ALTITUDE	105.69	(NM)
ECCENTRICITY	0.0035	
SEMI-MAJOR AXIS	3,537.26	(NM)
TRUE ANOMALY	357.33	(DEG)
PERIOD	87.95	(MIN)

*REFERENCED TO EQUATORIAL RADIUS (3,443.93 NM)

Figure 2-14

SL-2 S-IVB STAGE END CONDITIONS OF FLIGHT (AT ORBIT INSERTION)

FLIGHT TIME (t): ORBIT INSERTION	591.926	(SEC)
RADIUS (R):	21,417,867.	(FT)
ALTITUDE:	520,176.	(FT)
SPACE FIXED VELOCITY (V):	25,705.56	(FT/S)
SPACE FIXED FLIGHT PATH ANGLE (θ):	89.998	(DEG)
SPACE FIXED FLIGHT AZIMUTH (AZI):	55.941	(DEG)
EARTH FIXED FLIGHT AZIMUTH (AZE):	54.367	(DEG)
GEOCENTRIC DECLINATION (δ):	39.115	(DEG)
GEODETIC LATITUDE (ϕ):	39.303	(DEG)
LONGITUDE (λ): (WEST)	65.418	(DEG)
INCLINATION (i):	50.000	(DEG)
DESCENDING NODE ARGUMENT (θ):	154.675	(DEG)
INERTIAL RANGE ANGLE:	18.103	(DEG)

SPACE FIXED POSITION AND VELOCITY COMPONENTS

X_s	=	20,369,815.	(FT)
Y_s	=	456,777.	(FT)
Z_s	=	6,602,044.	(FT)
$\dot{X}_s$	=	-7,924.82	(FT/S)
$\dot{Y}_s$	=	-4.97	(FT/S)
$\dot{Z}_s$	=	24,453.69	(FT/S)

VEHICLE ATTITUDE ANGLES

PITCH ATTITUDE ANGLE (ϕ_p):	-99.996	(DEG)
YAW ATTITUDE ANGLE (ϕ_y):	-2.906	(DEG)
ROLL ATTITUDE ANGLE (ϕ_r):	.102	(DEG)

OSCULATING CONIC PARAMETERS

*PERIGEE ALTITUDE	80.99	(NM)
*APOGEE ALTITUDE	119.37	(NM)
ECCENTRICITY	0.0054	
SEMI-MAJOR AXIS	3,544.11	(NM)
TRUE ANOMALY	0.30	(DEG)
PERIOD	88.20	(MIN)

*REFERENCED TO EQUATORIAL RADIUS (3,443.93 NM)

Figure 2-15

Engine-Out Capability.

At any time during the S-IB stage boost powered flight after 3.0 sec, there exists the possibility of having one of the eight H-1 engines shut down prematurely. The mission profile must then be completed with the remaining seven engines. The S-IB stage steering commands are the body attitude Euler angles preprogrammed in the Launch Vehicle Digital Computer (LVDC) as functions of time only (i.e., the steering program is open-loop). The pitch attitude steering program (CHI) is designed to enforce a near-zero angle of attack time history throughout the high dynamic pressure regime of normal vehicle flight. At a preprogrammed time of approximately 10 seconds prior to outboard engine cutoff (OECO), the pitch steering command is arrested and remains constant for the remainder of the S-IB stage flight and for the first 29 sec of S-IVB stage flight.

Chi-Freeze Steering.

Subsequent to an engine failure, the vehicle guidance is switched to the chi-freeze steering mode. In the chi-freeze steering mode, the value of the commanded pitch attitude is frozen, upon engine failure, for an incremental duration and then the nominal program (displaced in time) is resumed until chi-arrest. The chi-freeze interval is a function of engine failure time as depicted in figure 2-17. Because extended periods of vertical or near-vertical flight are objectionable near the launch complex, the chi-freeze mode is inhibited during the first 30 sec of flight. For an engine failure during the inhibited period, the pitch attitude follows the nominal program until 30 sec of flight at which time the proper (i.e., as a function of engine out time) chi-freeze interval is initiated. For engine failures prior to 110 sec, chi-arrest is initiated as the nominal time plus the chi-freeze interval minus 20 sec. The deviation in the pitch attitude command history due to utilization of the chi-freeze steering mode for an H-1 engine failure at liftoff is illustrated in figure 2-18.

Section II Performance

THREE-SIGMA TRAJECTORY DISPERSION ENVELOPES (TYPICAL)

TRAJECTORY PARAMETER		THREE-SIGMA DISPERSION ENVELOPES			
		S-IB/IVB SEPARATION		ORBIT INSERTION	
		+RSS	-RSS	+RSS	-RSS
FLIGHT TIME (SEC)		4.26	2.94	11.75	10.67
ALTITUDE (FT)		8,494.	7,894.	1,722.	1,749.
SPACE FIXED VELOCITY (FT/SEC)		133.60	144.16	5.22	5.45
SPACE-FIXED FLIGHT PATH ANGLE (DEG)		2.281	2.008	0.018	0.018
GROUND RANGE (FT)		17,972.	14,088.	129,485.	126,663.
SPACE-FIXED POSITION VECTOR (FT)	X	8,502.	8,150.	44,347.	46,266.
	Y	12,602.	9,380.	16,749.	17,231.
	Z	21,962.	16,434.	134,708.	131,250.
SPACE-FIXED VELOCITY VECTOR (FT/SEC)	X	233.53	279.40	159.28	163.09
	Y	224.93	212.01	10.50	10.50
	Z	194.09	201.44	53.97	56.27
VEHICLE ATTITUDE (DEG)	PITCH	1.538	1.534	2.527	2.481
	YAW	1.522	1.521	2.515	2.280
	ROLL	3.648	3.648	▷	▷

▷ NOT APPLICABLE DUE TO APS CONTROL LIMIT OF APPROXIMATELY ONE DEGREE ERROR.

Figure 2-16

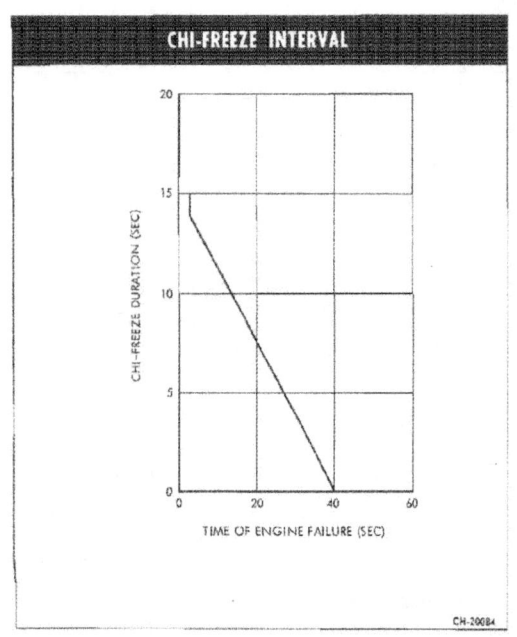

Figure 2-17

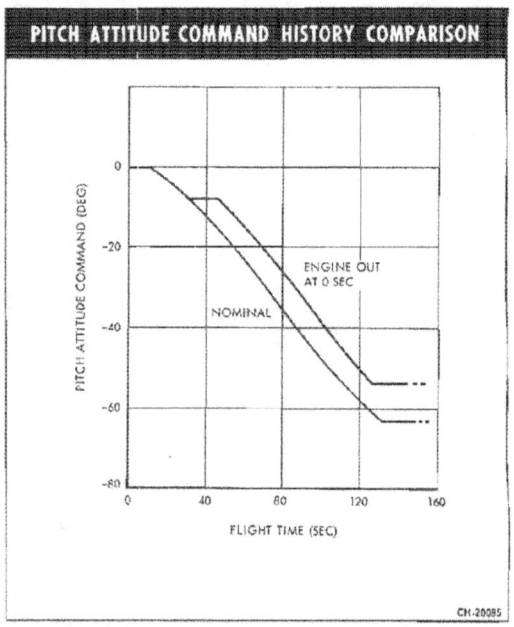

Figure 2-18

Performance Capability.

For a planar flight launch and nominal performance, the predicted weight-in-orbit capability for the SA-206 vehicle is 67,790 lbm. This provides approximately 3,400 lbm. of excess useable S-IVB stage propellant. This propellant may be utilized to compensate for low performance, perform S-IVB stage yaw steering to provide launch windows, or improve engine-out capability. Propellant required to compensate for three-sigma low performance (FPR) is 2,200 lbm. Propellant requirements to compensate for single H-1 engine failures are provided in figure 2-19. The launch window performance requirements are discussed in Section IX.

PROPULSION PERFORMANCE.

The predicted launch vehicle propulsion performance, as presented, is derived from computer simulations of the S-IB and S-IVB propulsion system performance using data from stage static tests, single engine acceptance tests, and previous flight tests.

S-IB STAGE.

Performance data of the S-IB stage is extracted from prediction data for the SL-2 mission. Stipulated criteria of the prediction are:

1) Launch from LC-39B of the AFETR;

2) Terminal delivery into an 81 x 120 NM phasing orbit.

Engine Simulation.

The predicted performance for the eight, 205K thrust H-1 engines was calculated from ground test data using a table of influence coefficients (partials) to estimate the performance with the pump inlet pressures and propellant densities expected during flight. Basically, the engines are characterized by Rocketdyne single engine acceptance test data modified by empirical factors derived from previous flight data. Since S-IB-6 is the first stage to use 205K thrust engines and no previous flight history exists, the magnitudes of the factors were influenced by the results of the stage static tests conducted at MSFC. The factors are somewhat smaller than would be used for 200K engine stages.

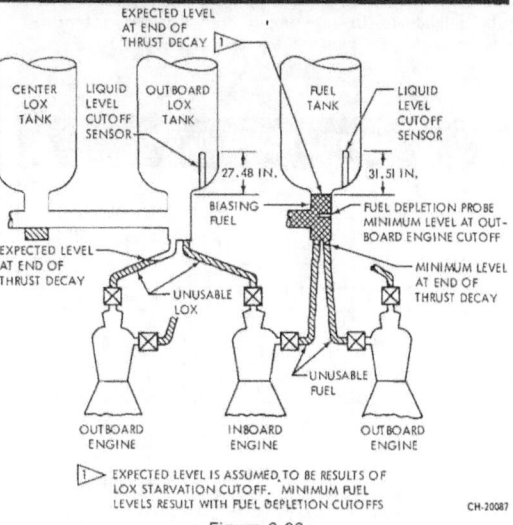

Figure 2-20

Pressurization and Propellant Data.

The predicted lox tank and fuel tank pressurization histories expected for flight were calculated from data obtained from the MSFC stage static tests.

The predicted fuel density was determined from temperature-density chemical analyses of the fuel that will be used and the expected fuel temperature at launch. The fuel temperature is based on the ambient temperature and an appropriate chilldown due to lox tank proximity expected for the time of year of launch.

The lox density was determined from the mean bulk temperature calculated from the wind velocities, ambient temperature, ambient pressure, and humidity expected at the time of launch.

Propellant Utilization.

The predicted amount of lox loaded is established as a full load to the minimum ullage of 1.5 percent. The predicted fuel load is calculated as the amount required to deplete all of the usable lox load with the performance expected during flight.

The predicted fuel load is not necessarily the amount of fuel loaded on the stage at the time of launch. Propellant load tables are furnished to KSC to provide fuel load data for the actual fuel temperature at launch time. No adjustments to the predicted lox loads are attempted.

The amount of lox trapped residual (approximately 2600 lbm) is determined as the lox in the suction lines to the main valves of the inboard engine, a few gallons trapped in the center lox tank sump, and approximately 70 gal in the outboard engines.

The amount of trapped fuel residual (3070 lbm) is determined as the amount of fuel remaining if the engines are cutoff by the fuel depletion sensors located in the sumps of F-2 and F-4 (figure 2-20).

Included in the fuel load is a 1550 lbm of fuel bias which provides a nominal fuel residual of 4620 lbm. The fuel bias is provided to minimize the amount of propellant residuals caused by deviations from predicted consumption ratios and loading inaccuracies.

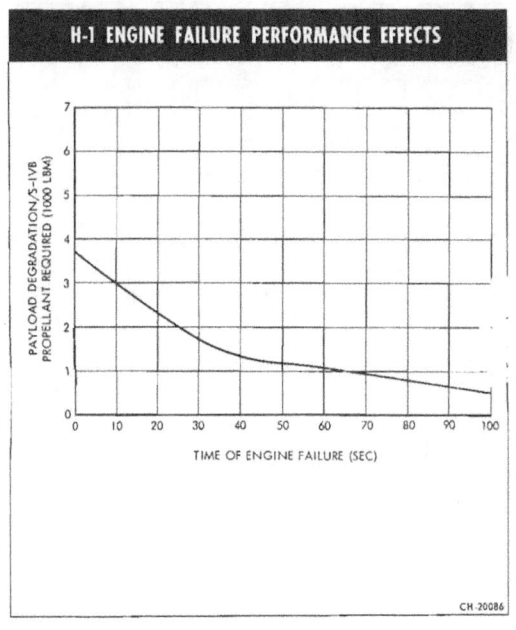

Figure 2-19

Section II Performance

S-IB Stage Predicted Performance.

Figure 2-21 shows the ambient pressure profile during powered flight. Individual S-IB stage propulsion system performance parameter variations are shown in figures 2-22 through 2-31.

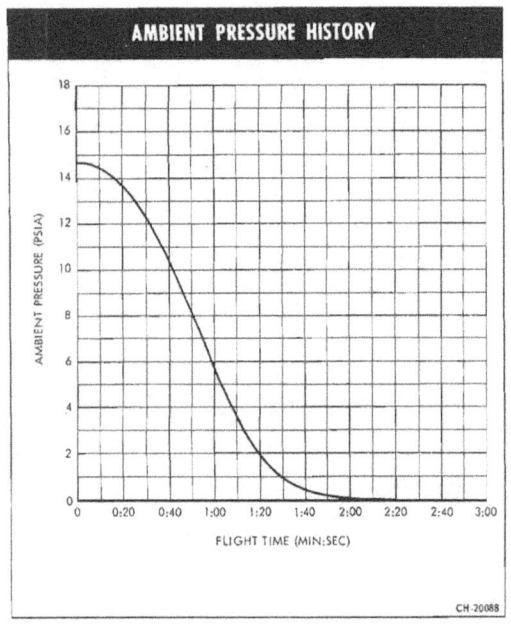

Figure 2-21

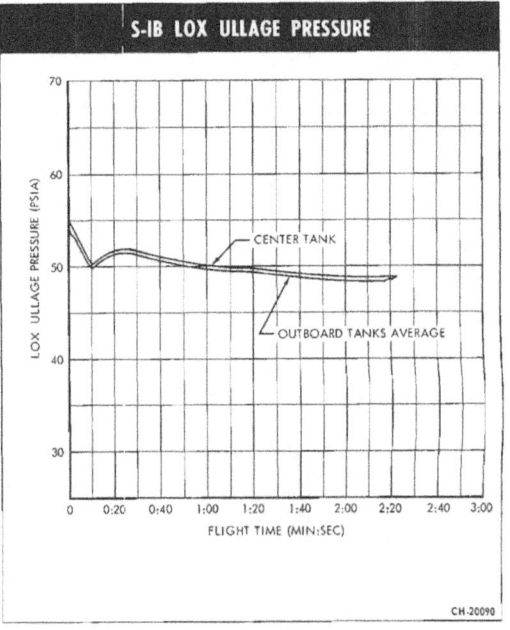

Figure 2-23

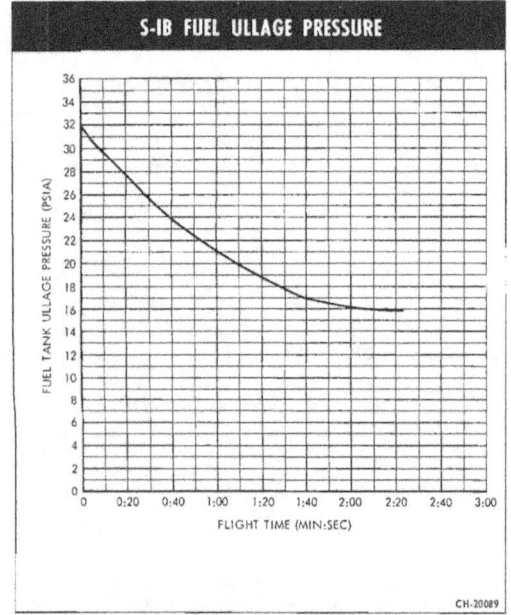

Figure 2-22

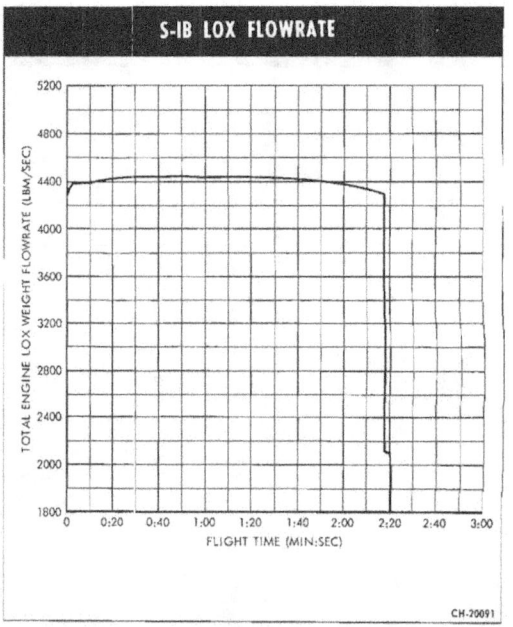

Figure 2-24

2-8

Section II Performance

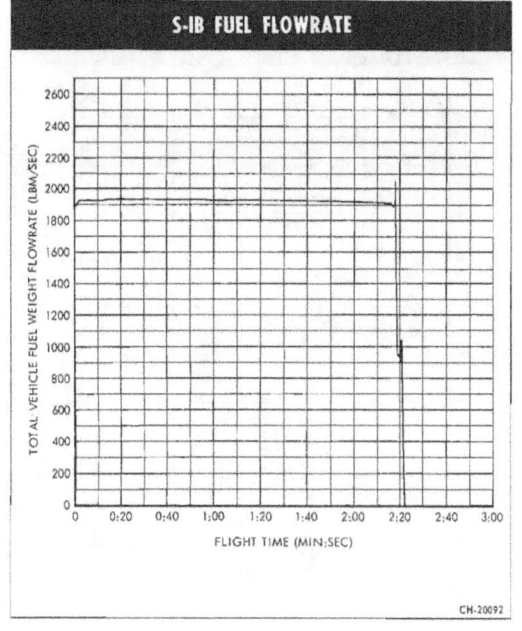

Figure 2-25

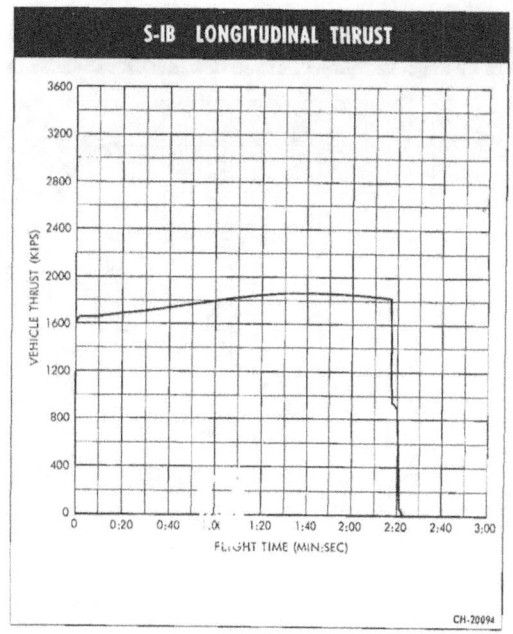

Figure 2-27

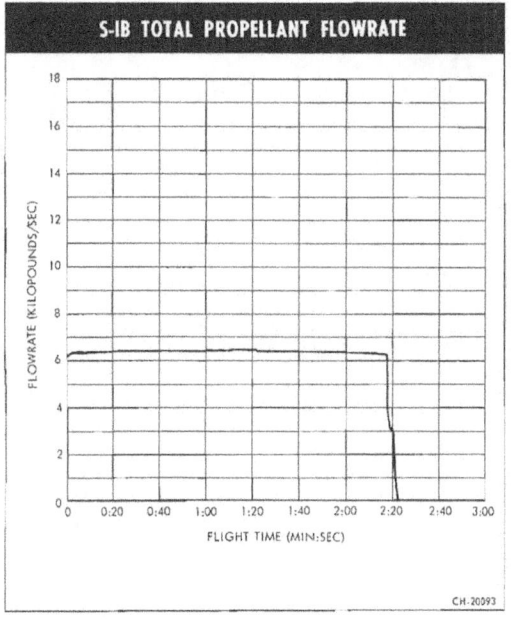

Figure 2-26

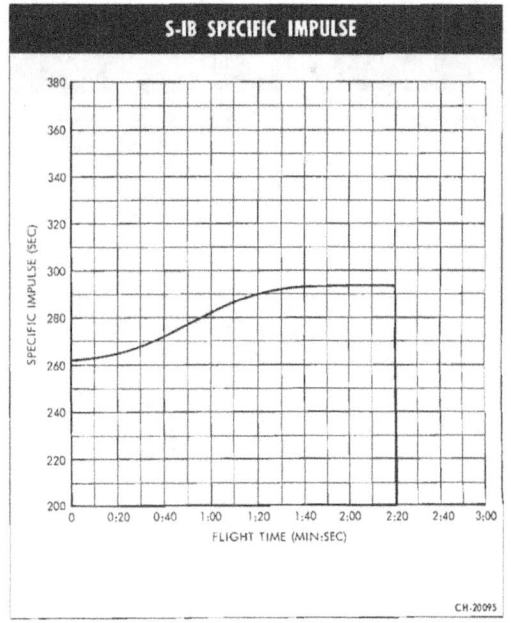

Figure 2-28

2-9

Section II Performance

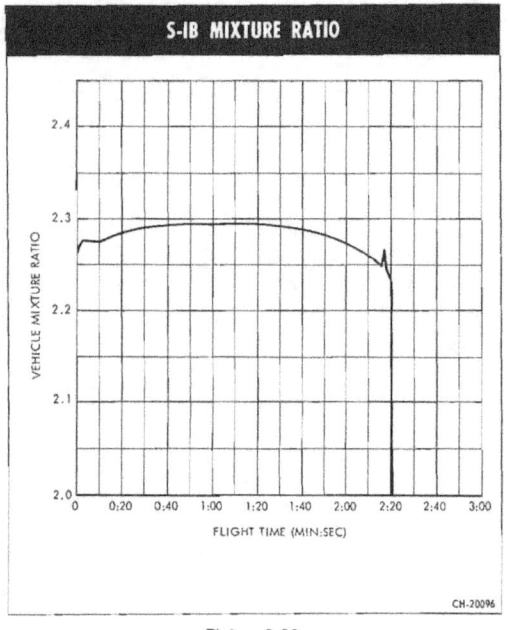

Figure 2-29

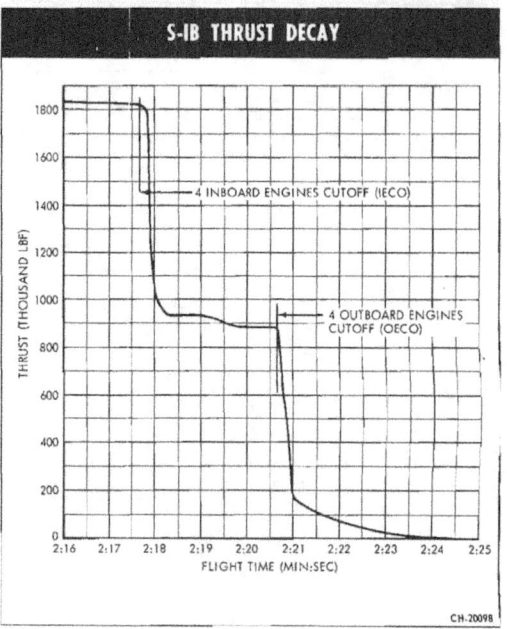

Figure 2-31

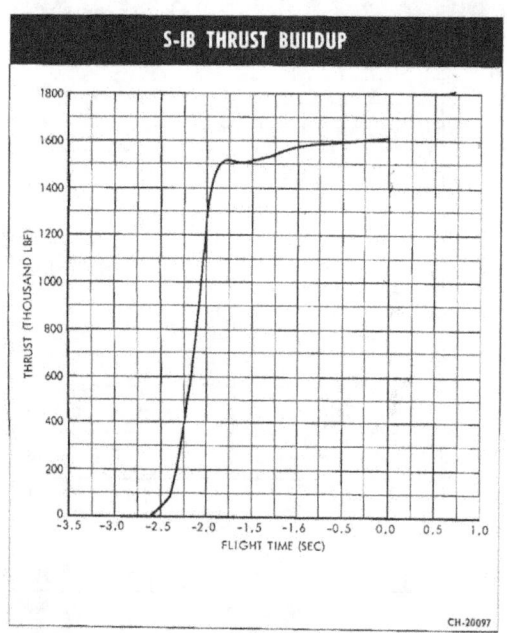

Figure 2-30

S-IVB STAGE.

The S-IVB stage data were extracted from reference propulsion flight performance predictions for a 81 by 120 NM orbital mission. This prediction is based on the following stage propellant utilization:

a. Total lox load of 195,972 lbm.

b. Total fuel load of 37,900 lbm.

c. Total lox consumed by engine and boiloff during flight of 192,621 lbm.

d. Total fuel consumed by engine and for tank pressurization during flight of 36,016 lbm.

e. Lox residual of 3,351 lbm.

f. Fuel residual of 1,884 lbm.

Figures 2-32 through 2-38 show stage propulsion system performance parameter variations.

Section II Performance

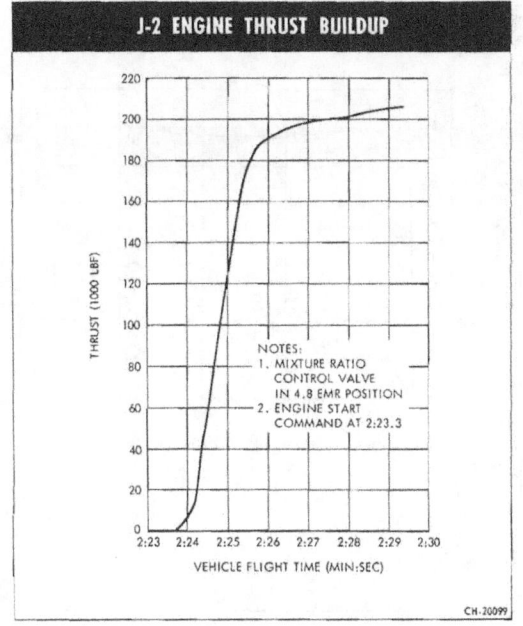

Figure 2-32

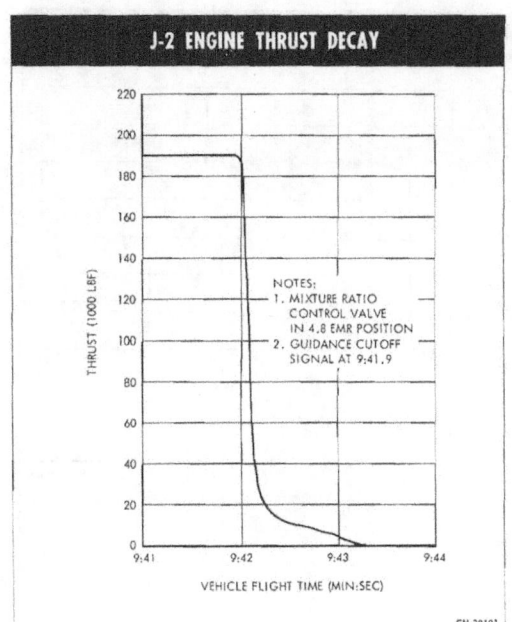

Figure 2-34

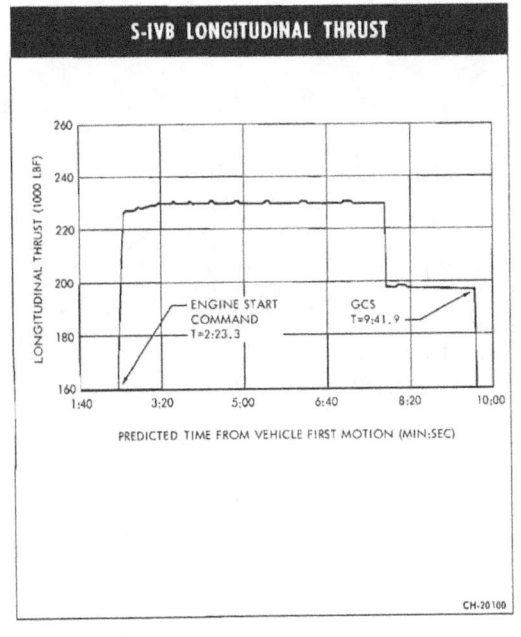

Figure 2-33

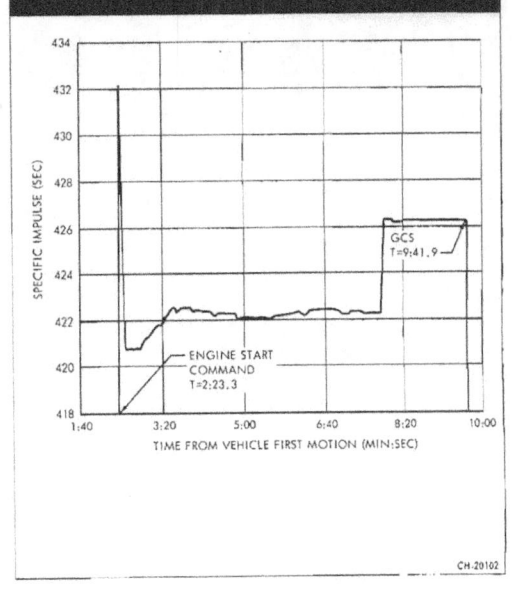

Figure 2-35

Section II Performance

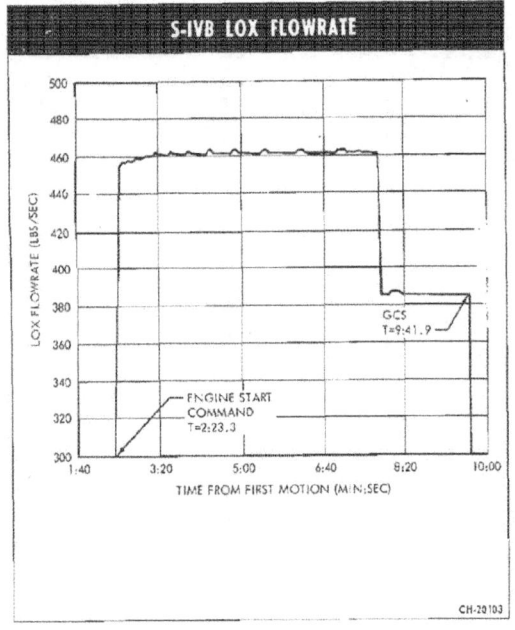

Figure 2-36

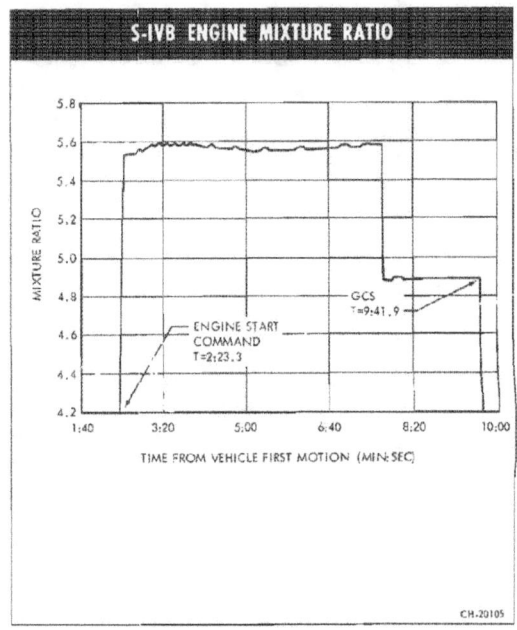

Figure 2-38

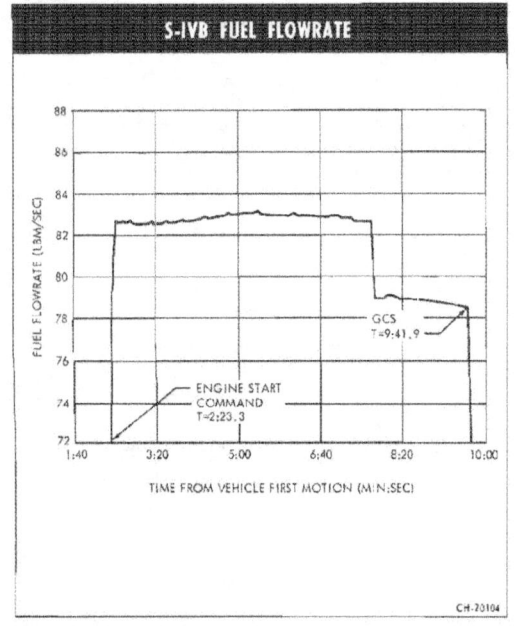

Figure 2-37

SEPARATION DYNAMICS.

The latest launch vehicle dynamics analyses indicate that no problem exists for S-IB/S-IVB stage separation motion. Potential problems considered are relative lateral motion of the J-2 bell and S-IVB aft interstage wall during physical separation, and S-IVB post separation controllability. The probability (cumulative distribution) of the J-2 bell clearing the S-IVB aft interstage wall in the case of a single retro failure in combination with stage separation tolerances is in excess of 3-sigma.

Both potential separation problems of J-2 bell interstage collision and S-IVB stage controllability are mainly affected (assuming no retro failures) by large aerodynamic moments or attitude rates existing at first stage boost flight termination. These two problems are minimized by first stage boost trajectory shaping. A nose-over maneuver initiated approximately 95 sec into flight and a tilt arrest initiated approximately 131 sec into flight results in acceptable levels of dynamic pressure, a small angle-of-attack, and attitude rates that are essentially zero at S-IB/S-IVB separation.

The main contributor to the physical separation of the S-IB stage from the S-IVB stage is the thrust of the four retro rockets. To

Section II Performance

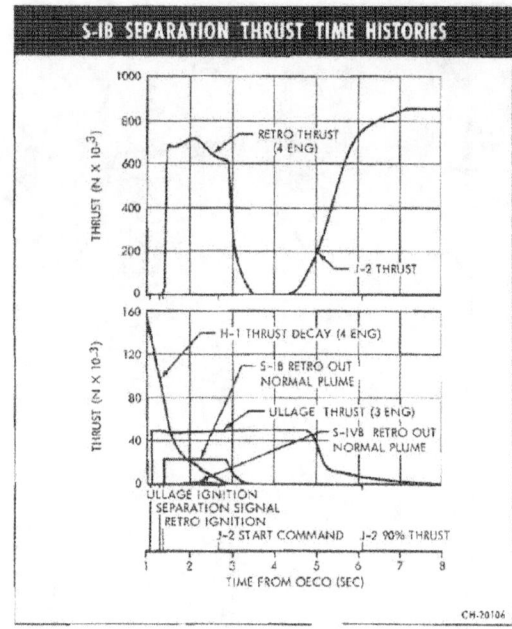

Figure 2-39

to ascertain successful retro-out staging probability. The quoted probabilities are defined by the probability law:

$$P = \sum_{i=1}^{4} P_i P_i^*$$

where: P ≡ probability of successful separation with one retro rocket failed

P_i ≡ probability that retro rocket number "i" is the one which failed

P_i^* ≡ probability of successful separation with retro rocket number "i" failed

The P_i^* probabilities quoted pertain to the cumulative distribution function. Each P_i^* is determined by root-sum-squaring the incremental lateral travel due to each tolerance with retro rocket number "i" failed. Those stage separation tolerances which have the greatest influence on S-IB/S-IVB relative lateral motion are those which create significant moments on the S-IB stage. Aerodynamic moments resulting from aerodynamic tolerances are not large enough on either stage to be significant contributors to a potential S-IB/S-IVB collision. The stage separation tolerances considered in the retro-out collision analysis are, therefore, retro rocket thrust variation ($\pm 13.28\%$), retro rocket thrust misalignment ($\pm 0.50°$), and S-IB lateral CG deviation (± 1.1 in.).

The single retro rocket failure results are presented in figures 2-42 and 2-43. Figure 2-42 gives the lateral clearance of the undeflected J-2 bell bottom with the S-IVB aft interstage (at interstage exit plane) for each of the four single retro rocket failures possible. These results are based upon all retro failures being simulated during an otherwise nominal separation subsequent to a nominal S-IB boost flight. The smallest lateral clearance is 0.434 meters which results when retro no. 3 fails. Figure 2-43 presents the J-2 bell lateral drifts in profile view for nominal and retro-out conditions. In addition 3 σ off-nominal drifts are presented for nominal and the worst case retro rocket failure condition (retro no. 3 out). As depicted in the figure 2-43 the probability (cumulative distribution) of the J-2 bell clearing the interstage for a single retro failure in combination with stage separation tolerances is in excess of 3σ.

The qα product (dynamic pressure times total angle-of-attack) at physical separation may be used as an indicator of S-IVB stage controllability after separation. Large qα products result in large dynamic responses in the S-IVB subsequent to physical separation. These post-separation transients may cause the J-2 engine to hardover against the 7-degree gimbal limit. However, vehicle response is such that the J-2 engine hardover exists for only a few seconds (in the worst case) and S-IVB control is maintained throughout flight.

The maximum S-IVB stage attitude and rate errors due to terminal boost flight conditions at separation are to be found in pitch. Figure 2-44 illustrates the peak pitch attitude errors after physical separation for the S-IB stage engine failures analyzed.

a very slight degree, the three ullage rockets also contribute to the physical separation. Proper phasing of the retro thrust with respect to the separation signal and H-1 thrust decay is necessary for successful staging and is shown in Figure 2-39. Impingement of the retro rocket plumes on the vehicle creates pressure distributions on the surface of the S-IB/S-IVB interstage and lower S-IVB stage. If a retro rocket fails to ignite, these pressure distributions then become asymmetric thereby causing imbalanced forces to act on the stages as shown in Figures 2-40 and 2-41. This imbalanced force condition constitutes a potential S-IB/S-IVB collision hazard. Figure 2-39 indicates that the S-IVB stage is without effective J-2 control thrust for approximately 4.5 sec after physical separation from the S-IB stage. It is during this time interval that S-IVB stage dynamic transients can become excessively large.

The S-IB/S-IVB relative motion resulting from each of four retro rocket failures in combination with stage separation tolerances, subsequent to a nominal S-IB boost flight, is analyzed in order

Section II Performance

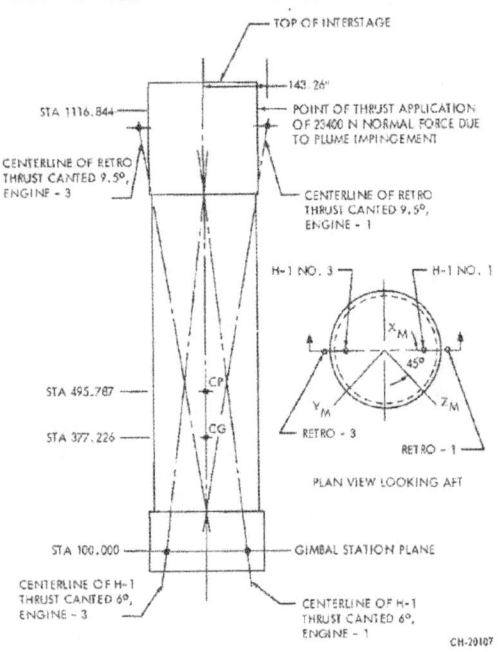

Figure 2-40

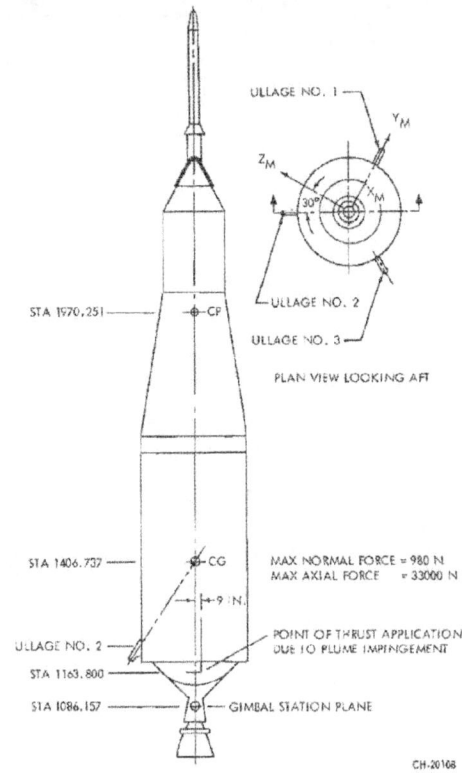

Figure 2-41

SINGLE RETROROCKET FAILURE STAGING ANALYSIS

RETRO FAILED	LATERAL CLEARANCE (METERS) [1>
NO. 1	.471
NO. 2	.469
NO. 3	.434
NO. 4	.435

[1> LATERAL CLEARANCE OF THE UNDEFLECTED J-2 BELL BOTTOM WITH THE S-IVB AFT INTERSTAGE AT THE INTERSTAGE EXIT PLANE.

Figure 2-42

One of the most important of the post physical separation responses is the peak pitch attitude rate. As indicated in figure 2-45, the envelope of the peaks of these pitch attitude rates was 4.36 deg/sec for engine no. 5 failure at liftoff. It can be seen that the 10 deg/sec EDS abort limit (see Section III) for post separation controllability is not violated by any of these acceptable malfunction modes.

J-2 engine gimbal deflections and time against the 7-degree limit stop are shown in figure 2-46. The upper graph shows, for example, that failure of engine no. 5 at liftoff causes the J-2 to be on the 7-degree gimbal limit for 2.4 sec, while with a failure of that engine subsequent to 4.5 sec, the engine does not reach this limit. The lower graph illustrates the maximum deflections in pitch for all failure times. It can be seen that engine no. 5 never reaches the 7-degree limit for a failure subsequent to 4.5 sec. Both graphs show that an engine no. 2 failure will never cause the J-2 to reach the gimbal limits.

Section II Performance

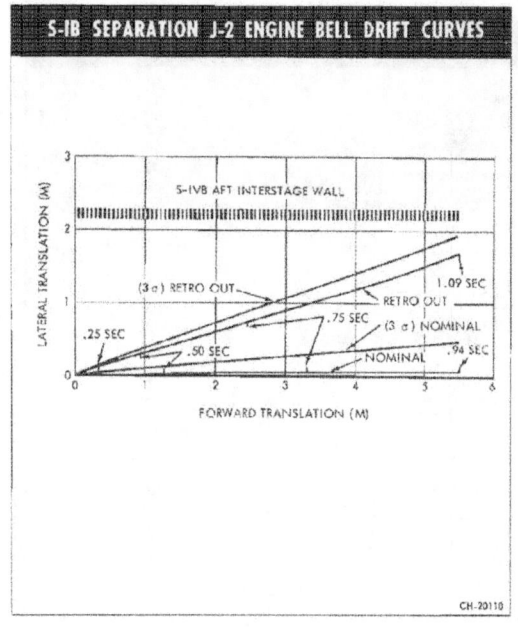

Figure 2-43

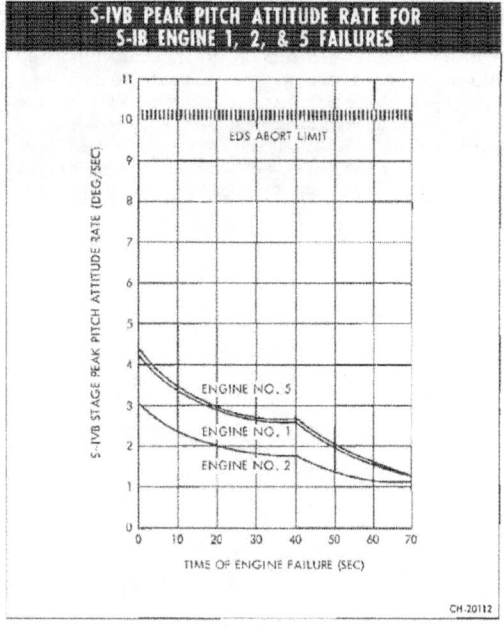

Figure 2-45

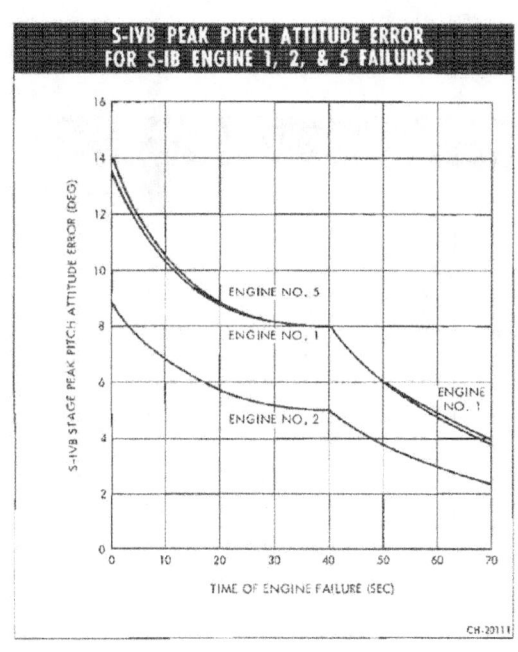

Figure 2-44

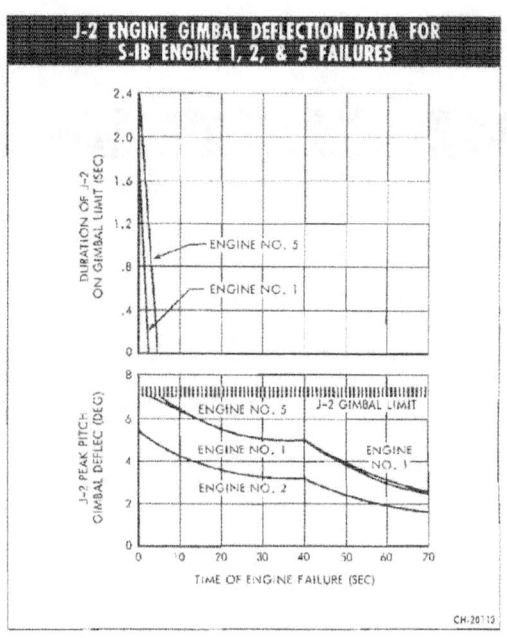

Figure 2-46

2-15

Section II Performance

MASS CHARACTERISTICS.

The following mass characteristics data were extracted from the latest final predicted mass characteristics studies, which are based on measured weights of the dry stages and hardware and propulsion systems performance predictions. Figure 2-47 lists the vehicle weight breakdown at liftoff. Figures 2-48 and 2-49 contain a weight breakdown at major events during S-IB and S-IVB powered flight, respectively. Figure 2-50 shows the total residuals at J-2 engine cutoff command.

SL-2 VEHICLE WEIGHT BREAKDOWN (LBM)

CM	13,500	
SM	17,500	
SLA PANELS	2,800	
SLA (FIXED)	1,500	
INSTRUMENT UNIT	4,298	
S-IVB STAGE INERT	24,465	
USEABLE S-IVB PROPELLANT (INCLUDES FPR)	3,727	
ORBIT INSERTION WEIGHT		67,790
J-2 THRUST DECAY PROPELLANT	121	
S-IVB CUTOFF WEIGHT		67,911
S-IVB PROPELLANT CONSUMED	227,637	
S-IVB APS PROPELLANT CONSUMED	6	
LES	9,350	
ULLAGE CASES	214	
S-IVB "90% THRUST" WEIGHT		305,118
S-IVB GH2 START TANK	4	
S-IVB BUILDUP PROPELLANT CONSUMED	565	
ULLAGE PROPELLANT CONSUMED	176	
S-IVB WEIGHT AT SEPARATION		305,863
S-IVB AFT FRAME HARDWARE	31	
S-IB/S-IVB INTERSTAGE	6,800	
S-IB DRY WEIGHT	84,600	
S-IB RESIDUALS AND RESERVES	10,718	
S-IVB DETONATION PACKAGE	5	
S-IVB FROST CONSUMED	100	
S-IB FROST CONSUMED	1,000	
S-IB SEAL PURGE CONSUMED	6	
S-IB FUEL ADDITIVE CONSUMED	27	
S-IB GEARBOX CONSUMPTION (RP-1)	706	
INBOARD ENGINE THRUST DECAY PRPT CONSUMED	2,162	
OUTBOARD ENGINE THRUST DECAY PRPT CONSUMED TO SEPARATION	1,672	
S-IB MAINSTAGE PROPELLANT CONSUMED	882,092	
VEHICLE LIFTOFF WEIGHT		1,295,762

Figure 2-47

SL-2 WEIGHT BREAKDOWN DURING FIRST STAGE FLIGHT (LBM)

DESCRIPTION	GROUND IGN	FIRST MOTION	IECO SIGNAL	OECO SIGNAL	SEP SIGNAL	SEP COMP	ETD
S-IB DRY	84,600	84,600	84,600	84,600	84,600	84,600	84,600
LOX IN TANKS	624,033	612,549	2,345	0	0	0	0
LOX BELOW TANKS	7,760	8,201	8,094	3,317	2,767	2,761	2,683
LOX ULLAGE (GOX)	32	80	2,619	2,651	2,653	2,653	2,652
FUEL IN TANKS	274,833	270,637	4,794	1,004	122	115	111
FUEL BELOW TANKS	4,826	5,758	5,758	5,292	5,080	5,064	4,827
FUEL ULLAGE (He)	5	8	58	59	59	59	59
FUEL PRESS. HELIUM SUPPLY	78	75	75	24	24	24	24
GN2	15	15	9	9	9	9	9
ORONITE	33	33	6	6	6	6	6
HYDRAULIC OIL	28	28	28	28	28	28	28
ICE	1000	1000	0	0	0	0	0
S-IVB AFT FRAME	0	0	0	0	0	31	31
TOTAL S-IB	997,243	982,984	108,336	96,990	95,348	95,350	95,030
S-IB/S-IVB INTERSTAGE	5,738	5,738	5,738	5,738	5,738	5,738	5,738
RETRO-PROPELLANT	1,062	1,062	1,062	1,062	1,062	1,062	1,062
TOTAL S-IB/S-IVB INTERSTAGE	6,800	6,800	6,800	6,800	6,800	6,800	6,800
FIRST VEHICLE STAGE	1,004,043	989,784	115,136	103,790	102,148	102,150	101,830
S-IVB LOADED	257,051	257,031	256,951	256,951	256,951	0	0
VIU	4,298	4,298	4,298	4,298	4,298	0	0
LES	9,350	9,350	9,350	9,350	9,350	0	0
CM	13,500	13,500	13,500	13,500	13,500	0	0
SERVICE MODULE	17,500	17,500	17,500	17,500	17,500	0	0
SLA WITH RING	4,300	4,300	4,300	4,300	4,300	0	0
TOTAL VEHICLE	1,310,042	1,295,783	421,035	409,689	408,047	102,150	101,830

Figure 2-48

Section II Performance

SL-2 WEIGHT BREAKDOWN DURING SECOND STAGE FLIGHT (LBM)

DESCRIPTION	OECOS	SEP SIGNAL	J-2 IGNITION	90% THRUST	J-2 CUTOFF	J-2 ETD
FIRST VEHICLE STAGE	103,790	102,148	0	0	0	0
S-IVB DRY	21,950	21,950	21,950	21,950	21,950	21,950
LOX IN TANK	195,576	195,576	195,576	195,152	2,951	2,891
LOX BELOW TANK	395	397	397	397	397	367
LOX ULLAGE (GOX + He)	17	17	17	17	347	347
He IN SPHERES	320	320	320	320	190	190
LH₂ IN TANK	37,887	37,887	37,887	37,745	1,833	1,812
LH₂ BELOW TANK	58	58	58	58	58	48
LH₂ ULLAGE (GH₂ + He)	136	136	136	137	412	412
AFT FRAME	31	31	0	0	0	0
DETONATION PACK	5	5	0	0	0	0
ULLAGE ROCKET PROP.	176	176	105	0	0	0
ULLAGE ROCKET CASES	214	214	214	214	0	0
APS PROPELLANT	132	132	132	132	126	126
APS He	3	3	3	3	3	3
HYDRA HYDRAULIC OIL	15	15	15	15	15	15
GN2	3	3	3	3	3	3
ENVIRONMENTAL CONT	14	14	14	14	14	14
He PNEUMATICS	12	12	12	12	12	12
GH₂ START	5	5	5	1	1	1
TOTAL S-IVB	256,951	256,951	256,844	256,170	28,312	28,191
VIU	4,298	4,298	4,298	4,298	4,298	4,298
LES	9,350	9,350	9,350	9,350	0	0
CM	13,500	13,500	13,500	13,500	13,500	13,500
SERVICE MODULE	17,500	17,500	17,500	17,500	17,500	17,500
SLA WITH RING	4,300	4,300	4,300	4,300	4,300	4,300
TOTAL VEHICLE	409,689	408,047	305,792	305,118	67,910	67,789

CH-20116

SL-2 RESIDUALS AT J-2 CUTOFF COMMAND (LBM)

TOTAL RESIDUALS AT J-2 ENGINE CUTOFF COMMAND		6,362
TOTAL LOX		3,348
THRUST DECAY	60	
IN TANK UNUSABLE	0	
IN TANK USABLE	3,087	
TRAPPED IN SUMP	0	
TRAPPED IN LINES	63	
DISSIPATED AT ETD (BELOW VALVE)	30	
TRAPPED IN ENGINE	108	
TOTAL FUEL		1,891
THRUST DECAY	21	
IN TANK UNUSABLE (INCL. LH₂ BIAS)	1,172	
IN TANK USABLE	640	
TRAPPED IN SUMP	0	
TRAPPED IN LINES	38	
DISSIPATED AT ETD (BELOW VALVE)	10	
TRAPPED IN ENGINE	10	
LOX ULLAGE (GOX + He)		347
FUEL ULLAGE (GH₂ + He)		412
SERVICE ITEMS		364
APS PROPELLANT	126	
HYDRAULIC OIL	15	
GN₂	3	
ENVIR. CONTROL FLUID	14	
HELIUM PNEUMATICS	12	
HELIUM IN SPHERES	190	
GH₂ START TANK	1	
HELIUM APS	3	

PROPELLANT UTILIZATION

	LOX	FUEL
TOTAL LOADED	(195,973)	(37,945)
THRUST BUILDUP	424	141
MAINSTAGE	192,002	35,636
THRUST DECAY	60	21
DISSIPATED AT ETD (BELOW VALVE)	30	10
CONVERTED TO GAS	199	277
REMAINING AT ETD	3,258	1,860
IN TANK	2,891	1,812
BELOW TANK	367	48

CH-20117

Figure 2-50

Section II Performance

TRACKING COVERAGE.

Tracking, command, and communications systems, and telemetry coverage data were extracted from the latest launch vehicle operational flight trajectory tracking analysis, for the SA-207 vehicle. During the launch phase, overlapping coverage is provided by sites at MILA, Bermuda and Newfoundland. Figure 2-51 shows a map trace of this launch phase coverage. Figure 2-52 contains a summary of orbital tracking coverage of the S-IVB/IU with orbital traces on a world map. The tracking and communications network shown on this figure is currently planned to be available for Saturn IB flights in the Skylab program.

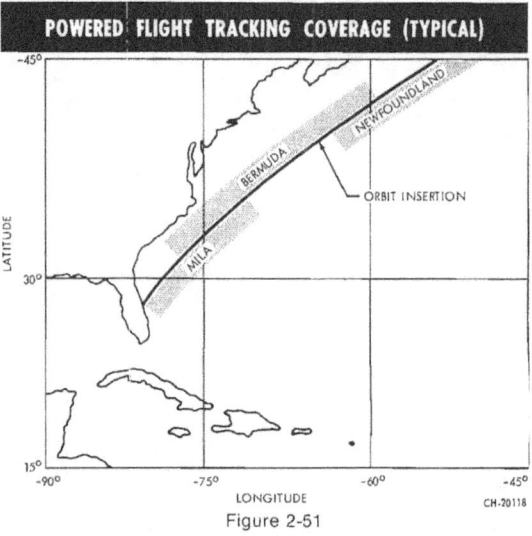

Figure 2-51

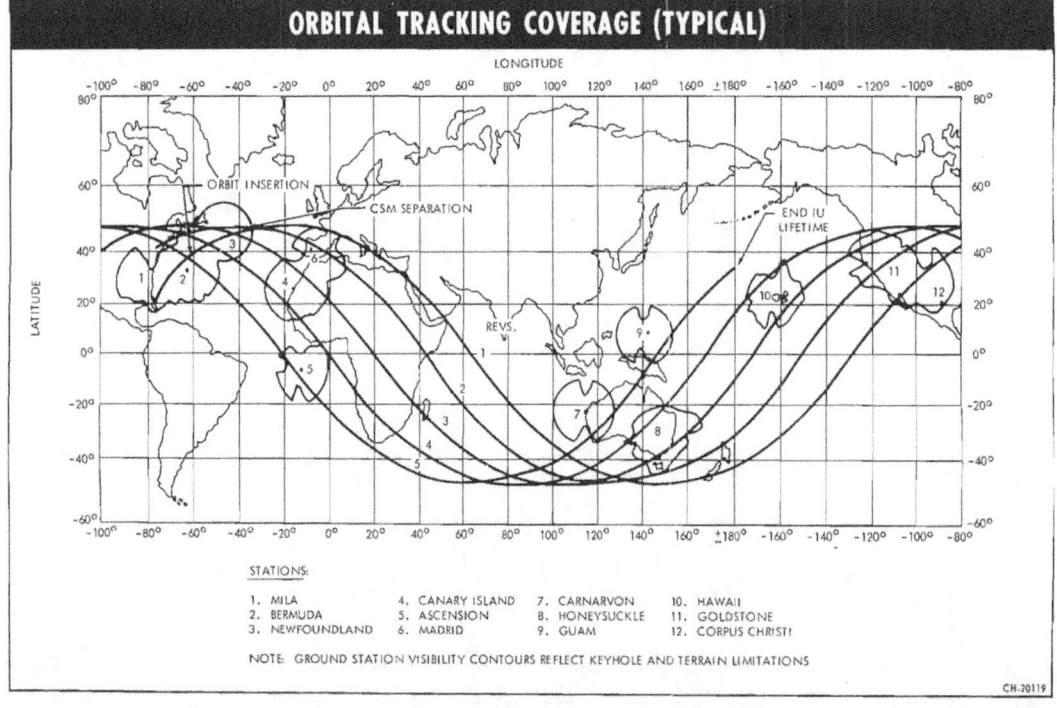

Figure 2-52

SECTION III

EMERGENCY DETECTION AND PROCEDURES

TABLE OF CONTENTS

Launch Vehicle Monitoring And Control	3-1
Launch Vehicle Monitoring Displays	3-1
Launch Vehicle Normal Controls	3-10
Launch Vehicle EDS Controls	3-11
Abort Controls	3-12
Abort Modes And Limits	3-15

LAUNCH VEHICLE MONITORING AND CONTROL.

The spacecraft is equipped with a number of displays and controls which permit monitoring the launch vehicle conditions and controlling the launch vehicle under normal and emergency conditions. Many of these displays and controls are related to the Emergency Detection System (EDS). The displays implemented for EDS monitoring were selected to present as near as possible those parameters which represent the failures leading to vehicle abort. Whenever possible, the parameter was selected so that it would display total subsystem operation. Manual abort parameters have been implemented with redundant sensing and display to provide highly reliable indications to the crewmen. Automatic abort parameters have been implemented triple redundant, voted two-out-of-three, to preclude single point hardware or sensing failures causing an inadvertent abort. The types of displays have been designed to provide onboard detection capability for rapid rate malfunctions which may require abort. Pilot abort action must, in all cases, be based on two separate but related abort cues. These cues may be derived from the EDS displays, ground information, physiological cues, or any combination of two valid cues. In the event of a discrepancy between onboard and ground based instrumentation, onboard data will be used. The EDS displays and controls are shown in figure 3-1. As each is discussed it is identified by use of the grid designators listed on the border of the figure.

LAUNCH VEHICLE MONITORING DISPLAYS.

FLIGHT DIRECTOR ATTITUDE INDICATOR.

There are two Flight Director Attitude Indicators (FDAI's), each of which provides a display of Euler attitude, attitude errors and angular rates. Refer to figure 3-1, Q-45 and J-60, for locations of the FDAI's on the MDC and to figure 3-2 for details of the FDAI's. These displays are active at liftoff and remain active throughout the mission, except that attitude errors are not displayed during S-IVB flight. The FDAI's are used to monitor normal launch vehicle guidance and control events. The roll and pitch programs are initiated simultaneously at +10 seconds. The roll program is terminated when flight azimuth is reached, and the pitch program continues to tilt-arrest. IGM initiate will occur approximately one second after LET jettison during the S-IVB stage flight. The FDAI ball displays Euler attitude, while needle type pointers across the face of the ball indicate attitude errors, and triangular pointers around the periphery of the ball display angular rates. Attitude errors and angular rate displays are, clockwise from the top, roll, pitch and yaw, respectively. Signal inputs to the FDAI's are switch selectable and can come from a number of different sources in the spacecraft. This flexibility and redundancy provides the required attitude and error backup display capability. Excessive pitch, roll, or yaw indications provide a single cue that an abort is required. Additional abort cues will be provided by the FDAI combining rates, error, or total attitude. Second cues will also be provided by the LV RATE light (R-50), LV GUID light (R-52), physiological sensations and MCC ground reports.

LV ENGINES LIGHTS.

The eight LV ENGINES lights on (S-51, figure 3-1) indicate that each corresponding S-IB stage engine is below 90 percent nominal thrust. The engine light cluster also provides indications of the launch vehicle staging sequence. Physical separation of stages is indicated with all engine lights off (after normal S-IB cutoff). The no. 1 light will come on again 2.4 sec after OECO and go off when the S-IVB engine exceeds 65 percent nominal thrust. For abort decisions, the on indication is considered zero thrust for the corresponding engine and off is 100 percent thrust. Each S-IB engine light and sensing circuit is redundant and constitutes a single "warning" cue of a possible abort situation. However, it is not an abort cue in itself. Upon ground verification of a single engine failure, the ABORT SYSTEM 2 ENGINE OUT switch should be moved to OFF (if T > + 15 sec). A second engine light is not necessarily a second cue for immediate abort. For this failure case (two non-adjacent engine lights on), secondary abort indications will be provided by the FDAI, LV RATE light, and ground information. The Saturn IB vehicle has limited capability to continue the mission with the loss of one or two engines depending upon the time of the second failure. Mission rules will prescribe the timelines after which flight will be continued with one or two engines out. Simultaneous illumination of two or more engine lights at any time during S-IB flight is sufficient cue for immediate abort except at normal IECO or OECO. During S-IVB burn the no. 1 engine light on is a single abort cue.

Two engine out automatic aborts are active until manually deactivated by the crew. Deactivation times will be prescribed by mission rules. Two engine automatic abort disabling by the flight crew is backed up by the launch vehicle sequencer prior to IECO and is not reactivated.

LV RATE LIGHT.

The LV RATE light on (R-50, figure 3-1) is the primary cue from the launch vehicle that preset overrate settings have been exceeded. It is a single cue for abort, while secondary cues will be provided by FDAI indications, physiological cues, or ground information. Automatic LV rate aborts are enabled automatically at liftoff (with EDS and LV RATES AUTO switches enabled in SC) and are active until manually deactivated by the crew. EDS auto abort deactivation times will be governed by mission rules. The automatic LV rate abort capability is also deactivated by the launch vehicle sequencer

3-1

Section III Emergency Detection and Procedures

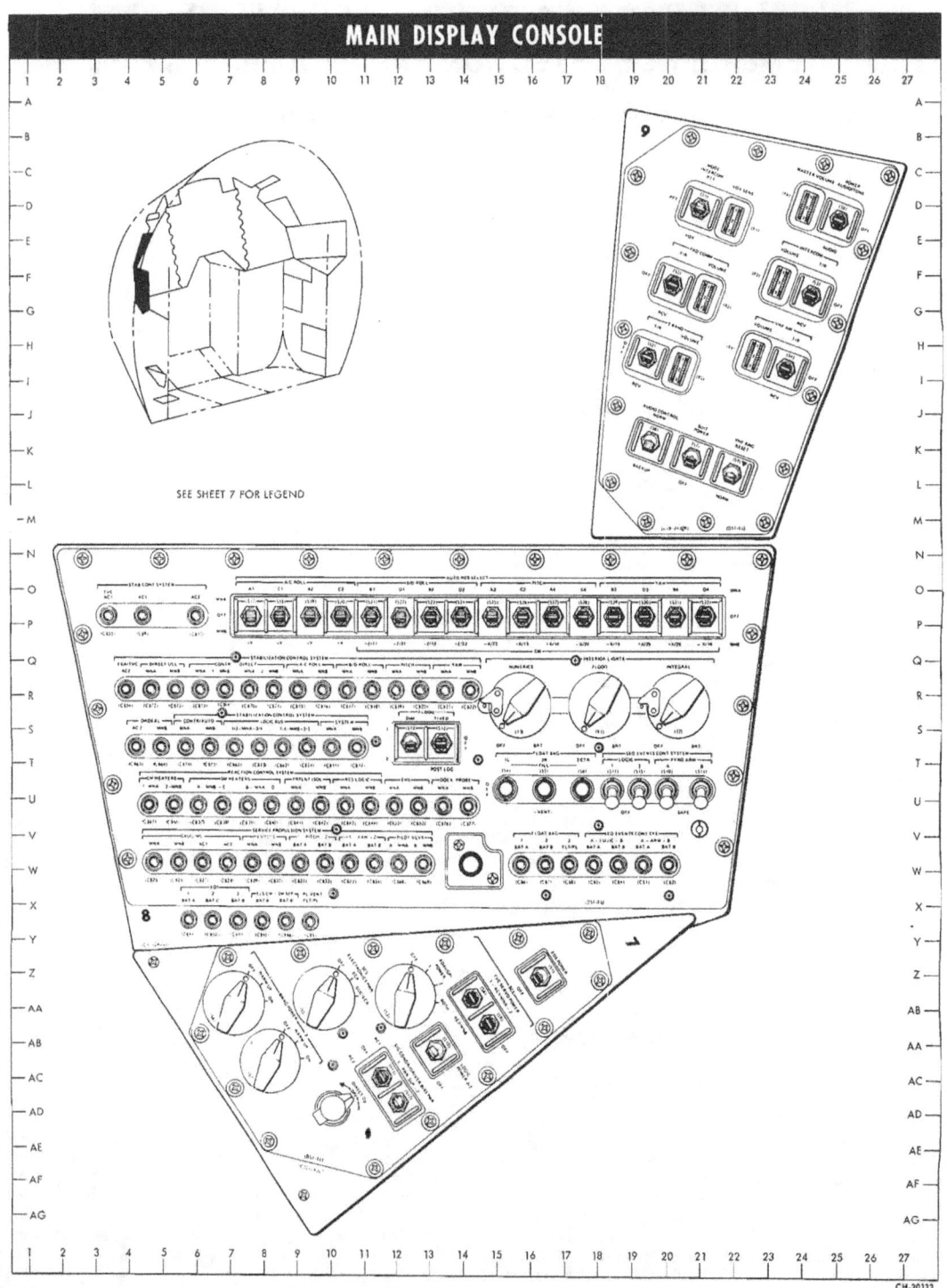

Figure 3-1 (Sheet 1 of 7)

Section III Emergency Detection and Procedures

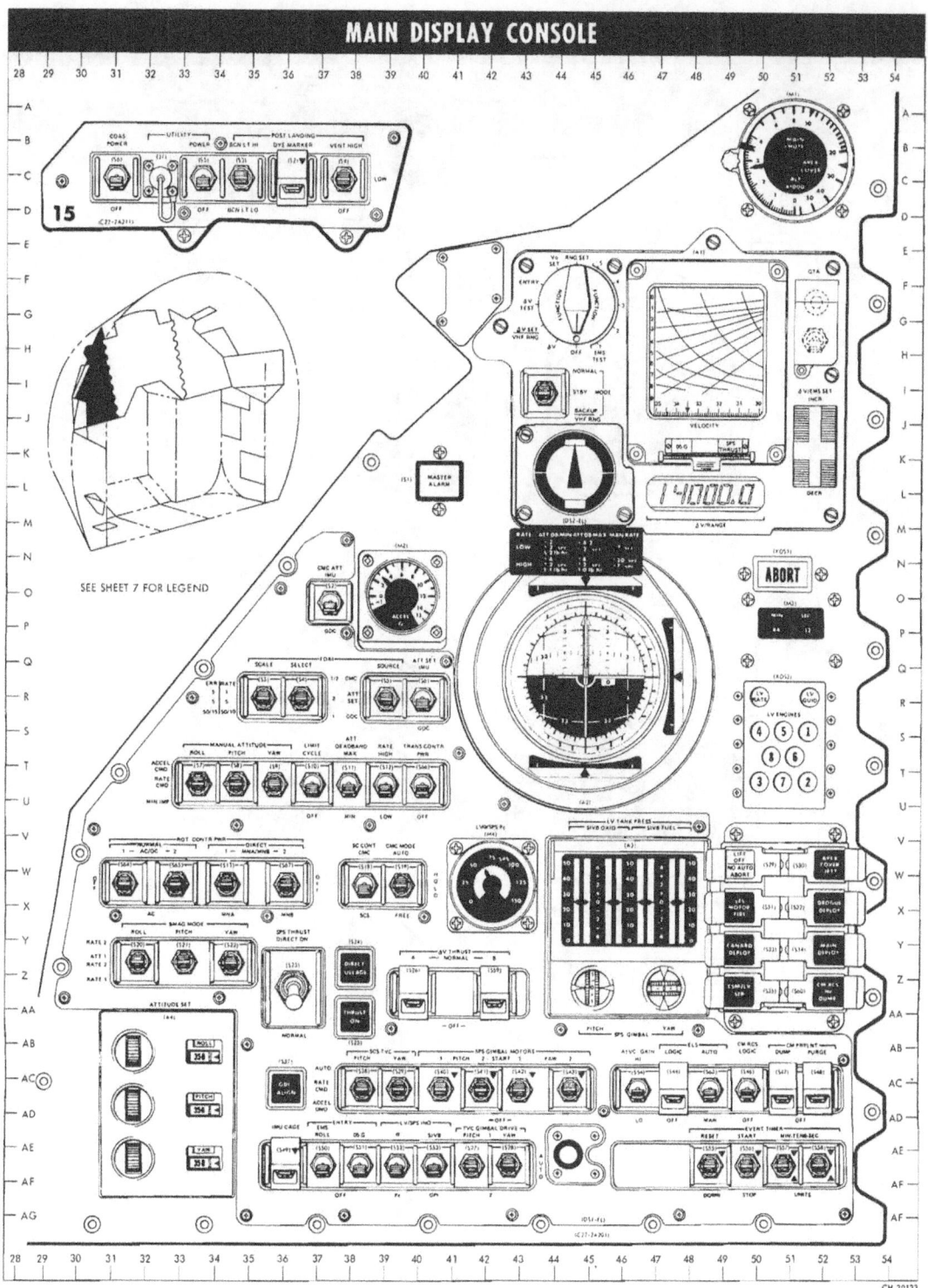

Figure 3-1 (Sheet 2 of 7)

Section III Emergency Detection and Procedures

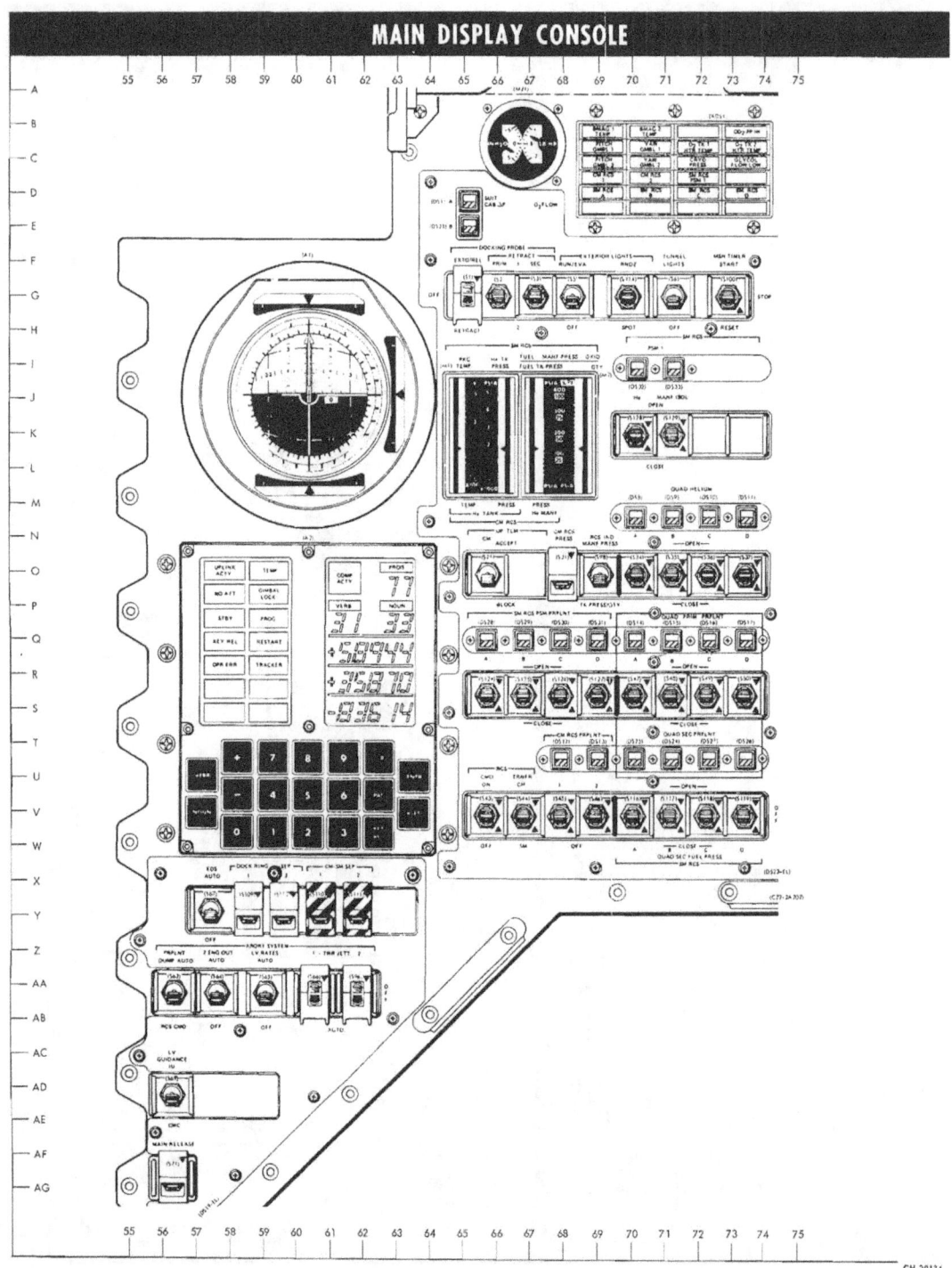

Figure 3-1 (Sheet 3 of 7)

Section III Emergency Detection and Procedures

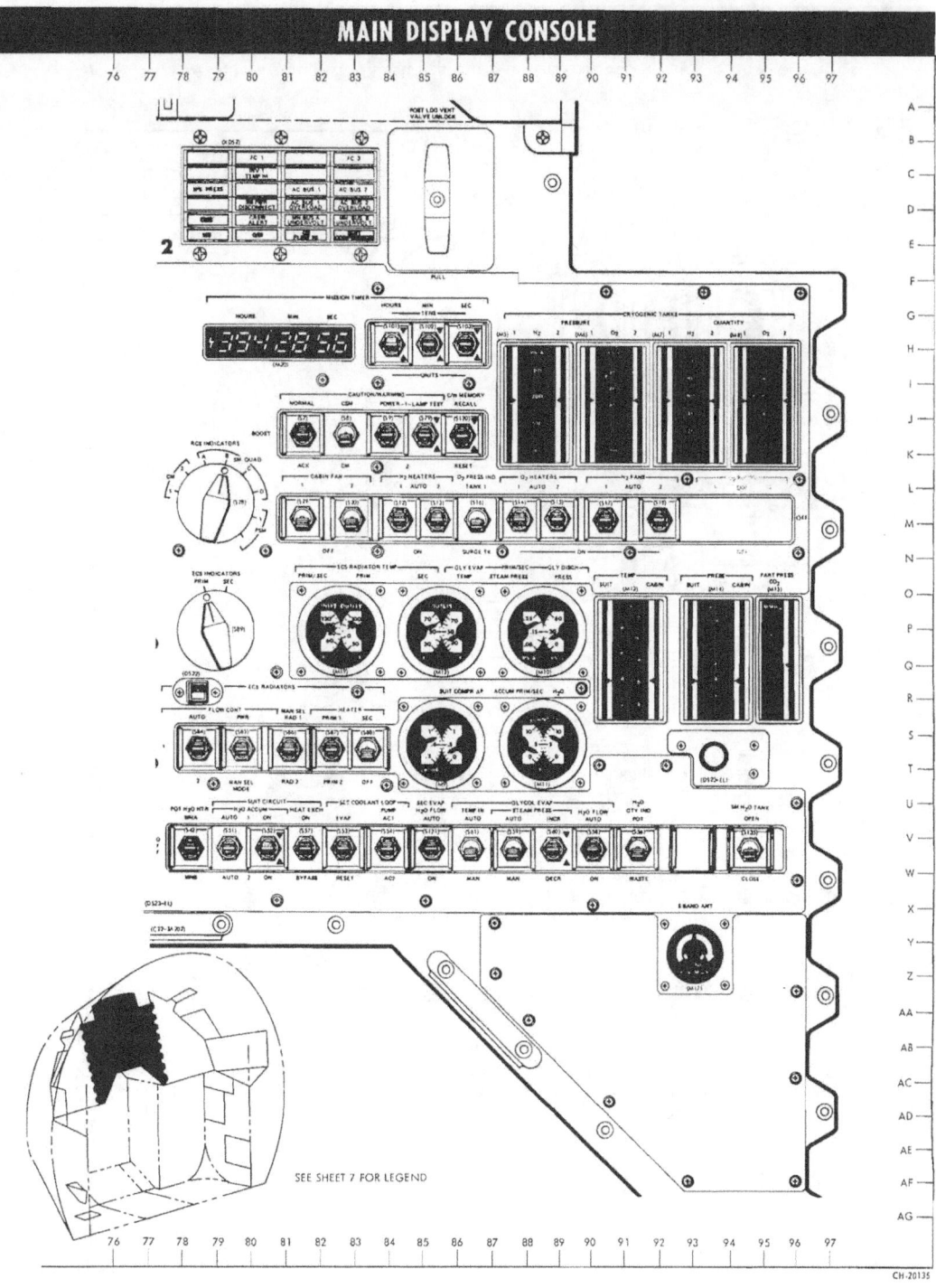

Figure 3-1 (Sheet 4 of 7)

Section III Emergency Detection and Procedures

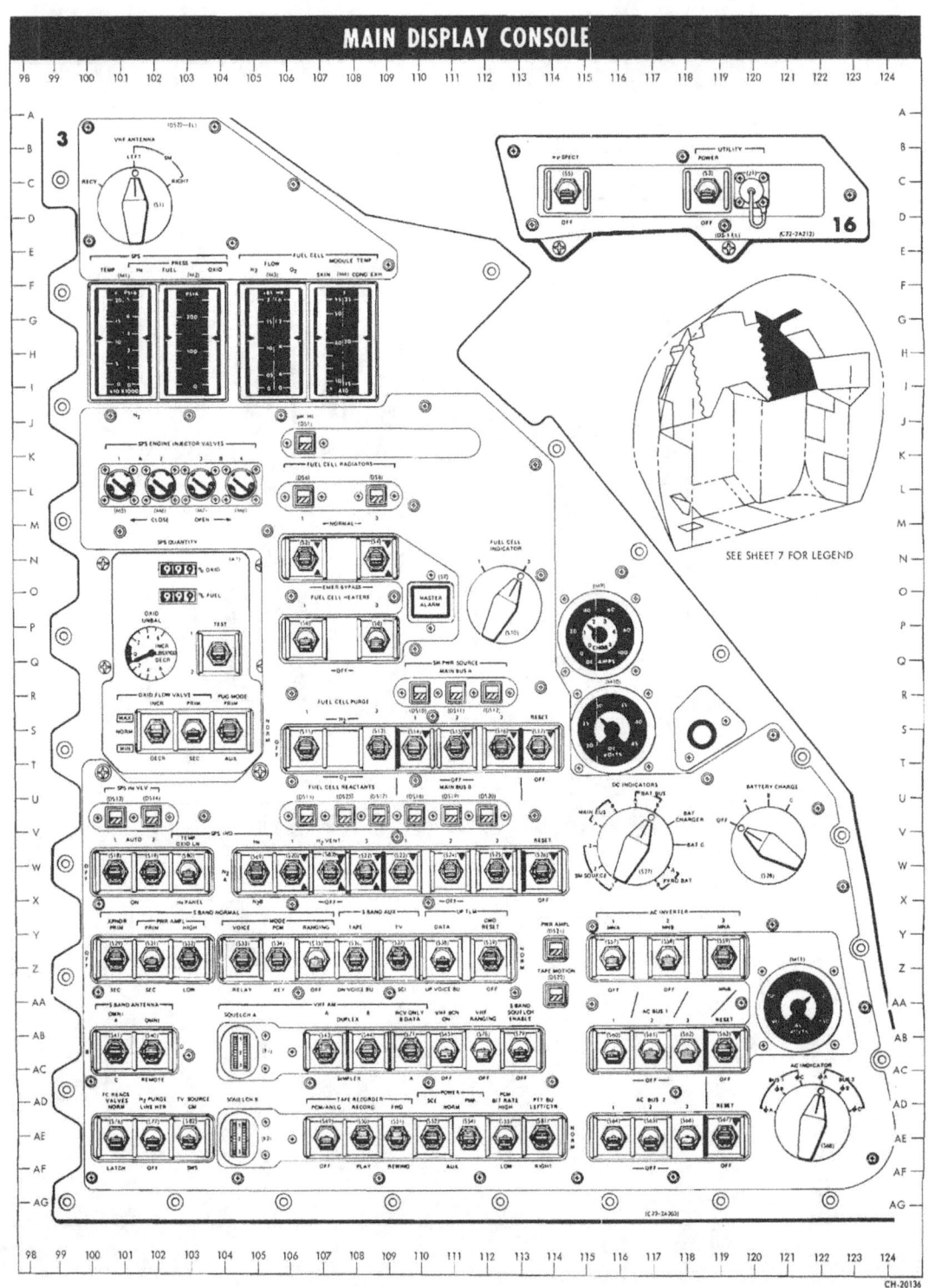

Figure 3-1 (Sheet 5 of 7)

Section III Emergency Detection and Procedures

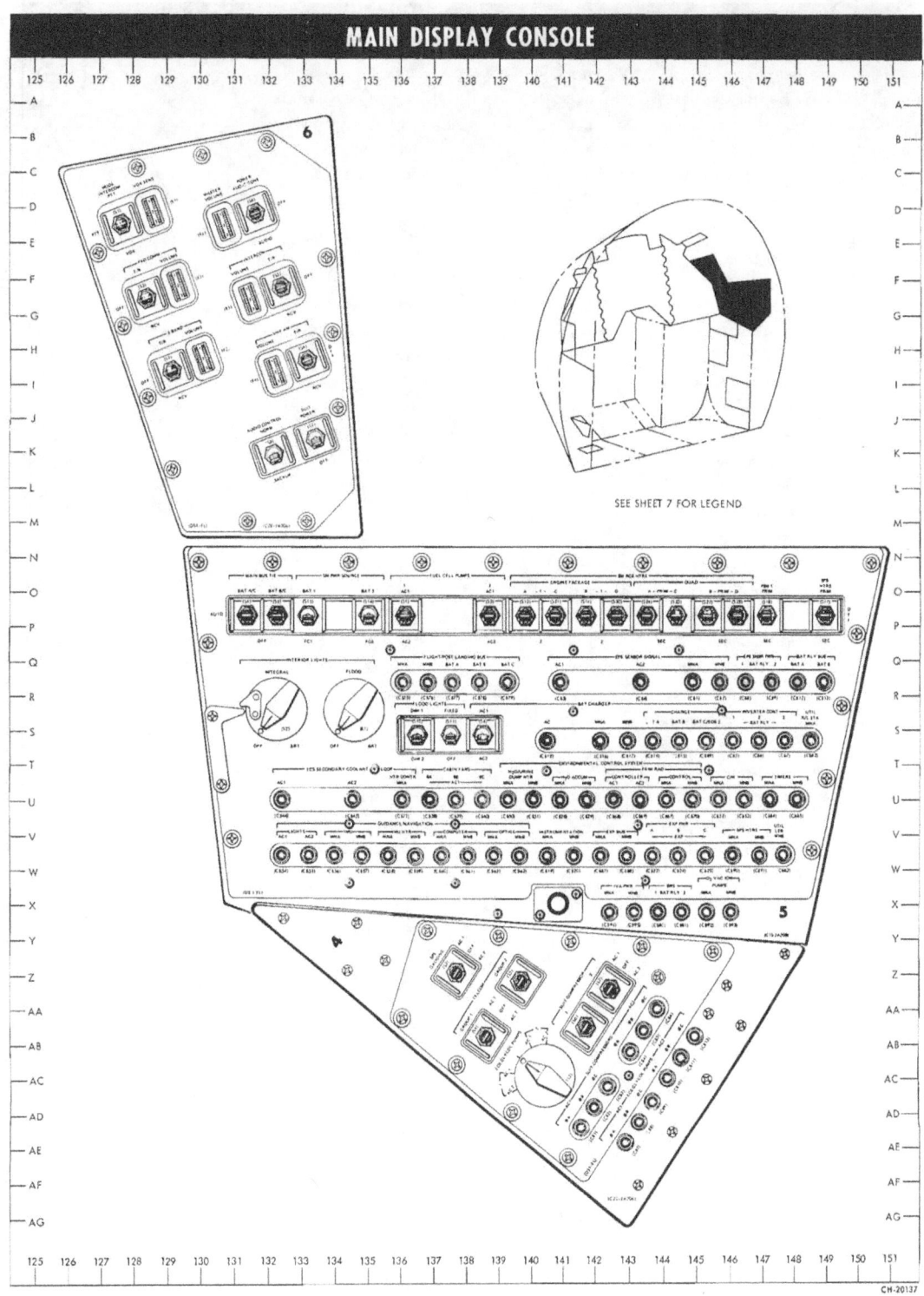

Figure 3-1 (Sheet 6 of 7)

Section III Emergency Detection and Procedures

MAIN DISPLAY CONSOLE LEGEND

ABORT LIGHT	N-51	IMU CAGE SWITCH	AE-36
ABORT SYSTEM SWITCHES	AA-59	LES MOTOR FIRE PUSHBUTTON	X-49
ACCELEROMETER	O-40	LIFTOFF/NO AUTO ABORT LIGHTS	W-50
ALTIMATER	C-51	LIMIT CYCLE SWITCH	T-37
APEX COVER JETT PUSHBUTTON	W-52	LOGIC POWER SWITCH	AB-13
ATTITUDE SET CONTROL PANEL	AD-32	LV ENGINE LIGHTS	S-51
ATT DEADBAND SWITCH	T-38	LV GUID LIGHT	R-52
ATT SET SWITCH	R-40	LV RATE LIGHT	R-50
ATVC GAIN SWITCH	AC-46	LV RATES SWITCH	AA-59
AUTO RCS SELECT SWITCHES	O-14	LV/SPS IND SWITCHES	AE-40
BMAG MODE SWITCHES	Z-32	LV TANK PRESS GAUGES	X-46
BMAG PWR SWITCHES	AA-9	$LV\alpha /SPS\ P_c$ INDICATOR	W-42
CANARD DEPLOY PUSHBUTTON	Y-49	MAIN DEPLOY PUSHBUTTON	Y-52
CMC ATT SWITCH	O-37	MAIN RELEASE SWITCH	AG-56
CMC MODE SWITCH	W-39	MANUAL ATTITUDE SWITCHES	T-34
CM PRPLNT SWITCHES	AC-51	MASTER ALARM LIGHT	L-40, O-110
CM RCS LOGIC SWITCH	AC-50	PRPLNT DUMP SWITCH	AA-56
CM RCS He DUMP PUSHBUTTON	Z-52	RATE SWITCH	T-39
CM RCS PRESS SWITCH	O-68	REACTION CONTROL SYSTEM C/B	U-8
CM/SM SEP SWITCHES	Y-61	ROLL STABILITY INDICATOR	L-45
CORRIDOR INDICATORS	K-45, L-45	ROT CONTR PWR SWITCHES	W-33
CSM/LV SEP PUSHBUTTON	Z-49	SC CONT SWITCH	W-38
DIRECT O_2 SWITCH	AD-10	SCS ELECTRONICS POWER SWITCH	AA-10
DIRECT ULLAGE PUSHBUTTON	Z-38	SCS TVC SERVO POWER SWITCHES	AA-15
DOCK RING SEP SWITCHES	Y-59	SCS TVC SWITCHES	AC-39
DROGUE DEPLOY PUSHBUTTON	X-52	SERVICE PROPULSION SYSTEM C/B	W-10
DSKY PANEL	R-60	SIG COND/DRIVER BIAS POWER SWITCHES	AC-12
EDS BATTERY C/B	X-5	SPS THRUST LIGHT	K-49
EDS POWER SWITCH	Z-17	SPS THRUST SWITCH	Z-36
EDS AUTO SWITCH	Y-57	SPS GIMBAL THUMBWHEEL CONTROLS	AA-46
ELS/CM-SM SEP C/B	X-9	SPS GIMBAL MOTORS SWITCHES	AC-42
ELS SWITCHES	AC-48	STABILIZATION CONTL SYS C/B	P-5, R-10, S-9
EMS FUNCTION SWITCH	G-45	THRESHOLD INDICATOR (.05 G LIGHT)	K-48
EMS MODE SWITCH	I-44	THRUST ON PUSHBUTTON	AA-38
ENTRY SWITCHES	AE-38	TRANS CONTR SWITCH	T-40
EVENT TIMER INDICATOR	P-51	TVC GIMBAL DRIVE SWITCHES	AE-42
EVENT TIMER SWITCHES	AE-50	TWR JETT SWITCHES	AA-61
FDAI	J-60, Q-45	UP TLM SWITCH	O-66
FDAI CONTROL SWITCHES	R-37	2 ENG OUT SWITCH	AA-58
FDAI/GPI POWER SWITCH	AA-12	ΔV THRUST SWITCHES	Z-41
GDC ALIGN PUSHBUTTON	AC-36	ΔV/EMS SET SWITCH	K-52
GUIDANCE SWITCH	AC-56	ΔV/RANGE COUNTER	L-48
G-V PLOTTER	H-49		

NOTE:
THE PANEL INDICATORS AND SWITCHES ASSOCIATED WITH THE EDS AND LAUNCH VEHICLE CONTROL ARE LISTED FOR EASE IN LOCATION. REFER TO THE TEXT FOR DISCUSSION OF SWITCH AND INDICATOR FUNCTIONS.

Figure 3-1 (Sheet 7 of 7)

Section III Emergency Detection and Procedures

prior to IECO and is not active during S-IVB flight. After the automatic abort system is deactivated the LV RATE light is used only to indicate that preset LV overrate settings have been exceeded. The light is a single cue for abort. Secondary cues will be provided by the FDAI, physiological cues, or ground communications. The LV preset overrate settings are:

Pitch and Yaw	5 ± 0.5 deg/sec	Liftoff to J-2 ignition
	10 ± 0.5 deg/sec	J-2 ignition to J-2 cutoff
Roll	20 ± 0.5 deg/sec	Liftoff to J-2 cutoff

The LV RATE light will come on at any time during first or second stage flight if the LV rates exceed these values.

Note

The LV RATE light may blink on and off during normal staging.

LV GUID LIGHT.

The LV platform (ST-124M-3) is interrogated every 40 ms for the correct attitude. Three or more excessive attitude discrepancy readings per second, in any one axis, will cause the system to switch to the coarse resolvers. Fifteen or more excessive attitude discrepancy readings per second on the coarse resolvers, in any one axis, will inhibit the LV attitude change commands being sent to the flight control system. The flight control system will then hold the last acceptable command.

A signal is sent from the LVDA to activate the LV GUID light (R-52, figure 3-1) at the same time the flight control commands are inhibited. It is a single cue for abort. Second cues will be provided by the LV RATE light (only when the automatic abort system is on) and by the FDAI, ground information, or both.

The LV GUID light ON is a prerequisite to spacecraft takeover of the Saturn during launch vehicle burn phases. See subsequent paragraph on GUIDANCE SWITCH for a further discussion of spacecraft takeover.

LIFTOFF/NO AUTO ABORT LIGHTS.

The LIFTOFF and NO AUTO ABORT lights (W-50, figure 3-1) are independent indications contained in one switch/light assembly. The LIFTOFF light ON indicates that vehicle release has been commanded and that the IU umbilical has ejected. The spacecraft digital event timer is started by the same function. The LIFTOFF light is turned OFF at S-IB IECO.

The NO AUTO ABORT light ON indicates that one or both of the spacecraft sequencers did not enable automatic abort capability at liftoff. Automatic abort capability can be enabled by pressing the switch/light pushbutton. If the light remains ON one or both of the automatic abort circuits failed to energize. The crew must then be prepared to back up the automatic abort manually. The NO AUTO ABORT light is also turned OFF at S-IB IECO.

WARNING

If the NO AUTO ABORT pushbutton is depressed at T-0 and a pad shutdown occurs, a pad abort will result.

ABORT LIGHT.

The ABORT light (N-51, figure 3-1) can be illuminated by ground command from the Flight Director, the Mission Control Center (MCC) Booster Systems Engineer, the Flight Dynamics Officer, the Complex 39 Launch Operations Manager (until tower clearance

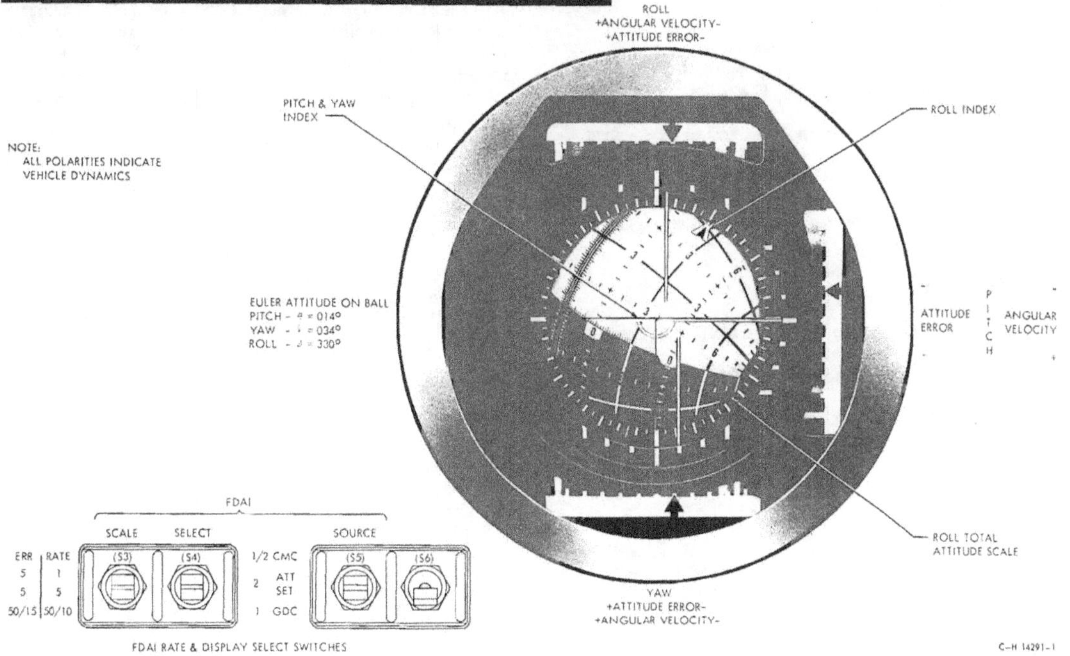

Figure 3-2

Section III Emergency Detection and Procedures

at +10 sec), or in conjunction with range safety booster engine cutoff. The ABORT light ON constitutes one abort cue. An RF voice abort request constitutes one abort cue.

Note

Pilot abort action is required prior to receipt of an ABORT light or a voice command for a large percentage of the time critical launch vehicle malfunctions, particularly at liftoff and staging.

ANGLE OF ATTACK METER.

The angle of attack (Qa) meter (W-42, figure 3-1) is time shared with service propulsion system (SPS) chamber pressure. The Qa display is a pitch and yaw vector summed angle-of-attack/dynamic pressure product (Qa). It is expressed in percentage of total pressure for predicted launch vehicle breakup (breakup limit equals 100%). It is effective as an information parameter only during the high Q flight region from +50 sec to +1 min 40 sec. Except as stated above, during ascent, the Qa meter provides trend information on launch vehicle flight performance and provides a secondary cue for slow-rate guidance and control malfunctions. Primary cues for guidance and control malfunctions will be provided by the FDAI, physiological cues, and/or MCC callout.

Nominal angle of attack meter indications should not exceed 25%. Expected values based on actual winds aloft will be provided by MCC prior to launch.

ACCELEROMETER.

The accelerometer (O-40, figure 3-1) indicates longitudinal acceleration/deceleration. It provides a secondary cue for certain engine failures and is a gross indication of launch vehicle performance. The accelerometer also provides a readout of G-forces during reentry.

ALTIMETER.

Due to dynamic pressure, static source location, and instrument error the altimeter (C-51, figure 3-1) is not considered to be an accurate instrument during the launch phase. The primary function of the altimeter is to provide an adjustable reference (set for barometric pressure on launch date) for parachute deployment for pad/near pad LES aborts. However, the aerodynamic shape of the CM coupled with the static source location produces errors up to 1800 feet. Therefore, the main parachutes must be deployed at an indicated 3800 feet (depends on launch day setting) to ensure deployment above 2000 feet true altitude.

EVENT TIMER.

The event timer (P-51, figure 3-1) is a critical display because it is the primary cue for the transition of abort modes, manual sequenced events, monitoring roll and pitch program, staging, and S-IVB insertion cutoff. The event timer is started by the liftoff command which enables automatic aborts. The command pilot should be prepared to manually back up its start to ensure timer operation. The event timer is reset to zero automatically with abort initiation.

MASTER ALARM LIGHT.

There are three MASTER ALARM lights, one on main display panel 1 (L-40, figure 3-1), one on main display panel 3 (O-110) and one in the lower equipment bay. The three MASTER ALARM lights ON alert the flight crew to critical spacecraft failures or out-of-tolerance conditions identified in the caution and warning light array. After extinguishing the alarm lights, action should be

initiated to correct the failed or out-of-tolerance subsystem. If crew remedial action does not correct the affected subsystem, then an abort decision must be made based on other contingencies. Secondary abort cues will come from subsystem displays, ground verification, and physiological indications.

Note

The Commander's MASTER ALARM light (L-40) will not illuminate during the launch phase, but the other two MASTER ALARM lights can illuminate and the alarm tone will sound.

LV TANK PRESS GAUGES.

The LV TANK PRESS Gauges (X-46, figure 3-1) indicate the S-IVB tank pressures. The two left-hand pointers indicate S-IVB oxidizer pressure. The two right-hand pointers indicate S-IVB fuel tank pressure until LV/spacecraft separation.

LAUNCH VEHICLE NORMAL CONTROLS.

GUIDANCE SWITCH.

The GUIDANCE switch is a two position guarded toggle switch with the two positions being IU and CMC (AC-56, figure 3-1). The switch controls a relay in the IU which selects either the IU or the CMC in the spacecraft as the source of flight control attitude error signals for the LV. The normal position of the GUIDANCE switch is IU. Placing the switch in the CMC position permits spacecraft control of the LV under certain conditions.

Guidance Reference Failure Condition.

During the LV burn modes, when the LVDC recognizes a guidance reference failure and turns on the LV GUID light, the GUIDANCE switch can be placed in the CMC position. (The switch function is interlocked in the LVDC such that a guidance reference failure must be recognized before the CMC switch position will be honored.) With the switch in the CMC position and LVDC recognition of a guidance reference failure, the LV will receive attitude error signals from the CMC via the LVDC. During the S-IB burn phase the CMC provides attitude error signals based on preprogrammed polynomial data. During S-IVB burn phases the rotational hand control (RHC) is used to generate the attitude error signals. The amount of attitude error transmitted to the LVDC is a function of how long the RHC is out of detent. The RHC in this mode is not a proportional control. S-IVB engine cutoff must be executed with the RHC ccw ($<$ 3 sec) based on observation of the computed velocity display.

No Guidance Reference Failure Condition.

During the coast mode, T_4, the guidance reference failure is not interlocked with the GUIDANCE switch and the spacecraft can assume control of the LV any time the switch is placed in the CMC position. With the GUIDANCE switch in the CMC position and no guidance reference failure, the LVDC will function in a follow-up mode. When LV control is returned to the LVDC during T_3 attitude orientation will be maintained with reference to local horizontal.

EDS POWER SWITCH.

The EDS POWER switch (Z-17, figure 3-1) should be in the EDs power position during prelaunch and launch operations. The switch,

Section III Emergency Detection and Procedures

if placed in the OFF position results in an "EDS Unsafe" function. This function has the following effect on the launch countdown.

1. If the switch is turned OFF before automatic sequence, the automatic sequence will not be entered.

2. If the switch is turned OFF during the automatic sequence (starting at T-187 sec), but prior to T-30.0 sec, the countdown is stopped at T-30.0 sec and the countdown is recycled to T-24 min.

3. If the switch is turned OFF after T-30 sec, but prior to T-16.2 sec, the countdown is stopped at T-16.2 sec and the countdown is recycled to T-24 min.

4. If the switch is turned OFF after T-16.2 sec, but prior to T-3.1 sec (ignition command), the countdown is stopped at T-3.1 sec and the launch is recycled to T-24 min (or is scrubbed for that day).

5. If the switch is turned OFF after T-3.1 sec, but prior to T-50 ms (launch commit), the countdown is stopped immediately and the launch is scrubbed for that day.

6. After T-50 ms, the switch will not stop launch; however, the SC EDS power off will have the following effect on the mission.

a. The EDS displays will not be operative.

b. The Auto-Abort capability will not be enabled; however, manual abort can be initiated.

WARNING

If the SC EDS POWER switch is returned to the ON position after liftoff, an immediate abort may result, depending upon which relays in the EDS circuit activate first.

LAUNCH VEHICLE EDS CONTROLS.

EDS SWITCH.

The EDS switch is a two position toggle switch with the two positions being AUTO and OFF (Y-57, figure 3-1). Prior to liftoff the EDS switch is placed in the AUTO position so that an automatic abort will be initiated if:

1. A LV structural failure occurs between the IU and the CSM.

2. Two or more S-IB engines drop below 90% of rated thrust.

3. LV rates exceed 5 degrees per second in pitch or yaw or 20 degrees per second in roll.

The two engine out and LV rate portions of the auto abort system can be manually disabled, individually, by the crew (Normally at T+ 1 min 40 sec). The LV RATES and the 2 engine out is automatically disabled just prior to IECO.

DOCK RING SEP SWITCHES.

The DOCK RING SEP switches are a pair of two position guarded toggle switches (Y-59, figure 3-1). Their purpose is to provide a means of manually initiating final separation of the LM docking ring. During a normal entry or an SPS abort, the docking ring must be jettisoned by actuation of the DOCK RING SEP switches. Failure to jettison the ring could possibly hamper normal earth landing system (ELS) functions.

CM/SM SEP SWITCHES.

The two CM/SM SEP switches (Y-61, figure 3-1) are redundant, momentary ON, guarded switches, spring loaded to the OFF position. They are normally used by the command pilot to accomplish CM/SM separation prior to the reentry phase. These switches can also be used to initiate an LES abort in case of a failure in either the EDS or the translational controller. All normal post-abort events will then proceed automatically. However, the CANARD DEPLOY pushbutton (Y-49, figure 3-1) should be depressed 11 seconds after abort initiation, because canard deployment will not occur if the failure was in the EDS instead of the translational controller.

PRPLNT SWITCH.

The PRPLNT switch is a two position toggle switch with the two positions being DUMP AUTO and RCS CMD (AA-56, figure 3-1). The switch is normally in the DUMP AUTO position prior to liftoff in order to automatically dump the CM reaction control system (RCS) propellants, and fire the pitch control (PC) motor if an abort is initiated during the first 61 seconds of the mission. The propellant dump and PC motor are inhibited by the SC sequencer at 61 sec. The switch in the RCS CMD position will inhibit propellant dump and PC motor firing at any time.

ABORT SYSTEM—2 ENG OUT SWITCH.

The 2 ENG OUT switch is a two position toggle switch, the two positions being AUTO and OFF (AA-58, figure 3-1). The purpose of this switch is to enable or disable EDS automatic abort capability for a two engine out condition. Normal position of the switch is AUTO, which enables the EDS automatic abort capability. With the switch in OFF the EDS automatic abort capability is disabled.

ABORT SYSTEM—LV RATES SWITCH.

The LV RATES switch is a two position toggle switch, the two positions being AUTO and OFF (AA-59, figure 3-1). The purpose of this switch is to enable or disable EDS automatic abort capability for excessive LV rates. Normal position of the switch is AUTO, which enables the EDS automatic abort capability for excessive LV rates. Placing the switch in OFF disables the capability. The capability is disabled automatically just prior to IECO.

ABORT SYSTEM—TWR JETT SWITCHES.

There are two redundant TWR JETT guarded toggle switches (AA-61, figure 3-1). When these switches are placed in AUTO, explosive bolts and the tower jettison motor are fired to jettison the LET. Appropriate relays are also de-energized so that if an abort is commanded, the SPS abort sequence and not the LES sequence will occur.

MAIN RELEASE SWITCH.

The MAIN RELEASE switch (AG-56, figure 3-1) is a toggle switch guarded to the down position. It is moved to the up position to manually release the main chutes after the command module has landed. No automatic backup is provided. This switch is armed by the ELS LOGIC switch ON and the 10K barometric switches closed (below 10,000 feet altitude).

Note

The ELS AUTO switch must be in the AUTO position to allow the 14-sec timer to expire before the MAIN CHUTE RELEASE switch will operate.

ELS SWITCHES.

There are two ELS two position toggle switches (AC-48, figure 3-1). The left hand switch is guarded to the OFF position and should only be placed in the LOGIC position during normal reentry

Section III Emergency Detection and Procedures

or following an SPS abort, and then only below 45,000 feet altitude. If the ELS LOGIC and AUTO switches are activated at any time below 24,000 feet (pressure altitude), the landing sequence will commence, i.e., LES and apex cover jettison and drogue deployment. If activated below 10,000 feet altitude, the main chutes will also deploy. ELS LOGIC is automatically enabled following any manual or auto EDS initiated LES abort. It should be manually backed up if time permits.

WARNING

Do not use ELS LOGIC and ELS AUTO switches during normal launch. Activation of ELS LOGIC and ELS AUTO switches below 40,500 feet during ascent will initiate the landing sequence causing LES and apex cover jettison and deployment of drogue chutes.

The right hand switch is not guarded and has positions of AUTO and MAN. Its normal position is MAN until AUTO is required at +14 sec on mode 1A and 1B aborts or at 30,000 ft on high altitude aborts or entry to enable the automatic sequencing of the ELS during a CM descent period. If the switch is placed in MAN it will inhibit all automatic sequencing of the ELS.

CM RCS PRESS SWITCH.

The CM RCS PRESS switch is a two position guarded toggle switch (O-68, figure 3-1). Any time the CM is to be separated from the SM, the CM RCS must be pressurized. The normal sequence of events for an abort or normal CM/SM SEP is to automatically deadface the umbilicals, pressurize the CM RCS, and then separate the CM/SM. However, if the automatic pressurization fails, the CM RCS can be pressurized by the use of the CM RCS PRESS switch.

ABORT CONTROLS.

TRANSLATIONAL CONTROLLER.

The TRANSLATIONAL CONTROLLER, which is mounted on the left arm of the commanders couch, can be used to accomplish several functions. A manual LES abort sequence is initiated by rotating the T-handle fully ccw. This sends redundant engine cutoff commands to the LV (engine cutoff from the SC is inhibited during the first 30 sec of flight), initiates CM/SM separation, fires the LES motors, resets the SC sequencer and initiates the post abort sequence. For a manually initiated SPS abort, the ccw rotation of the T-handle commands LV engine cutoff, resets the SC sequencer and initiates the CSM/LV separation sequence.

Note

Returning the T-handle to neutral before the 3 sec expires results only in an engines cutoff signal rather than a full abort sequence.

CW rotation of the T-handle transfers control of the SC from the CMC to the SCS. The T-handle can also provide translation control of the CSM along one or more axes. The T-handle is mounted approximately parallel to the SC axis; therefore, T-handle movement will cause corresponding SC translation. Translation in the +X axis can also be accomplished by use of the direct ullage pushbutton; however, rate damping is not available when using this method.

SEQUENCER EVENTS—MANUAL PUSHBUTTONS.

These are a group of covered pushbutton switches (X-51, figure 3-1) which provide a means of manual backup for abort and normal reentry events which are otherwise sequenced automatically.

LES MOTOR FIRE Switch.

The LES MOTOR FIRE switch is used to fire the launch escape motor for an LES abort if the motor does not fire automatically. It is also a backup switch to fire the LET jettison motor in the event the TWR JETT switches fail to ignite the motor.

CANARD DEPLOY Switch.

The CANARD DEPLOY switch is used to deploy the canard in the event it does not deploy automatically during an abort.

CSM/LV SEP Switch.

The CSM/LV SEP switch is used as the primary means of initiating CSM/LV separation after the ascent phase of the mission. When the switch is pressed it initiates ordnance devices which explosively sever the SLA, circumferentially around the forward end, and longitudinally, into four panels. The four panels are then rotated away from the LV by ordnance thrusters. Upon reaching an angle of 45 degrees spring thrusters jettison the panels away from the SC. The same ordnance train separates the CSM/LV umbilical. The CSM/LV SEP switch is also used as a backup to initiate separation of the SLA when an SPS abort cannot be initiated from the TRANSLATIONAL CONTROLLER. The +X translation would have to be manually initiated under these circumstances.

APEX COVER JETT Switch.

The APEX COVER JETT switch is used to jettison the APEX COVER in the event it fails to jettison automatically during an abort or a normal reentry.

DROGUE DEPLOY Switch.

The DROGUE DEPLOY switch is used to deploy the drogue parachutes in the event they fail to deploy automatically 2 sec after the 24,000-foot barometric pressure switches close.

MAIN DEPLOY Switch.

The MAIN DEPLOY switch is used to deploy main parachutes in the event they fail to deploy automatically when the 10,000-foot barometric pressure switches close. This switch can also be used to manually deploy the main parachutes during mode 1A aborts.

CM RCS He DUMP Switch.

The CM RCS He DUMP switch is used to initiate depletion of the CM He supply if depletion does not occur normally as an automatic function during abort.

SERVICE PROPULSION SYSTEM (SPS) CONTROL.

The SPS provides primary thrust for major velocity changes subsequent to SC/LV separation and prior to CM/SM separation. The SPS is also used to accomplish mode III and IV aborts.

SPS Engine Start.

SPS engine ignition can be commanded under control of the CMC, the SCS or manually. For all modes of operation, the ΔV THRUST switch A (Z-40), or ΔV THRUST switch B (Z-42), or both, must be in the NORMAL position. (If double-bank operation is desired, Δ V THRUST switch B is moved to NORMAL 5 seconds or more following SPS ignition.) Ullage is normally provided by the THC (+X translation) and backup is by DIRECT ULLAGE pushbutton (Z-38, figure 3-1). The DIRECT ULLAGE pushbutton is a momentary switch and must be held depressed until the ullage maneuver is complete. It does not provide rate damping. The SPS THRUST light (K-49) will illuminate when the engine is firing. In the CMC mode THRUST ON is commanded as a result of internal computations. Prerequisites are SC CONT switch (W-38) in CMC position and THC in neutral (except for +X translation). In the SCS mode, the SC CONT switch must be in the SCS position, or the THC rotated cw. SPS ignition is commanded by pressing the THRUST ON pushbutton (AA-38). Prerequisites are +X translation (from THC or DIRECT ULLAGE pushbutton) and ΔV/RANGE counter > 0 (L-48). Once SPS ignition has occurred in this mode, +X translation and THRUST ON commands can be removed and ignition is maintained until a THRUST OFF command is generated. In the manual mode the SPS THRUST switch (Z-36) is placed in DIRECT ON. Ignition is maintained until a THRUST OFF command is generated.

WARNING

The SPS THRUST switch is a single-point failure with the ΔV THRUST switches in the NORMAL position.

SPS Engine Shutdown.

In the CMC mode, normal engine shutdown is commanded by the CMC as a result of internal computations. In the SPS mode, engine shutdown can be commanded by the EMS ΔV counter running down to 0 or by placing the ΔV THRUST switches (both) to OFF. In the manual mode, shutdown is commanded by placing the ΔV THRUST switches (both) to OFF.

Thrust Vector Control.

Four gimbal motors control the SPS engine position in the pitch and yaw planes; two motors for each plane. These motors are activated by the SPS GIMBAL MOTORS switches (AC-42, figure 3-1).

Note

The motors should be activated one at a time due to high current drain during the start process.

Control signals to the gimbal motors can come from the CMC, SCS or the RHC. Gimbal trim thumbwheels (AA-46) can also be used to position the gimbals in the SCS ΔV mode. The TVC GIMBAL DRIVE switches (AE-42) are three position toggle switches. Their purpose is to select the source and routing of TVC signals. The switches are normally in the AUTO position.

STABILITY CONTROL SYSTEM (SCS).

The SCS is a backup system to the primary guidance navigation and control system (PGNCS). It has the capability of controlling rotation, translation, SPS thrust vector and associated displays. Switches which affect the SCS are discussed in the following paragraphs.

AUTO RCS SELECT Switches.

Power to the RCS control box assembly is controlled by 16 switches (O-14, figure 3-1). Individual engines may be enabled or disabled as required. Power to the attitude control logic is also controlled in this manner, which thereby controls all attitude hold and/or maneuvering capability using SCS electronics (automatic coils).

Note

The automatic coils cannot be activated until the RCS ENABLE is activated either by the MESC or manually.

DIRECT Switches.

Two DIRECT switches (W-35, figure 3-1) provide for manual control of the SM RCS engines. Switch 1 controls power to the direct solenoid switches in rotational controller 1 and switch 2 controls power to the direct solenoid switches in rotational controller 2. In the down position switch 1 receives power from MNA and switch 2 receives power from MNB. In the up position both switches receive power from both MNA and MNB. Manual control is achieved by positioning the rotational control hardover to engage the direct solenoids for the desired axis change.

ATT SET Switch.

The ATT SET switch (R-40, figure 3-1) selects the source of total attitude for the ATT SET resolvers as outlined below.

Position	Function	
UP	IMU	Applies inertial measurement unit (IMU) gimbal resolver signal to ATT SET resolvers. FDAI error needles display differences. Needles are zeroed by maneuvering SC or by moving the ATT SET dials.
DOWN	GDC	Applies GDC resolver signal to ATT SET resolvers. FDAI error needles display differences resolved into body coordinates. Needles zeroed by moving SC or ATT SET dials. New attitude reference is established by depressing GDC ALIGN button. This will cause GDC to drive to null the error; hence, the GDC and ball go to ATT SET dial value.

MANUAL ATTITUDE Switches.

The three MANUAL ATTITUDE switches (T-34, figure 3-1) are only operative when the SC is in the SCS mode of operation.

Position	Description
ACCEL CMD	Provides continuous RCS firing as long as the rotational controller is out of detent.
RATE CMD	Provides proportional rate command from rotational controller with inputs from the BMAG's in a rate configuration.
MIN IMP	Provides minimum impulse capability through the rotational controller.

LIMIT CYCLE Switch.

The LIMIT CYCLE switch (T-37, figure 3-1), when placed in the LIMIT CYCLE position, inserts a psuedo-rate function which provides the capability of maintaining low SC rates while holding the SC attitude within the selected deadband limits (limit cycling). This is accomplished by pulse-width modulation of the switching amplifier outputs. Instead of driving the SC from limit-to-limit with high rates by firing the RCS engines all the time, the engines are fired in spurts proportional in length and repetition rate to the switching amplifier outputs. Extremely small attitude corrections could be commanded which would cause the pulse-width of the resulting output command to be of too short a duration to activate the RCS solenoids. A one-shot multivibrator is connected in parallel to ensure a long enough pulse to fire the engines.

RATE and ATT DEADBAND Switch.

The switching amplifier deadband can be interpreted as a rate or an attitude (minimum) deadband. The deadband limits are a function of the RATE switch (T-39, figure 3-1). An additional deadband can be enabled in the attitude control loop with the ATT DEADBAND switch (T-38, figure 3-1) See figure 3-3 for relative rates. The rate commanded by a constant stick deflection (proportional rate mode only) is a function of the RATE switch position. The rates commanded at maximum stick deflection (soft stop) are shown in figure 3-4.

Section III Emergency Detection and Procedures

ATTITUDE DEADBAND SWITCH POSITIONS

RATE SWITCH POSITION	RATE DEADBAND °/SEC	ATT DEADBAND SWITCH POSITION	
		MINIMUM	MAXIMUM
LOW	±0.2	±0.2°	±4.2°
HIGH	±2.0	±4.0°	±8.0°

Figure 3-3

MAXIMUM PROPORTIONAL RATE COMMAND

RATE SWITCH POSITION	MAXIMUM PROPORTIONAL RATE COMMAND	
	PITCH AND YAW	ROLL
LOW	0.7°/SEC	0.7°/SEC
HIGH	7.0°/SEC	20.0°/SEC

Figure 3-4

EMS FUNCTION SWITCH OPERATION

OPERATIONAL MODE	SWITCH SELECTION	SWITCH POSITION	DESCRIPTION
ΔV MODE	START AT ΔV AND ROTATE CLOCKWISE.	ΔV	CORRECT PORTION FOR SPS THRUST MONITORING (ΔV DISPLAY).
		ΔV SET / VHF RNG	ENABLES USE OF EMS/ΔV SET SWITCH TO SLEW ΔV/RANGE DISPLAY TO INITIAL CONDITION FOR ΔV TEST AND SPS THRUST MONITORING. PROVIDES VHF RANGING INFORMATION FOR ΔV/RANGE DISPLAY.
		ΔV TEST	VERIFIES CORRECT OPERATION OF: 1. SPS THRUST LAMP 2. ΔV DISPLAY (AND COUNTDOWN ELECTRONICS). (SEE ΔV SET POSITION ABOVE.) 3. THRUST-OFF COMMAND
SELF TEST AND ENTRY MODE	START AT NO. 1 AND ROTATE COUNTERCLOCKWISE	NO. 1	TEST EMS FOR DECELERATION <.05G. (NO LAMPS ILLUMINATED).
		NO. 2	DECELERATION >.05G (.05G LAMP SHOULD ILLUMINATE.)
		NO. 3	DECELERATION <.262G. 1. .05G LAMP ILLUMINATES IMMEDIATELY. 2. TEN SECONDS LATER BOTTOM LAMP ON RSI ILLUMINATED. 3. ENABLES SLEWING OF ΔV/RANGE DISPLAY.
		NO. 4	EMS SYSTEM TEST. 1. ΔV/RANGE DISPLAY DRIVES TO 0 ± 0.2 IN 10 SECONDS. 2. VELOCITY SCROLL DRIVES RIGHT TO LEFT. 3. G SCRIBE DRIVES DOWN TO 9G IN 10 SECONDS. 4. .05G LAMP ON.
		NO. 5	DECELERATION >.262G. 1. ILLUMINATES .05G LAMP IMMEDIATELY. 2. TEN SECONDS LATER TOP LAMP ON RSI ILLUMINATED. 3. G SCRIBE DRIVES UP TO 0.28 ± 0.01G. 4. ENABLES SLEWING SCROLL TO 37,000 FPS.
		RNG SET	ENABLES SLEWING ΔV/RANGE DISPLAY TO INITIAL CONDITION USING EMS/ΔV SET SWITCH. G SCRIBE DRIVES VERTICALLY TO 0 ± 0.1G.
		VO SET	ENABLES SLEWING VELOCITY SCROLL TO INITIAL COUNTDOWN USING EMS/ΔV SET SWITCH.
		ENTRY	OPERATIONAL POSITION FOR EMS ENTRY DISPLAY FUNCTIONS.
		OFF	DEACTIVATES EMS EXCEPT FOR SPS THRUST ON LIGHT AND ROLL ATTITUDE INDICATOR.

Figure 3-5

Section III Emergency Detection and Procedures

SC CONT Switch.

The SC CONT switch (W-38, figure 3-1) selects the spacecraft control as listed below:

Position	Description
CMC	Selects the G&N system computer controlled SC attitude and TVC through the digital auto-pilot. An auto-pilot control discrete is also applied to CMC.
SCS	The SCS system controls the SC attitude and TVC.

BMAG MODE Switches.

The BMAG MODE switches (Z-32, figure 3-1) select displays for the FDAI using SCS inputs.

Position	Description
RATE 2	BMAG set no. 2 provides rate damping and the rate displays on the FDAI.
ATT 1	BMAG set no. 1 is uncaged providing attitude hold and attitude error display on the FDAI while,
RATE 2	Set no. 2 provides rate damping and the rate display.
RATE 1	BMAG set no. 1 provides rate damping and the rate displays on the FDAI.

ENTRY MONITOR SYSTEM (EMS).

The EMS provides displays and controls to show automatic primary guidance control system (PGNCS) entries and ΔV maneuvers and to permit manual entries in the event of a malfunction. There are five displays and/or indicators which monitor automatic or manual entries and four switches to be used in conjunction with these displays.

ENTRY EMS ROLL Switch.

The ENTRY EMS ROLL switch (AE-37, figure 3-1) enables the EMS roll display for the earth reentry phase of the flight.

ENTRY .05 G Switch.

Illumination of the .05 G light (K-48, figure 3-1) is the cue for the crew to actuate the .05 G switch (AE-38). During atmospheric reentry (after .05 G), the SC is maneuvered about the stability roll axis rather than the body roll axis. Consequently, the yaw rate gyro generates an undesirable signal. By coupling a component of the roll signal into the yaw channel, the undesirable signal is cancelled. The .05 G switch performs this coupling function.

EMS FUNCTION Switch.

The EMS FUNCTION switch (G-45, figure 3-1) is a 12 position mode selector switch, used as outlined in figure 3-5.

EMS MODE Switch.

The EMS MODE switch (I-44, figure 3-1) performs the following functions in the positions indicated:

NORMAL—Enables EMS accelerometer.

STBY—Inhibits operation in all but ΔV SET, RNG SET, and Vo SET positions of FUNCTION switch.

BACKUP VHF RNG

1. A manual backup to automatic .05G trigger circuits that starts scroll drive and RANGE integrator display drive circuits. Also backup to TVC MODES for velocity monitoring.

2. Does not permit negative acceleration pulses into countdown circuits.

3. Enables VHF ranging information to be displayed on ΔV/RANGE display.

Threshold Indicator (.05 G Light).

The threshold indicator (.05G light) (K-48, figure 3-1) provides the first visual indication of total acceleration sensed at the reentry threshold (approximately 290,000 feet). Accelerometer output is fed to a comparison network and will illuminate the .05 G lamp when the acceleration reaches .05 G. The light will come on not less than 0.5 sec or more than 1.5 sec after the acceleration reaches .05 G and turns off when it falls below .02G (skipout).

Corridor Indicators.

By sensing the total acceleration buildup over a given period of time, the reentry flight path angle can be evaluated. This data

APOLLO ABORT MODES

PERIOD	MODE	DESCRIPTION	NOTE
PAD TO 1:01	MODE IA	LET LOW ALT	(1)
1:01 TO 100,000 FEET (1:50)	MODE IB	LET MED ALT	(1)
100,000 FEET TO LET JETT (1:50) (2:51)	MODE IC	LET HIGH ALT	(1)
LET JETT TO $\Delta R = -1134$ NM (2:51) (9:32)	MODE II	FULL-LIFT	(1) (2)
$\Delta R = -1134$ NM TO $\Delta R = -450$ NM (9:32) (9:49)	MODE IIIA	FULL-LIFT SPS POSIGRADE	(1) (2)
$\Delta R = -450$ NM TO INSERTION (9:49) (10:01)	MODE IIIB CSM NO GO/SLV LOFTED	SPS RETRO FULL-LIFT	(1) (2)
COI CAPABILITY TO INSERTION (9:47) (10:01)	MODE IV CSM GO	SPS TO ORBIT	(1) (3)

NOTES
(1) EVENT TIMES (MINUTES, SECONDS) ARE APPROXIMATIONS
(2) ΔR = CMC SPLASH ERROR WITH HALF-LIFT. $\Delta R = -450$ NM AND FULL LIFT WILL LAND SC AT TARGET.
(3) FOR POSITIVE h AND S-IVB CUTOFF BEYOND THE 5 MINUTE TO APOGEE LINE (CREW CHART) AN APOGEE KICK MANEUVER WOULD BE RECOMMENDED FOR THE MODE IV.

Figure 3-6

Section III Emergency Detection and Procedures

is essential to determine whether or not the entry angle is steep enough to prevent superorbital "skipout." The two corridor indicator lights (K-45 and L-45, figure 3-1) are located on the face of the roll stability indicator (L-45). If the acceleration level is greater than 0.262 G at the end of a ten second period after threshold (.05 G light ON), the upper light will be illuminated. It remains ON until the G-level reaches 2 G's and then goes OFF. The lower light illuminates if the acceleration is equal to or less than 0.262 G at the end of a ten second period after threshold. This indicates a shallow entry angle and that the lift vector should be down for controlled entry, i.e., skipout will occur.

Roll Stability Indicator.

The roll stability indicator (L-45, figure 3-1) provides a visual indication of the roll attitude of the CM about the stability axis. Each revolution of the indicator represents 360 degrees of vehicle rotation. The display is capable of continuous rotation in either direction. The pointer up position (0 degrees) indicates maximum lift-up vector (positive lift) and pointer down (180 degrees) indicates maximum lift-down vector (negative lift).

G-V Plotter.

The G-V plotter assembly (H-49, figure 3-1) consists of a scroll of mylar tape and a G-indicating stylus. The tape is driven from right to left by pulses which are proportional to the acceleration along the velocity vector. The stylus which scribes a coating on the back of the mylar scroll, is driven in the vertical direction in proportion to the total acceleration. The front surface of the mylar scroll is imprinted with patterns consisting of "high G-rays" and "exit rays." The "high-G-rays" must be monitored from initial entry velocity down to 4000 feet per second. The "exit rays" are significant only between the entry velocity and circular orbit velocity and are, therefore, only displayed on that portion of the pattern. The imprinted "high-G-rays" and "exit rays" enable detection of primary guidance failures of the type that would result in either atmospheric exits at supercircular speeds or excessive load factors at any speed. The slope of the G-V trace is visually compared with these rays. If the trace becomes tangent to any of these rays, it indicates a guidance malfunction and the need for manual takeover.

Δ V/RANGE Display.

The ΔV/RANGE display provides a readout of inertial flight path distance in nautical miles to predicted splashdown after .05G. The predicted range will be obtained from the PGNCS or ground stations and inserted into the range display during EMS range set prior to entry. The range display will also indicate ΔV (ft/sec) during SPS thrusting.

ABORT MODES AND LIMITS.

The abort modes and limits listed in figures 3-6 and 3-7 are based on a nominal launch trajectory. The nominal launch phase callouts are listed in figure 3-8.

Note

More specific times can be obtained from current mission documentation.

EMERGENCY MODES.

Aborts performed during the ascent phase of the mission will be performed by using either the Launch Escape System or the Service Propulsion System.

LAUNCH ESCAPE SYSTEM.

The Launch Escape System (LES) consists of a solid propellant launch escape (LE) motor used to propel the CM a safe distance from the launch vehicle, a tower jettison motor, and a canard subsystem. A complete description on use of the system can be found in the specific mission Abort Summary Document (ASD). A brief description is as follows:

Model IA Low Altitude Mode.

In Mode IA, a pitch control (PC) motor is mounted normal to the LE motor to propel the vehicle downrange to ensure water landing and escape the "fireball." The CM RCS propellants are dumped through the aft heat shield during this mode to preclude damage to the main parachutes. The automatic sequence of major events from abort initiation is as follows:

Time	Event
00:00	Abort
	Ox rapid dump
	LE and PC motor fire
00:05	Fuel rapid dump
00:11	Canards deploy
00:14	ELS arm
00:14.4	Apex cover jett
00:16	Drogue deploy
00:18	He purge
00:28	Main deploy

The automatic sequence can be prevented, interrupted, or replaced by crew action.

Mode IB Medium Altitude.

Mode IB is essentially the same as Mode IA with the exception of deleting the rapid propellant dump and PC motor features. The canard subsystem was designed specifically for this altitude region to initiate a tumble in the pitch plane. The CM/tower combination CG is located such that the vehicle will stabilize (oscillations of ± 30 degrees) in the blunt-end-forward (BEF) configuration. Upon closure of barometric switches, the tower would be jettisoned and the parachutes automatically deployed. As in Mode IA, crew intervention can alter the sequence of events if desired.

Mode IC High Altitude.

During Mode IC the LV is above the atmosphere. Therefore, the canard subsystem cannot be used to induce a pitch rate to the vehicle. If the LV is stable at abort, the LET is manually jettisoned and the CM oriented to the reentry attitude. This method provides a stable reentry but requires a functioning attitude reference. With a failed platform the alternate method will be to introduce a five degree per second pitch rate into the system. The CM/tower combination will then stabilize BEF as in Mode IB. The LES would likewise deploy the parachutes at the proper altitudes.

Mode II.

The Sm RCS engines are used to propel the CSM away from the LV. When the CSM is a safe distance and stable, the CM

Section III Emergency Detection and Procedures

is separated from the SM and maneuvered to a reentry attitude. A normal entry procedure is followed from there.

SERVICE PROPULSION SYSTEM.

The Service Propulsion System (SPS) aborts utilize the Service Module SPS engine to maneuver to a planned landing area, or boost into a contingency orbit. The SPS abort modes are:

Mode IIIA.

The SPS engine is used for a posigrade maneuver to fly over the cold water in the north Atlantic and land at a predetermined point. The duration of the SPS burn is dependent on the time of abort initiation. Upon completion of the burn, normal entry procedures will be followed.

Mode IIIB.

The SPS engine is used to slow the CSM combination (retrograde maneuver) so as to land at a predetermined point in the Atlantic Ocean. The length of the SPS burn is dependent upon the time of abort initiation. Upon completion of the retro maneuver, normal entry procedures will be followed.

Mode IV.

The SPS engine can be used to make up for a deficiency in insertion velocity up to approximately 1300 feet per second. This is accomplished by holding the CSM in an inertial attitude and applying the needed ΔV with the SPS to acquire the acceptable orbital velocity. If there is no communication with MCC, the crew can take over manual control and maneuver the vehicle using onboard data.

ABORT LIMITS

RATES

1. PITCH AND YAW
   LIFTOFF (T + 0) TO T + 1 MIN 40 SEC   5 ±0.5 DEG/SEC
   T + 1 MIN 40 SEC TO S-IVB CUTOFF   10 ±0.5 DEG/SEC
2. ROLL
   LIFTOFF TO S-IVB CUTOFF   20 ±0.5 DEG/SEC

PLATFORM FAILURE

1. DURING S-IB POWERED FLIGHT THE TWO CUES FOR PLATFORM FAILURE REQUIRING AN IMMEDIATE SWITCHOVER ARE:

    a. LV GUID LT - ON

    b. LV RATE LT - ON

2. AFTER LV RATE SWITCH DEACTIVATION THE PRIMARY CUE IS:

    LV GUID LT - ON

    THE SECONDARY CUES ARE:

    a. FDAI ATTITUDE

    b. LV RATES

    c. GROUND CONFIRMATION

AUTOMATIC ABORT LIMITS (LIFTOFF UNTIL DEACTIVATION AT T + 1 MIN 40 SEC)

1. RATE   PITCH - YAW   5 ±0.5 DEG/SEC
               ROLL   20 ±0.5 DEG/SEC
2. ANY TWO ENGINES FAIL
3. CM TO IU BREAKUP

S-IB ENGINE FAILURE (SUBSEQUENT TO AUTO ABORT DEACTIVATE AT T + 1 MIN 40 SEC)

1. SINGLE ENGINE FAILURE   CONTINUE MISSION.
2. SIMULTANEOUS LOSS TWO OR MORE ENGINES   ABORT IF LV CONTROL IS LOST.

S-IVB ENGINE FAILURE   ABORT (MODE IC, II, III, OR IV)

S-IVB DIFFERENTIAL TANK PRESSURE LIMITS

ΔP (ORBITAL COAST) $LH_2 > LO_2 = 26$ psid
                              $LO_2 > LH_2 = 36$ psid
                              $LO_2 > 50$ psia

Figure 3-7

Section III Emergency Detection and Procedures

NOMINAL LAUNCH PHASE VOICE CALLOUTS (BOOST TO ORBIT)

PROG TIME†	STA	ACTION/ENTRY *REPORT	OPTION/EVENT
-00:03	LCC	IGNITION*	UMBILICAL DISCONNECT
+00:01	LCC	LIFTOFF*	CMC TO P11
			DET & MET START
	CDR	CLOCK START*	
00:10	LCC	CLEAR TOWER*	ABOVE LAUNCH TOWER
00:12	CDR	ROLL & PITCH START*	ROLL AND PITCH PROGRAM START
00:30	CDR	ROLL COMPLETE*	ROLL PROGRAM COMPLETE
01:01	MCC	MODE 1B*	
	CMP	PRPLNT DUMP-RCS CMD	
00:50	CDR	MONITOR α TO T+1:40	
00:55	CMP	MONITOR CABIN PRESSURE DECREASING	IF NO DECREASE BY 17,000 FEET
			DUMP MANUALLY
01:40	CMP	EDS AUTO-OFF*	NO AUTO ABORT LIGHT-ON
		EDS ENG-OFF	
		EDS RATES-OFF	
		KEY V82E, N62E	
01:50	MCC	MODE 1C* (BASED ON 100,000 FT)	
02:10	MCC	GO/NO GO FOR STAGING*	
	CDR	GO/NO GO FOR STAGING*	SYSTEMS STATUS
02:19	CDR	INBOARD OFF*	ENGINE LIGHTS NO. 5, 6, 7, & 8 ON
			LIFTOFF LIGHT-OUT
			NO AUTO ABORT LIGHT-OUT
02:22	CDR	OUTBOARD OFF*	ENG LIGHTS 1 THRU 4-ON
02:23	CDR	S-IB/S-IVB STAGING*	ENG LIGHTS OUT
02:24		S-IVB IGNITION COMMAND	ENG LIGHT NO. 1 ON
02:27	CDR	S-IVB 65%	ENG LIGHT NO. 1 OUT
02:51	CMP	TWR JETT (2)-ON* (IF TFF >1+20)	TOWER JETTISON
	MCC	MODE II*	
	CMP	α/PC-PC	
	CDR	MAN ATT (PITCH)-RATE CMD	
02:52	CDR	GUIDANCE INITIATE*	IGM START
04:00	CDR	REPORT STATUS*	
	MCC	TRAJECTORY STATUS*	
05:00	CDR	REPORT STATUS*	
06:00	CDR	REPORT STATUS*	
07:00	CDR	REPORT STATUS*	
		GMBL MOT (4)-START-ON	
		CHECK GPI (MOMENTARILY)	INSURE ANGLES CORRECT
08:00	CDR	REPORT STATUS*	
09:00	MCC	GO/NO GO FOR STAGING*	SYSTEMS STATUS
	CDR	GO/NO GO FOR STAGING	
09:32	MCC	MODE IIIA	
09:47	MCC	MODE IV	
09:49	MCC	MODE IIIB	
10:01	CDR	SECO*	ENG LIGHT NO. 1 ON MOMENTARILY
		INSURE ORBIT	
		KEY RLSE TO N44	
10:11	MCC	INSERTION*	

†EVENT TIMES ARE APPROXIMATIONS

Figure 3-8

SECTION IV
S-IB STAGE

TABLE OF CONTENTS

Introduction	4-1
Structure	4-1
Propulsion	4-7
Control Pressure System	4-33
S-IB Hydraulic System	4-37
Electrical	4-43
Instrumentation	4-52
Environmental Conditioning	4-56
Ordnance	4-59

INTRODUCTION.

The function of the S-IB stage is to boost the upper stages and spacecraft through a predetermined trajectory that will place them at the proper altitude and attitude, with the proper velocity, at S-IVB stage ignition. The major S-IB stage assemblies, figures 4-1 through 4-3, are the tail unit with eight fins and eight H-1 engines, the nine propellant tanks, the second stage adapter (spider beam unit), and associated mechanical and electrical hardware discussed under specific systems in this section. For a summary of S-IB stage data, see figure 4-4.

STRUCTURE.

Figure 4-5 shows the primary, load-carrying structural subassemblies of the S-IB stage combined with its tail unit heat and flame shields, engine flame curtains, lox and fuel tank firewalls, and second stage adapter seal plate. Separate figures show the unique design details of the tail unit heat shield, and also the spider beam and lox fitting reinforcements employed as a result of qualification testing. The stage structure was designed to provide a safety factor of 1.10 on yield and 1.40 on ultimate, with a dry stage weight of 85,745 lbm. The adequacy of the 1.40 safety factor (ultimate) has been demonstrated by all load-carrying structural subassemblies. From a reliability point of view, the stage structure is a simple, passive system, and its reliability prediction is based solely on whether or not strength will exceed load. The 1.40 ultimate safety factor, a conservative compilation of two-sigma and three-sigma loads and allowables, plus the complete analysis and test program demonstrates a reliability assessment several times greater than the reliabilities of the other stage systems. Thus, with respect to the rest of the onboard systems, the structure has been assumed to be 100 percent reliable within the performance limits established by the CEI specification, and is considered so for purposes of calculating total stage reliability. The principal functional requirement of the stage structure is to provide adequate tankage and framework to support the other flight systems and to provide adequate support of the upper stages, both on the pad and in flight. The evolutionary structural changes resulting from design analyses, test results, static firings, and flight performance data are summarized separately in the discussion pertaining to the individual structural elements.

PROPELLANT TANKS.

The nine propellant tanks that cluster to form the main body of the stage are modifications of proven designs from the Redstone and Jupiter vehicles and have performed successfully on all Saturn I and IB flights. The individual tanks are constructed of cylindrical sections built up of mechanically-milled, butt welded, aluminum alloy skin segments that are internally reinforced with rings to form a monocoque type of construction. The material used to construct the tanks is the now readily weldable 5456 aluminum alloy in the H343 temper. The use of this alloy, made possible by welding

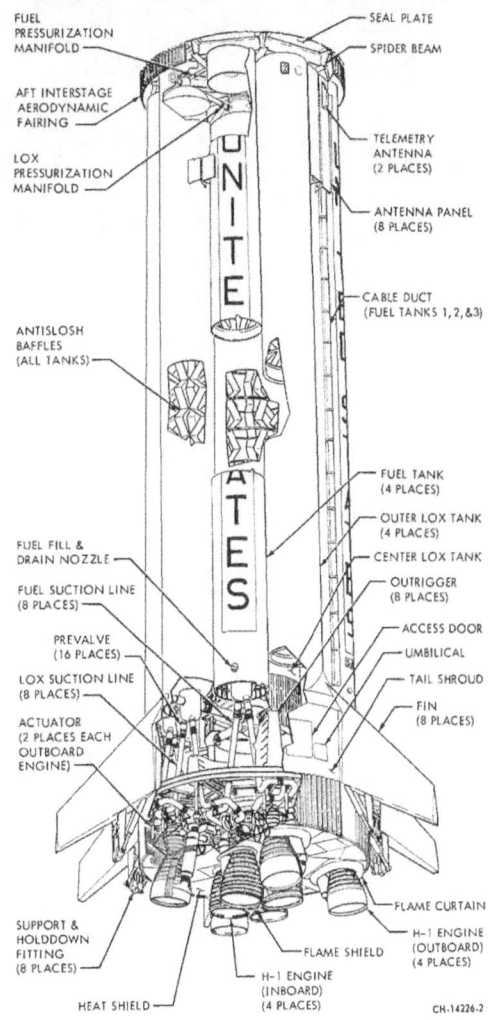

Figure 4-1

Section IV S-IB Stage

S-IB STAGE FORWARD DETAIL

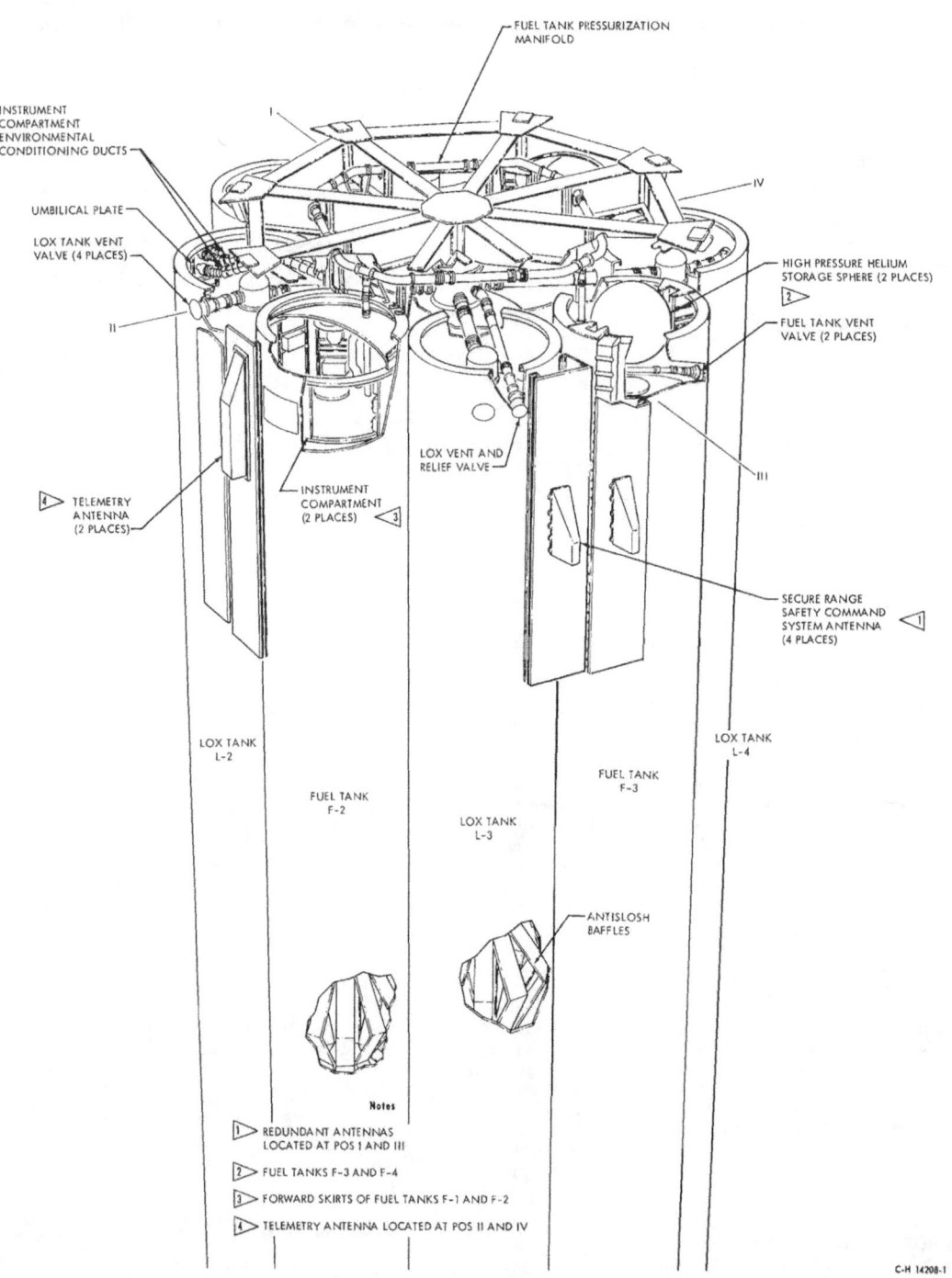

Figure 4-2

Section IV S-IB Stage

S-IB STAGE AFT DETAIL

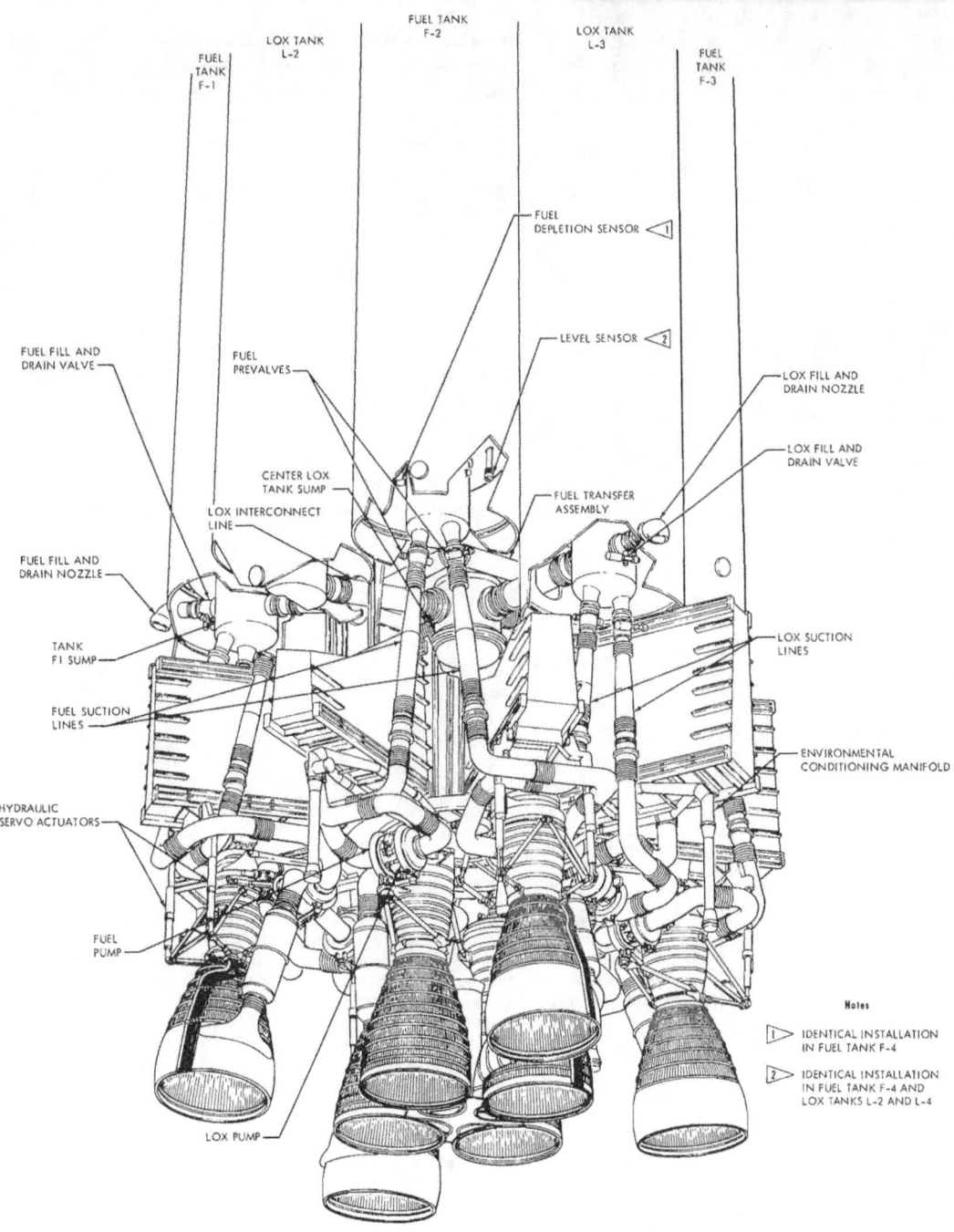

Figure 4-3

S-IB STAGE DATA SUMMARY

DIMENSIONS		**HYDRAULIC SYSTEM**	
LENGTH	80.2 FT	ACTUATORS (OUTBOARD ONLY)	2 PER ENGINE
DIAMETER		GIMBAL ANGLE	$\pm$ 8 DEG SQUARE PATTERN
AT PROPELLANT TANKS	21.4 FT	GIMBAL RATE	15 DEG/SEC IN EACH PLANE
AT TAIL UNIT ASSEMBLY	22.8 FT	GIMBAL ACCELERATION	1776 DEG/SEC^2
AT FINS	40.7 FT	**PRESSURIZATION SYSTEM**	
FIN AREA	53.3 FT^2 EACH OF 8 FINS	OXIDIZER CONTAINER	INITIAL HELIUM FROM GROUND SOURCE; S-IB BURN, GOX
MASS		FUEL CONTAINER	HELIUM
DRY STAGE	84,521 LB$_m$	OXIDIZER PRESSURE	
LOADED STAGE	997,127 LB$_m$	PREFLIGHT	58 psia
AT SEPARATION	95,159 LB$_m$	INFLIGHT	50 psia
ENGINES, DRY, LESS INSTRUMENTATION		FUEL PRESSURE	
INBOARD, PLUS TURNBUCKLES	2,003 LB$_m$ EACH	PREFLIGHT	17 psig
OUTBOARD, LESS HYDRAULICS	1980 LB$_m$ EACH	INFLIGHT	15 TO 17 psig
PROPELLANT LOAD	912,606 LB$_m$ (408,000 KG)	ULLAGE	
ENGINES		OXIDIZER	1.5%
BURN TIME	141 SEC (APPROX)	FUEL	2.0%
TOTAL THRUST (SEA LEVEL)	1.64 MLBf	**ENVIRONMENTAL CONTROL SYSTEM**	
PROPELLANTS	LOX AND RP-1	PREFLIGHT AIR CONDITIONING	AFT COMPARTMENT & INSTRUMENT COMPARTMENTS F1 & F2
MIXTURE RATIO	2.23:1 $\pm$ 2%	PREFLIGHT GN$_2$ PURGE	AFT COMPARTMENT & INSTRUMENT COMPARTMENTS F1 & F2
EXPANSION RATIO	8:1		
CHAMBER PRESSURE	702 psia	**ASTRIONICS SYSTEMS**	
OXIDIZER NPSH (MINIMUM)	35 FT OF LOX OR 65 psia	GUIDANCE	PITCH, ROLL, AND YAW PROGRAM THRU THE IU DURING S-IB BURN
FUEL NPSH (MINIMUM)	35 FT OF RP-1 OR 57 psia	TELEMETRY LINKS	FM/FM, 240.2 MHz; PCM/FM, 256.2 MHz
GAS TURBINE PROPELLANTS	LOX AND RP-1	TRACKING	ODOP
TURBOPUMP SPEED	6680 RPM	ELECTRICAL	BATTERIES, 28 Vdc (2 ZINC-SILVER OXIDE); MASTER MEASURING VOLTAGE SUPPLY, 28 Vdc TO 5 Vdc.
ENGINE MOUNTING			
INBOARD	32 IN. RADIUS, 3 DEG CANT ANGLE		
OUTBOARD	95 IN. RADIUS, 6 DEG CANT ANGLE	RANGE SAFETY SYSTEM	PARALLEL ELECTRONICS, REDUNDANT ORDNANCE CONNECTIONS.

Notes: ALL MASSES ARE APPROXIMATE.

Figure 4-4

advancements, allows for considerable tank weight reduction over the 5086 and 5052 alloys used to construct the Jupiter and Redstone tanks, respectively. Tank wall thickness varies from top to bottom in relation to stress distributions. Hemispherical bulkheads are welded to the forward and aft end of the cylindrical sections, and a sump is welded to the aft bulkhead. A pressurization and vent manifold is fastened to the forward bulkhead of each of the lox tanks. A cylindrical skirt reinforced with longerons is attached to the forward and aft bulkheads to complete a basic tank. The eight outer tanks are 70 in. in diameter and contain lox and fuel alternately. The center tank is 105 in. in diameter and contains lox. The center lox tank is bolted to the spider beam and is attached to the tail barrel with huck bolts. Ball and socket fittings attach the aft ends of the 70-in. lox tanks to the tail unit. Banjo fittings and studs rigidly secure the forward ends of the lox tank to the spider beam. The fuel tanks are supported by ball and socket fittings at the tail unit. During shipment, banjo fittings rigidly secure the fuel tanks to the tail unit; however, they are removed before flight. The forward ends of the fuel tanks are mounted to the spider beam unit by sliding pin connections, allowing the lox tanks to shorten due to thermal contraction when loaded. In summary, the major structural improvements incorporated into the propellant tanks are revised skin gages and reduced bulkhead and frame gages to agree more closely with stress levels, inversion of the aft dome manhold cover in the center tank, and the addition of GOX interconnect domes (with the related forward skirt cutouts) and a GOX pressurant diffuser. Also, the fuel tanks are painted white instead of black on SA-206 and subsequent vehicles for thermal reasons.

TAIL UNIT.

Primarily, the tail unit rigidly supports the aft ends of the propellant tank cluster and the vehicle on the launcher; mounts the eight engines and fins; and provides the thrust structure between the engine thrust pads and the propellant tanks. Other functions of the tail unit are to support the lower shroud panels, lox and fuel bay firewalls, heat shield support beam and panel assemblies, engine flame curtains, and the engine flame shield support installation. Unlike the propellant tank units, the tail unit is constructed with higher strength aluminum alloys of the 7000 series that are heat-treated to the T6 or the T-73 condition.

Section IV S-IB Stage

S-IB STAGE STRUCTURE

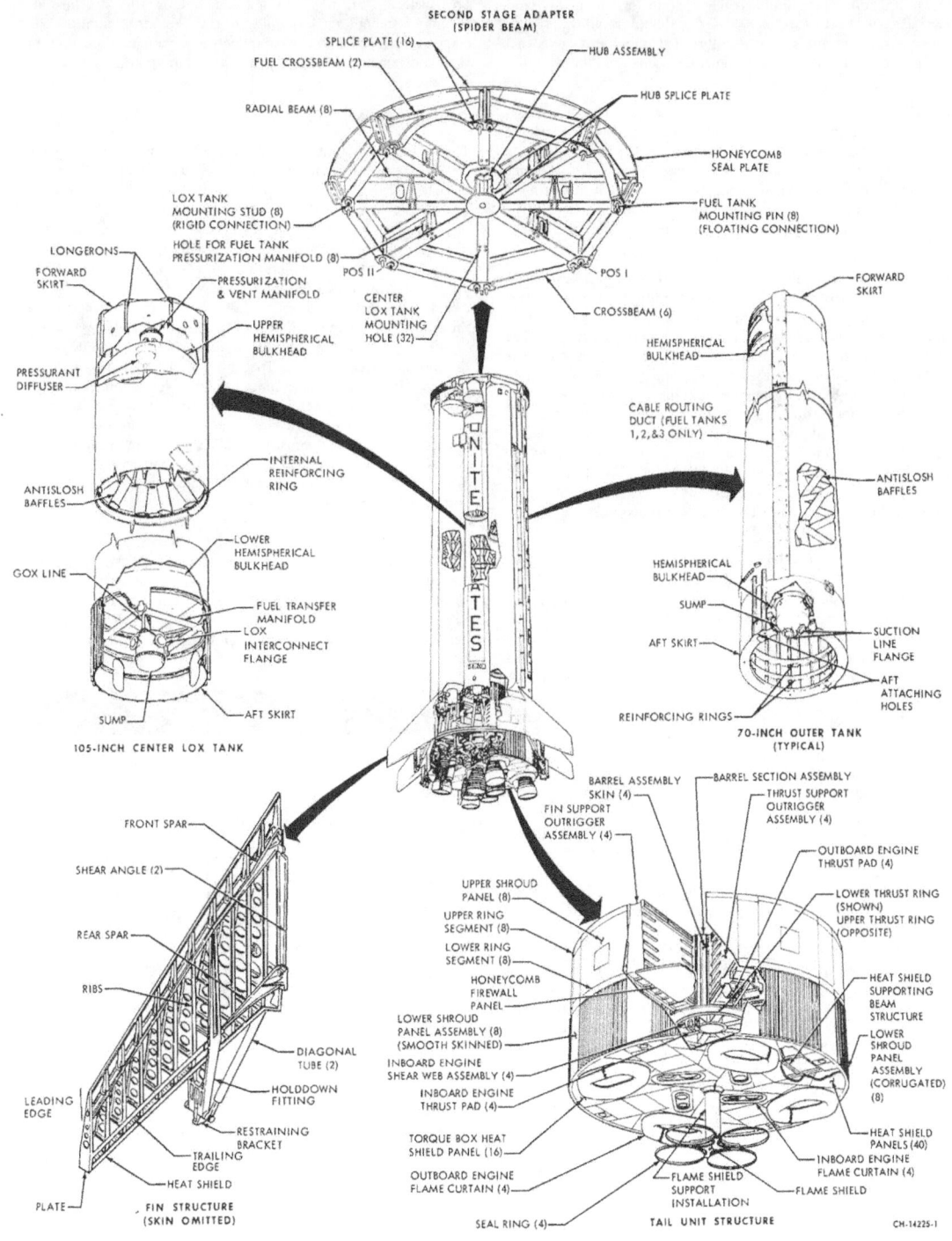

Figure 4-5

4-5

Section IV S-IB Stage

The 7000 series aluminum alloys used in the tail unit assembly are not recommended for welded applications because of low welding efficiencies. These alloys are used in unpressurized areas where the assembling is done with mechanical fasteners. The tail unit thrust structure configuration lends itself to this definition; and high strength, heat-treatable 7000 series aluminum alloy forgings, extrusions, plates and sheets are fabricated into components that are joined with mechanical fasteners to construct the tail unit assembly.

Because of the susceptibility of the high strength aluminum alloys to stress corrosion cracking, methods have been employed in the design and manufacture of the tail unit assembly which minimize the danger of failure due to stress corrosion cracking. Methods employed are: heat treatment to the T-73 condition, heat treatment after heavy machining operations, the use of closed die forgings, and the use of adequate final protective finishes.

The tail unit consists of a barrel assembly, 105 in. in diameter, that directly supports the center propellant tank, encloses the inboard engine thrust beams, and acts as the hub for the four thrust support outriggers and the four fin support outriggers. The four thrust support outriggers also act as fin support outriggers. The fin support outriggers are similar to the thrust support outriggers but differ mainly in that they have no thrust support beam or actuator support beam. The outer ends of the outriggers are spanned by upper and lower ring segments and eight upper shroud panels to form the basic thrust structure. Eight smooth and eight corrugated lower shroud panels are attached to the aft end of the thrust structure to form a compartment for the eight H-1 engines. Lox and fuel bay firewall panels are installed to cover the space between the outrigger assemblies and the space over the aft end of the barrel assembly.

A reinforcing beam structure is fitted into the aft end of the lower engine shroud assembly. The heat shield panels, engine flame curtains, and flame shield support installation are attached to the beam structure. Figure 4-6 shows the unique design details of the tail unit heat shield. Unlike most other honeycomb composites used in the vehicle utilizing phenolic cores that are adhesively bonded to face sheets and are limited by the upper and lower temperature constraints on the adhesive system, the heat shield honeycomb composite consists of both corrosion-resistant steel foil cores and thin face sheets that are joined by a brazing process. The 0.25-in. square-cell core is brazed to both the inner and outer face sheets, has a layer thickness of 1.00 in., and acts as the chief structural core member of the composite. The 0.50-in. square-cell core is brazed to only the outer face sheet, has a layer thickness of 0.25-in., and acts as the thermal insulation retaining member of the composite structure. M-31 insulation is trowled into the retaining core cells. Laboratory tests have generally demonstrated that, compared with adhesively bonded honeycomb composites, brazed honeycomb composites are over 100 percent greater in tensile strength, over 75 percent greater in core shear strength, over 20 percent greater in edgewise compression strength, and equal in flatwise compression strength. This heat shield design provides a lighter panel with increased stiffness which greatly improves the retention of the M-31 insulation material. Successful results of laboratory testing and static tests of the S-I-10 and S-IB-3 through S-IB-7 stages have fully qualified this heat shield panel design. In addition to the new configuration heat shield assembly, the tail unit incorporates new fin attachment fittings, gage reduction of sheet-metal and framing, and removal of the engine skirts from the lower shroud assembly.

FINS.

The eight fins, of semi-monocoque construction, provide aerodynamic stability in mid-region of first stage flight and support the vehicle on the ch pad prior to ignition and during the hold-down period ignition. A fin is fastened mechanically to each of the four thrust support outriggers and the four fin support outriggers. A heat shield is attached to the trailing edge to protect the fin from engine exhaust, and a plate is fastened to the tip of the fin between the leading edge and the heat shield. Skin panels are riveted to the ribs and spars, completing the structure and forming a smooth aerodynamic surface. The fins used on the S-IB stage are identical, and of a completely new configuration, replacing the arrangement of four large-fins and four stub-fins used on the S-I stages.

SPIDER BEAM UNIT.

The spider beam unit holds the propellant tank cluster together at the forward end and attaches the S-IB stage to the S-IVB aft interstage. The five lox tank units are rigidly attached to the spider beam while the fuel tank units are attached with sliding pin connections. Structurally, the spider beam consists of a hub assembly, to which eight radial beams are joined with upper and lower splice plates by mechanical fasteners. The outer ends of the radial beams are spanned by crossbeams and joined with upper and lower splice plates by mechanical fasteners. Like the tail unit thrust structure the spider beam is constructed of extrusions and fittings made of high strength, heat-treatable, aluminum alloys of the 7000 series that are heat-treated to the T6 condition. To form an aft closure for the S-IVB stage engine compartment, twenty-four honeycomb composite seal plate segments of approximately 0.05-in. thickness are fastened to the forward side of the spider beam. The seal plate honeycomb composite consists of 5052 aluminum alloy fail core material adhesively bonded to 7075-6 aluminum alloy face sheets to form thinner and lighter panels than those used on the S-I stages. During qualification testing of the S-IB stage spider beam, a failure of the lox tank fitting occurred. Figure 4-7 shows the radial beam reinforcing angle and bracket, crossbeam web stiffening brackets, and reinforced mounting stud flange incorporated to fix each of the eight lox tank fittings. Other changes incorporated into the spider beam design for the S-IB stages are the reduction of beam gages and the removal of retrorockets, 45-deg fairing, and radial beam tips. This unit is qualified and has performed successfully on all Saturn IB flights.

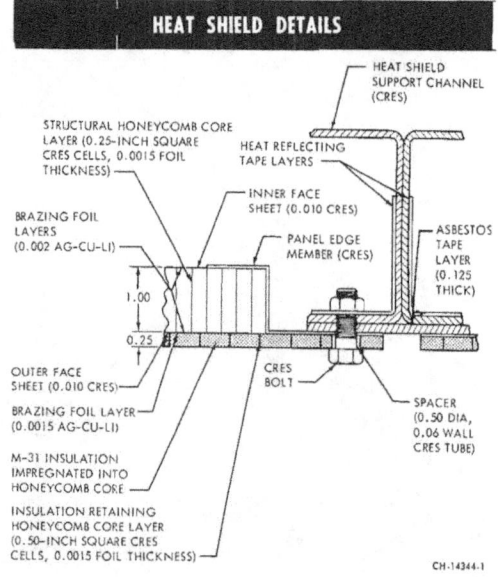

Figure 4-6

Section IV S-IB Stage

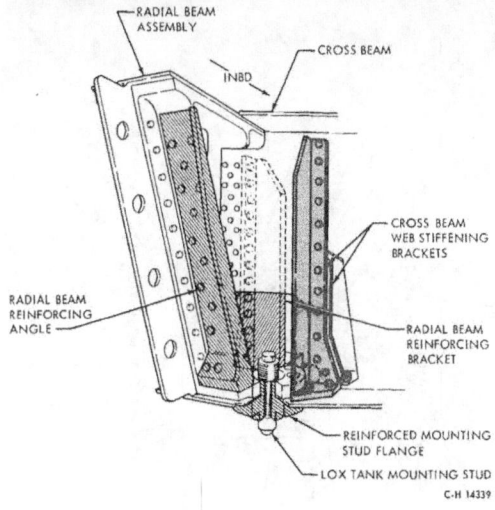

Figure 4-7

PROPULSION.

The S-IB stage propulsion system consists of an eight-engine cluster of H-1 engines that burn lox and RP-1 fuel to propel the Saturn IB vehicle during the first boost phase of powered flight. Propellant from the lox and fuel tanks feed the H-1 engines under tank pressure to assure the NPSH necessary for satisfactory engine operation. Boosters S-IB-1 through S-IB-5 used engines developing 200,000 lbf of thrust for a total stage thrust of 1,600,000 lbf. Boosters S-IB-6 and subsequent will use engines developing 205,000 lbf of thrust for a total stage thrust of 1,640,000 lbf. Four inboard engines are mounted 90 deg apart (at vehicle positions I, II, III, and IV) on a 32-in. radius from the vehicle longitudinal axis and are canted 3 deg outboard from the vehicle centerline. Four outboard engines are gimbal mounted 90 deg apart (at fin lines 2, 4, 6, and 8) on a 95-in. radius from the vehicle longitudinal axis. The engines cant outboard 6 deg from the vehicle centerline. Each of the eight engines is attached by a gimbal assembly to its thrust pad on the tail unit thrust structure. Inboard engine thrust pads are on the barrel assembly and outboard engine thrust pads are on the thrust support outriggers. Although the inboard engines do not gimbal for vehicle control the gimbal assemblies permit alignment of the engines to the thrust-structure; two turnbuckles used on each inboard engine, with the gimbal assembly, align and secure the engine in place. Two hydraulic actuators and a gimbal assembly secure each out-

board engine to the thrust structure. The actuators attach to an actuator support beam, which is part of the thrust support outrigger. The actuators, one mounted in the pitch plane and one in the yaw plane, gimbal the engine for vehicle attitude control. The engine gimbal centerline for both outboard and inboard engines lies in a plane perpendicular to the vehicle longitudinal axis at vehicle station 100 (figure 4-8). Canting the engines provides stability by directing the thrust vectors to common points on the vehicle longitudinal axis. The outboard engine thrust vectors intersect the longitudinal axis at vehicle station 1004, while the inboard engine thrust vectors intersect the longitudinal axis at vehicle station 711. The difference in cant angles and radii from vehicle centerline account for the two different intersect points. Directing the thrust vectors to the vehicle longitudinal axis reduces the possibility of excessive loading of the vehicle structure in the event of engine(s) failure during flight.

H-1 ENGINE.

The H-1 engine is a single-start, fixed-thrust, bipropellant rocket engine that burns RP-1 (MSFC-SPEC-342A) fuel and lox (MSFC-SPEC-399). Calibrated orifices installed in the high-pressure fuel, and at the lox and fuel inlets to the gas generator control valve, fix the propellant flowrates, which effect the fixed thrust. The engine

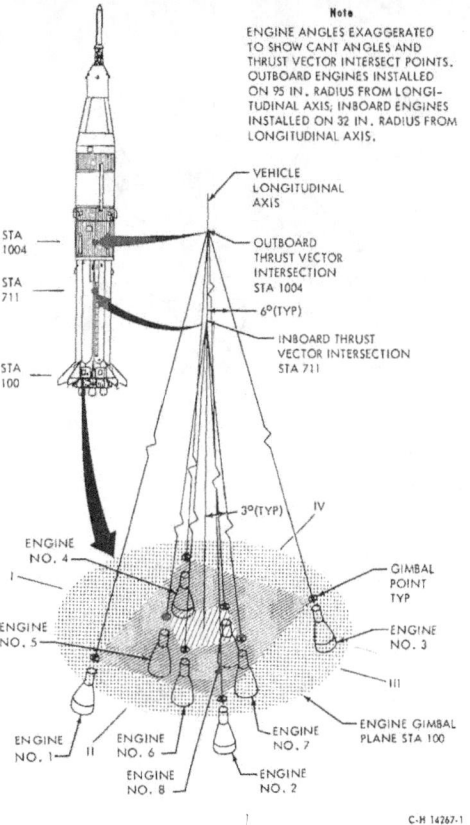

Figure 4-8

Section IV S-IB Stage

H-1 ENGINE

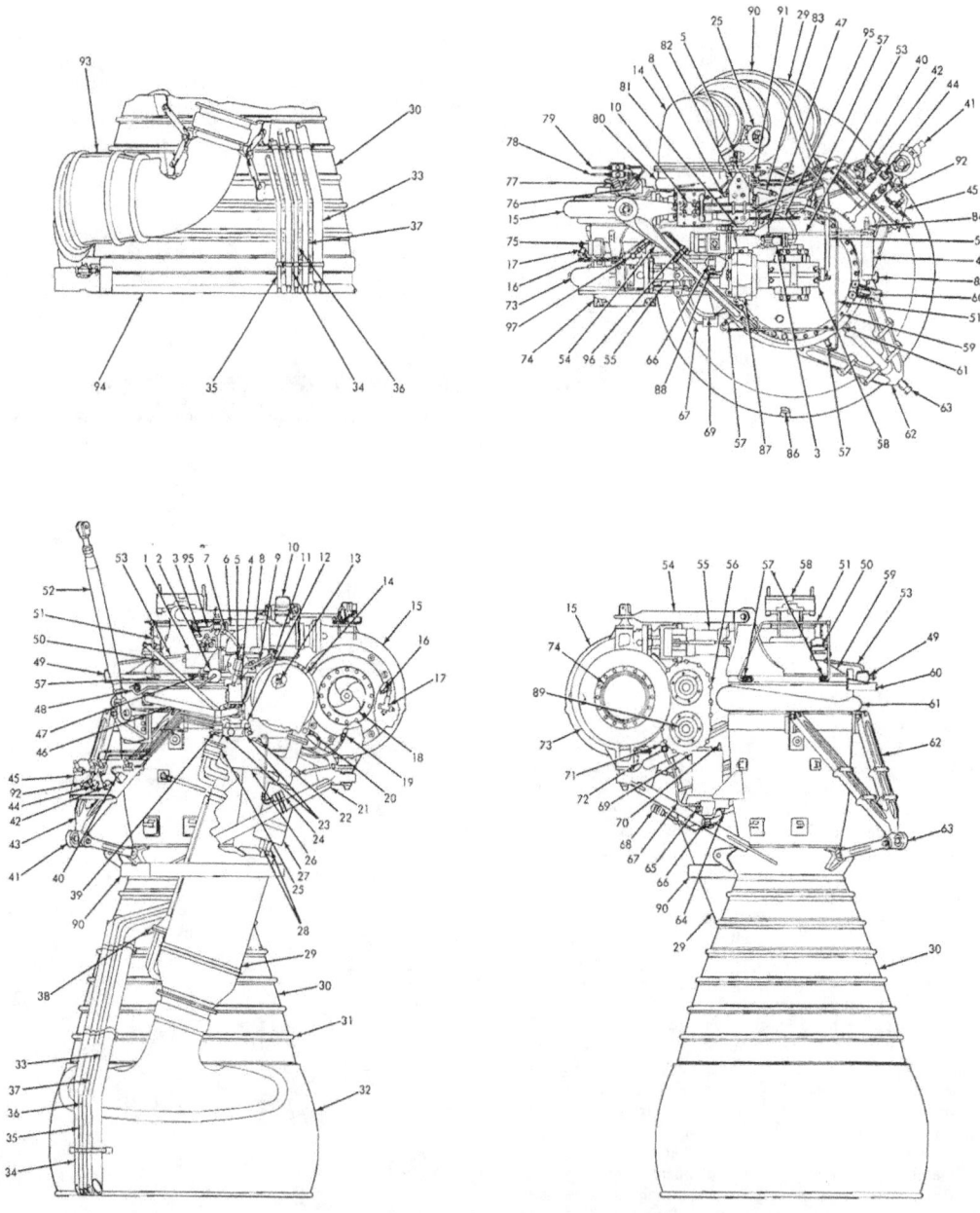

Figure 4-9 (Sheet 1 of 2)

LEGEND FOR H-1 ENGINE

1. MAIN FUEL VALVE
2. IGNITION MONITOR VALVE
3. IGNITER FUEL VALVE
4. LPGG CONTROL VALVE
5. CONAX VALVE
6. CONTROL-PRESS FUEL LINE
7. CONTROL-PRESS FUEL MANIFOLD (CONAX INLET)
8. HIGH-PRESS. FUEL DUCT
9. MLV CLOSING LINE
10. TOPS (3)
11. TOPS SENSING LINE
12. LPGG INJECTOR
13. MEAS T12 (TURBINE RPM)
14. TURBINE EXHAUST HOOD
15. FUEL PUMP VOLUTE
16. LOX PUMP SEAL PURGE PORT
17. GEARBOX PRESSURIZATION CHECK VALVE AND PORT
18. FUEL PUMP INLET
19. MEAS D20 (LOWER GEARCASE LUBE PRESS. SENSING PORT)
20. MEAS D14 (TURBINE INLET PRESS SENSING PORT)
21. AFT TURBOPUMP MOUNT SUPPORT
22. MEAS C9 (LPGG COMBUSTION CHAMBER TEMP)
23. SQUIBLESS IGNITERS
24. MEAS C9 ZONE BOX
25. HEAT EXCHANGER GOX OUTLET
26. LPGG COMBUSTION CHAMBER
27. SOLID PROPELLANT GAS GENERATOR
28. SPGG INITIATORS
29. HEAT EXCHANGER
30. THRUST CHAMBER
31. STIFFENING BAND AND TENSION RING
32. ASPIRATOR (OUTBOARD ENGINE ONLY)
33. LUBE DRAIN
34. LOX DRAIN LINE EXTENSION
35. LOX SEAL CAVITY DRAIN (SECONDARY)
36. IGNITION MONITOR VALVE DRAIN
37. FUEL DRAIN MANIFOLD DRAIN
38. HEAT EXCHANGE INLET MANIFOLD ASSY
39. FUEL JACKET FILL Q-D COUPLING
40. MEAS D20 (LOWER GEARCASE LUBE PRESS. TRANSDUCER)
41. FUEL ACTUATOR ATTACH POINT (OUTBOARD ONLY)
42. MEAS D14 (TURBINE INLET PRESS. TRANSDUCER)
43. FUEL ACTUATOR OUTRIGGER ASSY (OUTBOARD ONLY)
44. MEAS XD35 (GG LOX INJECTOR PRESS. TRANSDUCER)
45. TRANSDUCER PANEL
46. FUEL BOOTSTRAP LINE
47. IMV INLET LINE
48. GG AND IMV CONTROL LINE
49. HYPERGOL CONTAINER
50. LOX BOOTSTRAP LINE
51. THRUST CHAMBER FUEL INJECTOR PURGE LINE
52. TURNBUCKLE ASSY (INBOARD ENGINES ONLY)
53. IGNITER FUEL LINE
54. TURBOPUMP FORWARD MOUNT ASSY
55. HIGH-PRESS. LOX DUCT
56. ACCESSORY DRIVE ADAPTER
57. TC FUEL INJECTOR PURGE CHECK VALVES (3)
58. GIMBAL ASSEMBLY
59. LOX DOME
60. MEAS D1 (COMBUSTION CHAMBER PRESS. TRANSDUCER)
61. FUEL INLET MANIFOLD
62. LOX ACTUATOR OUTRIGGER ASSY (OUTBOARD ONLY)
63. LOX ACTUATOR ATTACH POINT
64. MEAS XC89 (GEARCASE LUBE TEMP THERMOCOUPLE)
65. FABU FILL Q-D COUPLING
66. FUEL DRAIN Q-D COUPLING
67. FABU DISCHARGE LINE
68. MEAS XC89 ZONE BOX
69. FUEL ADDITIVE BLENDER UNIT
70. FABU FULL INDICATOR
71. TURBOPUMP LUBE FILTER
72. LUBE DRAIN MANIFOLD
73. LOX PUMP VOLUTE
74. LOX PUMP INLET
75. HEATER COVER (LOX BEARING NO. 1)
76. MEAS D13 (LOX PUMP INLET PRESS. TRANSDUCER OUTBOARD ONLY)
77. MEAS D12 (FUEL PUMP INLET PRESS. TRANSDUCER OUTBOARD ONLY)
78. MEAS D12 SENSING LINE (PORT LOCATED ON SUCTION LINE)
79. MEAS D13 SENSING LINE (PORT LOCATED ON SUCTION LINE)
80. TURBINE
81. MLV OPENING LINE
82. CUSTOMER CONNECT PURGE PANEL
83. GG LOX INJECTOR PURGE CHECK VALVE
84. HYPERGOL INSTALLED SWITCH ASSY
85. HYPERGOL CARTRIDGE LOCK PIN
86. FUEL JACKET DRAIN PLUG (3)
87. MAIN LOX VALVE
88. LUBE FUEL INLET LINE
89. AUXILIARY DRIVE PAD
90. HEAT SHIELD ASSY
91. GG FUEL INJECTOR PRESS. SENSING PORT
92. MEAS DJ4 (GG FUEL INJECTOR PRESS. TRANSDUCER)
93. EXHAUST DUCT (INBOARD ENGINES ONLY)
94. FUEL RETURN MANIFOLD (INBOARD AND OUTBOARD ENGINES)
95. CONTROL-PRESS. FUEL MANIFOLD (IFV INLET)
96. MEAS C1 ZONE BOX
97. MEAS C1 (LOX PUMP BEARING NO. 1 TEMP THERMOCOUPLE)

Figure 4-9 (Sheet 2 of 2)

has a regeneratively-cooled thrust chamber with propellant feed and control components clustered around the forward end of the combustion chamber (see figure 4-9). A turbopump, driven by a gas turbine through a gear train, delivers fuel and lox under high pressure and at high flow rates to the combustion chamber. During operation, engine control is a function of fuel turbopump discharge pressure. This control-pressure fuel is manifolded from the high-pressure fuel duct upstream of the main fuel valve (MFV) to the propellant feed valve actuators. A solid propellant gas generator (SPGG) spins the turbine to start the H-1 engine and a liquid propellant gas generator (LPGG), using bootstrap fuel and lox, sustains engine operation by supplying large volumes of gas to the turbine. Each engine has its own turbine exhaust system for expended turbine gases. A heat exchanger in the turbine exhaust system converts lox to GOX for S-IB stage lox tank inflight pressurization.

Characteristics.

Inboard engines designated H-1C and outboard engines designated H-1D have basically the same physical characteristics, except inboard engines use a partial aspirator, or exhaust duct, for exhausting turbine gases, while the outboard engines use a peripheral aspirator on the engine thrust chamber to control the turbine exhaust gas flow. The outboard engines have outriggers mounted on the thrust-chamber for hydraulic actuator attachment, while inboard engines use struts (turnbuckles) attached to thrust-chamber stabilizing lugs for installation on the tail unit thrust structure. The H-1C and H-1D engines weigh 1942 and 2003 lbm (dry) respectively. At start, engine fluids increase the weight 220 lbm per engine. Approximately 166 lbm of fluids remain in each engine at cut-off. The engines measure 101.61 in. in length from the bottom of the thrust chamber to the gimbal assembly mounting face. The outboard engine aspirator extends an additional 1 in. below the thrust chamber.

H-1 Engine History.

H-1 engines have flown on all fifteen Saturn I&IB launch vehicles. The H-1 engine has received modifications during the Saturn program to uprate thrust and performance capabilities by the addition of injector baffles, a tapered fuel manifold, a Mark 3H turbopump, low fuel ΔP injector, and furnace-brazed, stainless steel thrust chamber. All inboard engines of the Saturn vehicles SA-1 through SA-202 exhausted turbine gases through exhaust fairings on the tail unit thrust structure. SA-203 and subsequent vehicles utilize inboard engines having integral exhaust ducts that exhaust turbine gases into the thrust chamber exit flow of the inboard engines. Throughout the Saturn program, the outboard engines have utilized aspirators for dispersing the turbine exhaust gases. H-1 engines, rated at 165,000 lbf each, propelled Block I vehicles SA-1 through SA-4 to test the concept of engine clustering and multi-tank propellant container. Engines rated at 188,000 lbf each propelled the Block II vehicles, SA-5 through SA-10, providing primary boost phase for orbiting boilerplate Apollo payloads and testing the S-IV stage. Engines rated at 200,000 lbf propelled SA-201 through SA-205. These engines provided the primary boost phase for testing the S-IVB stage, IU, and Apollo payloads which were on the Saturn V lunar vehicles. On the Skylab vehicles, the H-1 engines are rated at 205,000 lbf.

H-1 Engine Predicted Performance.

The H-1C and H-1D engines meet the requirements of H-1 Engine Model Specification R-1141dS. The engines must produce 205,000 ±2,000 lbf, an instantaneous impulse of 261 sec (min) and 263.4 sec (nom), a nominal chamber pressure of 652 psia, and a mixture ratio of 2.23±2 percent (O/F). The nominal fuel inlet pressure is 57 psia prior to ignition. Required NPSH during flight is 35 ft. The minimum required lox pump inlet pressure prior to ignition

Section IV S-IB Stage

is 80 psia with a required NPSH for flight of 35 ft. The engine must have an effective duration of 155 sec (min). Figure 4-10 presents the S-IB-6 propulsion predictions using static test data and Rocketdyne acceptance test data. These predictions may change slightly as parameters for individual missions are defined. Predicted inboard engine cutoff time is 2 min 17.7 sec into flight with the outboard engines cutoff time occurring 3 sec later. The reliability assessment of the H-1 engine during the AS-206 Design Certification Review was 0.99 at 0.869 confidence. Refer to Section II for vehicle performance.

Loading Limitations.

The engine and its structural mounts, while meeting the gimbaling requirements, must operate without deformation or failure under the following conditions: (1) flight loading 8.0 G parallel to the direction of flight and 0.5 G perpendicular to the direction of flight; (2) flight loading 4.0 G parallel to direction of flight and 1.0 G perpendicular to flight direction. The engine is designed to withstand a minimum of 1.5 times the forces resulting from all combinations of the above loading conditions or 4.0 G handling loads applied in any direction.

H-1 Engine Servicing.

To prepare the engines for firing, the fuel additive blender unit (FABU) is serviced with extreme-pressure additive ST0140RB0013, the thrust chamber fuel jacket is serviced with RP-1 fuel, and the installation of the LPGG squibless igniters, the SPGG, the SPGG initiators, the Conax valves, and the hypergol cartridges is completed.

FABU Fill. A portable unit services the FABU with 105 in.^{3} (min) of extreme-pressure additive ST0140RB0013 per engine. The extreme-pressure additive, maintained at $120\pm10°$ F, enters the FABU through a quick-disconnect on the base of each unit. A plunger-type indicator on top of the FABU extends when the unit is full.

Thrust Chamber Fuel Jacket Fill. At T-1 day, 15 hr, each thrust chamber fuel jacket is prefilled with 12 ± 0.5 gal of RP-1 fuel. Filling each chamber reduces the time delay of fuel entering the combustion chamber after the main fuel valve opens. Ambient RP-1 enters the thrust chamber through a quick-disconnect on the GG fuel bootstrap line.

H-1 Engine Ordnance Devices Installation. Installation of ordnance devices begins prior to RP-1 tanking, with the Conax valves installation, followed on T-1 day by installation of the squibless igniters, solid propellant gas generators, and SPGG initiators. Prior to installation the components receive checkouts for insulation resistance, pin-to-case checks, visual inspections, and electrical checks of the bridge-wires in the initiators and Conax valves. The harnesses for the electro-explosive devices also receive electrical checkout for stray voltages and for continuity. The SPGG initiator circuits are tested while the ignition switch on the S-IB networks panel is first, in the ARM position, and second, in the SAFE position. Sixteen squibless igniters, two per engine, are installed in the LPGG combustors just below the injector. Eight solid propellant gas generators, one per engine, attach to the LPGG combustor SPGG attach flanges. Two initiators for redundancy are installed in each SPGG. Eight Conax valves, one per engine, attach to the MLV closing control manifolds located on the main fuel valves. After checkout of the harness assemblies with no discrepancies noted, the electrical connectors are connected to the Conax valve (two per valve), squibless igniters, and SPGG initiators. Two additional electrical connectors are connected to the position indicator on each Conax valve. These connections do not interface with the ordnance charges.

Hypergol Cartridge Installation. Installation of the hypergol cartridge in the engines occurs on T-1 day. The cartridge containing 6 in.^{3} of triethylaluminum is inserted into the hypergol container on the injector. A lockpin secures the cartridge in place. The HYPERGOL INSTALLED detector switch senses cartridge installation and provides a corresponding signal to the S-IB firing prep panel. O-ring seals on the cartridge prevent leakage of fuel during engine operation. Burst diaphragms contain the hypergol until ruptured by fuel pressure during engine start.

Engine Purges.

Four ambient GN_2 purges supplied by valve panel no. 10 in the ML prevent accumulation of contaminants in engine components. The purges initiated during countdown continue until overcome by engine internal pressures during start.

Lox System Bypass Purge. Lox system bypass purge, required anytime thrust chamber exit covers are removed from the engines when the stage is on the launch pad, prevents entrance of contaminants into the lox dome from the thrust chamber. GN_2 from a common manifold that supplies the purge to all engines, enters the lox dome through the unitized check valve and the heat exchanger lox supply line. The purge flowrate is $5(+1.5, -3.3)$ scfm per engine at an engine interface pressure of 2 to 8 psig. A poppet-type check valve in the unitized check valve of each engine prevents lox flow into the manifold during engine operation. Switch S19, LOX DOME BYPASS PURGE, on S-IB firing prep panel initiates the low flowrate purge. Feedback from the lox dome bypass valve in valve panel no. 10 illuminates the LOX DOME BYPASS OPEN indicator on the S-IB firing prep panel. Pressure switches upstream from the lox dome bypass valve illuminate the LOX DOME PURGE ON indicator on the S-IB firing prep panel and the firing panel. Pressure switches on the lox dome bypass supply line illuminate the LOX DOME LOW PRESS. indicator on the S-IB firing prep panel.

Lox System Purge. At T-28 sec, the launch sequencer initiates a high flowrate lox system purge, 100 ± 20 scfm per engine at an engine interface pressure of 95 ± 10 psig, which is superimposed on the lox system bypass purge. The purge prevents contamination of the lox dome from a fuel-rich cutoff in the thrust chamber if abort is initiated after engine ignition but before liftoff. Lox system purge can be initiated manually by the LOX DOME PURGE switch on the S-IB firing prep panel.

Gas Generator Lox Injector and Thrust Chamber Fuel Injector Purges. At T-28 sec the launch sequencer also initiates the gas generator (GG) lox injector manifold purge and the thrust chamber (TC) fuel injector purge. The GG lox injector purge prevents solid propellant gas generator combustion products from entering the lox injector manifold during engine start. The purge prevents contamination of the gas generator lox injector and controls turbine inlet temperature spikes if abort is initiated after engine ignition but before liftoff.

Purge flowrate is 76 ± 7 scfm per engine at an engine interface pressure of 255 ± 25 psig. Lox pressure buildup in the manifold terminates the purge. The TC fuel injector purge prevents lox from entering the injector fuel ports during engine start. Flowrate is 238 ± 35 scfm per engine at an engine interface pressure of 487.5 ± 62.5 psig. Increasing fuel pressure in the injector, as a result of turbopump acceleration, terminates the purge. The purge pressure must be vented prior to initiating an abort after engine ignition but prior to liftoff.

Poppet-type check valves at each engine prevent reverse flow through the two purge manifolds when engine operating pressures exceed the purge pressures.

Purge Controls. GAS GENERATOR LOX INJECTOR PURGE and THRUST CHAMBER FUEL INJECTOR PURGE switches on the S-IB firing prep panel permit manual actuation of the purge. Pressure switches on the purge supply lines in valve panel

Section IV S-IB Stage

S-IB-6 PREDICTED SEA-LEVEL REFERENCE VALUES AT 30-SEC FLIGHT TIME (PRELIM) [3]

ENG NO.	THRUST (1000 LB)	SPECIFIC IMPULSE (SEC)	LOX FLOWRATE (LB/SEC)	FUEL FLOWRATE (LB/SEC)	CHAMBER PRESS. (PSIA)	ENGINE MIXTURE RATIO (O/F)	TURBOPUMP SPEED (RMP)
NOM	205.00	263.60	536.86	240.74	704.71	2.2300	6,691.3
1	206.00	262.72	541.66	242.43	709.32	2.2343	6,754.3
2	205.33	262.82	540.01	241.26	709.27	2.2383	6,690.8
3	205.49	262.82	541.54	240.30	710.03	2.2536	6,687.8
4	205.68	262.39	541.71	242.19	705.52	2.2367	6,749.9
5	206.27	263.26	542.69	240.85	709.08	2.2532	6,771.1
6	205.47	262.85	541.41	240.30	706.45	2.2530	6,694.4
7	205.49	263.55	541.18	238.51	703.84	2.2690	6,695.7
8	205.17	263.28	539.91	239.39	701.11	2.2554	6,674.0
AVG.	[1] 205.61	262.96	[2] 541.26	240.65	--	[2] 2.2491	--

[1] THRUST ALONG LONGITUDINAL AXIS
[2] INCLUDES FUEL USED AS LUBRICANT
[3] AMBIENT PRESSURE 14.7 PSIA, LOX DENSITY 70.79 LBM/FT^3, LOX PUMP INLET PRESSURE 65 PSIA, FUEL PUMP INLET PRESSURE 57 PSIA

Figure 4-10

no. 10 provide feedback signals to indicators, GG LOX INJECTOR PURGE and TC FUEL INJECTOR PURGE ON, on the S-IB firing prep panel and the firing panel. Each of the four purge switches on the S-IB firing prep panel has three positions: OFF, AUTO, and ON. The LOX DOME PURGE, GG LOX INJECTOR PURGE, and TC FUEL INJECTOR PURGE switches must be in the AUTO position for final countdown as a prerequisite for PURGES ARMED signal, an interlock for FIRING COMMAND. PURGED ARMED is monitored on firing panel. The ALL ENGINES RUNNING signal closes the TC fuel injector purge valve and opens the supply regulator dome vent valve to effect regulator closure. Redundant vent valves in the thrust chamber purge vent valve panel open, venting the 550-psig purge supply line to atmosphere before umbilicals disconnect. The COMMIT signal closes the lox system purge valve and the GG lox injector purge valve by removing +1D116 bus power. In the event of an abort before liftoff, the cutoff command reinstates the lox system purge and GG lox injector purge to prevent vapors from entering the lox components during the fuel-rich shutdown. The lox system purge is required for a minimum of 15 sec after engine cutoff, then the lox system bypass purge continues until the thrust chamber exit covers are reinstalled. The GG lox injector purge is required for a minimum of 10 min after engine cutoff. Supply pressures and flowrates are the same as preflight requirements.

Engine Gearbox Pressurization and Lox Pump Seal Purge. Engine gearbox pressurization and purge, and lox pump seal purge begin with pressurization of the S-IB stage control pressure system and continue throughout propellant tanking, engine start, launch, and powered flight. GN_2 flowing into the gearbox provides a constant purge during countdown. During engine operation the GN_2 pressurizes the gearbox to prevent lubricant foaming at high altitudes and to hold the gearcase lube drain relief valve open precluding flooding of the gearcase with lubricant. GN_2 from the 750-psig control pressure system is orificed to a 1.85(+0.6, -0.8) scfm per engine flowrate, which maintains the gearbox pressure between 2 and 10 psig. The lox pump seal purge, applied in the area between lox and lube seals, keeps the lox and lube seals leakage separated to prevent formation of a highly explosive hydrocarbon gel in the turbopump. The GN_2 pressure forces any seal leakage overboard through separate drain lines. The lox pump seal purge supply tees off the gearbox pressurization line and is orificed to purge the lox seal at a 1.35±0.6 scfm per engine flowrate. In the event of an abort before liftoff the lox pump seal purge is required to maintain separation of lox and lubricant leakage and continues until the pump returns to the ambient temperature. The gearbox pressurization purge supply maintains gearbox pressure between 2 and 10 psig.

H-1 Engine Operation.

At T-3 sec, the ignition sequencer issues the IGNITION commands to start the engines in pairs, opposite inboard engines 5 and 7, and 6 and 8; and opposite outboard engines 2 and 4, and 1 and 3. The engines start in this order, 100 msec apart per pair, to reduce stress on the vehicle during engine start. The electrical command fires the redundant SPGG initiators on each SPGG. See figures 4-11 and 4-12. The SPGG propellant begins burning, producing hot gases that accelerate the turbine and ignite the two liquid propellant gas generator squibless igniters. Through a gear train, the turbine drives the turbopump causing discharge pressures to increase against the closed propellant valves. During tanking operations fuel and lox filled the suction lines, the pump sections

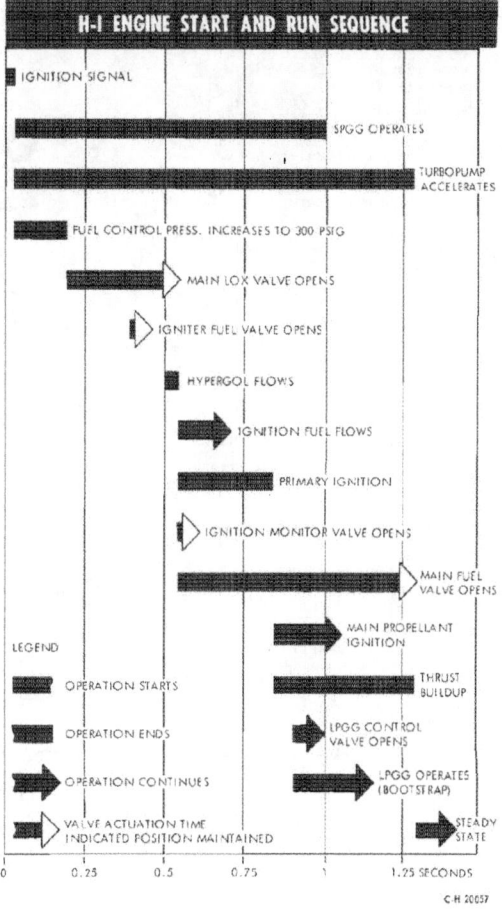

Figure 4-11

Section IV S-IB Stage

H-1 ENGINE START AND RUN OPERATION

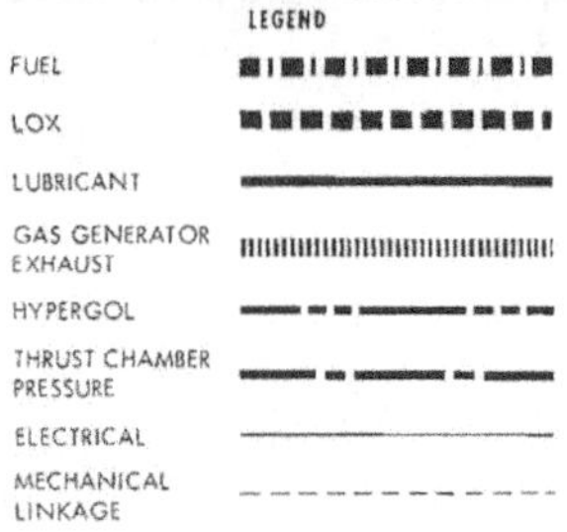

LEGEND

FUEL	■ ▪ ■ ▪ ■ ▪ ■ ▪ ■ ▪ ■ ▪
LOX	▬ ▬ ▬ ▬ ▬ ▬ ▬ ▬
LUBRICANT	───────────
GAS GENERATOR EXHAUST	⫼⫼⫼⫼⫼⫼⫼⫼⫼⫼⫼⫼⫼⫼
HYPERGOL	─ ─ ─ ─ ─ ─
THRUST CHAMBER PRESSURE	─ ·· ─ ·· ─ ··
ELECTRICAL	─ · ─ · ─ ·
MECHANICAL LINKAGE	─ ─ ─ ─ ─

① IGNITION COMMAND AT T-3 SEC FIRES SOLID PROPELLANT GAS GENERATOR (SPGG) INITIATORS.

② SPGG IGNITES PRODUCING GASES TO DRIVE TURBOPUMP TURBINE AND IGNITE SQUIBLESS IGNITERS.

③ TURBINE STARTS FUEL AND LOX PUMPS.

④ PUMPS DELIVER PROPELLANTS UNDER PRESS. TO MAIN VALVES.

⑤ FUEL ENTERS FUEL ADDITIVE BLENDER UNIT (FABU) AND MIXES WITH EXTREME-PRESSURE ADDITIVE. MIXTURE LUBRICATES TURBOPUMP GEAR TRAIN.

⑥ CONTROL PRESS. FUEL OPENS MAIN LOX VALVE (MLV). LOX FLOWS INTO ENGINE COMBUSTION CHAMBER.

⑦ MLV MECHANICALLY OPENS IGNITER FUEL VALVE (IFV) WHEN MLV OPENS 50-70 DEG. CONTROL PRESS. FUEL FLOWS TO HYPERGOL CONTAINER AND IGNITION MONITOR VALVE INLET.

⑧ FUEL BURSTS TWO DIAPHRAGMS IN HYPERGOL CARTRIDGE. HYPERGOL FLOWS THROUGH FUEL INJECTOR INTO COMBUSTION CHAMBER.

⑨ IGNITION. HYPERGOL AND LOX IGNITE SPONTANEOUSLY.

⑩ TC FUEL INJECTOR MANIFOLD PRESSURE RESULTING FROM INCREASED COMBUSTION CHAMBER PRESS. SHUTTLES IGNITION MONITOR VALVE (IMV) ALLOWING FUEL TO OPEN MAIN FUEL VALVE (MFV).

⑪ FUEL FLOWS THROUGH MFV INTO THRUST CHAMBER FUEL JACKET AND THROUGH FUEL INJECTOR INTO COMBUSTION CHAMBER TO SUSTAIN IGNITION.

⑫ TC FUEL INJECTOR MANIFOLD PRESSURE RESULTING FROM INCREASED COMBUSTION CHAMBER PRESS OPENS LIQUID PROPELLANT GAS GENERATOR (LPGG) CONTROL VALVE. BOOTSTRAP FUEL AND LOX ENTER LPGG COMBUSTOR. SPGG GASES AND SQUIBLESS IGNITERS IGNITE BOOTSTRAP PROPELLANT.

⑬ SQUIBLESS IGNITERS BURN APPROX 2 SEC TO ENSURE BOOTSTRAP PROPELLANT IGNITION.

⑭ LPGG GASES CAUSE TURBOPUMP TO ACCELERATE TO OPERATIONAL SPEEDS.

⑮ THRUST OK PRESS. SWITCHES (TOPS) SENSE FUEL PRESS. AND SEND THRUST OK SIGNALS TO IU EDS DISTR VOTING LOGIC CKTS THAT GENERATE 'ALL ENGINES RUNNING' SIGNAL & A COMMIT INTERLOCK. LOGIC CKT OUTPUTS OPERATE 'L/V ENGINES' LIGHTS ON COMMAND MODULE MAIN DISPLAY CONSOLE.

⑯ EIGHT 'L/V ENGINE' LIGHTS EXTINGUISH AS RESPECTIVE ENGINES ATTAIN SATISFACTORY THRUST.

⑰ HEAT EXCHANGER CONVERTS LOX TO GOX FOR S-IB LOX TANKS INFLIGHT PRESSURIZATION.

⑱ TURBINE EXHAUST GASES EXIT OUTBOARD ENGINES THROUGH PERIPHERAL ASPIRATORS AND INBOARD ENGINES THROUGH PARTIAL ASPIRATORS.

⑲ LOX SEAL PURGE AND GEARBOX PRESSURIZATION FROM CONTROL PRESS. SYS CONTINUES THROUGHOUT S-IB STAGE OPERATION.

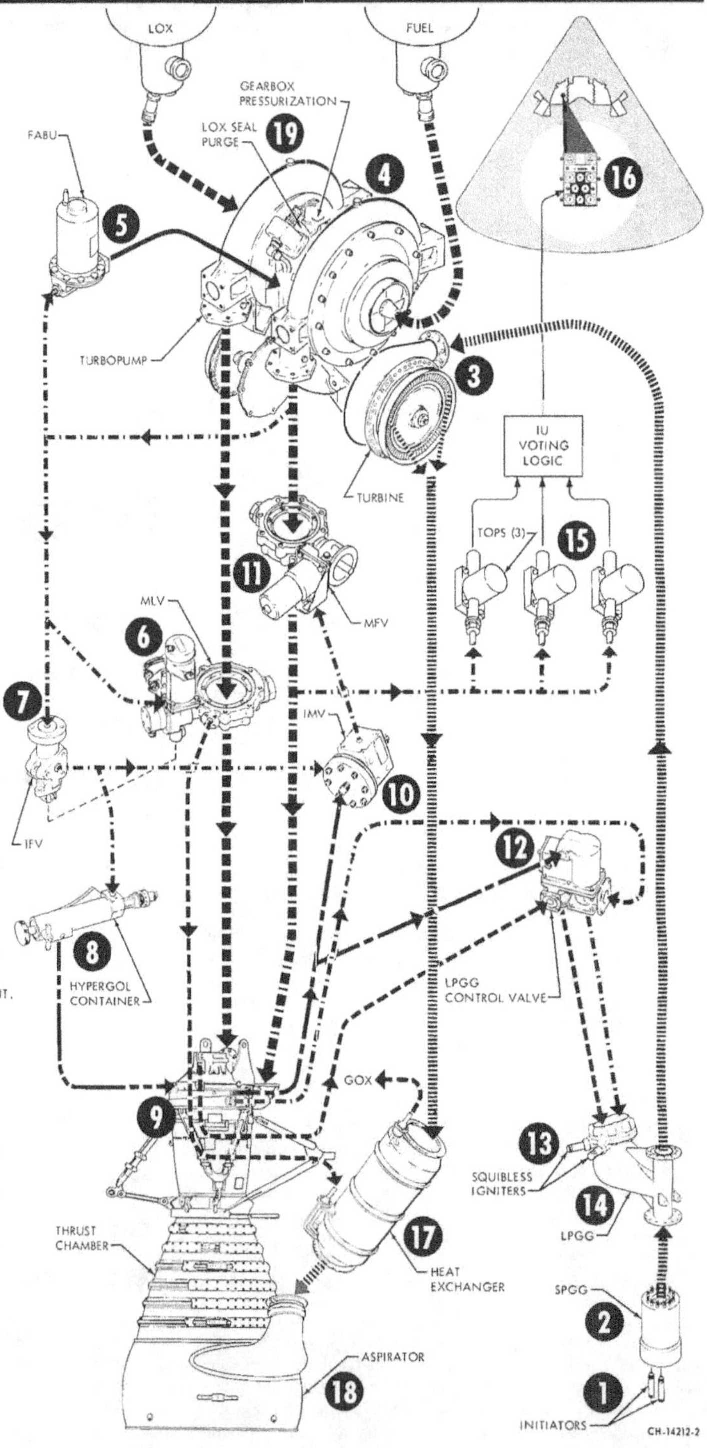

Figure 4-12

4-12

of the turbopump, and both high-pressure propellant ducts down to the closed gates of the main fuel and lox valves. Control-pressure fuel tapped off the high-pressure fuel duct flows through a series control line to the main lox valve (MLV), igniter fuel valve (IFV), fuel additive blender unit (FABU), and to the Conax valve. When fuel pressure increases to 70-150 psig the FABU opens to permit lubricant flow to the turbopump bearings and gears. As the turbopump accelerates and fuel pressure increases to 300±50 psig, control-pressure fuel opens the MLV permitting lox to enter the lox dome and gas generator lox bootstrap line. A cam on the MLV gate shaft mechanically opens the IFV when the MLV has traveled 50 to 70 degrees open, which permits fuel flow to the ignition monitor valve (IMV) and to the hypergol cartridge. Inlet and outlet burst diaphragms in the hypergol cartridge break at 300±25 psig and the control-pressure fuel forces the pyrophoric fluid through seven igniter fuel ports in the injector into the combustion chamber. The hypergol ignites spontaneously with lox entering the combustion chamber through the injector lox ports. Propellant ignition causes a pressure increase in the combustion chamber, which is sensed by the IMV. The IMV shall not open at 12±0.3 psig, and must open at 22±0.3 psig. Nominal IMV opening pressure is 15 psig. The IMV permits control-pressure fuel from the IFV to flow to the main fuel valve (MFV) opening actuator resulting in MFV opening. The pressure required to open the MFV is 350±50 psig.

Fuel, under turbopump pressure, flows through the MFV to the gas generator fuel bootstrap line, through the thrust chamber fuel jacket, through the injector manifold, and into the combustion chamber where main propellant ignition occurs. Increasing fuel pressure resulting from main propellant ignition and sensed at the thrust chamber (TC) fuel injector manifold opens the liquid propellant gas generator (LPGG) control valve. The control valve fuel poppet opens at 105±20 psig; the lox poppet opens at 200±20 psig.

Bootstrap propellants enter the LPGG with a slight lox lead and are ignited by SPGG hot gases and the two squiblets igniters. Gases from the LPGG cause the turbopump acceleration to continue until rated thrust is attained. Calibrated orifices in the fuel high-pressure duct inlet and in the LPGG lox and fuel bootstrap lines control the propellant flow, thus allowing the engine to operate at rated thrust.

Three thrust OK pressure switches (TOPS) on each engine sense fuel pressure downstream from the MFV to indicate satisfactory engine thrust. The switches actuate at 785±15 psig, indicating satisfactory thrust attained. TOPS outputs are inputs to voting logic circuits in the IU EDS distributor. The EDS distributor provides the ALL ENGINES RUNNING signal, when all engines have attained satisfactory thrust. Engine running indications monitored on LCC panels, are also monitored on the main display console (MDC) in the command modules. The MDC L/V ENGINES lights extinguish when the engines attain rated thrust. Gearbox pressurization and lox pump seal purge, supplied by the S-IB stage control pressure system continues throughout flight. Sensors installed on each engine provide information on engine conditions during flight to the S-IB stage telemetry system. See H-1 Engine Measuring. See figures 4-13 and 4-14 for H-1 engine lines and orifice summaries.

H-1 Engine Cutoff.

The launch vehicle digital computer (LVDC) issues the engine cutoff commands in two steps through the S-IB stage switch selector. The first command, issued approximately 2 min 17.7 sec into flight, shuts down the four inboard engines. Approximately 3 sec later, the second command shuts down the four outboard engines. A Conax valve on each engine effects engine cutoff. See figures 4-15 and 4-16. The cutoff command fires redundant explosive actuators which open the Conax valve by shearing redundant metal diaphragms in the valve body. Control pressure fuel flows through

H-1 ENGINE LINES SUMMARY

LINE	SIZE (IN. DIA.)	FLOWRATE (NOMINAL)
FUEL H. P. DUCT	3.25 ID	241.0 lb/sec
LOX H. P. DUCT	3.375 ID	537.4 lb/sec
MFV CONTROL	1/4	--
MLV OPENING	1/2	--
MLV CLOSING	3/8 (ID)	--
IGNITER FUEL	3/8	1.4 lb/sec
IMV SENSING	1/4	--
LUBE FUEL INLET	1/2	--
FABU DISCHARGE	3/8	0.63 lb/sec
GG & IMV CONTROL	3/8	--
FUEL BOOTSTRAP	1 1/4	13.52 lb/sec
LOX BOOTSTRAP	3/4	4.61 lb/sec
TOPS SENSING	1/4	--
TURBINE EXHAUST	--	18.13 lb/sec
HEAT EXCHANGER INLET (LOX)	3/4	3.0 lb/sec
THRUST CHAMBER		
FUEL	--	227.5 lb/sec
LOX	--	532.8 lb/sec
GEARBOX PRESSURIZATION AND LOX PUMP SEAL PURGE	1/4	3.2 scfm

Figure 4-13

H-1 ENGINE ORIFICE SUMMARY

ORIFICE	SIZE (IN. DIA)
FUEL DISCHARGE ▷	2.680
LOX BOOTSTRAP ▷	0.356
FUEL BOOTSTRAP ▷	0.700
GEARBOX PRESSURIZATION	0.013
LOX PUMP SEAL PURGE	FLOWRATE OF 1.7 scfm AT 750 psig, 70°F
HEAT EXCHANGER (3)	0.101
FUEL BLEED	0.060
MLV OPENING	0.116
FABU OUTLET	0.147
LOX BOOTSTRAP (FIXED)	0.400

▷ ENGINE CALIBRATING ORIFICES (NOMINAL SIZE)

Figure 4-14

the Conax valve to the MLV closing actuator. MLV closure stops lox flow to the combustion chamber and the LPGG causing thrust chamber pressure and turbopump speed decay. Closing the MLV first in the cutoff sequence permits a fuel-rich cutoff to prevent a temperature spike in the LPGG and thrust chamber. The MLV permits the IFV to close mechanically and shut off control-pressure fuel to the IMV and hypergol inlet. The MFV closes by spring tension when its actuation pressure decays below 145(+40, −50) psig, shutting off fuel flow to the thrust chamber and LPGG. Fuel pressure decay causes TOPS deactuation, which closes the prevalves and removes the thrust OK indications to the IU EDS voting logic circuits. Loss of thrust OK signals causes the L/V engines lights on the MDC to illuminate, indicating to the astronauts that the engines have cut off. L/V ENGINES −5, −6, −7, and −8 lights will illuminate first, indicating inboard engines cutoff. Approximately 3 sec later, L/V ENGINES −1, −2, −3, and −4 lights will illuminate, indicating outboard engines cutoff. The lights will remain on until S-IB/S-IVB separation. When fuel pressure in the fuel injector manifold decays below 105±20 psig the LPGG control valve closes by spring pressure. The pressure decay also permits the FABU to close, shutting off lubricant flow to the turbopump gearbox. The FABU valve closes when the inlet pressure is 50-90 psig. Engine thrust decays to zero in approximately 3.5 to 4.5 sec after receipt of cutoff command. During flight, if an

4-13

Section IV S-IB Stage

engine malfunctions and turbopump discharge fuel pressure falls between 790 and 740 psia, TOPS deactuation on that engine will initiate engine cutoff. See Electrical Sequencing for additional information on engine shutdown.

H-1 Engine Cutoff Commands.

During flight, any of three basic cutoff modes will initiate H-1 engine cutoff. Those modes are: malfunction cutoff, normal 4-by-4 cutoff, and range safety command cutoff. The normal 4-by-4 cutoff is the planned engine cutoff mode, where the inboard engines shut down, followed by outboard engine shutdown 4 sec later. The malfunction mode will shut down individual engines while the range safety command will shut down all engines simultaneously to effect a zero-thrust condition. The range safety commands originate from the ground stations and are used in the event of vehicle deviation from planned trajectory or other emergencies endangering the launch facility or the mission. All other cutoff commands originate in the vehicle.

Normal 4-by-4 Engine Cutoff. The switch selector starts the engine cutoff sequence by enabling the four propellant level sensors at 2 min 11.2 sec into flight (times are predicted S-IB-6 flight times which may change as mission parameters are defined). See figure 4-18. One of the sensors will actuate at 2 min 14.7 sec (T_2) and provide an input to the LVDA/LVDC to start the cutoff sequence. At 2 min 17.5 sec (T_2 + 3.0 sec) the LVDC will issue the inboard engines cutoff command through the S-IB stage switch selector. The cutoff command fires both squibs in the Conax valves on the inboard engines to start the cutoff operation. See H-1 Engine Cutoff. At 2 min 19.0 sec (T_2 + 4.5 sec), a switch selector command groups together the outputs from the voting logic of the thrust OK pressure switches on each outboard engine. If any outboard engine experiences lox starvation and shuts down, deactuation of the TOPS on that engine will command the remaining three engines to cut off. At 2 min 19.5 sec (T_2 + 5.0 sec) the switch selector enables the fuel depletion sensors. This command permits the fuel depletion sensors to issue the cutoff command if the fuel drops below the sensor level in the tank sumps to prevent the engines from shutting down because of fuel starvation. The engines must shut down with a fuel-rich mixture to prevent excessive temperatures in the thrust chamber and gas generator. Either lox starvation or fuel depletion cutoff will establish the vehicle staging time base (T_3). The switch selector issues a backup cutoff command at 2 min 21.7 sec (T_3 + 0.1) to ensure engine shutdown for staging. Simultaneous cutoff of all outboard engines is necessary to prevent attitude deviations that would endanger the launch vehicle during staging.

One Engine-Out Capability. The S-IB stage has no engine-out capability during the first 3 sec of flight to preclude the possibility of a catastrophic vehicle/tower collision. At 3 sec into flight (T_1 + 3 sec), the switch selector initiates a command to enable a one-engine-out capability. Failure of one engine, resulting in TOPS deactuation, will initiate cutoff for that engine. Shutdown of that one engine will immediately remove bus power from the TOPS voting logic circuits of the other engines, thereby inhibiting the cutoff circuits for the remaining seven engines regardless of their performance (figure 4-17). This one-engine-out capability will remain in effect until 10 sec into the flight (T_1 + 10 sec) to ensure maximum available thrust to lift the vehicle clear of the pad area.

Multiple-Engine-Out Capability. Ten sec into flight (T_1 + 10 sec) the switch selector disables the one-engine-out bus (figure 4-17), which in turn enables the bus power to the TOPS voting logic circuitry. If an engine (or engines) fails, thrust decay in that engine will cause TOPS deactuation, which fires the Conax valve to complete cutoff for that engine. For the remainder of flight, until TOPS grouping command, each engine can shut down independently.

Range Safety Cutoff. Any time during flight that the vehicle deviates beyond acceptable limits of the intended trajectory or becomes a hazard, the range safety officer (RSO) can destroy the vehicle. As a requirement of the Air Force Eastern Test Range, thrust of liquid propellant vehicles must be reduced to zero before destroying the vehicle. The RSO issues two commands to destroy the vehicle. The first command shuts off all engines simultaneously and also charges the propellant dispersion system EBW firing units. The second command triggers the firing units to detonate the explosive shaped charges (See Ordnance). The first command fires the squibs in all Conax valves simultaneously, thereby initiating eight engine cutoff (figure 4-17). Thrust decays to zero within 3.5 to 4.5 sec.

Thrust Chamber.

The thrust chamber consists of a gimbal assembly, an oxidizer dome, an injector and hypergol container, and a thrust chamber body. The oxidizer dome, installed over the injector, encloses the forward end of the thrust chamber. The thrust chamber receives propellants under turbopump pressure, mixes and burns the propellants, and imparts a high velocity to the expelled combustion gases to produce thrust.

Gimbal Assembly. The gimbal assembly, mounted on the oxidizer dome, secures the thrust chamber to the thrust pad of the tail unit assembly. The gimbal is essentially a universal joint mounted on thrust vector alignment slides. The gimbal assemblies on the inboard engines are functional only to the extent of aligning the thrust vector during engine installation. The outboard engine gimbal assemblies permit the hydraulic actuator to change the thrust vector to satisfy the guidance and attitude corrections. The S-IB stage outboard engine gimbal assemblies permit an 11.31-deg angular displacement of the geometric thrust vector from the normal plane of the gimbal bearing axis (with both actuators fully extended or retracted, either in-phase or out-of-phase). A single actuator,

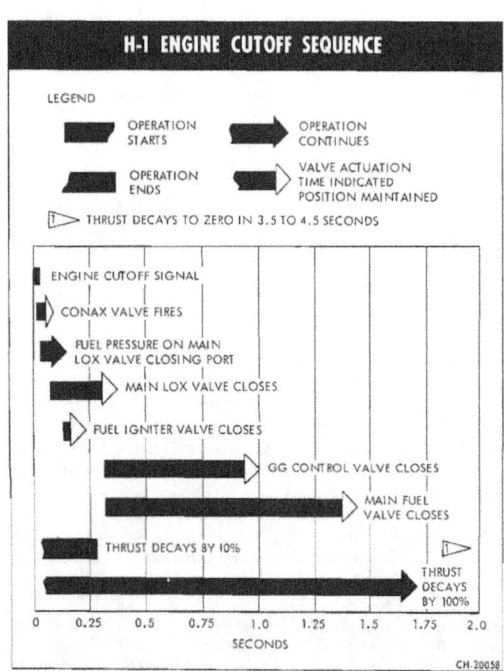

Figure 4-15

4-14

H-1 ENGINE CUTOFF OPERATION

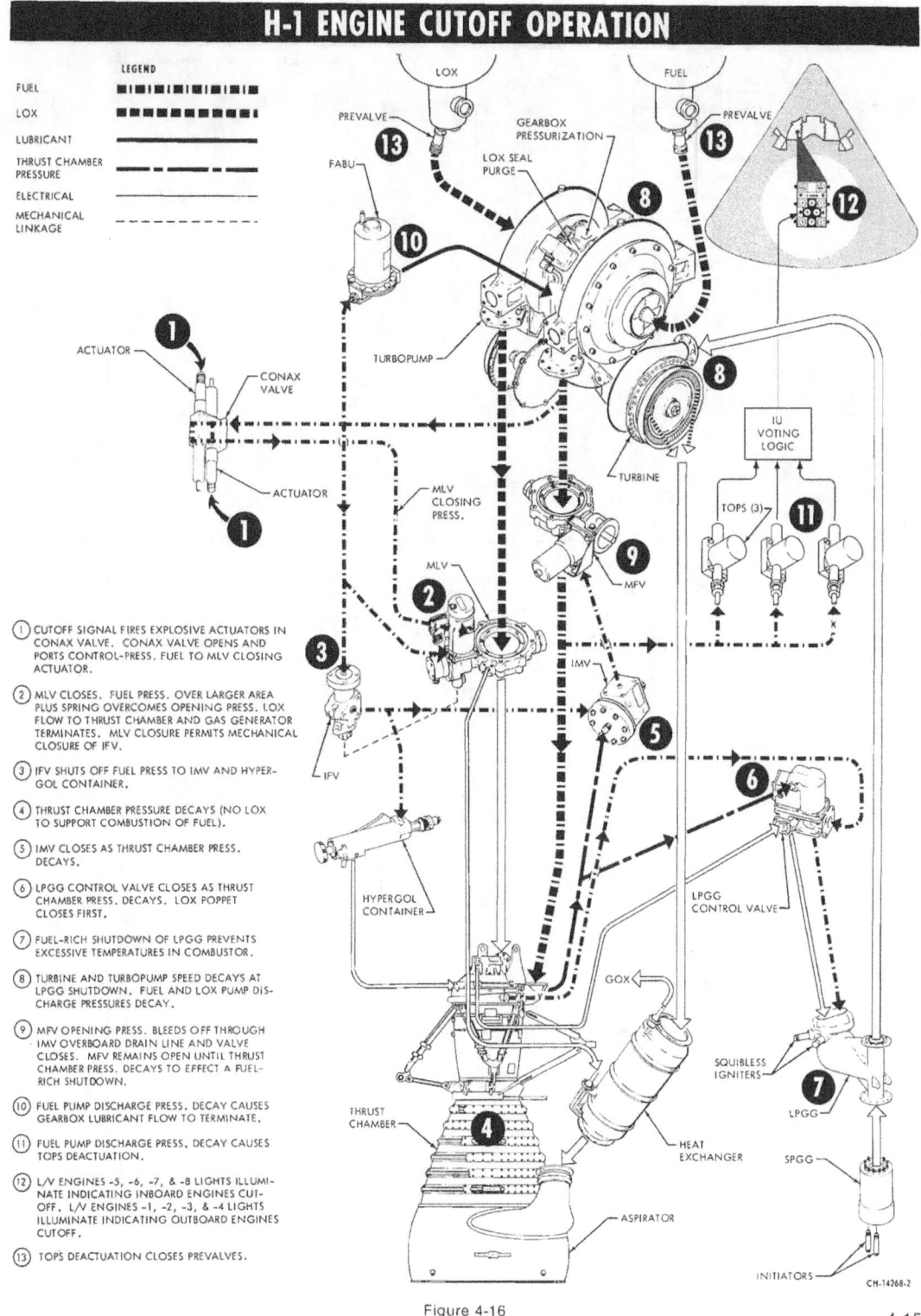

Figure 4-16

Section IV S-IB Stage

H-I ENGINE CUTOFF LOGIC

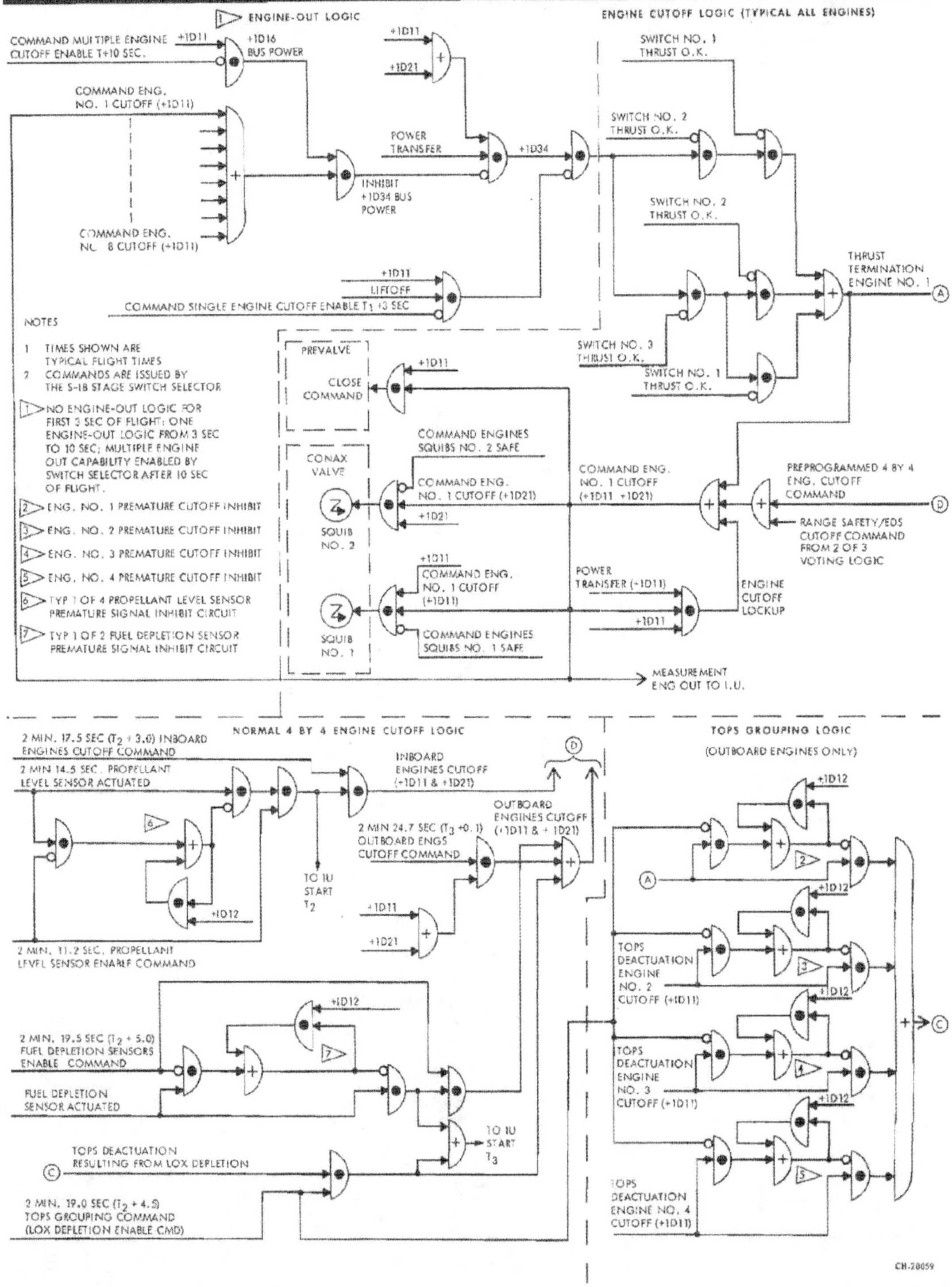

Figure 4-17

Section IV S-IB Stage

either fully extended or fully retracted, permits an 8-deg angular displacement of the geometric thrust vector. The gimbal assembly design will permit a 14.8-deg (max) geometric thrust vector angular displacement using two actuators, or a 10.5-deg (max) displacement using one actuator. The actuators determine the limits of the angular displacement for the S-IB stage outboard engines. The alignment slides on both inboard and outboard engines provide lateral positioning of the engine geometric thrust vector with respect to the engine centerline.

Oxidizer Dome. The oxidizer dome directs lox from the turbopump into the injector, provides a mount for gimbal and thrust chamber assembly, and transmits engine thrust to the vehicle structure. Bootstrap lox for the gas generator is supplied from the oxidizer dome.

Injector. The injector (figure 4-18) receives lox and fuel and injects the propellants into the combustion area in a fuel-on-fuel and lox-on-lox pattern to insure satisfactory combustion. Twenty-one concentric passage rings distribute the fuel and lox through angled orifices to attain the like-on-like impingement pattern. Fuel flows through the outermost ring and in each alternate ring. For increased combustion stability, copper baffles mounted on the injector divide the injector face into six equal areas around a center hub. Passages in the baffles that correspond to orificed holes in the lox and fuel rings permit propellant flow through the baffles into the combustion area. Hypergol flows through seven passages in the injector to fuel housings brazed into the injector face (one in each baffle compartment). Fuel from the thrust chamber body enters the injector through a ceramic-coated screen located around the injector periphery and lox enters the injector from the forward side, which is enclosed by the oxidizer dome. The injector transmits thrust forces, produced by combustion zone pressure acting upon the injector, to the oxidizer dome and subsequently to the vehicle structure. See figure 4-19 for injector characteristics.

Thrust Chamber Body. The thrust chamber body is a de Laval structure (100 percent bell) and consists of a converging combustion chamber, throat section, and diverging section through which combustion gases are expanded and accelerated. Longitudinal stainless steel tubes joined together by furnace brazing and retained by external rings and tension bands form the thrust chamber body. This type construction permits regenerative cooling during engine operation by fuel flow through the tubes. A tapered fuel manifold attaches to the top of the thrust chamber and provides equal flow

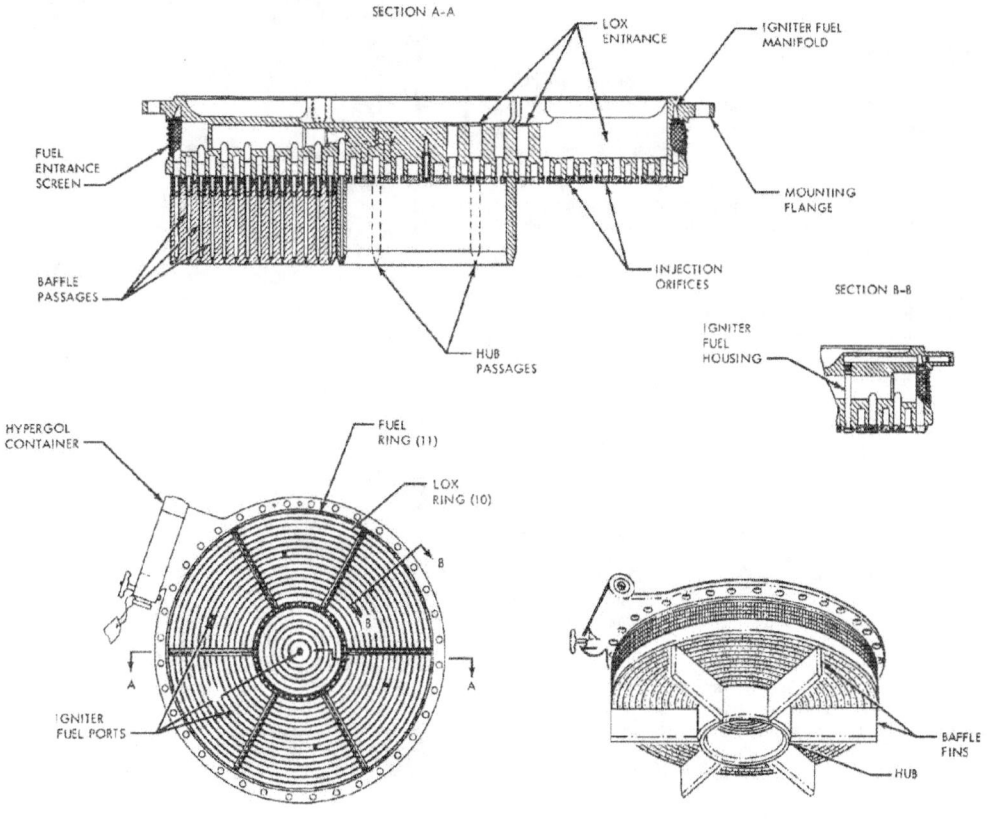

Figure 4-18

THRUST CHAMBER INJECTOR CHARACTERISTICS

ITEM	DESCRIPTION	
MATERIAL	347 STAINLESS STEEL CRES WITH COPPER RINGS	
FABRICATION METHOD	FURNACE BRAZED	
PLATE AREA	332 in.^{2}	
PLATE DIAMETER	20.55 in.	
SCREEN MATERIAL	18-8 cres CERAMIC-COATED 10-mesh (0.059 in. dia)	
OXIDIZER PRESS. DROP	122.2 psi AT 535.0 lb/sec	
FUEL PRESS. DROP	85.4 psi AT 235.0 lb/sec	

ITEM	LOX	FUEL
TOTAL NO. ORIFICES	1137	1394
TOTAL ORIFICE AREA	10.182 in.^{2}	7.044 in.^{2}
IMPINGEMENT DISTANCE	0.521 in.	0.300 in.
IMPINGEMENT ANGLE	40 deg	40 deg
NO. FILM COOLANT ORIFICES	0	84
NO. BAFFLE FIN COOLANT ORIFICES	42	42
NO. HUB COOLANT ORIFICES	0	44
PERCENT FILM COOLANT	0	2.1

Figure 4-19

THRUST CHAMBER BODY CHARACTERISTICS

ITEM	DESCRIPTION
CHAMBER AREA (INJECTOR END)	332 in.^{2}
CHAMBER DIAMETER (INJECTOR END)	20.56 in.
THROAT AREA, A_t	204.35 in^2
THROAT DIAMETER	16.13 in.
EXIT AREA	1634.3 in^2
EXIT DIAMETER	45.62 in.
NOZZLE AREA EXPANSION RATIO	8:1
OVERALL LENGTH	86.15 in.
TUBE WALL THICKNESS	0.012 in.
NUMBER OF TUBES	292
CHARACTERISTIC LENGTH $L = \dfrac{V_c}{A_t}$	39.10 in.
COMBUSTION CHAMBER VOLUME V_c	4.62 ft^3

Figure 4-20

THRUST CHAMBER OPERATING CHARACTERISTICS

PARAMETER	RATING
SEA LEVEL THRUST	204,300 lbf
SEA LEVEL SPECIFIC IMPULSE	268.9 sec
TOTAL PROPELLANT FLOWRATE	760.3 lb/sec
MIXTURE RATIO	2.342 : 1 (O/F)
LOX FLOWRATE	532.8 lb/sec
FUEL FLOWRATE	227.5 lb/sec
INJECTOR END CHAMBER PRESS.	701.8 psia
NOZZLE STAGNATION PRESS.	652.5 psia
CHARACTERISTIC VELOCITY (C*)	
C* INJECTOR END PRESS.	6,069 ft/sec
C* NOZZLE STAGNATION PRESS.	5,647 ft/sec
C* EFFICIENCY (NOZZLE)	97.20 %
THRUST COEFFICIENT (C_f)	
C_f INJECTOR END PRESS.	1.425
C_f NOZZLE STAGNATION PRESS.	1.532
C_f EFFICIENCY (NOZZLE)	101.4 %
RATIO OF INJECTOR END PRESS. TO NOZZLE STAGNATION PRESS.	1.080
JACKET PRESSURE DROP:	
AT 225 lb/sec	135.0 psig
AT 227.4 lb/sec	138.0 psig

Figure 4-21

through each down tube. This even flow through the cooling tubes increases the thrust chamber life. The return manifold encloses the tubes at the nozzle base and directs fuel flow through up-tubes to the fuel injector. See figures 4-20 and 4-21 for thrust chamber characteristics.

Propellant Feed System.

The propellant feed system consists of the turbopump, fuel and lox high-pressure ducts, main lox valve, main fuel valve, igniter fuel valve, check valves, and orifices. This system supplies the propellant, at the prescribed flowrates and pressures, to the thrust chamber and gas generator.

Turbopump. A turbine-driven, dual-pumping unit turbopump delivers fuel and lox at high-pressure and high flowrates to the engine combustion chamber. The turbopump (figure 4-22) consists of an oxidizer pump, fuel pump, reduction gearbox, accessory drive adapter, and gas turbine. The turbopump mounts on the side of the thrust chamber and requires only two short, high-pressure ducts connecting the volute outlets to the main propellant valves. This installation assures minimum pressure drop from pumps to valves. Bootstrap propellants from turbopump discharge flow to the LPGG for turbine operation. The high-speed gas turbine drives the turbopump through a series of reduction gears that drive the main shaft. In each pump, axial-flow inducers increase the pressure at the impeller inlet (thus requiring a low NPSH), and radial-flow hollow-vaned impellers and integral diffusers increase propellant flow-rate and pressures. Stationary diffuser vanes inside the pump provide uniform pressure distribution, reduction of fluid velocity around the impellers, and reduction of fluid turbulence in the pump volutes. Balance ribs on the inboard side of the impellers hydraulically balance the axial thrust on the pump shaft. During operation the fuel additive lubricates and cools the turbopump gears and bearings. The lox and fuel single-entry, centrifugal pumps mount back-to-back, one on each side of the gearbox. Bolts secure the fuel pump to the gearbox while radially-inserted steel pins secure the lox pump to the gearbox. The steel pins permit the lox pump housing to expand and contract during extreme temperature changes without distortion or misalignment. To prevent formation of an explosive environment, four lines from the turbopump drain expended lubricant and fuel and lox leakage overboard. Two lube drain ports manifold into a single line that incorporates a lube drain relief valve. This valve also maintains the gearbox pressure between 2 and 10 psig. Two lox lines drain any lox leakage from the lox cavity downstream from the primary lox seal. One line drains any fuel and lubricant leakage past the lube seal into the fuel drain manifold, which dumps the fuel overboard through a single line.

Gearbox. The gearbox contains the gear train that provides the turbine differential speed to drive the main pump shaft (figure 4-22). All gears are of full depth configuration. The intermediate and pinion gears contain inner races for the roller bearings. The main pump shaft utilizes a roller bearing and a ball bearing. The ball bearing restricts axial movement of the shaft. Lube passages cast within the gearbox walls direct the fuel and Oronite mixture through jets onto the bearings and gears. The jets apply lubricant to the disengaging side of the gears to prevent hydraulic lock. An accessory drive pinion, integral with the intermediate gear, drives the two counter-rotating accessory drive gears. The lower accessory

Section IV S-IB Stage

H-I ENGINE TURBOPUMP

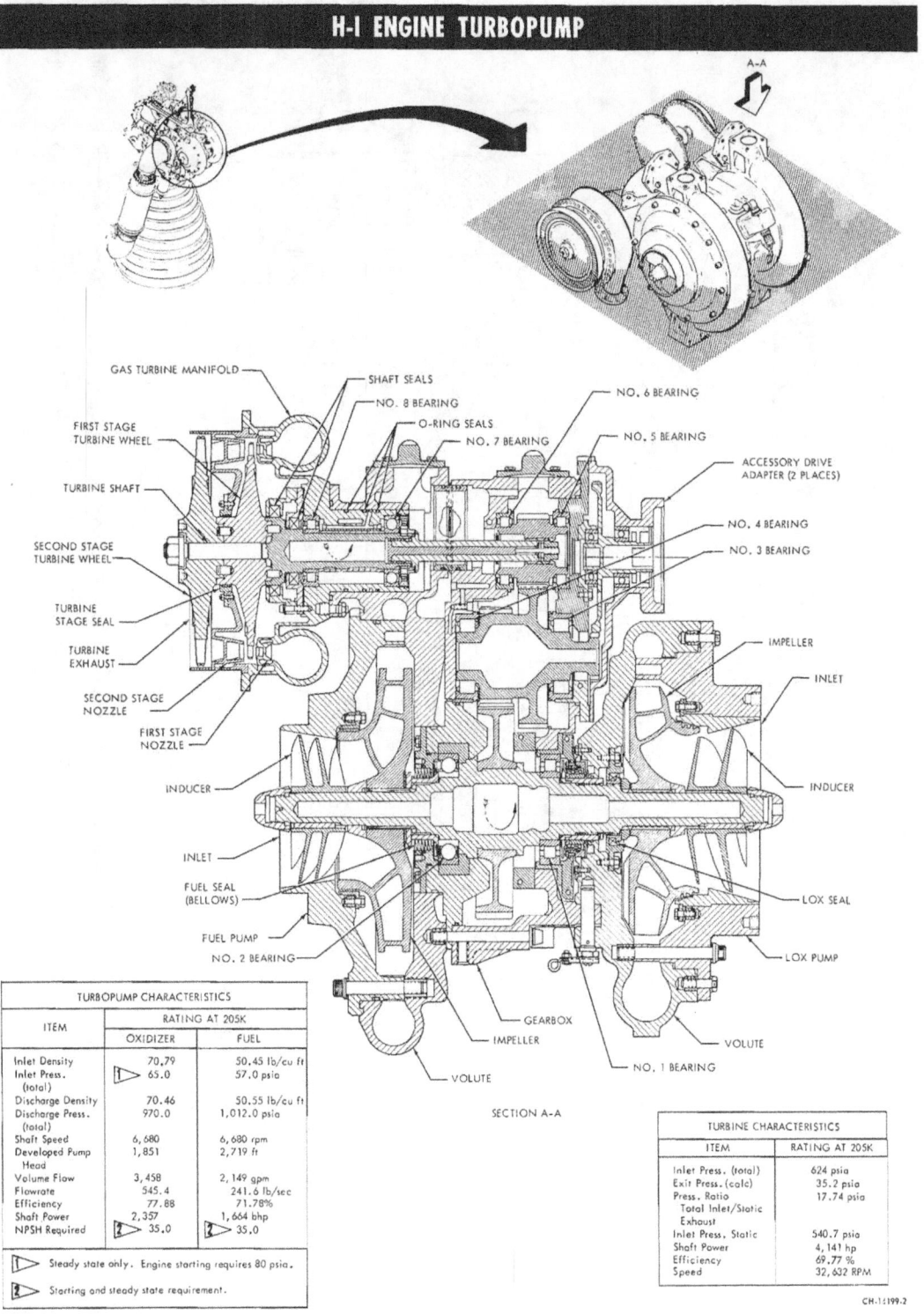

Figure 4-22 (Sheet 1 of 2)

4-19

Section IV S-IB Stage

H-1 ENGINE TURBOPUMP

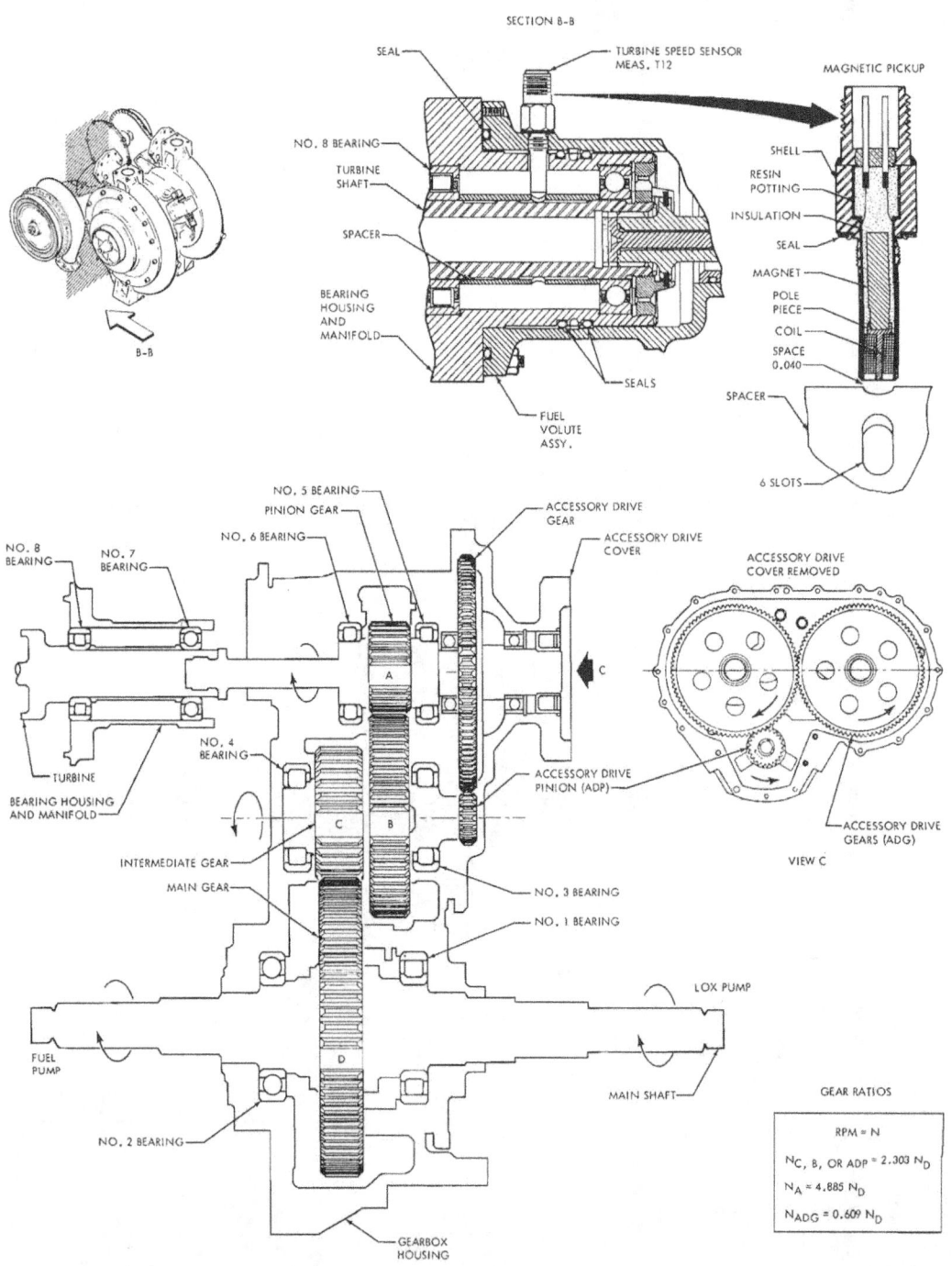

Figure 4-22 (Sheet 2 of 2)

Section IV S-IB Stage

drive gear operates the outboard engine hydraulic pump. Calrod heaters, one attached to the inner surface of the gearbox and one integral with the no. 1 bearing retainer, maintain the temperature of the lox pump main shaft bearing (no. 1) between 90 ± 5 and $110\pm8°F$. The heaters receive 115V, 60 Hz power at T-8 hr (before lox loading begins). The COMPONENT HEATERS lamp on the S-IB networks panel illuminates at power application to all eight turbopump heaters. A current sensor in each heater supply circuit provides an input to DEE-6 that the heater is operative. Power remains applied to the heaters until the ALL ENGINES RUNNING signal deenergizes the heater power contactor. Measurement XC1 on each engine senses the lox pump main shaft bearing temperature and provides a telemetry input signal. See H-1 Engine Measuring.

Gas Turbine. The gas turbine is an impulse, two-stage, pressure-compounded unit (figure 4-22). The turbine bolts to the fuel pump housing and consists of an inlet manifold, first- and second-stage turbine wheels and nozzles, a turbine shaft, and a splined quill shaft that connects the turbine shaft to the high-speed pinion gear. The turbine shaft inboard bearing (no. 7) is a split race ball bearing, and the outboard bearing (no. 8) is a roller bearing. Carbon ring shaft seals prevent hot gas leakage into the bearings. Gases from the LPGG enter the gas turbine manifold and flow through the first stage nozzle to the first stage turbine wheel. The gases then pass through the second stage nozzle, increasing in velocity, and through the second stage turbine wheel. The gases exhaust the turbine into the turbine exhaust system and ultimately into the thrust chamber exit flow. The turbine stage seal prevents the gases from bypassing the second-stage nozzles. An electrical tachometer using a single-element magnetic pickup senses the turbine shaft speed (figure 4-22). This measurement, T12, is telemetered to ground receiving stations and recorded for later evaluation of turbine performance. See H-1 Engine Measuring. Six slots in a spacer around the turbine shaft rotate through the magnetic field of the pickup, thus generating the shaft speed signal.

Propellant Ducts. High-pressure ducts connect the lox and fuel pump outlets with the main lox and main fuel valves on the thrust chamber. Bellows sections in the welded duct assemblies allow for expansion and contraction or movement between the turbopump and thrust chamber. The fuel bellows will accommodate a combined simultaneous deflection of 0.5-in. offset and 0.25-in. axial displacement, plus 1 deg of flange angulation while operating at 1124-psig maximum internal pressure. The lox joint bellows will deflect ±3 deg in all planes to allow a 0.25-in. parallel offset and withstand ±0 deg, 20 min of torsional rotation while operating at 1195-psig maximum internal pressure and at $-300°$ F. Two plates, a gimbal ring, and two links bear structural load imposed on the lox joint. The fuel duct diameter is 3.25 in. and the lox duct is 3.375 in. in diameter. See figure 4-23.

Main Lox Valve. The main lox valve (MLV) controls the flow of oxidizer to the thrust chamber. The normally closed, balanced butterfly type gate valve is installed between the lox high-pressure propellant duct and the oxidizer dome. Control-pressure fuel under turbopump discharge pressure actuates the valve. During engine start, increasing fuel pressure (300 ± 50 psig minimum) overcomes actuator spring pressure and opens the valve. A piston-crank assembly rotates the gate shaft to open or close the valve. A control cam mounted on the gate shaft mechanically opens the igniter fuel valve, which is installed on the MLV. A thermostatically controlled, $400\pm60W$ blanket heater on the actuator prevents lubricants and packings from freezing due to lox temperatures. The heater maintains the temperature at $100\pm15°$ F. GSE supplies 115V, 60 Hz power to the actuator heaters. The COMPONENT HEATERS lamps on the AC module and S-IB network panel indicate power application to the heater power buses. The heater power indication also serves as an input to the events display panel. A

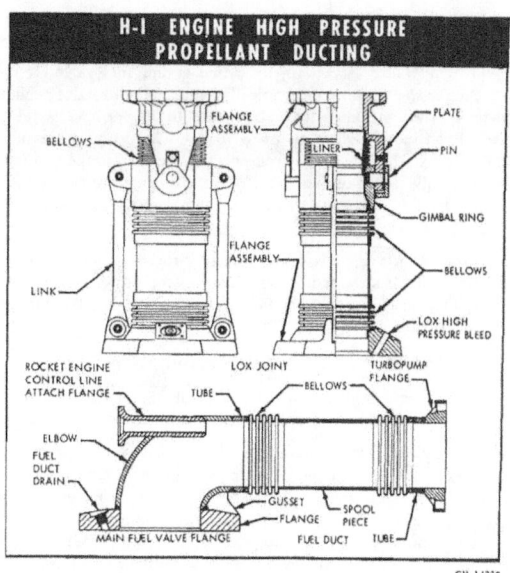

Figure 4-23

current sensor in each heater circuit provides an input to DEE-6 indicating that the heater is in operation. The ALL ENGINES RUNNING signal removes heater power just prior to umbilical disconnect. The heaters do not operate during flight. A pyrotechnically actuated valve (Conax Valve) closes the MLV to effect engine cutoff by porting control-pressure fuel to the closing side of the MLV actuator. The fuel pressure acting on a larger surface area of the actuator piston plus the spring pressure overcomes the fuel pressure at the opening port and the valve closes. Drain lines port lox leakage, from past the gate shaft seal and from the MLV linkage housing, into the thrust chamber exhaust stream. Additionally, two lines drain leakage of control-pressure fuel from past the actuator piston seal and from around the connecting rod into the fuel drain manifold and into the exhaust stream of the engine. These drain systems prevent accumulation of fuel and lox leakage in the engine compartment thus preventing a possible fire. A port on the downstream side of the MLV gate housing supplies lox to the heat exchanger for lox tanks inflight pressurization.

Main Fuel Valve. The normally closed, gate-type main fuel valve (MFV), installed between the fuel high-pressure propellant duct and the fuel inlet manifold, controls RP-1 flow to the thrust chamber. The valve remains closed during the engine start sequence until ignition of hypergol and igniter fuel with lox has been detected. The increase in chamber pressure shuttles the ignition monitor valve, which then ports control-pressure fuel from the igniter fuel valve to the opening port of the MFV actuator. Fuel pressure of 350 ± 50 psig operates a piston-crank assembly, which rotates gate shaft to open the valve. The MFV remains fully open during engine operation and closes after the MLV closes during engine shutdown to effect a fuel-rich engine cutoff. The fuel-rich cutoff prevents excessive temperatures in the engine thrust chamber. During engine shutdown the ignition monitor valve closes when the chamber pressure decays and ports the MFV opening actuator fuel overboard allowing the actuator spring to close the MFV gate. Three drain lines connected to the MFV port fuel leakage into the fuel drain manifold and ultimately into the engine exhaust stream. These lines drain leakage from the gate shaft seal and from the actuator housing seal and piston housing seal. A port

4-21

on the downstream side of the gate housing supplies fuel pressure for thrust OK pressure switches actuation.

Igniter Fuel Valve. The normally closed igniter fuel valve (IFV) attached to and operated mechanically by the main lox valve begins to open when the main lox valve opens to 50 deg and is fully open when the lox valve opens to 70 deg. Control-pressure fuel flows through the IFV to the hypergol container and to the ignition monitor valve for main fuel valve opening. Spring pressure closes the igniter fuel valve at engine cutoff when the main lox valve closes.

A drain line prevents any control-pressure fuel leakage from past the IFV seals from entering the MLV linkage housing. The line drains the leakage through the fuel drain manifold into the engine exhaust stream.

Gas Generator and Control System.

The gas generator and control system controls engine start sequencing and supplies power to drive the turbopump.

Solid Propellant Gas Generator. A solid propellant gas generator (SPGG) installed on the liquid propellant gas generator (LPGG) combustor of each engine, starts the engine by supplying large volumes of gas to spin the turbine. Upon receipt of the engine ignition command, two initiators installed in the SPGG ignite pellets in an igniter inside the SPGG and the pellets in turn start the grain propellant burning. As the grain burns, SPGG internal pressure increases until a diaphragm bursts at 600 to 700 psig and releases the gases through an orifice to the turbine. The turbine drives the turbopumps, through a gear train, and liquid propellants begin flowing in the engine. The SPGG continues to burn for approximately 100-200 msec after bootstrap fuel and lox enter the liquid propellant gas generator, thus igniting the liquid propellants.

SPGG Initiator. The SPGG initiator is an electrically activated pyrotechnic squib that ignites the SPGG. A 500-Vac, 2-A ignition command impulse applied to a bridgewire in the initiator ignites the pyrotechnic charge. To ensure against inadvertent firing of the initiator, a cold cathode trigger diode in the initiator circuit prevents current through the bridgewire until the voltage exceeds 250 Vac. A 100 ohm resistor in each electrical leg protects the bridgewire from high level static electric discharge. Functioning time is 15 msec (max). The pyrotechnic material and bridgewire firing circuit components are housed in a case made of 1018 steel bar material. The housing has two sets of threads, one for installing the initiator in the SPGG and one for attaching electrical cables.

Liquid Propellant Gas Generator. The liquid propellant gas generator produces combustion gases during steady-state operation to drive the two-stage turbine, which supplies power through a gear reduction train to drive the propellant pumps. The LPGG consists of a gas generator control valve, injector assembly, a combustor, and two squibless igniters. Propellants entering the LPGG are ignited by hot gases produced by the solid propellant gas generator and squibless igniters during engine start. The hot gases from the SPGG ignite the squibless igniters prior to liquid propellant entry into the LPGG. The igniters burn for 2.5 to 3 sec to ensure liquid propellant ignition. See figure 4-24 for LPGG characteristics.

Gas Generator Control Valve. The normally closed gas generator control valve contains two poppets that admit fuel and lox bootstrap propellants into the gas generator combustor during engine operation. TC combustion pressure actuates the control valve by repositioning a piston that opens the fuel poppet first. A yoke integral with the piston opens the lox poppet. Fuel poppet cracking pressure is 105±20 psig, lox poppet cracking pressure is 200±20 psig, and fully-open operating pressure for the control valve is 275±25 psig. Bootstrap fuel flowrate is 13.52 lbm/sec while bootstrap lox flowrate is 4.61 lbm/sec. A bellows assembly enclosing the lox poppet stem

LPGG CHARACTERISTICS

PARAMETER	RATING
TOTAL FLOWRATE	18.13 lb/sec
MIXTURE RATIO	0.341 (o/f)
GG COMBUSTION PRESS. INJECTOR END	646.3 psia
GG COMBUSTOR TEMP	1,198°F
FLOWRATE:	
LOX	4.61 lb/sec
FUEL	13.52 lb/sec
PRESS DROP ACROSS INJECTOR:	
LOX	93 psid
FUEL	90 psid

Figure 4-24

and closure spring, and seals on the actuator piston, prevent leakage of fuel and lox into the control valve actuator. A drain line ports any fuel leakage into the valve actuator to the fuel drain manifold where it is dumped overboard into the engine exhaust stream. The control valve design ensures a fuel-rich cutoff to prevent excessive temperature buildup in the combustor resulting in turbine burning. Spring pressure closes the control valve at engine cutoff when the thrust chamber pressure decays.

Gas Generator Injector. Fuel and lox from the gas generator control valve enter the injector and flow through passages that provide a uniform mixture-ratio of 0.341 (lox/fuel). The injector cavity design permits an oxidizer lead into the combustor during start to prevent detonation. From the injector, two fuel streams impinge on a single lox stream. The injector uses 44 impingement points. Fuel entering the combustor through 36 holes around the periphery of the impingements provides film coolant for the injector. During countdown the GG lox injector receives an ambient GN_2 purge to prevent entrance of SPGG contaminants into the injector. See Engine Purges.

Combustor. The bootstrap propellants burn in the combustor and exit to the gas turbine. Two squibless igniters installed in the combustor just below the GG injector assure propellant ignition during start. The combustor is a welded assembly with flanges for installation of the SPGG and for attachment to the gas turbine. Operating temperature is 1198±16° F and operating pressure is 646.3±4.9 psia.

Squibless Igniters. Two squibless igniters installed in the injector mounting flange on the combustor burn for 2.5 to 3 sec after their ignition by the solid propellant gas generator. They ensure ignition of fuel and lox if the SPGG burn should have expired before bootstrap propellant entry into the combustor. A 2-A link wire provides a monitoring capability for engine premature ignition. The sixteen igniter circuits (two per engine) are series connected to the ML. If any of the link wires break or burn through, for any reason, a PREMATURE IGNITION lamp on the S-IB firing panel will illuminate and the PREMATURE IGNITION SAFE lamp on the S-IB stage networks panel will extinguish. An emergency cutoff command will be initiated automatically when the PREMATURE IGNITION switch is in the ARM position. The link wire has no function in the ignition of the squibless igniter. A 7.5-Ω resistor in the series link wire circuit removes the possibility of igniting an igniter by the monitor circuit power.

Ignition Monitor Valve. The ignition monitor valve (IMV) opens the main fuel valve when ignition has been achieved in the combustion chamber. The IMV mounts below the MFV actuator and interfaces with the MFV opening port by an adapter. The three-way, normally closed valve has four ports; control-pressure fuel inlet (from IFV) and outlet (to MFV actuator), combustion chamber pressure sensing port, and a drain port. When combustion chamber pressure reaches 15±0.5 psig, the valve shuttles, closing off the drain port, and permits control-pressure fuel flow into the MFV actuator. The IMV remains open throughout H-1 engine operation. During engine shutdown the valve closes under spring pressure. The drain port opens and dumps the MFV opening actuator fuel overboard through a drain line permitting the MFV to close.

Thrust OK Pressure Switches. Three normally open, two-position pressure switches on each engine, sense fuel pressure downstream of the main fuel valve. Each hermetically sealed pressure switch contains a single-pole, double-throw switch with positive actuated snap-action electrical contacts. A checkout port enables CALIPS checkout testing without disconnecting engine system lines or pressurizing the fuel inlet manifold. The electrical outputs of the switches indicate satisfactory fuel pressure as a component of satisfactory engine thrust. The switches actuate when fuel pressure downstream from the main fuel valve reaches 800±15 psia (indicating approximately 90 percent engine thrust attained). Switch deactuation occurs between 45 psi (max) and 25 psi (min) below the actual actuation pressure. During checkout, pressure applied to the switches from the CALIPS console actuates the switches at 800±45 psia for first actuation and 800±30 psia for second and third actuation. Deactuation pressures during checkout are 15 to 65 psi below actuation pressure for the first cycle and 20 to 45 psi for second and third cycle. During the ignition sequence operation (T-3 to T-0) the IU EDS distributor monitors the thrust OK pressure switches for thrust buildup. Voting logic for each three-pressure-switch-group on each engine determines when all engines have attained approximately 90 percent of rated thrust and provides an ALL ENGINES RUNNING signal to the program distributor in the ML. The logic circuits remove power from the L/V engine indicator lamps on the command module main display console, indicating that the engines are running and have attained approximately 90-percent thrust level. If all engines are running at time for commit, T-0, the COMMIT signal will command the launch vehicle release circuits. If, however, the ALL ENGINES RUNNING signal is not present at time for commit, cutoff will occur automatically. ALL ENGINES RUNNING indications are monitored on the S-IB firing panel. The THRUST OK pressure switches provide discrete inputs to the ground computer systems and the telemetry system.

Hypergol Container. The hypergol container is an integral part of the thrust chamber injector (figure 4-18). The cylindrical housing accommodates a 6-in.³ hypergol cartridge of triethylaluminum (figure 4-25) and a HYPERGOL INSTALLED detector switch. O-ring seals on the cartridge prevent leakage during engine operation. Two diaphragms, one at the inlet and one at the outlet of the cartridges, contain the hypergol until fuel pressure, 300±25 psig, burst them during engine start. Hypergol flows, under turbopump fuel pressure, through the igniter fuel manifold to seven passages that direct hypergol to igniter fuel ports in the injector. Ignition occurs when the hypergol contacts lox. Fuel flows through the hypergol container and igniter fuel ports during engine operation. Insertion of the cartridge into the container actuates the detector switch, which provides an output to the S-IB firing preparation panel. HYPERGOL INSTALLED indicator lamps illuminate indicating all eight engines have received the hypergol cartridges. The switches also provide inputs to the digital event evaluator.

Conax Valve. The two-way, normally closed Conax valve effects engine cutoff by directing control-pressure fuel to the main lox valve closing actuator (figure 4-16). Two pyrotechnic actuators installed into a single body connect the inlet and outlet ports by driving a ram through metal membranes separating the ports. Actuation of either or both pyrotechnic actuator assemblies will allow fuel flow through the valve. Position indicators installed into the valve body opposite the actuators provide monitoring of the valve position. When the actuator fires, the ram strikes a plunger, which in turn breaks a link wire in the indicator. All sixteen position indicators (two per engine) are connected in series to ground monitoring circuits. With all link wires intact, no ground indication will be monitored. However, if any link wire breaks, a red ANY CONAX FIRED lamp will illuminate on the S-IB firing panel in the LCC. As a safety precaution, no electrical connections exist between the position indicator and the pyrotechnic actuator. To prevent inadvertant application of power or stray voltage from firing the Conax valve, a ground "Conax valve safe" command, constituted by deenergized emergency cutoff and ignition command relays, energizes four vehicle relays that ground actuator firing circuits. Either the T-5 sec signal or an emergency cutoff command disables the ground "Conax valve safe" command. In the vehicle, normally closed contacts of the command engine cutoff relays also keep the actuator firing circuits at ground potential until the guidance computer or RSO commands engine cutoff or if the thrust ok pressure switches on an engine initiate cutoff. The command engine cutoff relays reposition their contacts and apply power to the Conax valve actuator. Bridgewires detonate the actuator pyrotechnics. The valves open, and control-pressure fuel flows to the main lox valves closing actuators.

Exhaust System.

The H-1 engine exhaust system (figure 4-9) ducts the fuel-rich turbine exhaust gases overboard into the thrust chamber exit flow stream. A welded stainless steel turbine exhaust hood ducts the gases into a heat exchanger. A bellows section with an integral liner in the turbine exhaust hood permits movement of the system due to heating. The heat exchanger, also a welded stainless steel shell, houses a helix-wound four-coil system. Exhaust gases heat the coils. Lox, under turbopump pressure, flows through a unitized check valve into three of the coils and is converted to gox for lox tank inflight pressurization. An orifice in each of three coil inlets controls the lox flowrate. The fourth coil is not used. Exhaust gases exit the heat exchanger into a turbine exhaust duct on inboard engines, or an aspirator on outboard engines. The curved stainless steel turbine exhaust duct directs the exhaust gases into the thrust chamber exit flow stream of the inboard engines. The aspirator, a welded Hastelloy C-shell assembly installed on the periphery of the outboard engine nozzle, extends below the thrust chamber exit. The forward end of the aspirator is welded to a channel band approximately 20 in. forward of the fuel return manifold. The aft end of the aspirator is not secured. A 0.440-in. clearance between the fuel return manifold and the aspirator permits the turbine exhaust gases to escape into the thrust chamber exit flow stream.

Lubrication System.

The fuel additive blender unit (FABU) eliminates the requirement for a lube-oil tank, pressurization, equipment, plumbing, and controls. Extreme-pressure additive, Rocketdyne ST0140RB0013, blended with RP-1 fuel lubricates and cools turbopump gearbox components (figure 4-22). A thermostatically controlled 300±30 W heater maintains the correct viscosity of the additive by controlling the temperature between 12±4° F and 130±4° F. GSE supplies 115V, 60Hz power to the heater for preflight additive conditioning.

PHYSICAL PROPERTIES OF TRIETHYLALUMINUM	
MOLECULAR WT.	114.5
FREEZING POINT	-53°F
BOILING POINT	381°F
SPECIFIC GRAVITY	0.84
DENSITY	52.3 lb/ft³
WT. PER GALLON	7.02 lb
ENERGY RELEASE	18,300 Btu/lb
FLAME TEMP.	1,200°F

Figure 4-25

Section IV S-IB Stage

FUEL ADDITIVE HEATER POWER ON lamp on the AC module and FUEL ADDITIVE HEATERS lamp on S-IB networks panel illuminate at heater power application. A current sensor in each heater circuit provides an input to DEE-6 that the heater is operative. The ALL ENGINES RUNNING signal removes heater power just before the umbilicals disconnect. Lubricant flow begins when control-pressure fuel increases to 70 to 150 psig. Fuel flows through a 35-mesh inlet strainer made of Monel and pressurizes the additive, and aligns a spool containing a metering orifice with the additive outlet. The additive then flows through a 100-mesh Monel outlet screen and blends with the fuel (2.75±0.75 percent by volume). The mixture leaves the FABU and enters the turbopump through a 40 (nom) to 75 (max) micron-mesh filter. The filter element is made of 18-8 CRES stainless steel. Lubricant consumption rate is 5 to 6 gpm. After injection into the turbopump gearbox, the lubricant drains overboard through the lube-drain relief valve and drain lines that extend down the engine thrust chamber exterior and dump into the engine exhaust stream. At engine cutoff the decaying fuel control pressure permits spring closure of the FABU, which shuts off lubricant flow to the gearbox.

Electrical System.

The H-1 engine electrical system consists of armored and unarmored electrical harnesses that interface with engine components and stage and ESE circuitry. Harnesses that have flight as well as preflight functions are armored to prevent damage to the conductors that may compromise engine operation or cause failure of the mission. Harnesses that have preflight functions only do not have armor. Those preflight functions are heater operations (FABU, turbopump bearing no. 1, and MLV actuator), auxiliary hydraulic pump operation, Conax valve position indications, squibless igniter link monitoring (premature ignition), hypergol cartridges installed indications, and start commands to the SPGG initiators. The armored flight harnesses transmit thrust ok signals, hydraulic servoactuator commands and feedback, and cutoff commands to the Conax valves.

H-1 Engine Measuring.

The H-1 engine measuring systems monitor 13 conditions on each outboard engine and 11 conditions on each inboard engine. This information is telemetered to ground receiving stations through the S-IB stage PCM/DDAS assembly and RF assembly P1 during flight. During checkout, telemetry is received by coaxial cable from the PCM/DDAS assembly. Each measurement has a number comprised of a letter representing a parameter, measurement number within the parameter, and a dash number indicating the stage unit location (Example: C1-1). Three parameters of engine measurements are temperature, C; pressure, D; and RPM, T. Engine unit numbers are 1 through 8 respectively to engine location. See figure 4-26 for measurement numbers, titles, and other pertinent information and figure 4-9 for measurement locations. Output signals from measurement transducers XC1, C9, XC54, XC89, D1, and T12 are routed through measuring racks 9A516 for engines 1 and 5, 9A520 for engines 2 and 6, 9A526 for engines 3 and 7, and 9A530 for engines 4 and 8. The measuring racks contain signal conditioning modules for each measurement input. The modules assure compatibility of the measurement signal with the TM multiplexer input requirement. After signal conditioning, measurement signals XC1, C9, XC54, XC89, and D1 are multiplexed by TM multiplexer 13A484. Measurement signals T12 are applied directly to TM assembly F1. Measurement signals D12, D13, and D14 do not require signal conditioning and are applied directly to TM multiplexer 13A440. Measurement signals D20, D34, and D35 do not require signal conditioning and are applied directly to TM multiplexer 13A484. Measurement signals XC1, C9, XC54, D1, and T12 require remote automatic calibration system (RACS) checkout. This permits checking the individual circuits for response and

H-1 ENGINE ANALOG MEASUREMENTS

NUMBER	NAME	RANGE
XC1-1 THRU -8	TEMP, LOX PUMP BEARING 1	-20 TO 200°C
C9-1 THRU -8	TEMP, GAS GEN CHAMBER	0 TO 1000°C
XC54-1 THRU -8	TEMP, LOX PUMP INLET	-185 TO -168°C
XC89-1 THRU -8	TEMP, GEAR CASE	0 TO 150°C
C540-1 THRU -8	TEMP, LOX SEAL DRAIN LINE 1	-185 TO -12°C
C541-1 THRU -8	TEMP, LOX SEAL DRAIN LINE 2	-185 TO -12°C
C542-1 THRU -8	TEMP, LOX SEAL DRAIN LINE 3	-185 TO -12°C
D1-1 THRU -8	PRESS, COMBUSTION CHAMBER	0 TO 800 PSIA
D12-1 THRU -4	PRESS, FUEL PUMP INLET	0 TO 100 PSIA
D13-1 THRU -4	PRESS, LOX PUMP INLET	0 TO 150 PSIA
D14-1 THRU -8	PRESS, TURBINE INLET	0 TO 800 PSIA
D20-1 THRU -8	PRESS, GEAR CASE	0 TO 200 PSIA
D34-1 THRU -8	PRESS, GG FUEL INJECTOR	0 TO 900 PSIA
XD35-1 THRU -8	PRESS, GG LOX INJECTOR	0 TO 900 PSIA
D53-2 AND -6	PRESS, LOX PUMP INLET	0 TO 150 PISA
E513-2 AND -6	VIBRATION, ENG THRUST BLOCK, LONG	-1 TO +5 G
T12-1 THRU -8	TURBINE RPM	0 TO 45K RPM

▷ "X" PREFIX INDICATES AUXILIARY DISPLAY

Figure 4-26

accuracy of signal transmission. The RACS control panel operator selects HI, LO, or RUN reference signals that correspond to predicted transducer signals and applies the signal to an individual circuit. Upon receipt of instructions from the RACS, the S-IB stage measuring rack selector (9A546) addresses the particular measuring rack and module to be checked. With the reference signal transmitted to the signal conditioning module, the module returns a signal corresponding to the predicted transducer output. This signal verifies calibration of the individual circuit. Two H-1 engine parameters are considered critical and have redline values which, if exceeded, will produce unsafe or unsatisfactory operations. Measurements XC89-1 through XC89-8, extreme-pressure additive temperature at each engine, must be within a 105° F to 160° F range. If the lube additive is outside this range, the additive consistency will cause improper mixture of fuel and additive resulting in improper turbopump lubrication. Measurements XC1-1 through XC1-8, turbopump bearing no. 1 temperature at each engine, must have a minimum temperature of 40° F at T-3 min. Redline value at ignition is 0° F, which is based on heat loss (assuming heater failure) at T-3 min. The position of each thrust ok pressure switch on each engine is telemetered back to the ground receiving station as event measurements. See figure 4-27. In addition to being recorded, these signals are monitored by ENG THRUST OK indicators, one for each pressure switch, on the EDS monitor panel. Inputs to the EDS monitor panel are received through DDAS. Prime interest time of these measurements is from liftoff until S-IB stage outboard engine cutoff. These measurements are fed directly from the thrust OK pressure switch to the Remote Digital Submultiplexer 9A700 and are telemetered through the PCM/DDAS assembly and RF assembly P1 to the ground stations. During checkout, thrust OK pressure switch positions are also monitored, through hard-wired connections, on ENG THRUST OK indicators on the EDS Preparation Panel. This panel also has an indicator for each

Section IV S-IB Stage

MEAS. NO.	TITLE
VK138-1	ENG 1, SW 1 THRUST OK
VK139-1	ENG 1, SW 2 THRUST OK
VK140-2	ENG 2, SW 1 THRUST OK
VK141-2	ENG 2, SW 2 THRUST OK
VK142-3	ENG 3, SW 1 THRUST OK
VK143-3	ENG 3, SW 2 THRUST OK
VK144-4	ENG 4, SW 1 THRUST OK
VK145-4	ENG 4, SW 2 THRUST OK
VK146-5	ENG 5, SW 1 THRUST OK
VK147-5	ENG 5, SW 2 THRUST OK
VK148-6	ENG 6, SW 1 THRUST OK
VK149-6	ENG 6, SW 2 THRUST OK
VK150-7	ENG 7, SW 1 THRUST OK
VK151-7	ENG 7, SW 2 THRUST OK
VK152-8	ENG 8, SW 1 THRUST OK
VK153-8	ENG 8, SW 2 THRUST OK
VK171-1	ENG 1, SW 3 THRUST OK
VK172-2	ENG 2, SW 3 THRUST OK
VK173-3	ENG 3, SW 3 THRUST OK
VK174-4	ENG 4, SW 3 THRUST OK
VK175-5	ENG 5, SW 3 THRUST OK
VK176-6	ENG 6, SW 3 THRUST OK
VK177-7	ENG 7, SW 3 THRUST OK
VK178-8	ENG 8, SW 3 THRUST OK

H-1 ENGINE EVENT MEASUREMENTS

Figure 4-27

pressure switch on each engine. Measurements D1-1 through D1-8 and VK138 through VK153 are flight control measurements that are monitored in real time at Mission Control Center in Houston.

The three lox seal drain line temperature measurements on each engine (C540, C541, and C542) are interlocked in the automatic countdown sequence between engine ignition and liftoff. If any two of the three measurements on any engine indicate the presence of lox (temperature below $-250°$ F) in the drain cavity, all engines will be cut off.

STATIC TEST.

S-IB-6 was static fired on June 23, 1966 (Test SA-36) for 35.58 sec duration and again on June 29, 1966 for 141.24 sec (Test SA-37). Performance and engine operation was satisfactory on Test SA-36. During Test SA-37, Engine 2 (H-7072) and Engine 4 (H-7075) experienced step decreases in power level at 105 sec and 25 sec respectively.

S-IB-6 is the first stage equipped with 205K engines. These engines are essentially the same as the previous 200K stages except for changes necessary to increase propellant flowrates and to allow the higher operating level. Specifically, the turbopump lox and fuel impellers were retrimmed for increased flowrates and the gas generator injector differential pressures were lowered.

Test SA-36.

This test was conducted at the Static Test Tower East (STTE) at MSFC, Huntsville, Alabama. Cutoff was initiated by the firing panel operator as scheduled. Duration from ignition command to Inboard Engine Cutoff (IECO) was 35.46 sec and to Outboard Engine Cutoff (OECO) was 35.58 sec. All engines performed within the 205,000±3000 lbf range (sea level reference conditions). No recalibration was necessary. Engine 6 (H-4069) operated at 207.4K sea level thrust, 2.8K higher than the Rocketdyne level.

Test SA-37.

This long duration test was also conducted on STTE. Time base two was initiated by uncovering of the low level sensor in Tank 02 at 135.38 sec after ignition command. At 3.2 sec after uncovering,

IECO was commanded by the switch selector. OECO was from lox depletion occurring 2.66 sec after IECO. Time from ignition command and actual event times are not necessarily representative of those expected for flight, but were altered to fit the static test conditions.

As mentioned in the initial paragraph, engines 2 and 4 exhibited step changes in thrust during SA-37. These shifts were noticed in thrust chamber combustion chamber pressure and amounted to approximately one percent.

In addition to the in-run shifts, all engines operated at a steady state power level lower than on Test SA-36. Exhaustive studies have failed to detect the reason for this overall reduction. The flight prediction has considered both the shift and the lower power level.

The step decreases have been observed in Rocketdyne tests and were found to be caused by changes in fuel flow distribution in the thrust chamber coolant tubes. The flow distribution change resulted in fuel system resistance increases. The addition of a baffle in the thrust chamber exit manifold successfully eliminated the shifts. However, the installation of the baffle is not scheduled until S-IB-13 engines.

See figure 4-28 for a summary of the 205K H-1 engine test history.

205K H-1 ENGINE TEST HISTORY

PROGRAM	NO. OF ENGINES TESTED	TEST DURATION (SECONDS)	NO. OF ENGINE TESTS
205K QUAL TEST	1	2,765	37
ACCEPTANCE TEST	97	28,657	364
R&D TEST PROGRAMS	27	62,373	843
STAGE STATIC TESTS	64	12,708	184
TOTAL		106,503	1,428

Figure 4-28

LOX SYSTEM.

The five S-IB lox tanks receive lox from the facility storage system through the fill-and-drain line storing it for consumption by the eight H-1 engines during boost phase (figure 4-29).

The lox system tanks consist of four outer units (O-1, O-2, O-3, and O-4) and a center unit (O-C), with a nominal system capacity of 66,277 gal and a minimum ullage volume of 1.5 percent. Sufficient ullage pressure is provided to ensure structural integrity of the lox tanks, and to maintain a net positive suction head of 35 ft at the lox pump engine inlet. The skin-milled, butt-welded aluminum alloy segments of the tank walls vary in thickness from top to bottom in relation to stress concentrations. The tank bulkheads are hemispherical, with skirts forward and aft providing space for pressurization and vent manifolds on the forward end. The aft skirts accommodate the sumps and interconnecting manifolds. Each of the outer tanks supplies lox to one inboard and one outboard engine. The center tank holds approximately 35 percent of the lox and is 105 in. in diameter by 678 in. long. The overall length is 750 in. Clustered around the center tank are four tanks, each having a capacity for 16 percent of the lox requirements. The outer tank dimensions are 70 in. in diameter by 678 in. long, between

Section IV S-IB Stage

S-IB LOX SYSTEM DIAGRAM

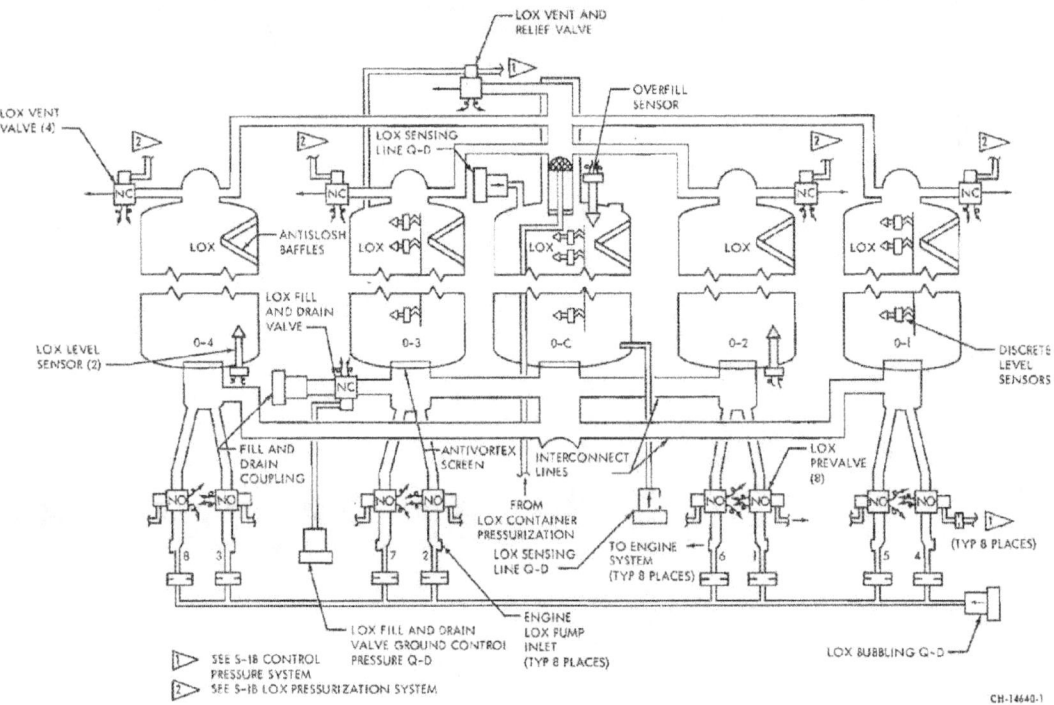

Figure 4-29

bulkheads, and an overall length with skirts of 747 in. Each tank has a capacity of 10,821 gal, but 1.5 percent of the capacity is reserved for ullage.

Lox Fill.

Initial Fill. The S-IB lox loading is an automatic operation controlled by the lox tanking computer. During the initial fill, the stage vent valves, the main fill valve, the slow fill valve and the pump discharge valve are opened to allow the lox to flow from facility storage to the stage tanks. As the lox is loaded, it is distributed equally to each tank by the manifold. The lox temperature, as it is loaded, remains almost constant for the entire fill operation between −285 and −297° F. The flowrate for this precool operation is approximately 500 gpm, for approximately 30 min. until 22-percent loading is obtained. The propellant tanking computer system (PTCS) generates a signal that terminates the precool operation and starts the main fill operation.

Main Fill. When the main fill valve, the pump discharge valve, and the S-IB fill-and-drain valve are opened the S-IB umbilical vent line drain valve is closed, the flowrate is increased to 14,250 lbm/min. This rate decreases when 95-percent load is reached in approximately 15 min. The topping, the last 5 percent of the 634,125-lbm load, is slow-fill loaded at a rate of 5260 lbm/min and at a pressure of 50 psig. This sequence is accomplished by closing the main fill valve. The 99-percent level signal causes the slow fill valve to close and the fill sequence is complete. The lox level is now maintained to the 100-percent level until 3 min and 7 sec before liftoff.

Lox Bubbling. Lox bubbling is initiated approximately 153 sec before liftoff and must not be terminated more than 100 sec before ignition. Helium is the medium for the bubbling action that is introduced into the lox suction lines at a ground regulated pressure of 225 psig, at ambient temperature and a flowrate of 3.263 lb/min. The helium gas flows upward through eight separate branch lines, each containing a metering orifice, and then into the pump inlet at the end of each suction line. This helium flow maintains subcooled lox at the turbopump inlets to prevent pump cavitation during engine start. The helium bubbles then rise through the suction lines and normally open lox prevalves and through the lox tank where it contributes to the ullage pressure.

Lox Tank Pressurization.

Prepressurization. The S-IB tank is automatically sequenced for pressurization after the lox bubbling from valve panel 2, while the stage is on the ground. Inflight pressurization is provided by gox converted from lox in heat exchangers located on the H-1 engines (figure 4-30). The prepressurization tank pressure is 55.3 to 57.7 psia at ambient temperature; the time required is 0.834 min. The pressurization is initiated by closing the lox vent valves and the lox vent-and-relief valve. Ground helium flows to the center lox tank and from there to the outer lox tanks through the upper interconnect lines and manifold. The tank pressure is controlled by the lox prepressurization switch between 55.3 and 58.5 psia. If the prepressurization switch fails, the lox vent and relief valve will mechanically open between 60 and 62.5 psia. The ground lox vent pressure switch will actuate at 67.5 psia and allow pneumatic

Section IV S-IB Stage

S-IB LOX PRESSURIZATION AND VENT SYSTEM DIAGRAM

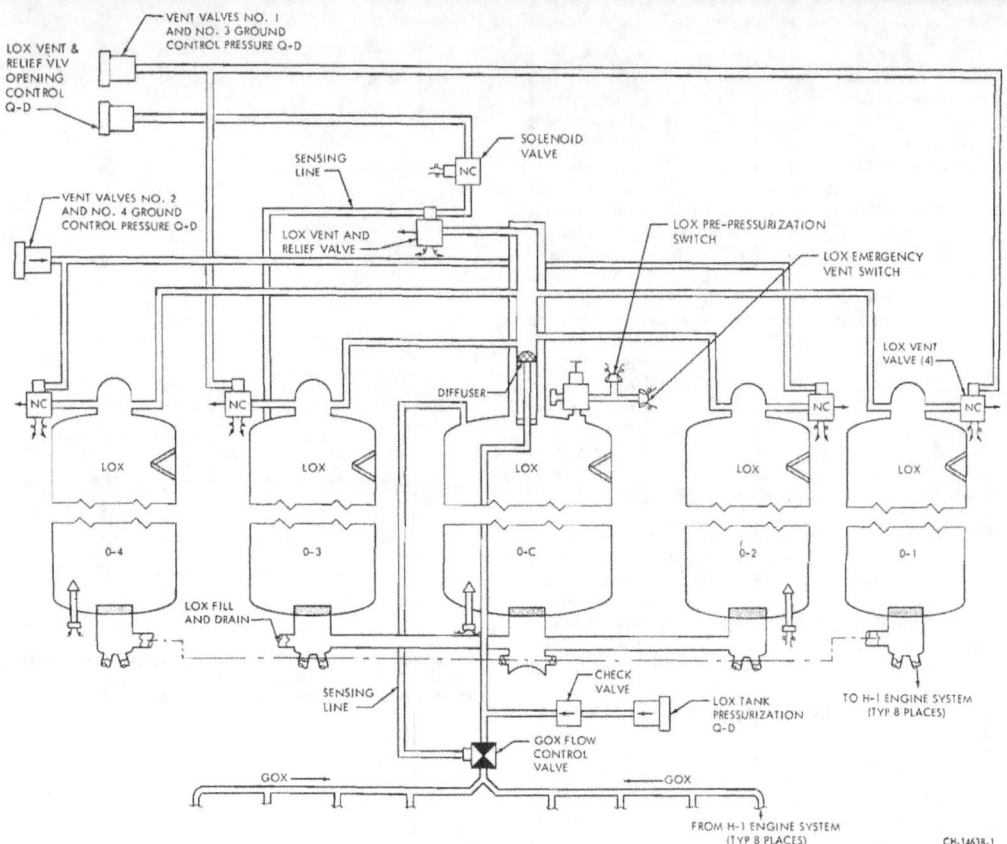

Figure 4-30

pressure from the ground GN_2 system to open the lox vent-and-relief valve.

Flight Pressurization. During the boost phase, lox is supplied to the engine heat exchangers. Gox from each heat exchanger flows into a common manifold, through the gox flow control valve, and into lox tank O-C. The outer lox tanks receive gox through the upper interconnect lines and manifold. The gox flow control regulator controls the flow and maintains a pressure of 50 psia. Overpressurization is prevented by the lox vent-and-relief valve.

Measurements.

Figure 4-31 lists the lox system flight measurements and indicates the information that is displayed.

Lox Characteristics.

Figure 4-32 lists the physical and chemical properties of lox, and figure 4-33 shows the oxygen vapor pressure curve.

Lox System Components.

Lox Fill and Drain Valve. The ball-rotor gate valve is located in the fill-and-drain line leading to the sump of lox tank O-3. The valve is spring-loaded to the closed position and is opened by

LOX SYSTEM MEASUREMENTS			
NUMBER	NOMENCLATURE	RANGE	DISPLAY
XD3-OC	PRESSURE, GOX, LOX TANK	+0 +100 PSIA	1
VK15-02	LOX LEVEL CUTOFF IND NO. 2	ON OFF	2
VK16-04	LOX LEVEL CUTOFF IND NO. 3	ON OFF	2
L500-0C	LOX LEVEL PROBE	0/+5 VDC	
L500-01	LOX LEVEL PROBE	0/+5 VDC	
L500-03	LOX LEVEL PROBE	0/+5 VDC	
1	AUXILIARY AND MCC-H DISPLAY		
2	ESE DISPLAY		

Figure 4-31

pneumatic pressure applied to the actuator assembly. The actuator is designed for an operating pressure of 750 psig, a proof pressure of 1125 psig, and a burst pressure of 1875 psig. The valve position is indicated by an electrical switch for an LCC readout. The closing

Section IV SIB Stage

and opening response time is 500 msec with a flow chamber pressure of 100 psig. The valve is designed to handle lox at ambient temperatures from −100 to 125° F with a nominal operating pressure of 150 psig, a proof pressure of 225 psig, and a burst pressure of 375 psig.

Lox Prevalve. The normally open, ball-rotor gate valves are located next to the sumps in each of the eight lox feed lines. The primary function is to stop the flow of lox to the engine and also provide back up capability to the engines main lox valves. The eight prevalves are operated by a signal from the TOPS deactuation so that prevalve closure is accomplished only after the engine has started thrust decay. The valve is spring-loaded and driven to the closed position by the pneumatic actuator; when the pressure is released, the valve returns to the normally open position. Gaseous nitrogen is the pneumatic medium operating at a pressure of 775±25 psig. The actuator is designed for a proof pressure of 1125 psig and a burst pressure of 1875 psig. The actuator operating temperature range is −65 to 125° F. A position indicator switch monitors the fully closed and the fully open position for an LCC readout. All prevalves must indicate OPEN to complete an interlock for start of automatic launch sequence. The valve response time for closing is 850±100 msec and for opening is 3,500 msec maximum under flow. The prevalve environmental temperature is −100 to +125° F at a nominal operating pressure of 150 psig, proof pressure of 225 psig, and a burst pressure of 375 psig.

Gox Flow Control Valve. Gox is accumulated from eight engine heat exchangers and its flow is regulated by the flow-control-valve as it passes to the dome of lox tank O-C for distribution and equal pressurization to each of the five lox tanks. The control valve is a modulating, spring-loaded, normally open, butterfly-type valve. The butterfly is controlled by a pressure-operated bellows that tends to close the butterfly. An aneroid sensor in the valve regulates the pressure flow to the bellows from lox tank O-C. Thus an increasing lox tank pressure tends to close the butterfly and a decreasing pressure tends to open it, thereby increasing the gox flow. A potentiometer monitors the position of the butterfly for LCC readout. The proof pressure of the flow chamber and the control pressure chamber is 750 psig. The bias pressure chamber is proofed at 470 psig. The main power bellows has a negative differential proof pressure of 305 psig and positive differential proof pressure of 420 psig. The proof pressure of the pilot valve is 70 psia. The operating pressure for the flow chamber and the control pressure chamber is 500 psig maximum. The bias pressure chamber operates at a maximum pressure of 305 psig. The main power bellows operate with a negative differential pressure of 202 psig, and a positive differential pressure of 280 psig. The pilot valve nominal operating pressure is 50 psia. The burst pressure of the flow chamber and control pressure chamber is 1250 psig. The bias pressure chamber burst pressure is 785 psig. Burst pressures of the main power bellows are a negative differential pressure of 505 psig and a positive differential pressure of 700 psig. The gox flow control valve operating temperature range is from 10 to 200° F.

Vent Valve. The four vent valves vent the pressure in the lox tanks during lox loading and replenishing operation. The normally closed, gate-type valves are spring-loaded and are opened by ground supply GN_2. A position indicator switch provides LCC readout for a fully open or a fully closed gate. A 110-Vac thermostatically controlled heater prevents the valve mechanism from freezing. The thermostat energizes at a minimum of 70° F and deenergizes at 145° F. The valve will open in 75 msec with a control pressure of 775±25 psig, and a minimum of 500 psig, while the flow chamber medium is in a temperature range of −250 to 250° F. The control actuator uses GN_2 and has a proof pressure of 1125 psig, with a burst pressure of 1875 psig. The valve flow chamber has an operating pressure of 60 psig, with a proof pressure of 90 psig and a burst pressure of 120 psig.

PROPERTIES OF LOX

COMMON NAME:
   LOX, LIQUID OXYGEN

CHEMICAL FORMULA:
   O_2

MOLECULAR WEIGHT:
   32.0

PHYSICAL PROPERTIES:

FREEZING POINT	−361.76°F
BOILING POINT	−297.4°F
CRITICAL TEMPERATURE	−181.04°F
CRITICAL PRESSURE	736.47 PSIA
LIQUID DENSITY	9.54 LBS/GAL
APPEARANCE	PALE BLUE, CLEAR LIQUID
ODOR	NONE
LIQUID TO GAS RATIO	1:862

CHEMICAL PROPERTIES:

STABLE AGAINST MECHANICAL SHOCK IN PURE FORM.

IMPACT SENSITIVE TO UNPREDICTABLE DEGREE IF CONTAMINATED, ESPECIALLY WITH ORGANIC MATERIALS.

MIXED WITH GREASE, OILS, PETROLEUM DERIVATIVE FUELS, ALCOHOL, ETC., IT FORMS A HIGHLY IMPACT SENSITIVE GEL WHICH MAY BE DETONATED BY SPARK OR FLAME AS WELL AS MECHANICAL SHOCK. THE EXPLOSIVE POTENTIAL OF THIS GEL HAS BEEN SHOWN TO BE APPROXIMATELY EQUIVALENT TO NITROGLYCERINE.

Figure 4-32

OXYGEN VAPOR PRESSURE

TEMP. °F	PSIA	
−361.76	0.00	FREEZING POINT
−297.4	14.70	BOILING POINT (1 ATM.)
−181.04	736.47	CRITICAL POINT

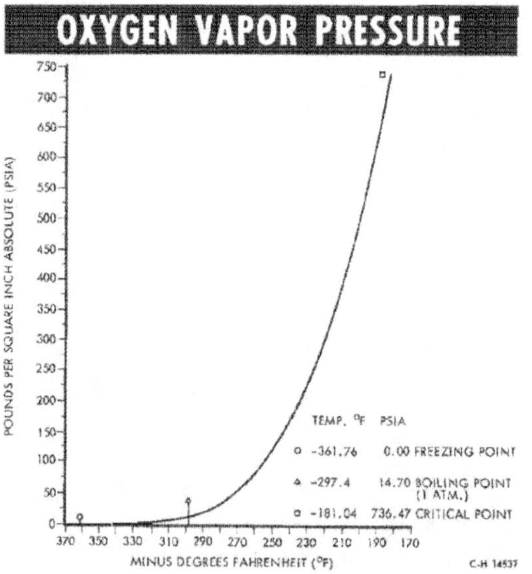

Figure 4-33

Vent and Relief Valve. The prime function of the vent-and-relief valve is to prevent over-pressurization of the lox tank should the pressurization system malfunction. The valve is a spring-loaded, normally closed, pilot-operated, in-line poppet-type valve. The valve vents the lox tank during the loading and replenishing operation and also relieves excessive pressure during preflight pressurization and after liftoff. The valve is operated by 775±25 psig GN_2 from the ground system or by He and gox pressure through a

Section IV S-IB Stage

sensing line connected between the valve pilot and the ullage area of tank O-3. The valve will crack at 60 psia minimum and will reseat at 59 psia. A hermetically sealed position indicator switch provides an LCC readout of the open and closed position of the valve. The flow chamber and sensing chamber have a design operating pressure of 63 psig, with a proof pressure of 95 psig and a burst pressure of 160 psig. The vent-and-relief valve will flow 16 lbm/sec (min) in the open position. The valve will move from fully closed to fully open position in 350 msec maximum with 750 psig pressure on the actuator and with 58.5 psia in the flow chamber. The maximum close cycle is 1.5 sec.

Pressure Transducer. Lox tank pressure is monitored at the prepressurization switch. The transducer is calibrated with ground source GN_2 through the calibration valve. The pressure range of the transducer is 0 to 100 psia, with a temperature range of −65 to +200° F, and an electrical resistance of 5000 ohms.

Calibration Valve. The calibration valve is used to calibrate the transducer. The nominal operating pressure is 3000 psig, with a proof pressure of 4500 psig and a burst pressure of 7500 psig.

Operating temperature range of the valve is −100 to 250° F.

Emergency Vent Switch. This pressure switch monitors only the preflight pressurization of the lox tank. Should the ullage pressure exceed 67.5±1.5 psia because the lox vent-and-relief valve relief function fails, the pressure switch will actuate a solenoid valve that opens the vent-and-relief valve thereby relieving the excessive pressure from the lox tank. The pressure switch deactuates at 63 psia to permit closing of the vent-and-relief valve. The switch is calibrated with ground source GN_2. It has a proof pressure of 110 psia and a minimum burst pressure of 175 psia. The temperature operating range of the switch is −65 to 165° F.

Prepressurization Switch. The prepressurization switch controls the ullage pressure of the lox tank, by actuating at 58.5 psia (max) closing down the pressure and by deactuating at 55.3 psia (min), permitting prepressurization to resume. This switch is disabled at liftoff. The switch is calibrated with ground source GN_2. It has a minimum proof pressure of 90 psig and a minimum burst pressure of 150 psig. The operating temperature range of the switch is from −65 to 165° F.

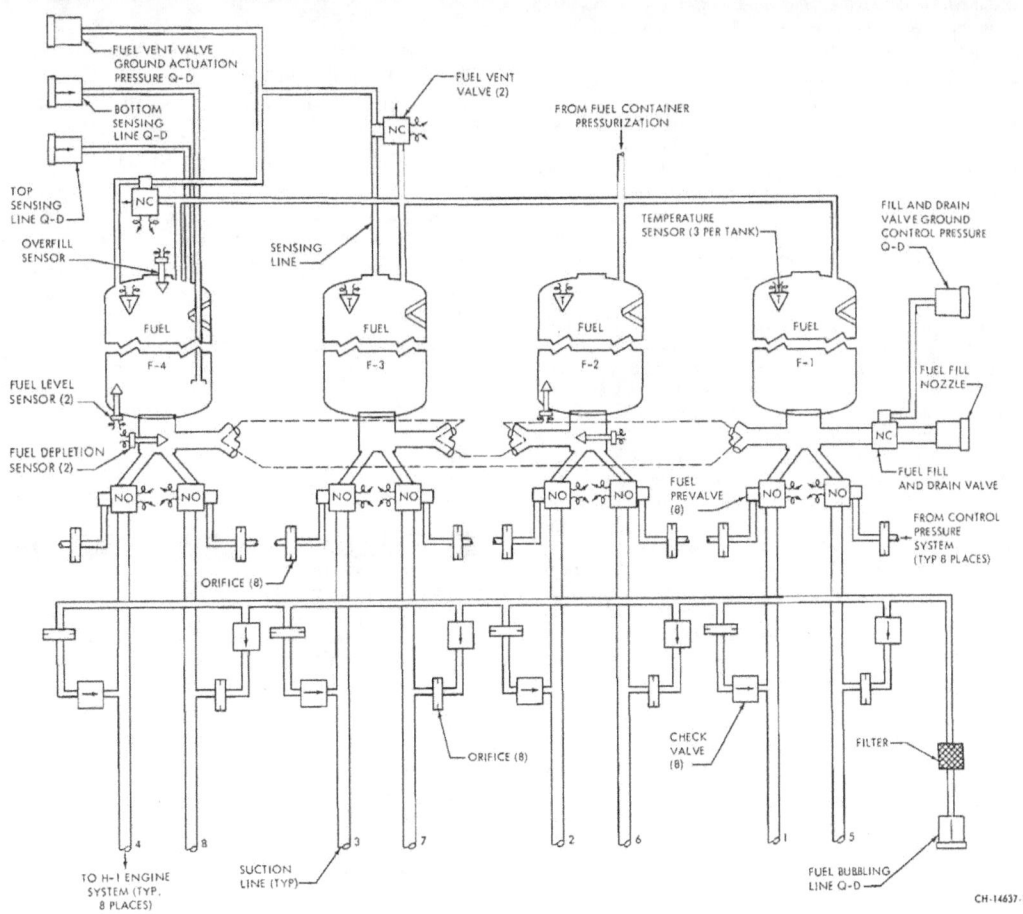

Figure 4-34

Section IV S-IB Stage

Engine Cutoff Lox Sensor. The engine cutoff lox sensors, located in the bottom of lox tanks O-2 and O-4, initiate the inboard engine cutoff when the lox is depleted to the level of the sensors. The operating temperature range of the sensors is −65 to 165° F. The linear accuracy of the sensors is plus 0.125 minus 2.00 in. of the static set point repeatable to ±0.125 in., when the level is decreasing at the rate of 5 in./sec. A heater and thermostat are incorporated into the sensor. The heater operates during prelaunch only on 110V, 60 Hz, single-phase power supply and is rated at 50 W.

Lox Overfill Sensor. The lox overfill sensor is located in the top of lox tank O-C, and actuates when the lox rises above the sensor during lox filling, thereby terminating the filling sequence. The operating temperature for the sensor is −65 to 165° F. The sensor has a thermostatically-controlled heater as an integral part that operates on 110V, single-phase power and is rated at 60 W.

FUEL SYSTEM.

The S-IB stage fuel system (figure 4-34) receives RP-1 fuel from the facility storage tanks, stores the fuel, and then supplies the fuel to the eight H-1 engines. The RP-1 fuel chemical and physical requirements are found in figure 4-35. The system consists of four fuel tanks (F-1, F-2, F-3, and F-4), tank pressurization components, distribution manifolds, control valves, switches, sensors, piping, interconnect lines, and the connecting hardware required to fill or drain the tanks, bubble fuel before flight, pressurize the tanks, and supply fuel to the engines.

The fuel tanks are interconnected at the top by a pressurization and vent manifold to ensure equal pressurization in all four tanks and to maintain the required net positive suction head (NPSH) at the engine fuel pumps. The manifold has two vent valves that vent the tanks in the event of overpressurization before flight, and also to vent the tanks during the fuel fill sequence. The system is not capable of becoming over-pressurized during flight due to the rapid fuel consumption. The fuel tank sumps are interconnected by a fuel transfer line assembly and interconnect lines that ensure a uniform fuel level in all four tanks and an equal distribution of fuel to the engines. In the event of an engine failure, the fuel normally consumed by the inoperative engine is supplied to the operating engines. Each tank supplies fuel to one outboard and one inboard engine through suction lines connected to the tank sump. A normally open prevalve connected between the tank sump and the suction line permits control of fuel flow from the tank sump to the engine. The prevalves provide a backup capability for fuel shutoff to the main fuel valve in the H-1 engines. During flight, two engine cutoff sensors, one each located in the bottom of tanks F-2 and F-4, generate a signal when the fuel decreases to their level that initiates inboard engine shutdown. Similar engine cutoff sensors in the lox system initiate inboard engine shutdown if lox depletion occurs prior to fuel depletion. Outboard engine shutdown occurs approximately 3 sec after inboard engine shutdown. The outboard engines are normally shut down simultaneously when engine thrust decay causes the outboard engines interconnected thrust OK pressure switches to deactuate. However, if fuel depletion occurs prior to lox depletion, two fuel depletion sensors, one each located in the sump of tanks F-2 and F-4, initiate outboard engine shutdown when the fuel level reaches the sensors. Three temperature sensors, located in each fuel tank, monitor the temperature of the fuel for fuel density calculations and electrically transmit the results to the fuel tanking computer prior to flight. An overfill sensor, located in the top of tank F-4 sends a signal to the tanking computer in the PTCS which terminates the fuel fill sequence in the event of tank overfill. Pressurized GN_2 (290 psig at ground regulator) is bubbled through each fuel suction line to agitate the fuel and aid in maintaining uniform fuel temperature in each tank. Fuel bubbling begins just before lox fill and continues until the start of fuel tank pressurization. The fuel tanks are pressurized with helium at approximately 2 min and 43 sec before launch until S-IB stage flight is completed. The tank pressure ensures the required fuel NPSH at the engine fuel pumps for engine starting and also to prevent the formation of a vacuum in the tanks as fuel is consumed during flight. Fuel tank pressurization starts at 29.6 to 32.4 psia and is maintained at this level until engine ignition. As the fuel is consumed, the tank pressure decreases to a minimum of 11.5 psia.

Fuel Fill.

The RP-1 fuel is stored in the facility storage tanks. Several days prior to transfer to the S-IB stage, the fuel is processed through a filter-separator unit that removes water and foreign matter that may have accumulated. The fuel is then transferred to the S-IB stage through a cross-country transfer line. Transfer of fuel from the facility storage tanks include all operations necessary to fill the S-IB stage fuel tanks. The operations include preparation for fuel transfer, manual fill, automatic fill, level adjust drain, and replenish. The manual fill and automatic fill operations are performed approximately 2 days before launch and the level adjust drain and replenish are performed on launch day.

Preparation for Fuel Transfer. Several major functions performed before transferring fuel are as follows:

a. The amount of fuel required for this particular mission is programmed into the PTCS.

b. The fuel filling mast located on the launcher is attached to the S-IB stage fuel fill and drain nozzle.

c. All closed hand valves in the transfer lines that permit fuel flow are opened.

d. Electrical power is applied to the RP-1 control panel and other necessary control components located in the LCC.

e. The pneumatic control pressures are made available at the facility pneumatic control console located in the fuel storage facility.

f. The S-IB stage fuel vent valves and the fuel fill-and-drain valve are opened using 750 psig GN_2 control pressure from the pneumatic control console.

Fuel Transfer to the S-IB Stage. Fuel transfer to the S-IB stage is accomplished as follows:

a. The manual mode of fuel transfer is initiated at the RP-1 control panel.

b. The solenoid valves in the transfer lines are each manually energized to the open position.

RP-1 CHEMICAL AND PHYSICAL REQUIREMENTS	
REQUIREMENTS	RP-1
GRAVITY °API – MINIMUM (MIN) (SPECIFIC (sp) GRAVITY, MAX)	42.0 (0.815)
GRAVITY °API – MAX (sp GRAVITY, MIN)	45.0 (0.801)
EXISTENT GUM, MILLIGRAMS (mg) PER 100 MILLILITERS (ml), MAX	7
POTENTIAL GUM, 16 HOURS AGING mg PER 100 ml, MAX	14
SULFUR, TOTAL, % WEIGHT, MAX	0.05
MERCAPTAN – SULFUR, % WEIGHT, MAX	0.005
FREEZING POINT, °F, MAX	−40
THERMAL VALUE: HEAT OF COMBUSTION BTU/lb, MIN	18,500
VISCOSITY, CENTISTOKES AT −30°F, MAX	16.5
AROMATICS, VOLUME %, MAX	5.0
OLEFINS, VOLUME %, MAX	1.0
SMOKE POINT, MILLIMETERS, MIN	25.0
COPPER STRIP CORROSION, ASTM CLASSIFICATION, MAX	1
FLASH POINT, MIN	110°F

Figure 4-35

Section IV S-IB Stage

RP-1 FUEL SYSTEM PRESSURIZATION DIAGRAM

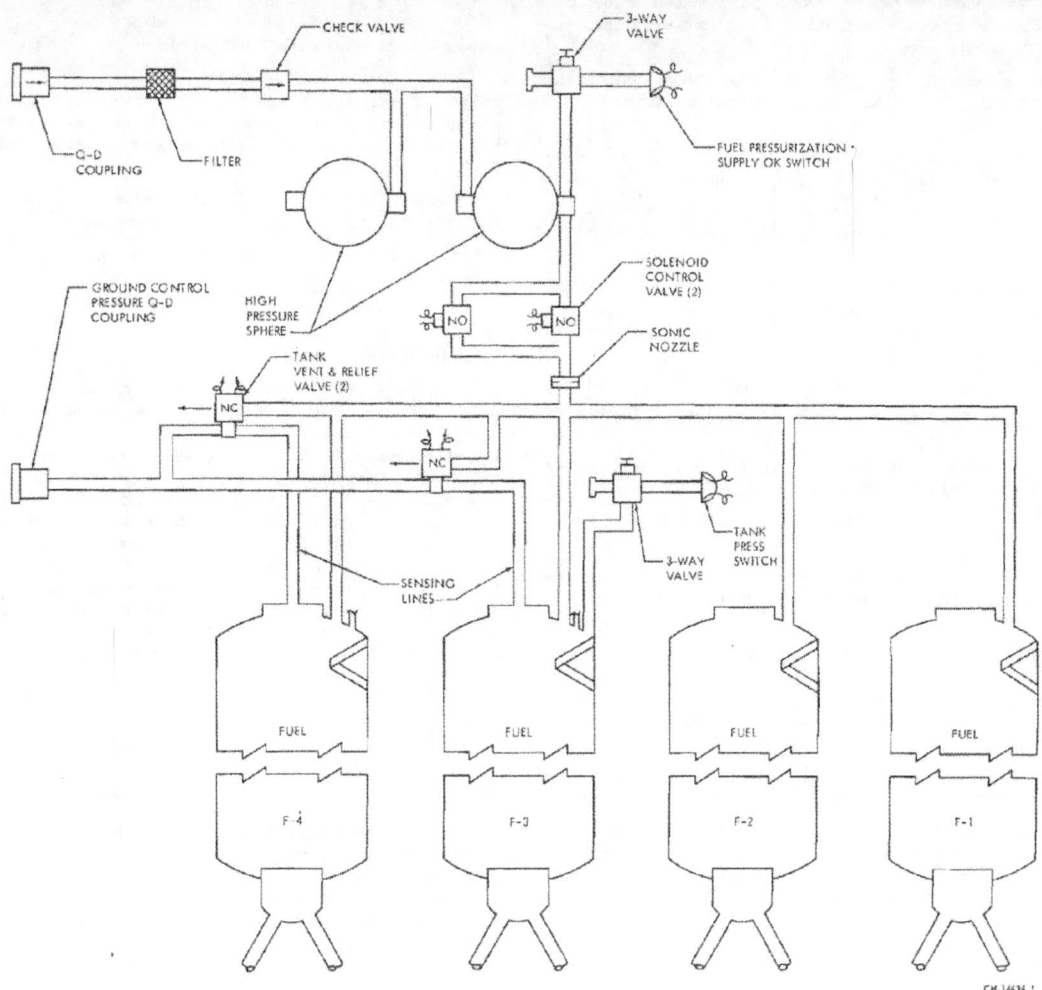

Figure 4-36

oid valves that route the helium from the spheres to the sonic nozzle. This switch is located in the top of tank F-3, and closes the solenoid valves when the pressure in the tanks exceeds 32.4 psia. In the event that tank pressure exceeds 35.7 psia, the fuel vent and relief valves mechanically vent the excess pressure to the atmosphere. At liftoff the electrical circuit to the solenoid valves and pressure switch is disconnected, and uninterrupted pressure from the spheres flows through the open solenoid valves and sonic nozzle into the tanks. As the fuel is depleted in S-IB stage flight, the tank pressure decreases to a minimum of 11.5 psia.

Measurements.

Figure 4-37 lists the fuel system flight measurements and indicates the information that is displayed.

Fuel System S-IB Stage Components.

Detailed descriptions and characteristics of the major components of the fuel system are presented in the following paragraphs.

Fuel Fill and Drain Valve. The fuel fill and drain valve controls filling and draining of the tanks and is installed in the fill and drain line leading to the sump of tank F-1. The ball-rotor gate is operated through a rack and pinion gear arrangement and is driven to the open position when pneumatic control pressure is applied to the actuator assembly. Simultaneously the rack and pinion arrangement compresses the return spring. When pneumatic control pressure is removed, the ball-rotor gate returns to the normally closed position. A position indicator switch monitors the fully open and fully closed positions for remote readout on the S-IB component test panel in the LCC.

Fuel Prevalve. There are eight prevalves, one each located between each suction line and the tank sump. The prevalves are normally open and provide backup capability to the main fuel valves in the H-1 engines for fuel cutoff during flight. The ball-rotor gate is operated through a rack and pinion gear arrangement and is driven to the closed position when pneumatic control pressure is

4-31

Section IV S-IB Stage

c. The fuel storage facility transfer pump is energized and fuel is pumped at approximately 200 gpm into the fuel tanks.

d. When the tanks are 15-percent full, as monitored in the LCC, the manual operation is terminated. This permits a leakage check of the transfer lines.

e. The automatic mode of fuel transfer is then initiated at the RP-1 control panel. The automatic fill functions include fast fill, slow fill, and line drain.

f. The necessary solenoid valves in the facility transfer lines are automatically energized to the open position.

g. The pump is energized and fuel is routed through the transfer line at a flow of approximately 2000 gpm into the S-IB stage fill-and-drain line and into the fuel tanks.

h. The tanks are filled to 98-percent full and then the tanking computer automatically initiates the slow fill sequence. The slow fill sequence restricts the fuel flow to 200 gpm. The percent of fill is monitored in the LCC.

i. The tanks are slow filled to approximately 102 percent, then the fill sequence is automatically terminated and the transfer line drain operations are automatically sequenced.

j. Fuel fill and drain valve control pressure is removed and the valve returns to the normally closed position.

k. Fuel in the facility transfer lines is returned to the RP-1 storage tanks.

l. Immediately after the line is drained, a 750-psig GN_2 transfer line and fuel fill mast purge is automatically sequenced.

m. After lox has been loaded and just prior to launch, a fuel level adjust drain operation is performed. When lox is in the S-IB lox tanks, heat transfer between the lox and fuel tank occurs and cools the fuel. The fuel density increases and the fuel load has to be adjusted to the required mass for a given mission.

n. The corrected fuel load is programmed into the facility tanking computer and the level adjust drain operations are initiated at the RP-1 components control panel.

o. The fuel fill and drain valve on the S-IB stage is opened and the excess fuel is drained into the facility transfer line until the fuel level in the S-IB stage tanks decreases to the 100 percent full level (see paragraph q). The fuel fill and drain valve returns to the normally closed position.

p. The facility transfer line is drained automatically and the GN_2 purge of the line and fuel fill mast is again performed.

q. In the event that the fuel tanks are less than 100 percent full, based on the latest fuel mass requirements, a replenish operation must be performed.

r. The fuel fill and drain valve and the fuel vent valves on the S-IB stage are again opened and fuel at 1000 gpm is routed through a fast fill valve and starts filling the transfer line. A 30-sec time delay is initiated and closes the fast fill valve when the transfer lines are full. The fuel is then routed through the slow fill line at 200 gpm into the S-IB stage tanks.

s. When the tanks are again 100-percent full, the replenish operation is terminated and the line drain and purge functions are automatically performed.

Fuel Drain From S-IB Stage To Facility Storage.

In the event of launch cancellation and the fuel has to be drained from the S-IB stage, a drain operation is performed as follows:

a. Drain operations are initiated at the RP-1 components control panel and drain functions are automatically sequenced.

b. The S-IB stage fuel tanks are pressurized to flight pressure from the facility helium source and valve panel no. 9 through the S-IB stage fuel tank pressurization components.

c. The S-IB stage fuel fill-and-drain valve and the necessary facility transfer line valves are opened and the fuel drains to the facility storage tanks.

d. When the fuel level decreases to 10 percent of the total fuel load, the tanking computer initiates a signal that sequences the line drain functions to return fuel in the facility transfer lines to the RP-1 storage tanks.

e. Immediately after the line is drained, a 750-psig GN_2 transfer line and fuel fill mast purge is automatically sequenced.

f. The fuel fill and drain valve is returned to the normally closed position.

Fuel Bubbling.

Pressurized GN_2 at approximately 135 psig is routed from valve panel no. 10 in the ML to the S-IB stage and bubbles the fuel in the fuel suction lines and the tanks. GN_2 flow into the suction lines and subsequently through the fuel tanks tends to maintain a uniform fuel temperature within each tank. Fuel bubbling is first initiated from the S-IB firing preparation panel in the LCC approximately 8 hr and 45 min before launch and is stopped during level adjust drain operation and then started approximately 10 min prior to launch and continues until fuel tank pressurization begins. The GN_2 is routed through a ring line manifold and branch lines to each fuel suction line. Each branch line is provided with an orifice to control GN_2 flow and a check valve to prevent reverse fuel flow into the GN_2 bubbling line. The bubbled GN_2 rises through the suction lines and the normally open prevalves to the ullage area of the fuel tanks and is vented through the open fuel vent valves.

Fuel Tank Pressurization.

The fuel tanks are pressurized (figure 36) with helium at 29.6 to 32.4 psia starting from approximately 2 min and 43 sec prior to launch and continuing until S-IB flight is completed. The tank pressure maintains a pressure head (NPSH) for starting the engine fuel pumps and provides structural support by preventing the formation of a vacuum in the tanks as fuel is depleted during flight. Pressurizing components of the S-IB stage include two high pressure storage spheres, solenoid valves, pressure switches, a sonic nozzle, and distribution lines. Prior to fuel loading, the two high pressure storage spheres are prepressurized to approximately 1600 psig. The helium is supplied from valve panel no. 9 through a Q-D coupling (service arm no. 1A) connected to the upper umbilical on the S-IB stage. The helium enters the stage and passes through a filter and check valve into the two storage spheres. During the tank pressurization sequence, the storage spheres are fully pressurized to 3000 psig and are continuously replenished if necessary prior to launch. From the spheres the helium is routed through two normally open solenoid valves and a sonic nozzle into the distribution lines to each tank. The sonic nozzle meters the helium flow and maintains a constant rate of flow into the tanks. Prior to launch, a pressure OK switch located on the outlet of one sphere monitors the sphere pressure and actuates and deactuates to cause a solenoid control valve in the facility source to shut off or open the supply. The switch actuates at 2965 ± 30 psia on increasing pressure to shut off the supply and deactuates when the sphere pressure decreases to 2835 psia. The FUEL PRESSURIZATION PRESSURE OK readout is in the LCC. Another pressure switch monitors tank pressure and controls the operation of the two solen-

4-32

Section IV S-IB Stage

NUMBER	NOMENCLATURE	RANGE	DISPLAY
XC179-F1	TEMPERATURE, FUEL	-0 +40°C	[1]
XC179-F2	TEMPERATURE, FUEL	-0 +40°C	[1]
XC179-F3	TEMPERATURE, FUEL	-0 +40°C	[1]
XC179-F4	TEMPERATURE, FUEL	-0 +40°C	[1]
XD2-F3	PRESSURE, HE, FUEL TANK	-0 +45 PSIA	[2]
VK17-F2	LEVEL CUTOFF, FUEL	ON OFF	[3]
VK18-F4	LEVEL CUTOFF, FUEL	ON OFF	[3]
L20-F1	LEVEL, FUEL, DISCRETE	ON OFF	
L20-F3	LEVEL, FUEL, DISCRETE	ON OFF	

[1] AUXILIARY DISPLAY
[2] AUXILIARY AND MCC-H DISPLAY
[3] ESE DISPLAY

FUEL SYSTEMS MEASUREMENTS

Figure 4-37

applied to the actuator assembly. Simultaneously the rack and pinion arrangement compresses the return spring. When pneumatic control pressure is removed the ball-rotor gate returns to the normally open position. A position indicator switch monitors the fully open and fully closed positions for remote readout on the S-IB firing preparation panel in the LCC. All prevalves in the OPEN position complete an interlock for the start of automatic launch sequence.

Engine Cutoff Fuel Sensor. Two engine cutoff fuel sensors initiate an electric signal to shut down the inboard engines when the fuel level in tanks F-2 or F-4 falls below the level of the sensors.

Fuel Depletion Sensors. Two fuel depletion sensors initiate an electrical signal to shut down the outboard engines when the fuel in tanks F-2 or F-4 is depleted below the level of the sensors.

Fuel Overfill Sensor. The fuel overfill sensor initiates an electrical signal to the PTCS to terminate the fuel fill sequence should the fuel level in the tanks rise to the level of the sensor. A FUEL OVERFILL readout is monitored on the S-IB component test panel in the LCC.

Fuel Temperature Sensor. Three temperature sensors in each of the four fuel tanks are located parallel to the longitudinal axis of the tanks. The temperature sensors monitor the temperature of the fuel and electrically transmit the results to the facility fuel tanking computer. The temperature data is used in density calculations for programming the correct fuel mass to be loaded into the S-IB stage fuel tanks.

High Pressure Storage Sphere. Two high pressure storage spheres are used to store the helium, 19.28 ft³ each, required for inflight fuel tank pressurization. The storage spheres are pressurized to 3000 psig from the facility source prior to launch and are maintained at this pressure until liftoff. After liftoff these spheres pressurize the fuel tanks during S-IB stage flight. Each of the two identical spheres is formed by welding together two fully-machined forged titanium hemispheres. The spheres are proof tested at 4650 psig through a temperature range from $-125°$ F to $200°$ F and are designed to withstand 6200 psig without rupture.

Solenoid Operated Control Valve. Two normally open solenoid operated control valves control the fuel tank pressurization prior to liftoff. A pressure switch located on top of tank F-3 senses the correct tank pressure and causes the solenoid valve to close or open to maintain tank pressure of 29.6 to 32.4 psia. At liftoff the switch is disconnected and the solenoid valves return to the normally open position. Helium then flows at the maximum flowrate through the sonic nozzle into the fuel tanks. During sphere fill or at any time prior to launch, the valves can be closed by operation of the FUEL PRES'G VALVES NO. 1 and NO. 2 switch on the S-IB component test panel in the LCC. During venting, prior to launch, an electrical signal from the facility closes the valves until tank pressurization is required again.

Pressure Switch. The pressure switch controls the operation of the solenoid operated control valves to maintain the fuel tank pressure at 32.4 psia (max) prior to liftoff. When the fuel tank pressure is below 29.6 psia (min) the switch causes the fuel pressurizing command circuit in the LCC to energize and the solenoid operated valves return to the normally open position and the helium is routed to the tanks. When the tank pressure increases to 32.4 psia (max), the switch causes the fuel tanks pressurized circuit in the ESE to energize and the solenoid operated valves close and the helium flow to the tanks is stopped. The switch also causes the FUEL PRESSURIZED indicator to light on the S-IB firing panel when the tank pressure is 29.6 to 32.4 psia. The switch has no inflight function.

Vent Valves. Two vent valves vent the fuel tanks during fill operations or when emergency venting is required. The normally closed, spring-loaded, poppet-type valves are opened by pilot valves. GN_2 from the facility source opens the valves during fuel filling operation. Pressure sensing lines from tanks F-3 and F-4 route ullage pressure to provide operating pressure for emergency venting. The valves are provided with position indicators for remote readout on the S-IB component test panel and the S-IB firing panel in the LCC.

Fuel Pressurization Supply OK Switch. The fuel pressurization supply OK switch monitors the helium stored in the two high pressure storage spheres. The switch is set to actuate at 2965 ± 30 psia increasing pressure and to deactuate at 2835 psia minimum as pressure decays. When the spheres are fully pressurized, the switch actuates and the power is removed from the facility control valve which shuts off the helium supply. Also, a FUEL PRESSURIZING PRESSURE OK readout signal is transmitted to the S-IB firing preparation panel in the LCC. Should the pressure decay and cause the switch to deactuate, the functions reverse to remove the pressure OK readout, apply power to the facility control valve, and recharge the storage spheres. The switch has no flight function.

CONTROL PRESSURE SYSTEM.

The S-IB stage control pressure system located in the aft skirt of fuel container F-3 stores 3100-psig GN_2 and supplies 750-psig regulated pressure for operation of fuel and lox prevalves, calorimeter purge, gearbox pressurization, and lox pump seal purge. Pressure switches and transducers monitor system conditions and provide inputs to the S-IB stage telemetry system. GN_2 from valve panel no. 10 enters the control pressure system through short cable mast 4 (figure 4-38) and pressurizes the 1-ft³ sphere. During preflight test activities, the control pressure system receives 50-psig GN_2 to purge the system. Adjustment of the regulator for outlet pressure of 40 to 50 psig and opening the gearbox pressurization and lox seal purge hand valve permits GN_2 flow through the systems. This operation continues for a minimum of 30 min before pressurizing the sphere to prelaunch operation pressure of 1500 psig. Approximately 5 hr before launch the sphere pressure is increased to 3100 psig. An orifice in the supply line in cable mast 4 valve panel controls the sphere pressurization rate. After the sphere pressure increases to 3100 psig, the 3000lb control pressure OK signal from the high pressure OK switch energizes a solenoid valve that bypasses the orifice to permit sufficient GN_2 flow to replenish the sphere. Replenish continues until the time for ignition signal at T-3 sec. The high pressure OK switch actuates at 2965 ± 30 psia and provides

4-33

Section IV S-IB Stage

S-IB STAGE CONTROL PRESSURE SYSTEM DIAGRAM

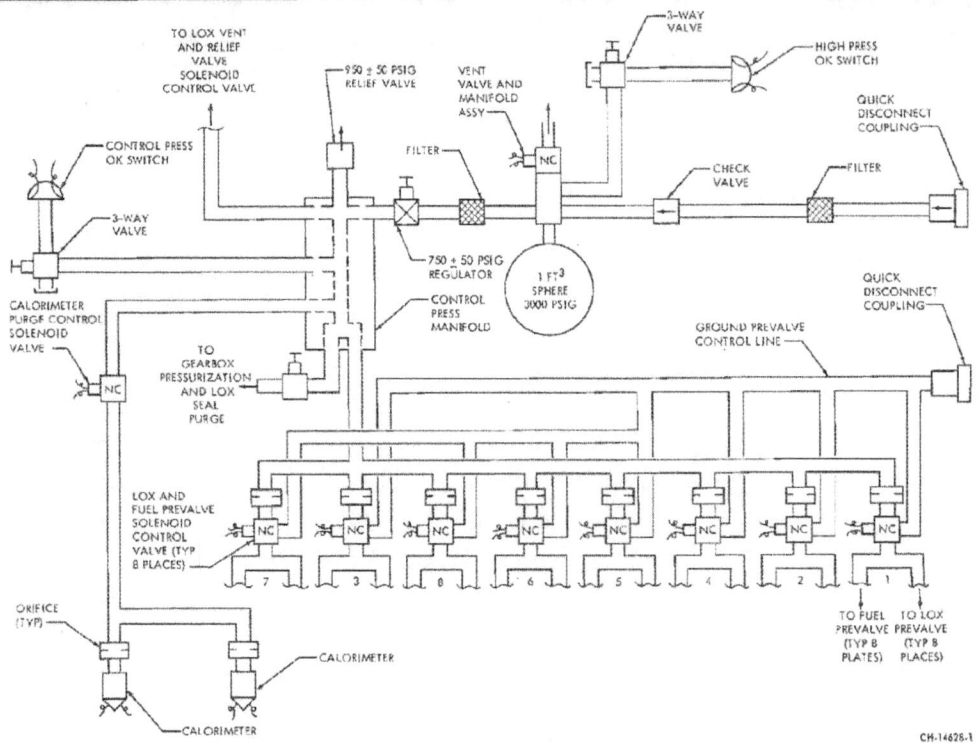

Figure 4-38

an output to illuminate the CONTROL 3000 LB OK indicator on the S-IB firing prep panel. After liftoff the switch has no function. Pressure transducer XD40-9 provides continuous monitoring of sphere pressure before liftoff and during flight. Measurement prefix X indicates that the signal is routed to KSC/LVO-1 for real-time analog recording. See figure 4-39. Redline value for XD40-9 is 3300 psia maximum.

Sphere pressure is regulated to 750±15 psig and distributed to the vehicle systems through the control pressure manifold. The control pressure OK switch actuates at 625±25 psig and provides an output that illuminates the CONTROL 750 LB OK indicator on S-IB firing prep panel. A 950±50 psig relief valve provides overpressure protection for the 750 psig system. Pressure transducers XD41-9 and XD42-9 provide continuous monitoring of the regulator output pressure before liftoff and during flight. XD41-9 and XD42-9 redline values are 710 psia minimum and 815 psia maximum. This information requires periodic monitoring from 3100-psig pressurization to the control pressure sphere until automatic sequence start.

TEST DATA.

The 1-ft³ fiberglass storage sphere is proofed at 5000 psig and has a minimum burst pressure of 6660 psig. The operating pressure is 3000 psig with an operating temperature of −65 to +160° F. The most critical items in the system are the bottle fill-and-vent valve, the 750-psig regulator, and the relief valve. Failure of the fill-and-vent valve to remain closed would result in a probable mission loss. Failure of the regulator or relief valve would result in a possible mission loss. The control pressure system components have been qualified and have flown previous Saturn flights. Prior to qualification testing, each component of the qualification test sample was individually proof tested. The sphere was pressurized hydrostatically; all other components were pressurized with helium. Each component was pressurized to 150 percent of normal operating pressure and maintained in that condition for 5 min. During the final 2 min of the 5 min period a leakage check was performed. The pressure was reduced to zero and an examination of the components for distortion was made. No discrepancies were detected.

CONTROL PRESSURE SYSTEM FUNCTIONS.

The calorimeter purge control solenoid valve may be energized by the CALORIMETER PURGE switch on the S-IB firing prep panel or, during countdown by the power transfer command. GN_2 flows through two orifices that reduce the flowrate to 0.218±0.022 lbm/min/calorimeter. Ambient purge prevents accumulation of combustion products and other contaminants on the calorimeter radiation window; purge continues throughout S-IB stage operation. The calorimeters measure heat flux in the engine area and relay the information through TM. The measurements are of prime interest from liftoff until S-IB stage impact. See figure 4-39.

Gearbox pressurization and lox pump seal purge begins when the control system receives initial pressure. A three-way valve permits the purge to be shut off when necessary during checkout. An orifice reduces the GN_2 flow to each engine to 0.23±0.09 lbm/min. The purge line tees and one branch becomes the lox pump seal purge with a flowrate of 1.35 scfm and the other branch becomes the gearbox pressurization with a flowrate of 1.85 scfm. A check valve in the gearbox pressurization line prevents reverse flow into the purge line. Gearbox pressure of 2 to 10 psig is maintained during flight. See H-1 Engine, for additional information.

The 750-psig control pressure closes the prevalves at H-1 engine cutoff, serving as a backup propellant flow termination device with the main propellant valves on the engine. A normally closed solenoid valve for each pair of normally open fuel and lox prevalves prevent closure until deactuation of TOPS on each engine. The solenoid control valve then opens and pressurizes the closing actuators of fuel and lox prevalves. See Lox System and Fuel System for additional prevalve information.

During flight, sphere pressure will decay because of continuous purging of the calorimeters, lox pump seals, and pressurization of the turbopump gear boxes. See figure 4-40 for predicted decay rates based on available pressure at liftoff.

CONTROL PRESSURE SYSTEM MEASUREMENT DATA

MEAS NO	TITLE	RANGE
C603-6	FLAME SHIELD TEMP (R. CAL)	0 TO 115 W/cm²
C609-3	HEAT SHIELD TEMP (R. CAL)	0 TO 45 W/cm²
XD40-9	CONTROL EQUIP SUPPLY PRESS	0 TO 3500 psia
XD41-9	CONTROL EQUIP REGULATED PRESS	0 TO 1000 psia
XD42-9	CONTROL EQUIP REGULATED PRESS	0 TO 1000 psia

Figure 4-39

SYSTEM FLIGHT HISTORY.

The S-IB-1 control pressure system sphere pressure at ignition was 2920 psi. Minimum acceptable pressure, determined by the deactuation setting of the high pressure OK switch, was 2862 psi. The regulated pressure ranged between 760 and 768 psi well within the specified redline value of 710 to 815 psi. The predicted sphere pressure decay compared favorably with the actual sphere pressure decay. Actual sphere pressure at 2 min 30 sec flight time was within 25 psi or 2.5 percent of predicted value. The S-IB-2 control pressure system utilized two 1-ft³ spheres because of increased pneumatic requirements for supplying purge to a fifth calorimeter. Sphere pressure at T-3 sec was 3160 psia. The regulated pressure was well within limits throughout countdown and flight. At T-10 sec, the regulated pressure was 768 psia and then varied between 760 and 770 psia during flight. The S-IB-3 sphere pressure at T-10 sec was 3040 psi. Regulated pressure ranged between 770 and 775 psi during flight. Two 1-ft³ spheres were used in the S-IB-3 control system to accommodate the additional purge flowrates for five calorimeters and one spectrometer. From T-10 sec until outboard engine cutoff, the pressure declined steadily to 1820 psi. Final sphere pressure was within 50 psi of predicted value. At approximately 2 min 20 sec into flight, a change in slope of the decay curves was caused by the pneumatic requirement for closing the prevalves. The regulated pressure was 770 psi by 2 min 30 sec into flight and remained within the range of 710 to 815 psi. S-IB-4 and S-IB-5 each utilized a single 1 ft³ sphere for control pressure storage. On each of these flights, the regulated pressure and gas usage rate remained within acceptable limits. The S-IB-4 regulated pressure varied between 770 and 785 psi; S-IB-5 varied between 759 and 754 psi.

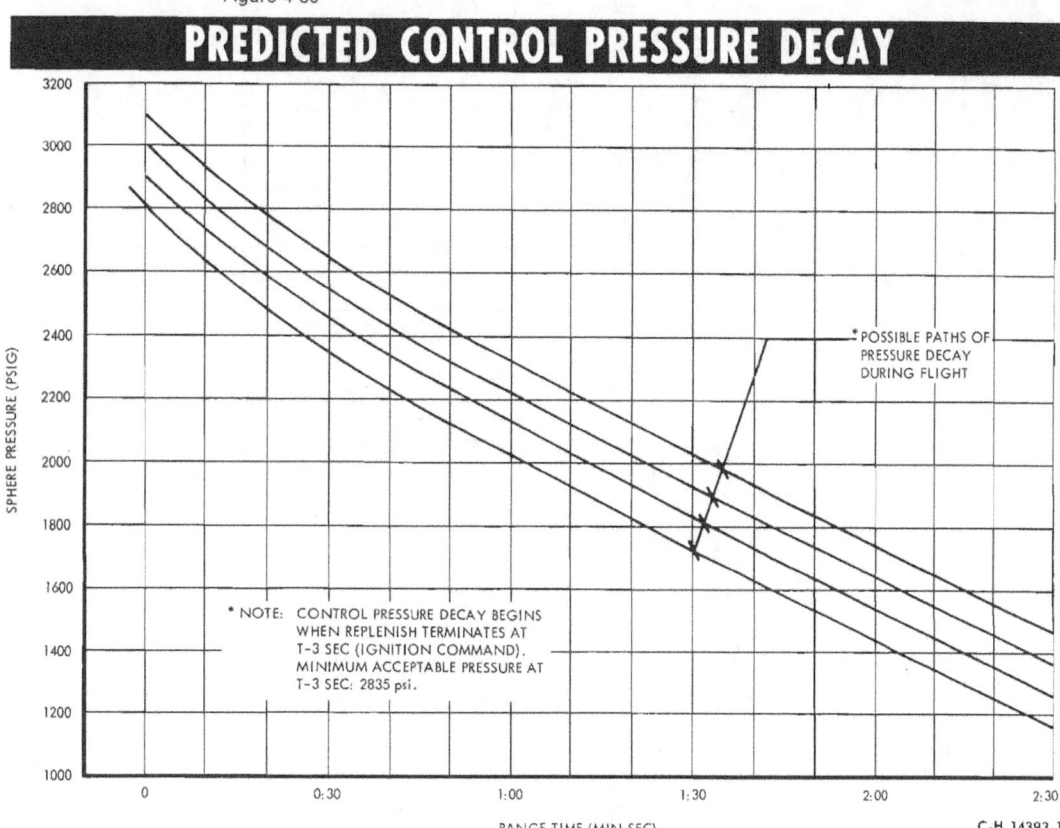

Figure 4-40

Section IV S-IB Stage

HYDRAULIC SYSTEM

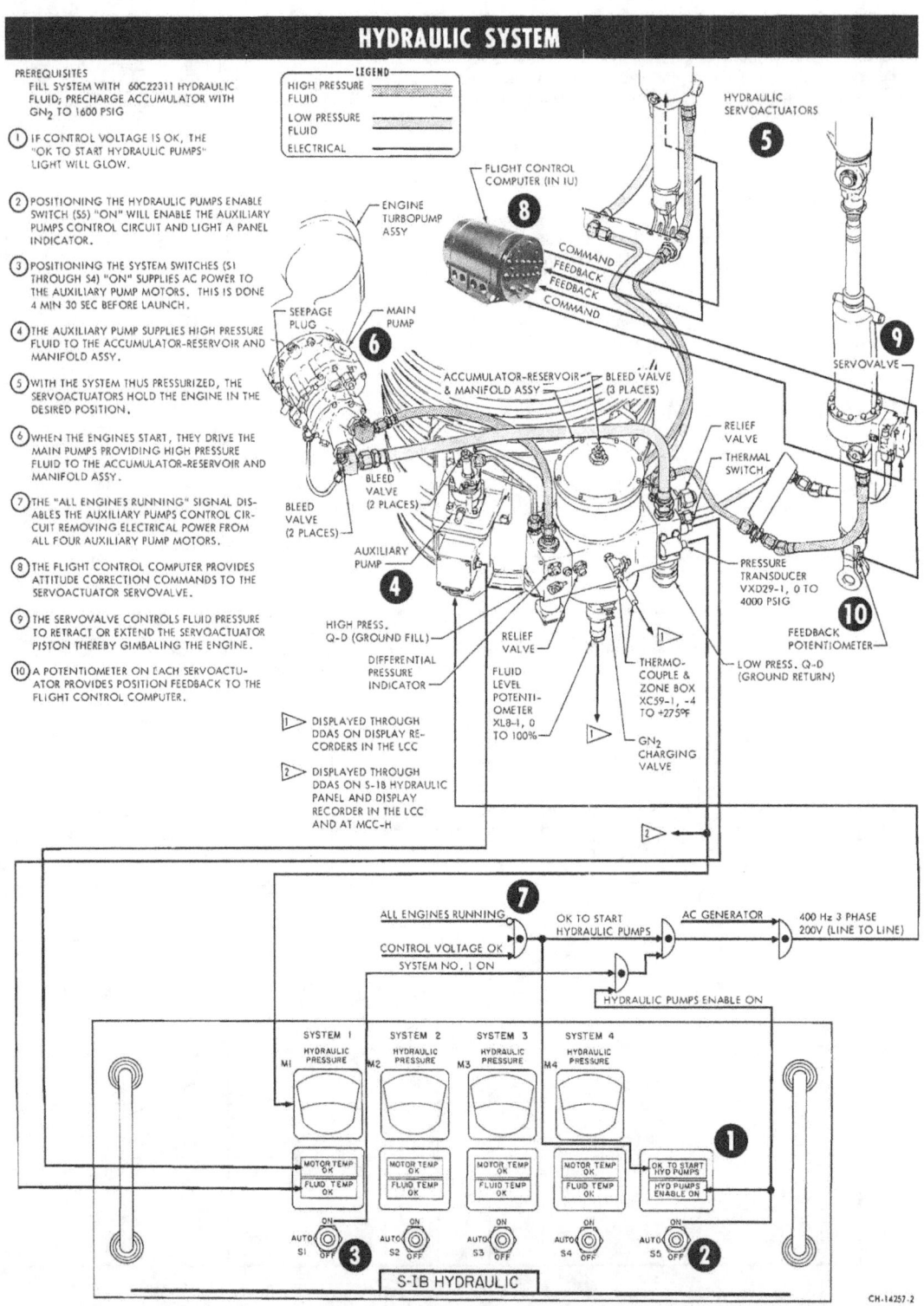

Figure 4-41

Section IV S-IB Stage

S-IB HYDRAULIC SYSTEM.

The flight control computer in the IU combines attitude error signals from the navigation and guidance systems with rate signals from control accelerometers and rate gyros and produces attitudes correction commands. Servoactuators on the outboard engines respond to these commands by gimballing the engines, thereby diverting the thrust vectors to correct control errors or execute programmed pitch and roll maneuvers. The gimbal direction and rate are proportional to the polarity magnitude of the command signal.

Each of the four outboard engines is equipped with an independent closed-loop hydraulic system. Each system consists of five major modules: accumulator, reservoir, and manifold assembly; main pump; auxiliary pump; pitch servoactuator; and yaw servoactuator. Tubing and flexible hoses interconnect these modules. Components within modules are ported through manifolds. This keeps external plumbing at a minimum reducing leakage problems and making the system less vulnerable. When the servoactuators extend and retract, the engine can be gimballed a maximum of ± 8 deg from the null position in both the pitch and yaw axes. In the null position, each engine is canted 6-deg outward from the vehicle centerline.

FILLING.

Normally the hydraulic system should not require filling and bleeding after the booster leaves the Michoud assembly plant. However, if the system is opened, filling and bleeding must be repeated as follows. Prior to filling the system with hydraulic fluid, the accumulator is charged to 1600 psig with GN_2 from the launcher pneumatic manifold through the GN_2 precharging valve. Each accumulator takes 0.251 lbm of nitrogen. Hydraulic fluid is then pumped into the system from the hydraulic servicer cart through the high pressure, quick-disconnect coupling. The system is then bled to the reservoir desired level as monitored by the fluid level potentiometer. Excess fluid returns to the servicer cart through the low pressure quick-disconnect coupling. After filling, the GN_2 precharge is bled down to 10 psig if the system is to remain inoperative for 10 days or more.

PRELAUNCH PREPARATION.

Just before operating the system for checkout or during countdown, the accumulator is charged with GN_2 to 1600 psig. If the pump motor temperature is under $350 \pm 18°$ F and the low pressure fluid temperature is under $200 \pm 10°$ F on all four systems the MOTOR TEMP OK and FLUID TEMP OK lights on the S-IB hydraulic panel will illuminate (figures 4-41 and 4-42). If the control voltage is OK, the OK TO START HYD PUMPS light will illuminate. To start the systems operating for checkout or countdown, the enable switch is placed on the ON position and the HYD PUMPS ENABLE ON light illuminates. Placing each system switch in the ON position supplies facility ac power to each auxiliary pump motor. The auxiliary pump can deliver 2.2 gpm at a minimum

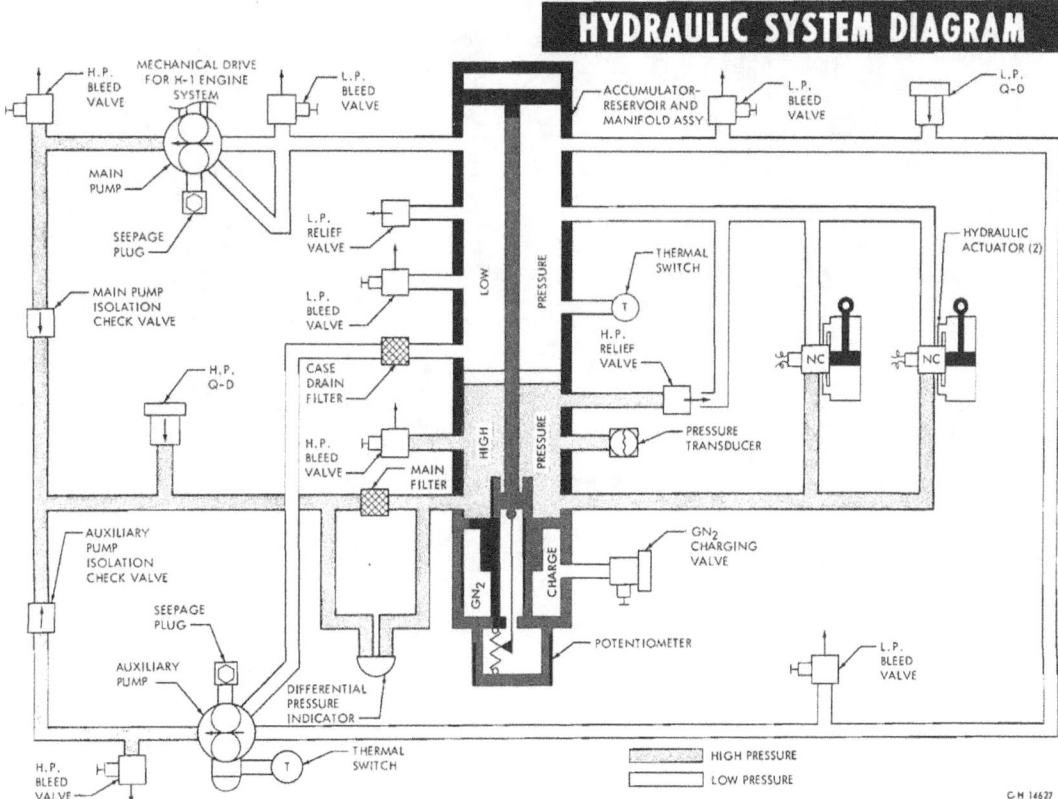

Figure 4-42

4-37

ACCUMULATOR-RESERVOIR AND MANIFOLD ASSEMBLY

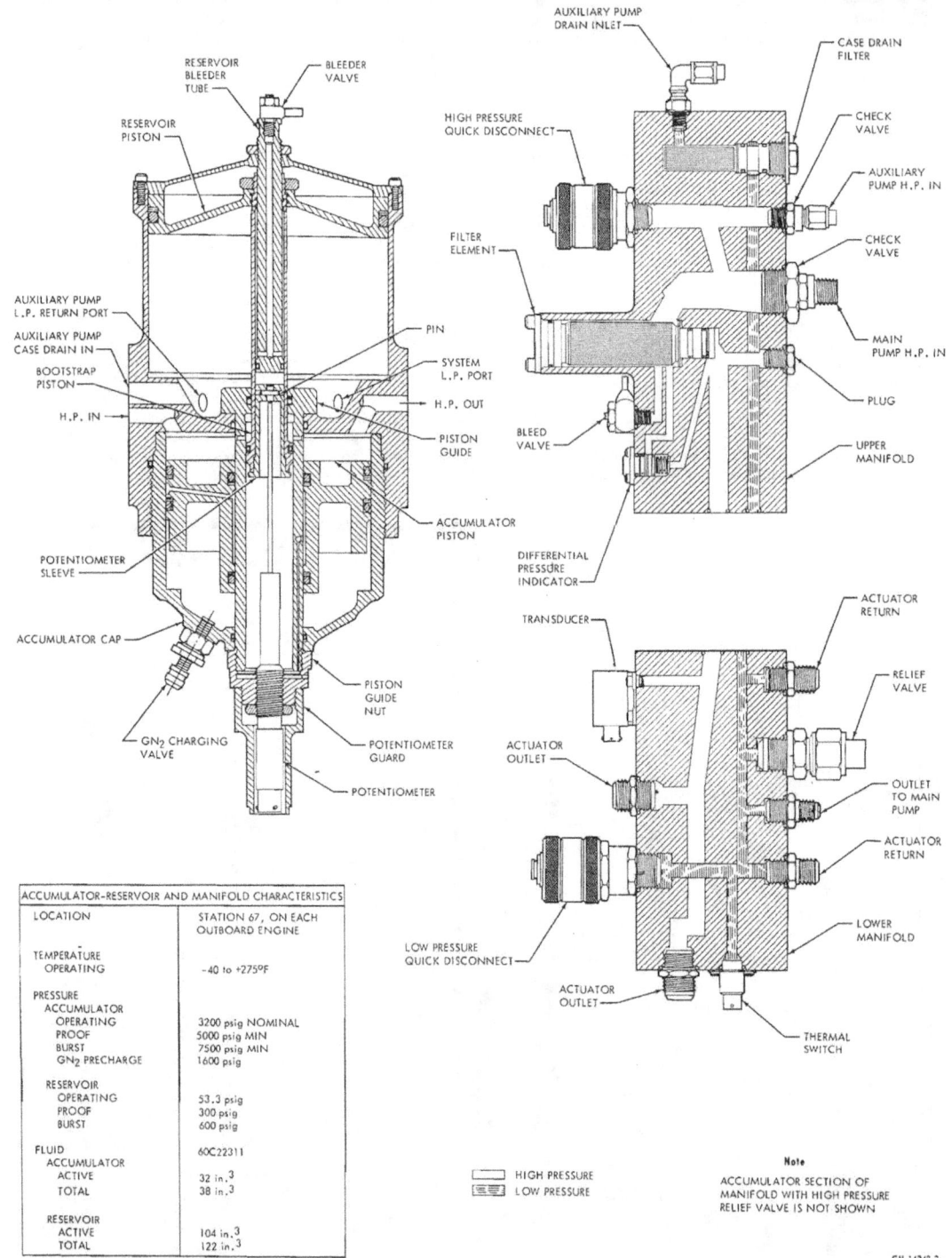

Figure 4-43

Section IV S-IB Stage

pressure of 2850 psig to allow system checkout and to hold the engines in the neutral position at ignition. As each engine ignites, its main hydraulic pump, driven by the engine turbopump, begins delivering high pressure fluid. After the ignition sequence, the all engines running signal removes power from the auxiliary pump motors.

OPERATION.

The main operating pump draws fluid from the low pressure, 53-psig, side of the accumulator-reservoir and increases the pressure to a maximum of 3250 psig. Inlet pressure is dependent on pump outlet pressure (60:1). When the auxiliary pump is operating, the pressures are 50 and 3050 psig (max). These maximums are no-flow pressures. In the complete system, operating pressure is less than this value. The fluid then passes through high pressure tube assemblies to the manifold assembly and into the high pressure portion of the accumulator-reservoir. From the opposite manifold portion through tube assemblies and flexible hoses, the fluid passes to the servoactuators. Electrical commands from the flight control computer in the IU signal the servovalves to throttle the fluid in a direction and at a rate corresponding to the polarity and magnitude of the command. The high pressure fluid flows to the proper side of the servoactuator piston causing it to extend or retract. Low pressure fluid returns to the low pressure side of the accumulator-reservoir through tube assemblies and flexible hoses and then through tube assemblies to the pump inlet.

ACCUMULATOR-RESERVOIR.

The accumulator portion, figure 4-43, stores some high pressure fluid during system operation to help meet sudden servoactuator demands and dampen pump surges. A floating piston allows the accumulator volume to vary according to the quantity of fluid stored at any instant by means of GN_2 precharge on the piston side opposite the high pressure fluid. The reservoir portion stores low pressure fluid to feed the pump inlets. A bootstrap piston in the reservoir has one piston surface acting on the low pressure fluid and another piston surface, one-sixtieth the area of the first, exposed to the high pressure fluid in the accumulator. The high pressure fluid forces the piston down on the low pressure fluid, maintaining reservoir pressure at 1/60th of accumulator pressure (47-53 psig) to help prevent pump cavitation. The volume of fluid in this unit and the pump induced pressure is sufficient to meet the maximum expected servoactuator demands with a safety factor of 1:6.

MANIFOLD.

One portion of the manifold, figure 4-43, contains the main system filter, which cleans the fluid as it flows from the pumps to the accumulator. Another filter cleans the auxiliary pump case drain fluid. Check valves protect against reverse flow in the outlet tubes of each pump. The differential pressure indicator shows, during prelaunch operations, when the main filter needs cleaning. Another portion of the manifold contains high and low pressure relief valves

MAIN PUMP

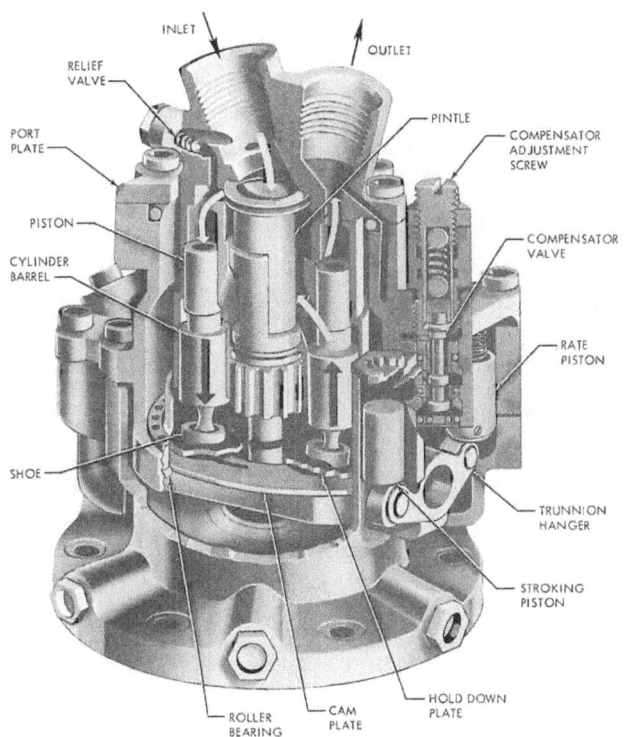

MAIN PUMP CHARACTERISTICS	
LOCATION	STATION 80
TEMPERATURE OPERATING	-65 to +275°F
PRESSURE ZERO FLOW FULL FLOW INLET	3200 ± 50 psig 2900 psig 7 to 50 psig
SPEED	4300 rpm
RATED FLOW	17 gpm @ 4300 rpm
LUBRICATION	SELF-LUBRICATING

Figure 4-44

to protect both portions of the system from overpressurization. The manifold contains the high and low pressure quick-disconnect couplings used to fill the system.

MAIN PUMP.

This two-stage, cam-actuated, variable displacement pump (figure 4-44) uses a sensing of output pressure to compensate for fluctuations, thereby maintaining a constant output pressure of 3200 psig. The pump is in effective operation by the time its cylinder barrel completes one revolution. Rotation of the cylinder barrel moves the seven dual-diameter pistons along the variable angle cam plate (wobble plate) causing the pistons to reciprocate within the cylinder block. The intake strokes of the pistons cause hydraulic fluid to be drawn through the inlet port into a fixed pintle and then into the primary stage of the pump. The discharge strokes of the large diameter portion of the piston pressurize the fluid to approximately 100 psig and route it along the outside of the pintle to the second stage. (The flow from the first stage is more than is required by the second stage. Excess fluid is diverted through a relief valve and is routed back to the pump inlet.) In the second stage, the smaller diameter portion of the pistons pressurizes the fluid to the operational level of 3200 psig. The high pressure fluid is discharged as the cylinder block rotates and aligns each piston with the high pressure outlet port.

Pump outlet can be varied by changing the angle of the cam plate. This is controlled by the rate piston and the stroking piston. Under high load conditions, the cam plate is held immobile in the maximum displacement position by the rate piston and pumping load. During operation these forces are opposed by the stroking piston, which receives control pressure fluid from the compensator valve. This normally closed valve is sensitive to pump discharge pressure and is screw-adjusted to maintain outlet pressure at the desired level.

During pump start, when rising discharge pressure approaches 3200 psig, the sensed force on the compensator valve spool overrides the preset valve adjustment and the spool is displaced, delivering high pressure fluid to the stroking piston. The stroking piston causes the trunnion hanger to rotate, and the cam plate moves to a reduced angle. As operating conditions stabilize, the cam plate will assume a relatively permanent position, subject only to minor changes when output pressures vary because of system demands. A hollow epoxy seepage plug, filled with sponge, absorbs any leakage around the drive shaft seals. The seepage plug threads into the base of the pump. A case drain line routes fluid that has escaped past the pistons and cylinder barrel back to the pump inlet. The main pump obtains its power from the engine turbopump gearcase.

AUXILIARY PUMP.

The auxiliary pump (figure 4-45) is a fixed-angle, nine-cylinder, pressure-compensated unit with variable delivery, controlled by a rotor, automatically rotated to meet system requirements of high pressure hydraulic fluid. The pressure capability is 3000±50 psig at zero flow to full flow with 2850 psig minimum. The flowrate

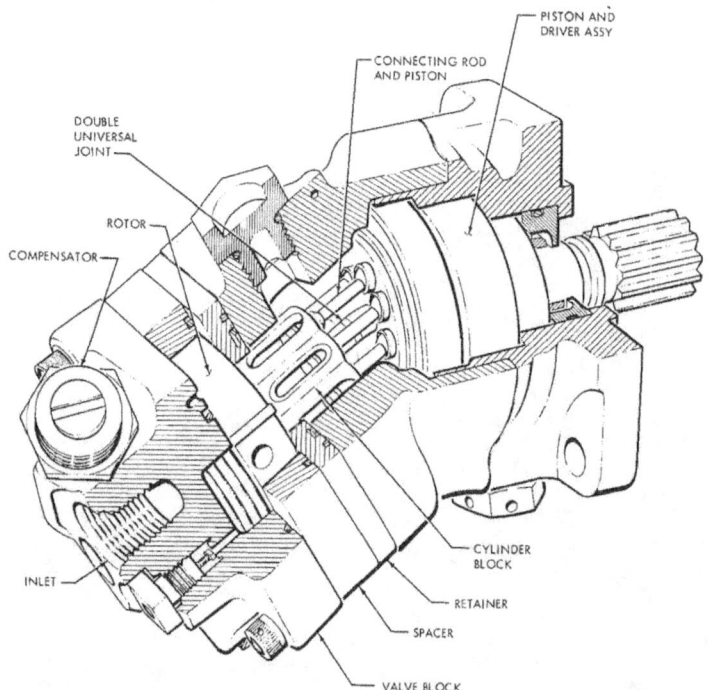

Figure 4-45

Section IV S-IB Stage

AUXILIARY PUMP MOTOR CHARACTERISTICS	
LOCATION	STATION 67
TEMPERATURE OPERATING	0 to 165°F
ELECTRICAL VOLTAGE	200 V LINE-TO-LINE, 400 Hz, 3 PHASE
SPEED	11,300 ± 5% rpm at 25 in-lb TORQUE
POWER RATING	3356 W (4.5 HP) MAX CONTINUOUS DUTY
SWITCH INDICATION	OPEN ABOVE 350°F CLOSE BELOW 310°F

Figure 4-46

at 11,000 rpm is 3.0 gpm minimum. During checkout the flowrate is limited to 2.2 gpm for 2 min duration by controlling the gimballing rate, thus preventing overheating the auxiliary pump motor. The pump is operated during the hydraulic system fill and purge operations, during leak checks, during engine gimballing precheck operations, and during launch countdown to prepressurize the system before engine ignition. The pump serves no flight function.

The auxiliary pump is powered by an electric motor (figure 4-46) fed from facility source of 200-V, 3-phase, 400 Hz. The motor is a 3-phase "Y" wound unit, capable of developing 11,300 rpm ±5 percent at 4.5-hp load. The pump and motor are mounted on opposite sides of a bracket located on the engine. The motor has an integral thermal switch that is interlocked in the motor start circuit and completes the circuit to the MOTOR TEMP OK indicator light on the hydraulic panel in the LCC.

SERVOACTUATOR ASSEMBLY.

Gimbal commands are received from the flight control computer to actuate the electrohydraulic servovalve in each actuator (figure 4-47). This action is used to control the outboard engines for pitch, yaw, and roll. Gimbal commands are accomplished by diverting the flow of high pressure hydraulic fluids against either side of the actuator piston. The system consists of a servovalve, a filter, a prefiltration valve, a manually operated servovalve bypass valve, an actuator position scale, a midstroke locking device, and valves for fluid sampling and system bleeding. Two servoactuator assemblies are mounted in a perpendicular plane that intersect at the longitudinal axis of each engine, and are mounted between stage superstructure and engine. The engines are moved, proportionally to the magnitude of the electrical input signals, as the actuators extend or retract, independently or simultaneously. The maximum gimbal angle is ±8 deg; yet past operations have not used over one-third of this capability. The stroke of the actuator is 9.560±.06 in., but is limited to 7.640±.06 in. To accomplish 1 deg of engine movement the piston moves 0.478 in. The operating pressure on these equal-area pistons is 3200 psig. The servovalve also completes the cycle by directing the return of low pressure fluid to the reservoir. The feed-back potentiometer sends a signal to the flight control computer corresponding to the piston movement, which is used to reduce the signal strength as desired actuator

HYDRAULIC SERVOACTUATOR CHARACTERISTICS	
LOCATION	STATION 76
TEMPERATURE	-65 TO +275°F
PRESSURE OPERATING PROOF BURST	3000 psig 4500 psig 7500 psig
EFFECTIVE PISTON AREA	5 in.²
LOAD RATED STALL	10,000 lbf 15,000 lbf
STROKE TOTAL LIMITED	9.560 ± 0.06 in. 7.640 ± 0.06 in.
LENGTH ADJUSTMENT	40 in. 0.5 in.
VELOCITY (input 12 MA diff) LOADED (10,000 lb) NO LOAD	8.75 + 0, -1.75 in/sec 15.0 + 0, -3.0 in/sec
LUBRICATION	SELF-LUBRICATING
SERVOVALVE TORQUE MOTOR 2 COILS COIL RESISTANCE DIFF INPUT SIGNAL (RATED) COIL CURRENT (MAX)	POLARITY OF SIGNAL DETERMINES COIL USED 1000 ± 50 OHMS 12 ± 0.12 MA 20 MA

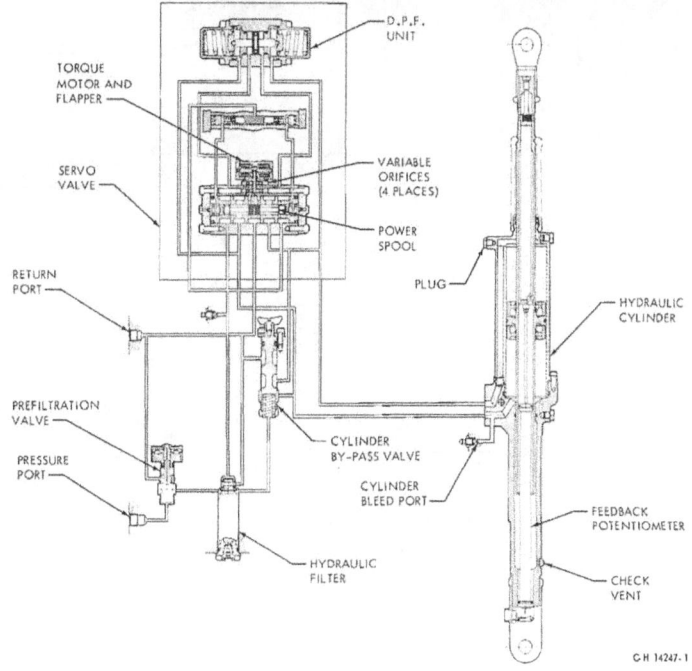

Figure 4-47

Section IV S-IB Stage

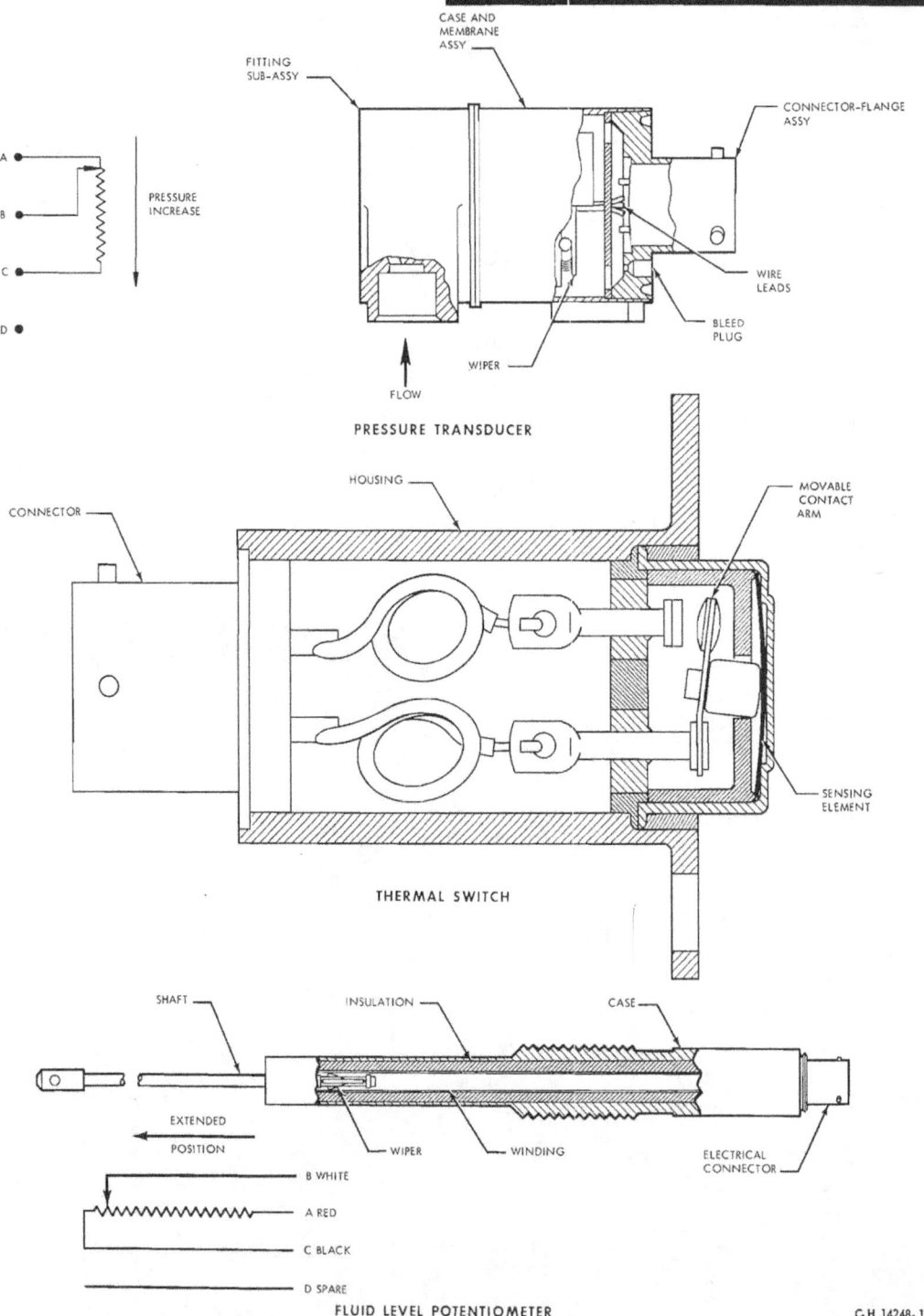

Figure 4-48

Section IV S-IB Stage

position is reached. The operations of the servoactuators are monitored at the engine deflection panel and the analog recorders in the LCC.

MONITORING DEVICES.

Various switches and indicators (figure 4-48) monitor the hydraulic system operation within the accumulator-reservoir and manifold assembly. High pressure relief valves protect the high pressure side of the system by allowing high pressure fluid to vent into the low pressure side. A low pressure relief valve protects the system from being overfilled. The pressure transducer, mounted on the manifold, monitors high pressure in the hydraulic system. The pressure sensed is converted to an electrical signal for the DDAS system and checkout. The DDAS system provides a readout of this function to a strip chart and a meter on the hydraulic panel in the LCC. The capability range of pressure for the transducer is zero to 4000 psig. A thermal switch monitors the hydraulic fluid temperature in the low pressure side of the system. When the fluid reaches normal operating temperature $200 \pm 10°$ F the bimetallic sensing element snap-actuates the contact arm and opens the circuit to an indicating lamp in the LCC. When the temperature decreases to $155 \pm 10°$ F the circuit will close and the indicator lamp will light. The reservoir is equipped with a fluid level potentiometer. The movement of the reservoir piston is detected to give an equivalent output voltage. The potentiometer is a single element, 2000 ± 100 ohm wire wound, linear translation-type unit and is internally mounted in the base of the accumulator-reservoir. The varying output voltage is processed into the telemetering system to provide a direct readout and recording in the LCC.

HYDRAULIC HOSES AND TUBING.

The main pump and actuator flexible hoses have the same general configuration but differ in detail construction. The outside cover is of stainless steel wire braid and an inner liner based on a teflon compound. The low pressure hose designed operating pressure is 50 psig. The low pressure hose was tested hydrostatically at 3000 psig and performed without leaks for 5 min. The high pressure hose has an additional carbon steel reinforcement braid between the inner and outer layers. The design operating pressure is 3200 psig. The high pressure hose was hydrostatically tested for 5 min without a leak at 6000 psig. The operating temperature range for both high and low pressure hose is $-65°$ F to $275°$ F.

The tube assemblies are made from seamless corrosion-resistant steel, 304 or 304L controlled by MSFC-SPEC-131. The tubing is flared in accordance with MC-146 and uses one sleeve and one nut on either end of the assembly. The nuts and sleeves are of corrosion-resistant steel. The nut used on high pressure applications are in compliance with MF-818 for precision flared fitting ends with interlocks.

ELECTRICAL.

Two independent bus networks (+1D11 and +1D21) distribute primary power as shown in figure 4-49. Source power to these buses originates from ground-base electrical support equipment during prelaunch operations and transfers to stage batteries 50 sec before liftoff. A separate battery supplies each bus network. Generally, one battery (D10) powers operational systems that draw high transient current and generate bus voltage transients. The other battery (D20) powers the measuring system and supplies parallel power for critical functions such as engine cutoff. This dual isolation feature ensures more than adequate electrical power. Primary power varies in the 27- to 30-Vdc range. Line loss is less than 2 Vdc from stage bus to stage load. A total of 13 cables interconnect the electrical system above and below the propellant tanks. Four cables are in the cableway on fuel tank F1; nine cables are in the cableway on fuel tank F2. Structure of the center lox tank provides a unipotential ground path (IDCOM). All electrical equipment incorporates design provisions in accordance with MIL-E-6051 (Electromagnetic Compatibility Requirements) to prevent electromagnetic interaction between electrical systems. Stage electrical equipment operates independently of power sources from other stages. However, some buses extend into the stage to form a feedback loop to indicate certain events. For example, the S-IVB stage powers a bus (+4D11) that indicates physical separation of stages and consequently initiates J-2 engine start logic; the S-IU stage powers buses (+6D91, +6D92, and +6D93) that loop the H-1 engine thrust OK pressure switches to the emergency detection system (EDS). There are other extra-stage buses, but none have drain on the S-IB stage electrical system since the requesting stage supplies them power.

BATTERIES.

Two batteries supply primary power to separate bus networks in the stage electrical system; both batteries are identical and discussion is limited to one unit. The battery is a manually activated storage unit containing 21 cells composed of silver oxide (pos) and zinc (neg) plates. The cell wiring arrangement permits selection of either 18, 19, 20, or 21 cells so that battery voltage is between 28 and 29.6 Vdc when measured on launch day under load conditions. The battery is packaged dry (without electrolyte) to extend shelf life; it is activated no longer than 168 hr before launch. The activation time was extended from 120 to 168 hr after test data confirmed the extension had no significant effect on battery performance. The activated battery is installed approximately 57 hr before liftoff. Physical and electrical characteristics of the battery are shown in figure 4-50.

Battery Construction.

Construction features protect the battery from damage and provide for internal pressure control as illustrated in figure 4-51. The battery box is composed of magnesium alloy for lightness and strength. It is coated for thermal emissivity effect and environmental protection. The cells are spaced with Neoprene shim-stock and potted to form a unitized cell block. A cover, molded of scotchcast and filler composition, protects the top of the cell block. The box cover adds support to the battery box unit and permits access to the cell block for inspection and cell activation. An o-ring gasket seals the battery box and cover. The seal aids primarily in maintaining the internal pressure as regulated by the pressure relief valve; it also prevents entrance of moisture and other foreign matter.

Blind Plug Selection.

The blind plug assembly (figure 4-51) is a jumper connector that completes the battery circuit when installed. The plug selects either 18, 19, 20, or 21 cells so that the battery output voltage (under load) is 28 to 29.6 Vdc. Plug selection is based on actual load test of the stage electrical system on launch day. A blind plug made for the testload range is selected as shown in figure 4-52. Initially, the blind plug is selected on the basis of the overall test (OAT) bus current measurements before battery installation. After installing the batteries (approximately 57 hr before liftoff), the stage is powered up and the bus current is rechecked. If the load current is the same as measured in OAT, the installed blind plug is used; if not, the proper blind plug is selected.

Activities.

The battery is manually activated with potassium hydroxide (KOH)

4-43

Section IV S-IB Stage

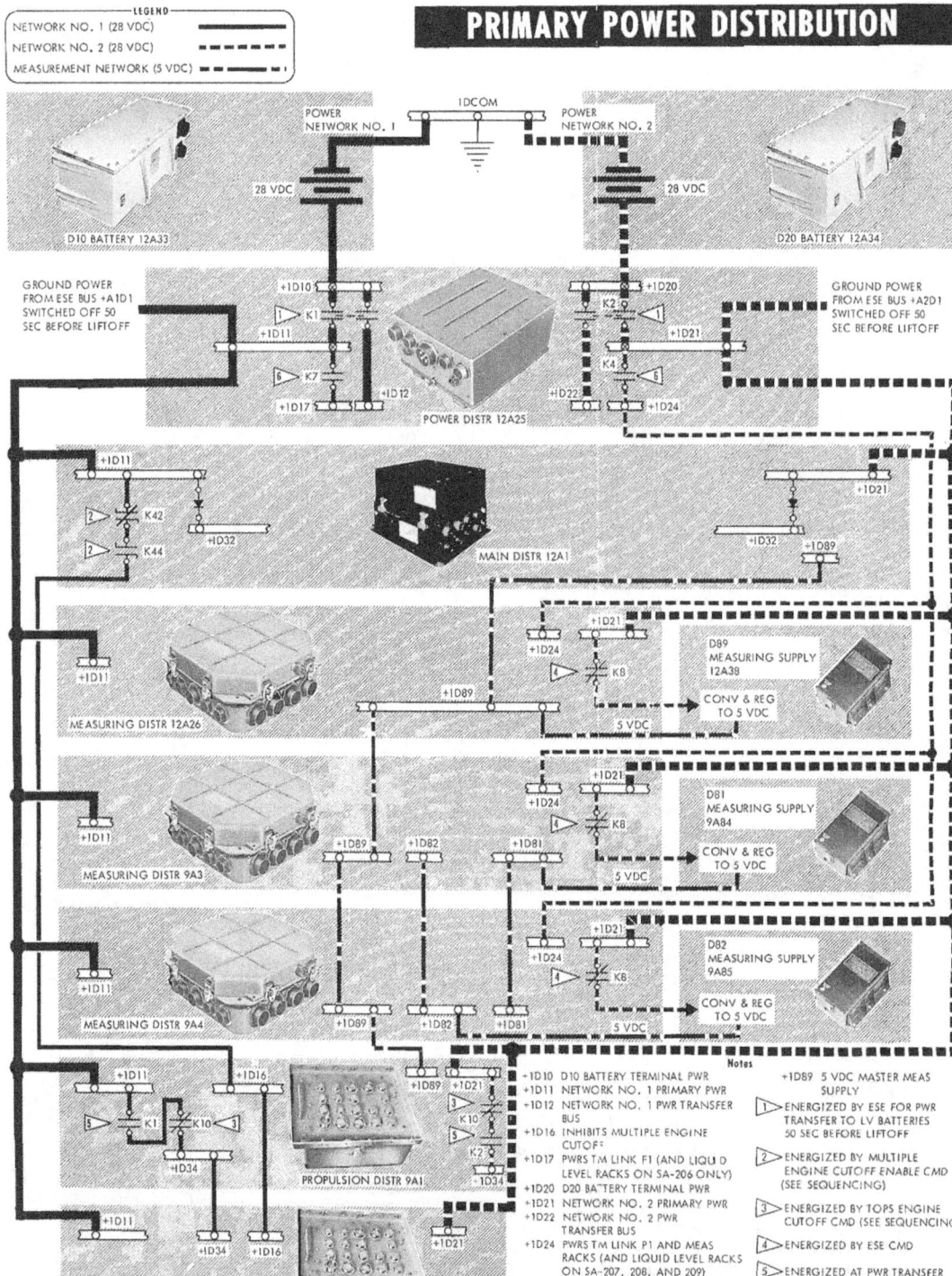

Figure 4-49

Section IV S-IB Stage

BATTERY CHARACTERISTICS

DIMENSIONS
LENGTH: 18.0 INCHES
WIDTH: 7.5 INCHES
HEIGHT: 7.0 INCHES

WEIGHT
DEACTIVATED: 52 POUNDS
ACTIVATED: 58 POUNDS

ELECTRICAL
VOLTAGE (UNLOADED): 38.3 TO 39.3 VDC (21 CELLS)
VOLTAGE (LOADED): 28 TO 29.6 VDC
POWER CAPACITY: 2000 AMP-MINUTES AT 100 AMPS

ELECTROLYTE
SOLUTION: POTASSIUM HYDROXIDE (KOH)
SPECIFIC GRAVITY: 1.40 ± 0.042 AT 77°F
VOLUME: 1890 ± 21 CM^3 PER BATTERY
DEGASSING PERIOD: 2 TO 4 HOURS

PRESSURIZATION
BATTERY BOX MEDIA: DRY NITROGEN (GN_2)
BATTERY BOX VENT: OPENS AT 26 PSID MAX;
CLOSES AT 10 PSID MIN
CELL VENT: OPENS AT 2 TO 6 PSID

TEMPERATURE
STORAGE: 68 TO 80°F
OPERATING: 68 TO 104°F

Figure 4-50

electrolytic solution having a specific gravity of 1.40±3 percent at 77° F. Each cell (figure 4-53) is activated with 90±1 cc of solution premeasured in individual containers. The battery chosen for flight use cannot be older than 36 mo, nor can it be activated longer than 168 hr before launch. There are two methods for adding electrolyte to the battery, drip activation and vacuum activation. Drip activation requires a rack that holds 21 individual activator cases, each containing 90 cm^3 of electrolyte. A drain needle, located centrally within each case, facilitates insertion into the respective cell filler holes. The rack is positioned to allow gravity feed into the individual cells. Vacuum activation also requires a rack with individual activator cases, but includes a vacuum pump as part of the assembly. Outlet filler tips, one for each cell, extend from the lower side of the activator case. Filler adapter screws thread into each cell filler port and connect flexible tubing to transfer to electrolyte. The vacuum system consists of a pump, gage, regulator valve, and relief valve to evacuate and fill the battery cells. After activation, each cell vent valve (part of filler cap assembly) is installed. The activated battery is subjected to a 10- to 20-A load for 15 to 30 sec to verify proper activation and performance; this test occurs 4 hr after installing the cell vent valves. The installed vent valves allow internal cell pressure to vent at 2 to 6 psid but prevent leakage of electrolyte regardless of battery position. The activated battery is stored at 68 to 80° F until ready for installation. Battery installation and the last case isolation test occur no sooner than 12 hr after installing the vent valves. Should the activated storage time exceed 168 hr, the battery must be replaced.

Internal Shunt.

A shunt is the means for measuring battery current during flight. Figure 4-52 shows the shunt connected in series with the negative lead of the battery output; monitoring connections are available to the telemetry system. The shunt design allows for an output voltage of 100±3 mVdc between pins A and B (J2) when the battery is delivering 75-A current at 77° F. The voltage drop across the shunt is proportional to battery load current.

Thermistor.

A thermistor in each battery (figure 4-52) provides battery temperature measurements. A calibration curve for this measurement is provided with each battery with a ±1% accuracy. Battery temperatures are continuously monitored after installation in the stage. An increase in battery internal temperature indicates that a cell failure is about to occur.

Predicted Performance.

Data from previous flights indicate that the stage power system is both adequate and reliable. For example, battery current averaged only 20A during the SA-205 flight. The D10 battery used 7.7% of the rated capacity while the D20 battery used 7.3 percent. Voltage output averaged 28 Vdc during this flight. A summation of the estimated voltage and current requirements for The SA-206 flight is profiled in figures 4-54 through 4-57, respectively. Analysis of the electrical load requirements for the flight shows that even a faulty current up to 60 A from either or both batteries would not hamper critical systems operation. Complete current loss from one battery would neither prematurely cut off the engines nor impair normal engine operation.

MEASURING SUPPLIES.

Measurements require voltage in the 0 to 5 Vdc range; three measuring supplies provide regulated voltage to instrumentation

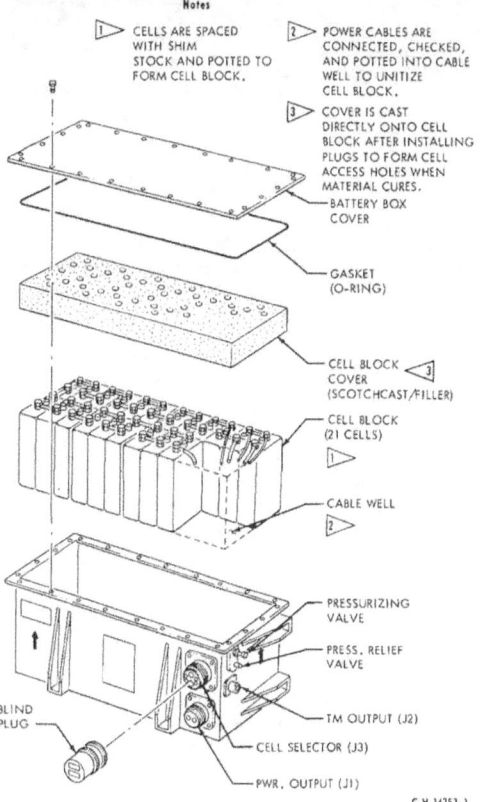

Figure 4-51

4-45

Section IV S-IB Stage

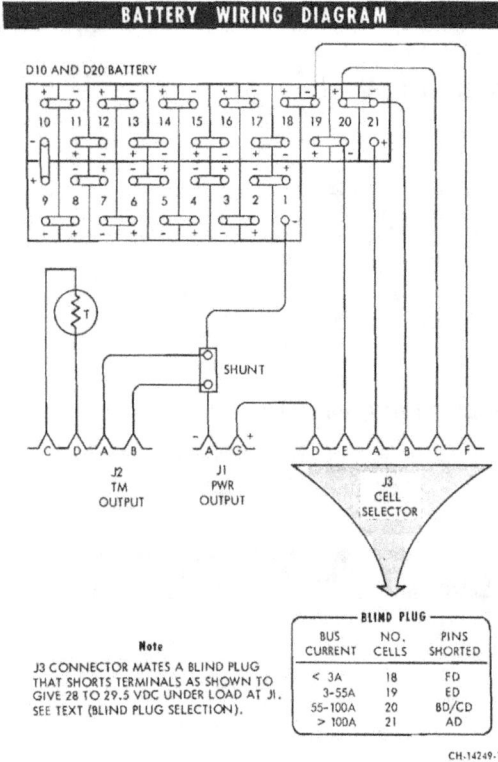

Figure 4-52

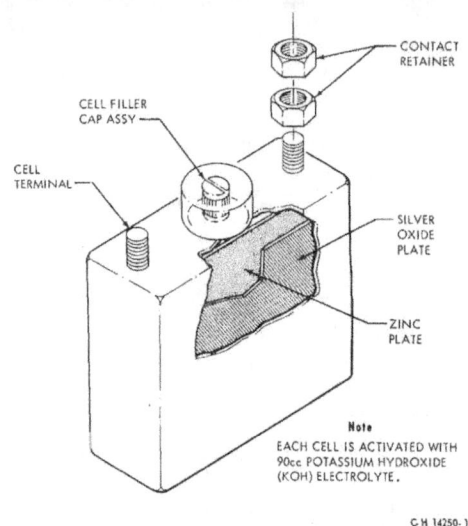

Figure 4-53

for this purpose. Known as the D81, D82, and D89 measuring supplies, the units each supply independent networks. Unit D89 powers instrumentation located above the firewall; the other two units (D81 and D82) provide power to instrumentation located below the firewall. Each unit receives 28-Vdc input from an associated measuring distributor. Solid-state circuitry converts this input to 5 Vdc±0.25 percent. This precise voltage returns to each measuring distributor for subsequent distribution to instrumentation.

DISTRIBUTION.

Seven distributors route electrical power (see figure 4-49); one power distributor, one main distributor, two propulsion distributors, and three measuring distributors. Cables transmit primary power from the batteries to the power distributor, establishing the two primary network buses. From this origin, secondary buses are established in the other distributors located in strategic areas (see Equipment Location) above and below the propellant tanks. Subsystems below the tanks interconnect through two propulsion distributors and the two remaining measuring distributors. Cabling reaches the distributors through two protective cableways; one on fuel tank F1, the other on fuel tank F2.

Power Distributor.

The power distributor routes battery power to auxiliary distributors that act as junction boxes and switching points. Battery terminal buses +1D10 and +1D20 (also ESE buses +A1D1 and +A2D1) are powered in this distributor. At 50 sec before liftoff, a command for power transfer from ground sources to stage batteries closes heavy relay contactors and establishes the two primary buses +1D11 and +1D21. Two pyrotechnic switch assemblies attached to the distributor ensure this transfer shortly after liftoff (see Power Transfer).

Main Distributor.

The main distributor supplies power mostly on command from the switch selector (see Sequencing) and contains relay circuitry that switches +1D11 and +1D21 power to operate or enable flight control components. It provides interface control for normal 4-by-4 engine cutoff (see H-1 Engine Cutoff) and EDS engines shutdown (see Section VI). Distributor circuitry inhibits multiple engine cutoff should one engine shut down between 3.0 and 10 sec of flight. Circuitry also prevents premature shutdown of one OB engine from affecting lox depletion cutoff. By enabling the propellant level sensors, it permits the separation EBW units to charge when one sensor actuates. The distributor triggers the EBW units on stage separation command. Since the main distributor performs many critical functions, its reliability has been closely checked. The most critical components in the unit are the electromechanical relays. Tests to determine contact chatter (open and close periods lasting 100 μsec) and the effect of intermittent output show that the distributor is both reliable and qualified to perform flight sequence and EDS functions.

Propulsion Distributor.

The two propulsion distributors contain circuitry to detonate the Conax valve squibs and subsequently cut off the engines upon command (see H-1 Engine Cutoff Commands). The distributors also operate the engine prevalve control valves. One distributor (9A1) controls propulsion functions on the OB engines, the other distributor (9A2) controls the IB engines. Prior to SA-202, only one propulsion distributor was used. Addition of the second unit

Section IV S-IB Stage

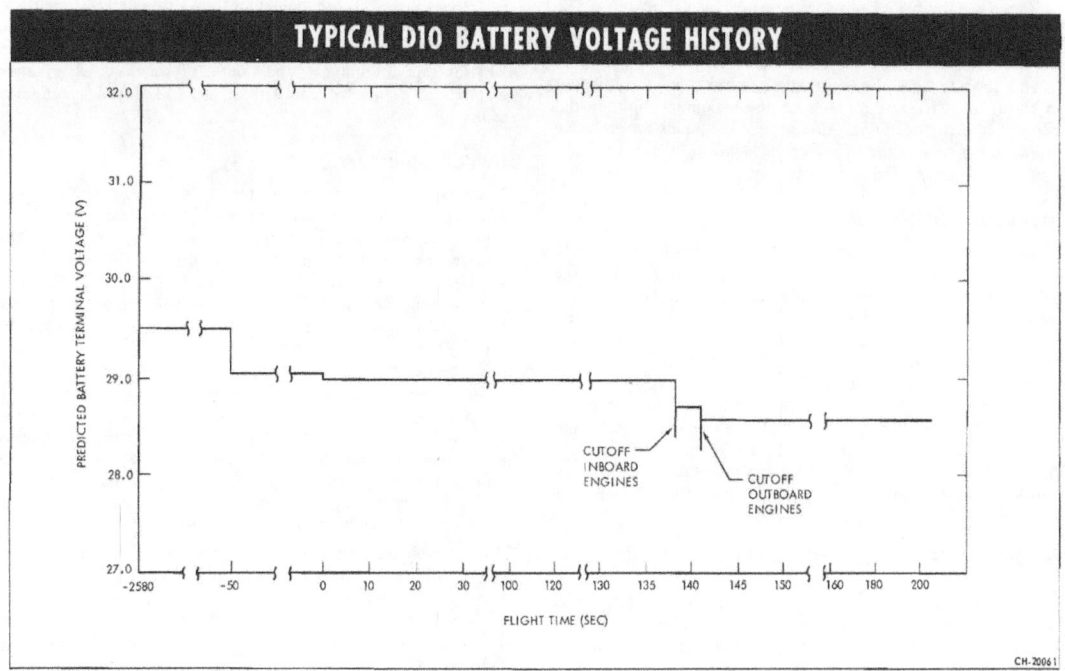

Figure 4-54

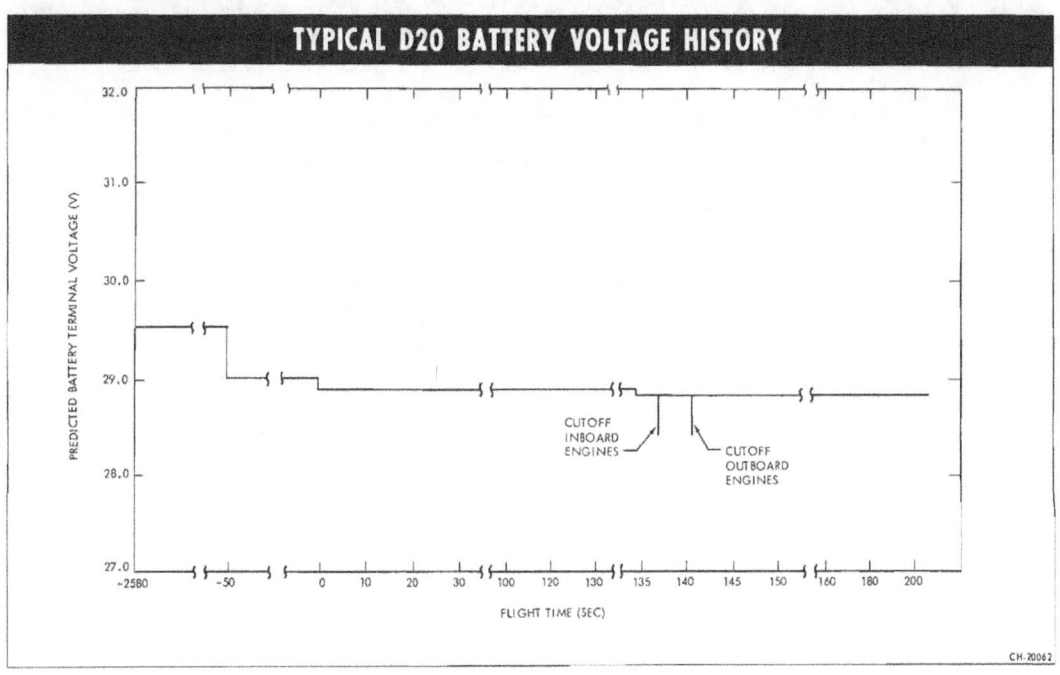

Figure 4-55

4-47

Section IV S-IB Stage

-ased reliability ... provides circuitry for a third thrust OK pressure switch on each engine. Each distributor now has 2-of-3 voting logic for thrust decay cutoff on each engine.

Qualification tests on the distributor showed some relay chatter in excess of 100 μsec duration, but the maximum duration did not impair circuit operation. Also, some wire strands broke during vibration but none of the effected circuits failed; flight distributors now have improved wiring arrangement.

Measuring Distributors.

Three measuring distributors route 28 Vdc to power signal conditioning equipment and 5 Vdc to energize measuring devices; the 5-Vdc source is also a reference and calibration signal for measurements and telemetry (see Instrumentation). Each distributor powers a 5-Vdc power supply (see Measuring Supplies) that returns the converted and regulated 5-Vdc source. The distributors supply three independent networks. One distributor (12A26), located above the propellant tanks, services the D89 network for measurements throughout the entire stage. The other two distributors (9A3 and 9A4), located below the tanks, service the D81 and D82 networks for engine measurements. With this arrangement, loss of one network would not impair critical measurements. Vibration tests have qualified all three distributors for flight use.

SEQUENCING.

A stored program in the launch vehicle digital computer controls inflight electrical sequencing; a switch selector is the stage communications link. The switch selector decodes incoming computer signals and activates the proper circuits in the main distributor. Figure 4-58 illustrates the results of the switch selector commands. These commands are sequenced in four time bases (T_n). A specific milestone event establishes each time base. If any one time base is not established, subsequent time bases cannot be started and the vehicle cannot continue the mission. Likewise, programmed safeguards prevent premature initiation of a time base. See Section VI (Navigation, Guidance, and Control) for a closer examination of program events sequenced from the LVDC and a definition of the time bases.

Switch Selector.

Each stage has a switch selector for communicating with the guidance and control system in the S-IU. Functionally, the switch selector receives coded commands from the LVDC (via the LVDA), decodes them, and activates the selected circuits. Use of a coded message decreases the number of interface lines and increases programming flexibility. The switch selector is divided into two functional sections to maintain power isolation between stages: the input section (relay circuits) receives power from the S-IU stage; and, the output section (decoding circuitry and drivers) receives power from the selected stage. The input and output sections couple through a diode matrix. This matrix decodes the 8-bit command input code and activates a PNP output driver, thus producing the switch selector command to stage system. The switch selector executes commands given by the 8-bit code or by its complement. The switch selector operates on positive logic, that is, +28 Vdc for a binary 1 and 0 Vdc for a binary 0.

A flight sequence command from the LVDA loads the switch selector register with a coded word command; the word format is shown in figure 4-59. Register bits 1 through 8 represent the functional sequence command. Bits 9 through 13 key the selected stage. Bit 14 resets all the relays in the switch selector when the LVDC receives faulty verification information. Bit 15 activates the addressed switch selector to read the command and produce the proper output.

The LVDC loads the switch selector register in two passes; bits 1 through 13 load during the first pass and, depending on the

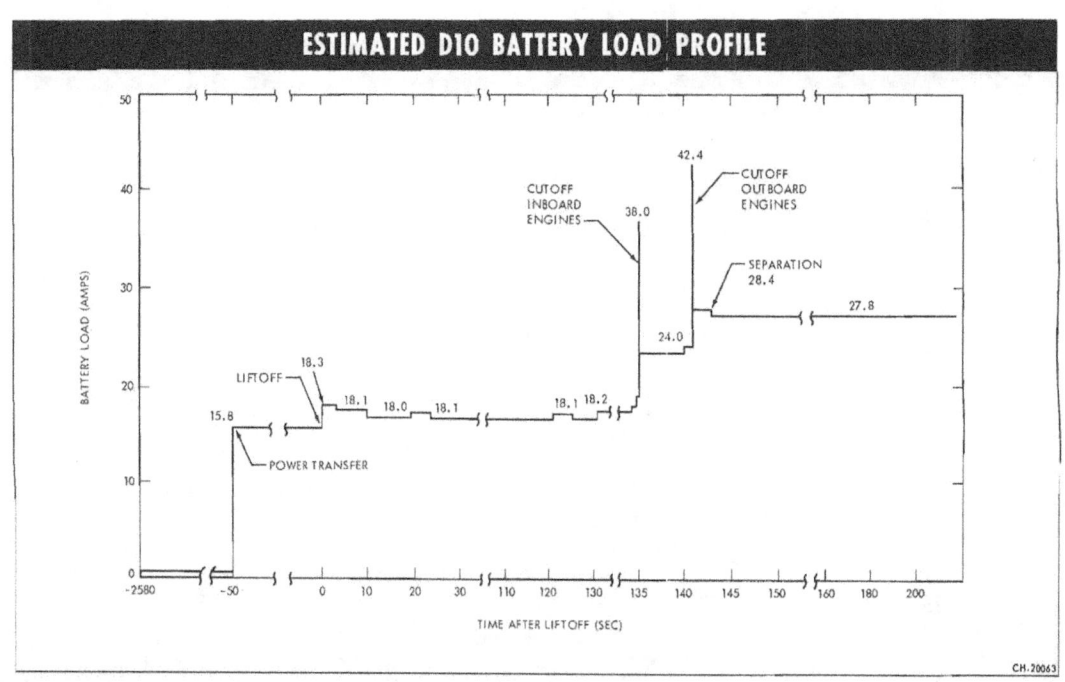

Figure 4-56

Section IV S-IB Stage

feedback code, either bit 14 or 15 loads during the second pass. After setting the 8-bit command, the received code complement is sent back to the LVDC through eight parallel lines. This feedback (verification) is returned to the digital input multiplexer of the LVDC and is subsequently compared with the original code in the LVDC. If the feedback agrees with the original code, a read command is given. If the feedback does not agree, a reset command is given (forced reset), and the LVDC reissues the 8-bit command in complement form.

POWER TRANSFER.

Primary power transfers from ML electrical supplies to internal stage batteries at 50 sec before liftoff. The command issues from the terminal countdown sequencer to the stage power distributor (12A25) at this time; however, relay logic in an ML program distributor restrains the command until confirmation signals indicate that the S-IB, S-IVB, and S-IU stages are ready for transfer. The command for transfer closes relay contacts connected directly between the stage batteries and network buses in the power distributor as shown in figure 4-60. Parallel transfer circuitry is provided when liftoff movement detonates two squib-actuated switches that complete a backup circuit. Gas pressure generated by the exploding bridgewire (see Ordnance) actuates ganged contacts in the switches that parallel contacts on the transfer relays (K1 and K2). The switches also maintain power on these relay coils to nullify contact chatter. When the switch assemblies are installed (1 day, 13 hr before launch) on the power distributor, a series circuit completes through an electrical loop in each switch and enables a safety switch installed signal to a display in the LCC (S-IB networks panel).

MEASUREMENTS.

Measuring instrumentation monitors switch selector operation, measurement bus voltage, primary network voltage, and battery current. These measurements are listed in figure 4-61, which shows the display ground station for each measurement.

HEATERS.

Cryogenic characteristics of the propellants (see Propulsion) require that certain components be heated to maintain operable temperatures. These components, which either immerse in the propellants or operate in the transfer system, are as follows:

a. Turbo Pump Bearing No. 1

b. Fuel Additive Blender Unit Gearcase

c. Main Lox Valves

d. Lox Vent Valves

e. Lox Overfill Sensor

f. Lox Level Sensors

g. Fuel Level Sensors

Each component has an integral resistance-type heater that is thermostatically controlled. The heaters receive power through a 115-V, 60-Hz, 3-phase, 4-wire system from ESE. Temperatures of the components are sensed by onboard thermistors. The resultant temperature indications are made available to the ground computer for evaluation. The heaters are turned on prior to loading cryogenic propellants approximately 8 hr before liftoff and are turned off at the ALL ENGINES RUNNING indication.

EQUIPMENT LOCATION.

Electrical power assemblies and instrumentation equipment are located in the upper and lower skirts of fuel tanks F1, F2, F3, and F4 as shown in figure 4-62. The upper portions of tanks F1 and F2 have extended skirts that house most of the equipment.

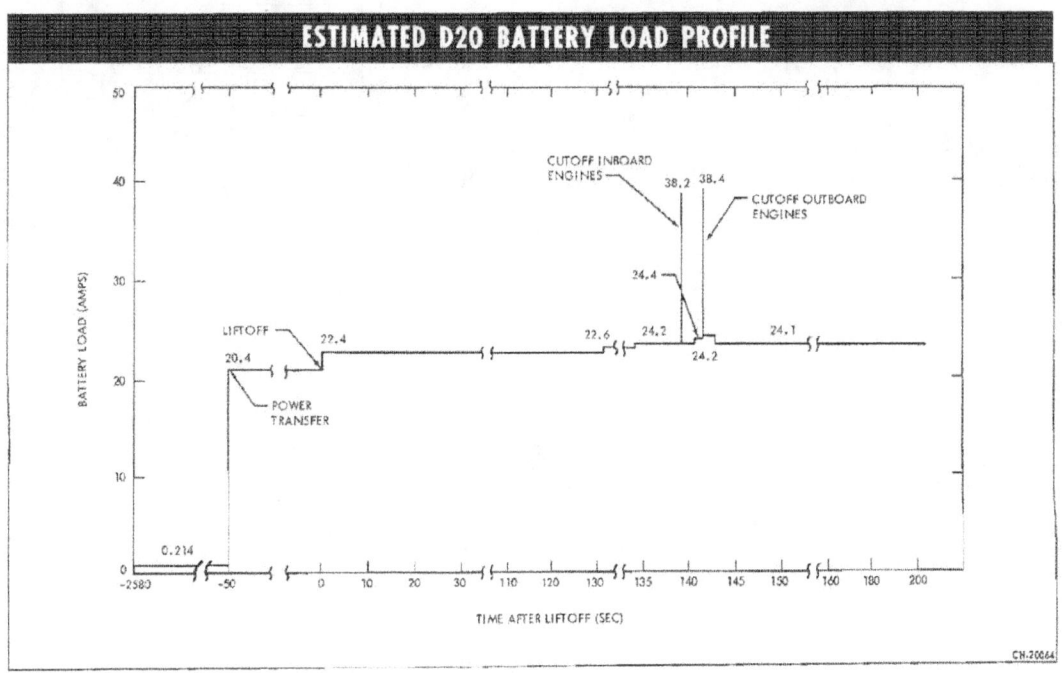

Figure 4-57

Section IV S-IB Stage

ELECTRICAL SEQUENCING

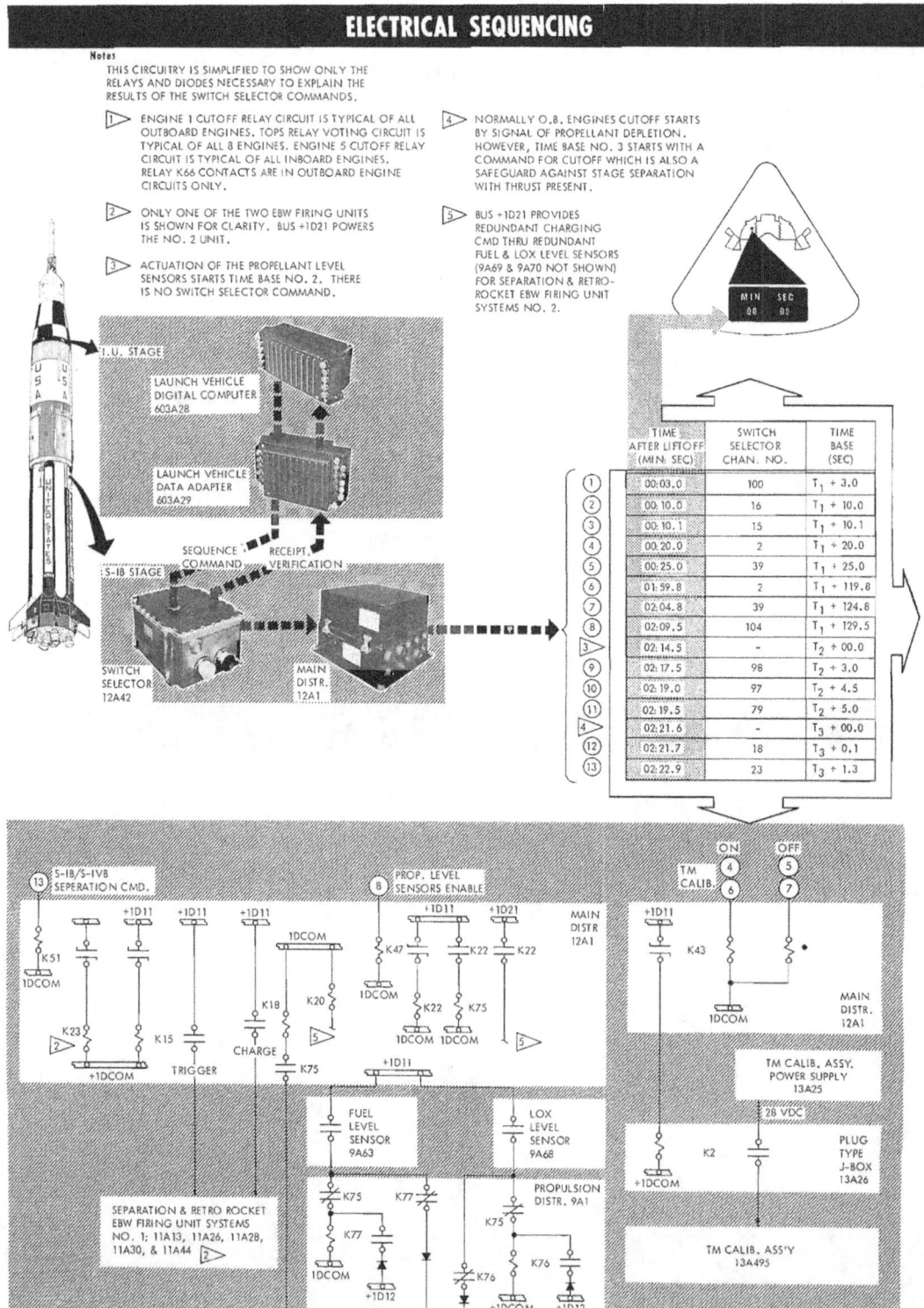

Figure 4-58 (Sheet 1 of 2)

Section IV S-IB Stage

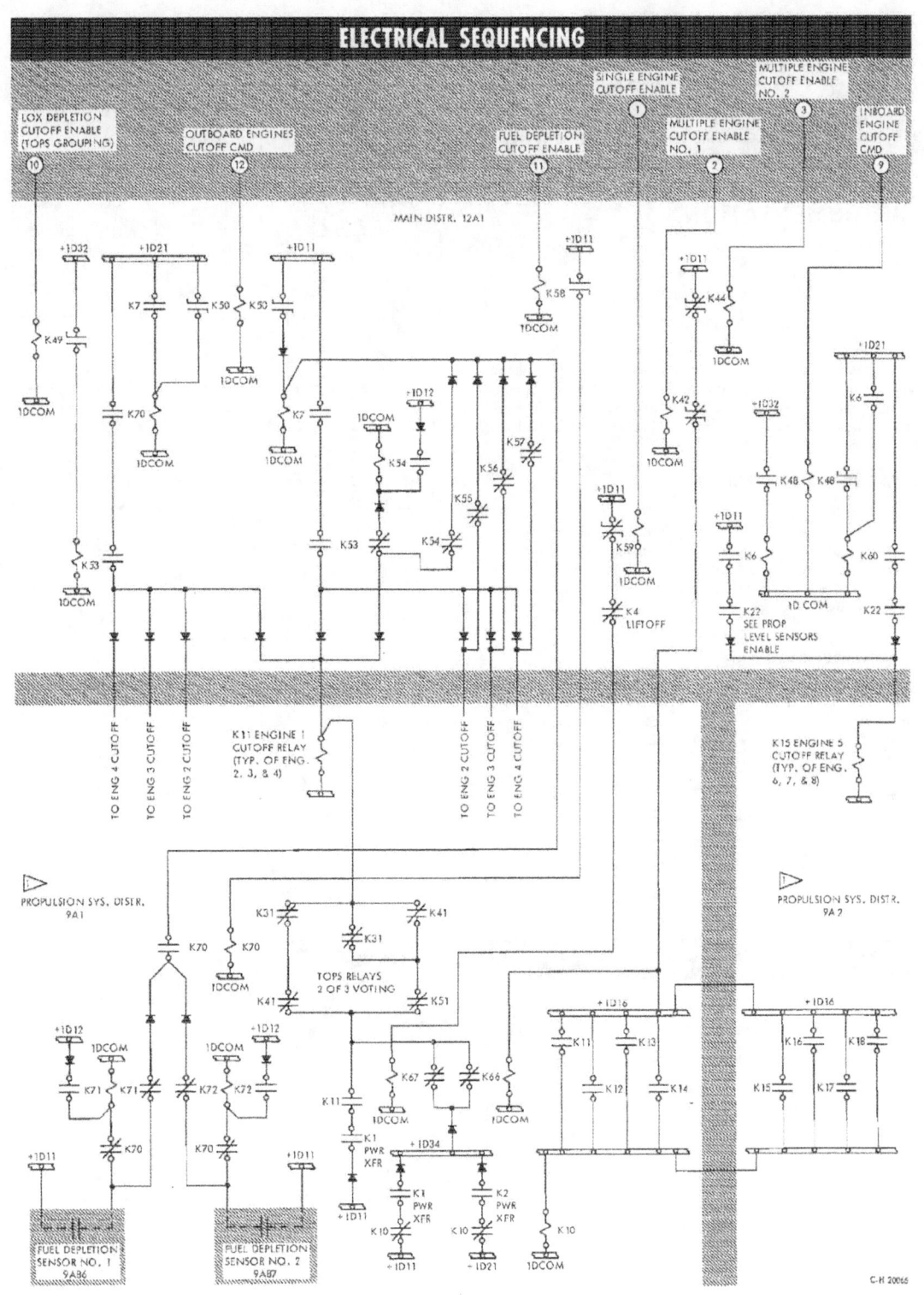

Figure 4-58 (Sheet 2 of 2)

Section IV S-IB Stage

A bulkhead seals this area above each tank and forms an air-tight compartment that is conditioned (see Environment Conditioning) for increased reliability of electrical components. Compartment no. 1 (F1) contains mostly electrical power equipment while compartment no. 2 (F2) contains mostly telemetry and tracking equipment. The lower portions of all tanks are open and house propulsion and measuring equipment. Electrical cabling reaches the upper and lower tank areas through two cableways, one on tank F1 and the other on tank F2.

INSTRUMENTATION.

The S-IB stage instrumentation system meet the following objectives:

a. Provide inflight data on critical subsystem performance

b. Provide prelaunch ground monitoring of stage subsystem operation

Measuring and telemetry subsystems comprise the S-IB stage instrumentation. The major instrumentation equipment is located on the interior of the forward and aft skirts of fuel tank F1, the forward and aft skirts of fuel tank F2 and the aft skirt of fuel tanks F3 and F4 (figure 4-62).

MEASURING SYSTEM.

The measuring system (figure 4-63) consists of transducers, signal conditioners located in the measuring racks, and measuring distributors. A total of 283 measurements are made on the S-IB stage of vehicle AS-206; 266 of these are telemetered to ground stations during flight. Prior to launch, 52 measurements not used in flight are transmitted to the block house by telemetry or hardwire connections.

Measurement Numbers.

A coding system is employed to provide identification for each measurement. The coding system used for measurement numbers describes the particular type of measurement and the area location on the S-IB stage. A typical measurement number is explained in figure 4-64.

Signal Conditioning.

Approximately 86 of the source signals in the measuring program are unsuitable for use by the telemetry system; therefore, signal conditioning is required to modify the measurements before delivery to the vehicle measuring distributors. Replaceable plug-in signal conditioning modules are installed in the measuring racks.

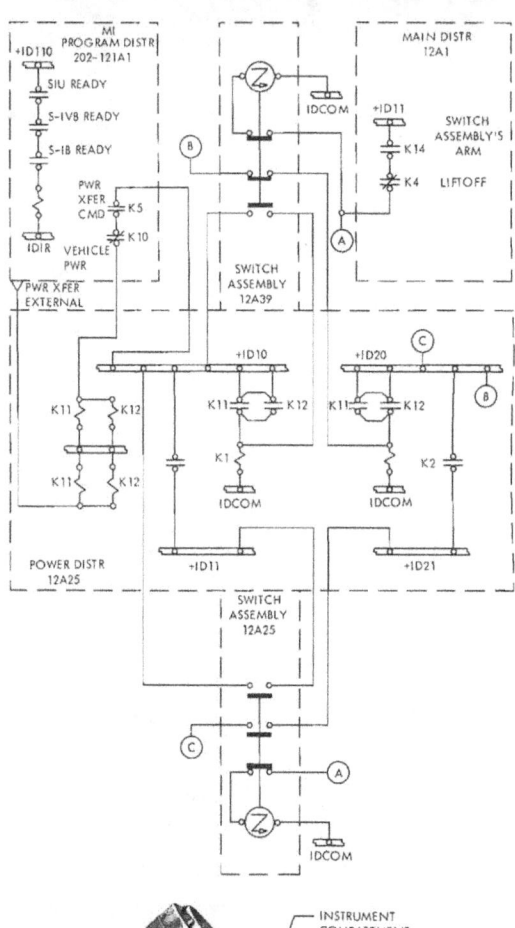

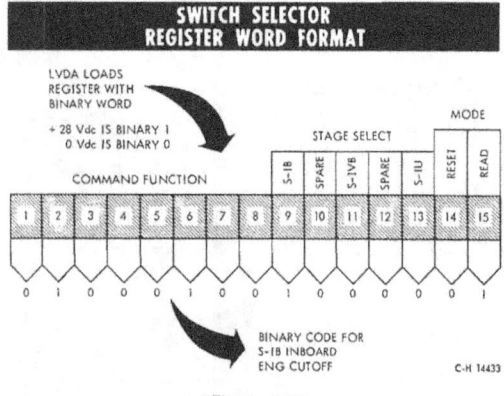

Figure 4-59

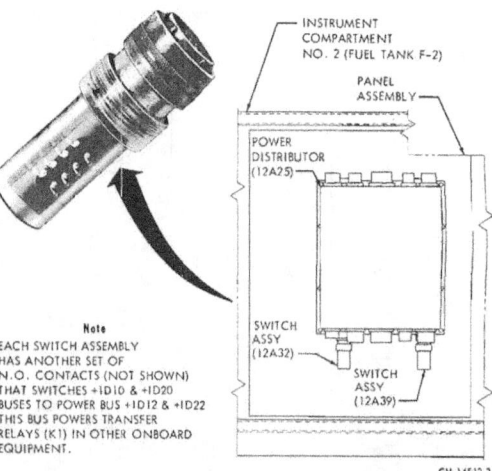

Figure 4-60

4-52

ELECTRICAL SYSTEM MEASURMENTS

MEASUREMENT		DISPLAY		
NO.	TITLE	ESE	AUX	MCC
VK1-12	SW SEL COUNT IND	X		X
VK115-12	SW SEL REG TEST	X		
VM1-9	MEAS VOLTAGE +1D81	X		
VM2-9	MEAS VOLTAGE +1D82	X		
VM9-12	MEAS VOLTAGE +1D89	X		
M16-12	D21 BUS VOLTAGE			X
M17-12	D11 BUS VOLTAGE			X
XM18-12	D10 BAT CURRENT		X	X
XM19-12	D20 BAT CURRENT		X	X

Figure 4-61

Six measuring racks are located in the S-IB stage. One rack is located in the aft skirt of each of the four fuel tanks and two racks are located in instrument compartment no. 1. The conditioned outputs of the measuring racks are sent to measuring distributors.

The S-IB stage contains three measuring distributors. Measuring distributor 9A4, located in the aft skirt of the F2 tank, controls direct measurements and routes the outputs of measuring racks 9A520 and 9A526, located in the aft skirt of tanks F2 and F3, to the proper telemetering system. Measuring distributor 9A3, located in the aft skirt of the F1 tank, controls direct measurements and routes the outputs of measuring racks 9A516 and 9A530, located in the aft skirt of tanks F1 and F4, to the proper telemetering system. Measuring distributor 12A26 located in instrument compartment no. 2 routes information signals and the outputs of measuring racks 12A439 and 12A440 to the proper telemetering system.

Checkout.

A remote automatic system permits a remote checkout of the flight instrumentation system. Each of the signal conditioning modules contains a printed circuit board that includes transducer simulation circuits for calibration testing of the module. Two relays are incorporated into the module: one relay for the upper (HI) end of the calibrated range of the measurement and one for the lower (LO) end. The run mode returns the measurement to its normal operating position. A control panel in the block house enables personnel to select the desired measurement module in the vehicle and the calibration mode (HI, LO, RUN). Any number of channels can be selected and energized in any of the three modes (HI, LO, RUN) individually, or in sequence. During simulated flight tests or launch countdown, a previously programmed computer may automatically operate the system.

A measuring rack selector (9A546) located in the aft skirt of tank F2 is used to decode measuring rack selection commands from the ESE, select the rack or racks to be calibrated, and pass on ESE calibration commands to the selected rack or racks. The measuring rack selector is used only during prelaunch checkout. Figure 4-65 lists the test specifications.

TELEMETRY SYSTEMS.

Measurements made on the S-IB stage are transmitted over two telemetry systems, GP1 and GF1 (figure 4-66). Components of the telemetry system are located in instrument compartment no. 1 (see figure 4-62) except the remote digital submultiplexer 9A700 which is located in the aft skirt of fuel tank F2. The two telemetry systems transmit data through a common antenna system consisting of multicoupler 13A494, power divider 13A470, and antennas 11A494 and 11A495. The antennas are installed on opposite sides of the stage at positions II and IV. A telemetry system calibrator, 13A495, supplies calibration voltages to the telemetry systems for prelaunch checkout and inflight calibration. The major components used in the telemetry systems have been qualified in previous flights. The momentary data drop-outs that have occurred in the past have been traced to antenna pattern null points resulting in insufficient signal strength, or ground station problems.

TELEMETRY SYSTEM GP1.

Telemetry system GP1 is a PCM/DDAS link that transmits realtime checkout data before launch, and measuring program information during flight. The telemetry system consists of Model 270 multiplexers 13A440 and 13A484, remote digital submultiplexer 9A700, Model 301 PCM/DDAS assembly 13A485, and RF assembly 13A486. The pulse-amplitude modulated (PAM) data from the Model 270 multiplexers are converted to digital words in the Model 301 assembly. The Model 301 assembly then encodes the digital words with a digital word from the remote digital submultiplexer into a data frame. The data frame is transferred to the RF assembly where it is converted to an RF signal. The RF signal is connected to the RF coupler for transmission to the antenna system. The data frame is also used to modulate a 600 kHz, voltage-controlled oscillator (VCO) the output of which is connected to the LCC through a coaxial cable for prelaunch checkout and monitoring.

REMOTE DIGITAL SUBMULTIPLEXER.

The RDSM provides additional data-handling capability to the pulse code modulation (PCM) telemetry system. The RDSM can accept a maximum of one hundred inputs, which are sampled sequentially in groups of ten. The ten outputs of the RDSM, containing the information provided by the 100 inputs, can therefore furnish the PCM telemetry system digital words made up of ten bits each. The assembly handles digital information only. Its inputs and outputs are set voltage levels that represent ON-OFF conditions or binary numbers. The sequential sampling action of the RDSM is controlled by timing signals from the Model 301 PCM/DDAS assembly.

MODEL 270 MULTIPLEXER.

The Model 270 Multiplexer consists of 23 submultiplexers, each of which gate ten data sources into a 30-channel main multiplexer. Four data channels are continuously supplied to the main multiplexer, the remaining three channels convey frame identification and synchronization pulses. The data frame output of the main multiplexer is supplied to a PAM amplifier through a gate. The PAM amplifier amplitude modulates a train of pulses in accordance with the sequential data input. The PAM pulse train is then supplied to the Model 301 assembly. The output gate, after the main multiplexer, allows a calibration signal to be substituted for data when calibration of the data channels is desired. An internal calibration generator produces a sequence of five calibrated voltages upon receipt of a calibrate command. Voltage levels of 0, 25, 50, 75, and 100 percent of maximum level are sustained for ten frames commencing with the master frame following receipt of the command.

MODEL 301 PCM/DDAS ASSEMBLY.

The PCM/DDAS scans the pulse amplitude modulation wavetrains from the Model 270 multiplexers in a programmed sequence and multiplexes these wavetrains into an analog-to-digital converter. The PAM samples in the wavetrain are encoded into a 10-bit digital word and stored in a 10-bit register. An output gate multiplexes the data in the register with data in the register of the remote digital submultiplexer. The serial data words are then supplied

Section IV S-IB Stage

EQUIPMENT LOCATION

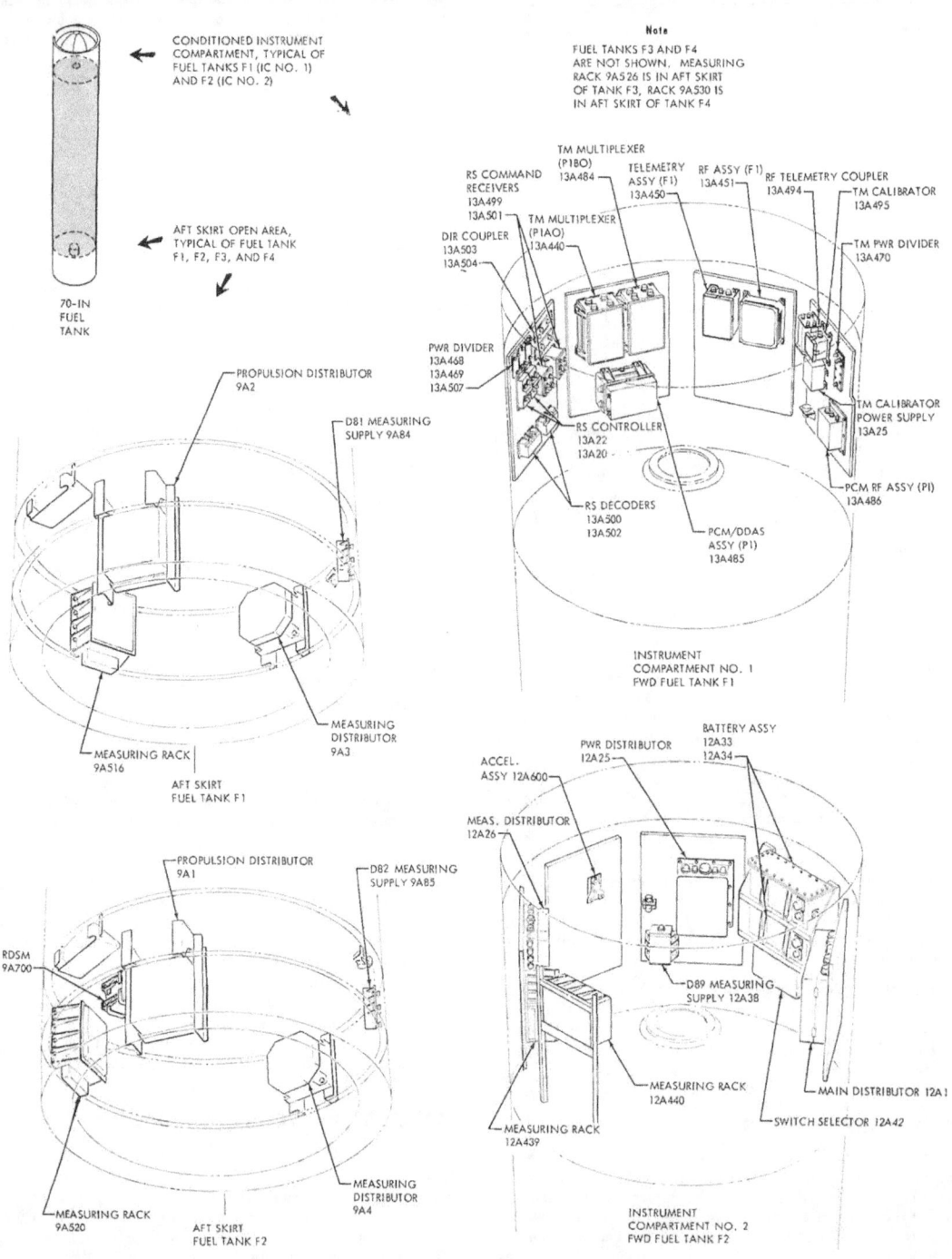

Figure 4-62

Section IV S-IB Stage

to the 600 kHz VCO and the RF assembly. The 600 kHz FM carrier comprises the DDAS output which is connected, through a coaxial cable, to the DDAS receiving equipment in the LCC. This signal is demodulated, demultiplexed, and decoded, to recreate the original measurements. These measurements can be displayed on meters, or recorded, or used as command verifications in the automatic checkout equipment.

RF ASSEMBLY—TELEMETRY LINK GP1.

The RF assembly consists of a filter, a frequency shift keyer, and a power amplifier. The digital data pulses are first filtered to reduce sideband components. The filtered pulses then modulate the frequency shift keyer. The FM signal is then amplified by the power amplifier and supplied to the RF coupler as a carrier of 256.2 MHz at a nominal 15 W with a modulation deviation of ±36 kHz. The data rate is 72 Kbps.

TELEMETRY SYSTEM GF1.

Telemetry System GF1 is a FM/FM link (frequency modulation/frequency modulation) which transmits measuring program information during flight. The telemetry system consists of Model CI-18 assembly 13A450 and RF assembly 13A451. The GF1 telemetry system is provided to convert analog measurement signals into proportional frequency-intelligent signals for subsequent modulation of an FM transmitter. All input signals must be preconditioned to a 0-Vdc to 5-Vdc range.

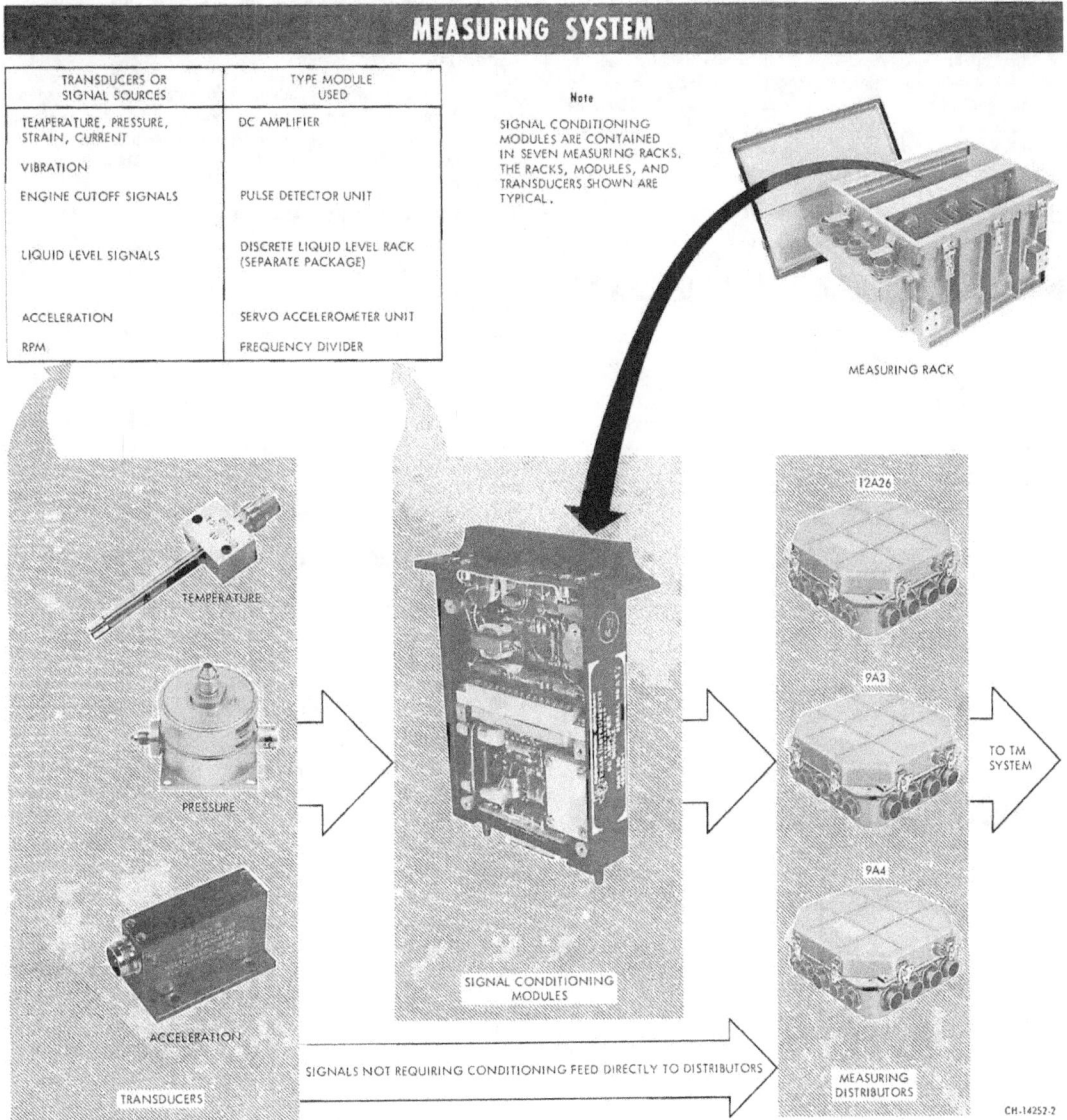

Figure 4-63

Section IV S-IB Stage

TELEMETRY COMPONENTS.

Model CI-18 Assembly.

The Model CI-18 assembly consists of 17 subcarrier oscillators operating at standard inter-range instrumentation group channels 2-18. Channel 1 is not used because the power supply of the guidance system in the IU operates at the same frequency. The subcarrier oscillators (SCO) are the units within the telemeter that convert the analog inputs into frequency-intelligent data. Each SCO is a voltage controlled oscillator that is set at a precise frequency range. The frequency output will shift up to the high limit as the input signal increases to 5 Vdc and shift down to the low limit as the input decreases to 0 Vdc. Thus, the output frequency of each SCO is set to be linearly proportional to an input of 0 to 5 Vdc. A gate at the input to the subcarrier oscillators allows a calibrating signal to be substituted for the data. Calibrating commands and voltages from the telemetry system calibrator are supplied through the program plug assembly to actuate the gate and check the operation of the channel. The program plug assembly is a wiring option that allows a channel to be preflight calibrated only, inflight calibrated only, or not calibrated at all. The plug is wired in accordance with the measuring program for the particular vehicle. The FM outputs of the SCO's are mixed in the mixer-amplifier and the composite signal is supplied to the RF assembly.

RF Assembly—Telemetry Link GF1.

The RF assembly uses the input signal to frequency modulate a VHF oscillator in the FM transmitter. The 1.5-W signal is raised to 20 W nominal by a power amplifier. The high power signal is then filtered to suppress harmonics and supplied to the telemetry system RF coupler.

RF Multicoupler.

The multicoupler uses tuned cavities to selectively pass the signals to be coupled. Each carrier input passes through a resonant cavity that is tuned to 1/4 wavelength of the carrier center frequency. The coupler provides 18-db isolation between adjacent frequencies. The carriers are then capacitively coupled together and routed to the power divider.

Power Divider.

The power divider equally divides the composite carrier signal and provides equal power to the two antennas.

Antennas.

The two telemetry antennas radiate the telemetry carrier to ground receiving stations. The antennas are mounted 180 deg apart on metal panels located on the upper portion of the propellant tanks. The antennas provide an omnidirectional pattern about the launch vehicle roll axis.

Telemetry Calibrator Assembly.

Periodic calibration of the telemetry systems establishes references for data reduction to increase the accuracy of data analyzation. Calibration commands and signals for preflight and inflight calibration of telemetry systems GP1 and GF1 are generated by the telemetry system calibrator. Preflight calibration is controlled from the LCC. Calibration is accomplished by switching the input of a data channel from the data source to the calibrator output. The calibrator supplies a precisely controlled voltage with steps at 0, 25, 50, 75, and 100 percent of channel capacity. When the calibration command is received, the calibrator will normally cycle to each step and return to zero. During preflight calibration, however, the calibrator output can be stopped at any desired step for alignment of data amplifiers or other signal conditioning devices.

ENVIRONMENTAL CONDITIONING.

INSTRUMENT COMPARTMENT.

Two instrument compartments located above fuel containers F-1 and F-2 require environmental conditioning during preflight operations only. A ground environmental control system (see Section VII) supplies the required conditioning medium through swing arm no. 1 quick-disconnect. Air is used as the conditioning medium from approximately 1 day and 8 hr prior to liftoff when electronic components are activated until 30 min prior to liquid hydrogen loading at 4 hr and 45 min prior to liftoff. GN_2 then provides the inert conditioning required for the remainder of preflight preparation. The conditioning medium is supplied to the instrument compartments (figure 4-67) through service arm 1A, precooling check valve, upper flexible tubing assembly, and a distribution manifold in each compartment. The medium is exhausted through a weldment near the bottom of each compartment, through the lower flexible tubing assembly, precooling check valve, swing arm ducting, and into the atmosphere. The compartment inlet temperature is controlled by a sensor which is located in the GSE supply duct and provides input to the S-IB forward ECS control panel in the LCC. A bleed orifice on compartment F-1 and the two precooling check valves maintain the desired environment during flight. At the end of S-IB powered flight for the AS-201 and AS-202 missions, pressure in instrument compartment F-2 was approximately 6 and 5 psi respectively. Compartment requirements during preflight cooling are listed in figure 4-68.

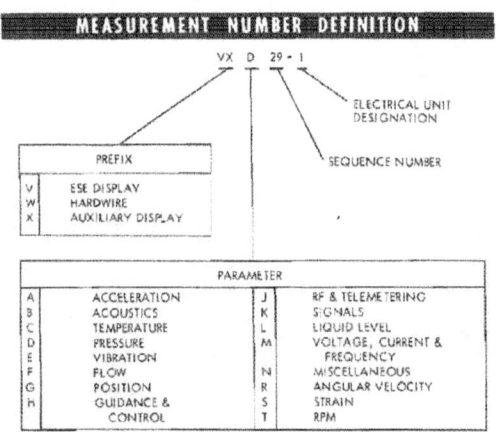

Figure 4-64

Figure 4-65

4-56

Section IV S-IB Stage

S-IB TELEMETRY SYSTEMS

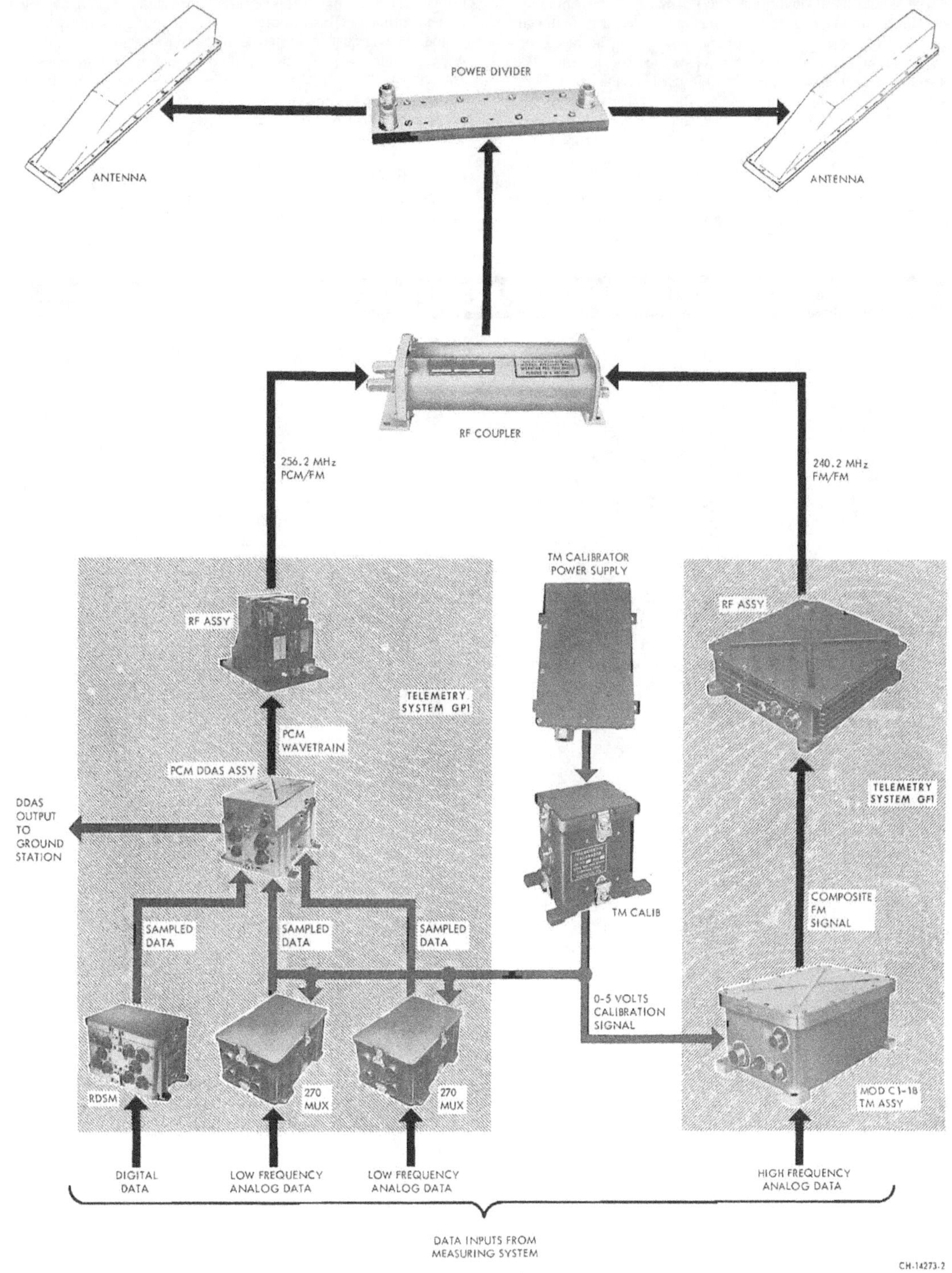

Figure 4-66

4-57

Section IV S-IB Stage

ENGINE COMPARTMENT.

Tail Unit Conditioning and Water Quench.

Four separate manifolds (figure 4-69) provide for distribution of conditioned air, GN_2, GN_2 deluge, or water quench to the engine compartments during preflight operations. During normal operation, the ground environmental control system provides air or GN_2 for compartment conditioning. Air is used as the conditioning medium until 30 min before lox loading. GN_2 is used for inert conditioning during the remainder of countdown. In the event of engine malfunction cutoff or fire in the engine compartment area, a cold GN_2 deluge is activated flooding the compartments with cold GN_2. The water quench is available as a backup in case of fire. Compartment conditioning requirements are presented in figure 4-70. The conditioning medium, GN_2 deluge, or water quench flows through a quick-disconnect coupling, pneumatic valve, and dispersal manifolds which direct the flow to the engine compartments and the area between the firewall and the center lox tank sump. Compartment inlet temperature is controlled by thermal probes that are located in the engine areas and provide temperature input to the S-IB aft ECS control panel in the LCC. A pneumatic valve in each manifold is held open during system operation by 750 psig GN_2 control pressure routed through short cable mast no. 4 from the launch complex (see Section VII) and closes at liftoff to prevent exhaust gases from entering the engine compartments during flight.

Fire Detection.

The fire detection system consists of 32 thermocouple sensors (connected in four loops of eight sensors) located on the aft thrust

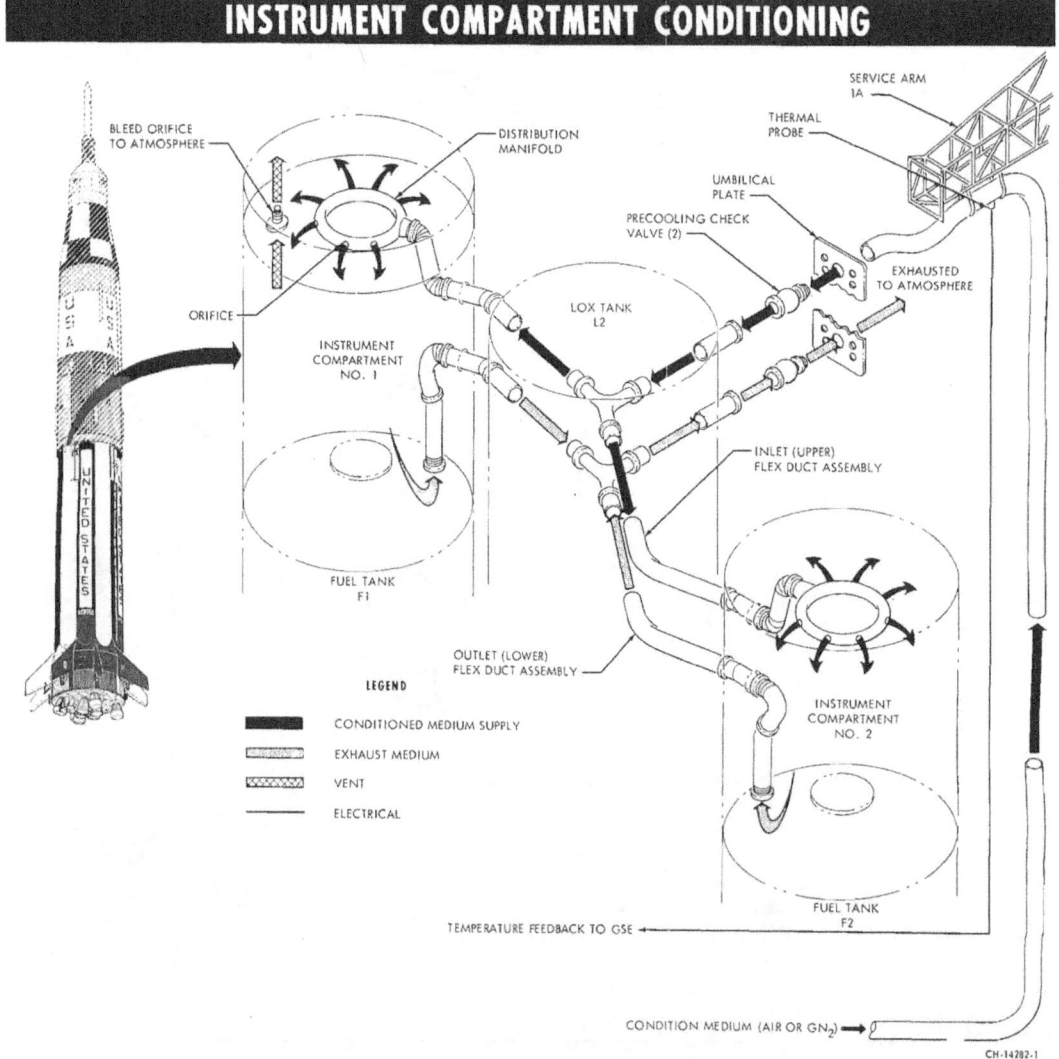

Figure 4-67

4-58

Section IV S-IB Stage

INSTRUMENT COMPARTMENT CONDITIONING REQUIREMENTS

FLOW (Lbs/Min)	COMPARTMENT INLET CONDITIONS			COMPARTMENT & TEMP. RANGE (Degrees F)	PROBE DATA	
	PRESSURE (In. of H_2O)	TEMP. (Degrees F)	HUMIDITY (Gr/Lb of Air)		LOCATION	SETTING (Degrees F)
45	15	75 ± 5	0 - 37	#1, 32-140 #2, 68- 84	DUCT (GSE)	75 ± 3
NOTE: REQUIREMENTS FOR AIR AND GN_2 ARE SAME.						

Figure 4-68

structure and the firewall substructure. The sensors monitor temperature rise-rates in critical engine areas prior to liftoff. The sensor output is displayed on a recorder in the LCC where it is continuously monitored by a redline observer. If the display indicates presence of a fire (a temperature rise of 60° F/sec for 0.5 sec), the observer notifies the test conductor who initiates a launch scrub and orders the cold GN_2 deluge turned on. A water quench may be manually initiated if necessary to extinguish fire.

HAZARDOUS GAS DETECTION SYSTEM.

To detect the presence of hydrogen and oxygen gases, the atmosphere in the S-IB stage engine compartment, the S-IB lox tank skirts, the S-IVB forward skirt, and the S-IVB aft interstage is monitored during prelaunch operations. This is accomplished through a hazardous gas analyzer located in the ML that draws gas samples from the various compartments through a mass analyzer. See figure 4-71.

ORDNANCE.

Ordnance components used on the Saturn IB vehicle initiate various operations necessary for proper stage function. Additionally, ordnance installed on the S-IB stage propellant containers provides a propellant dispersion capability for use if the vehicle becomes a safety hazard during the boost phase of flight.

H-1 ENGINE ORDNANCE.

Four types of ordnance devices either initiate or shut down the H-1 engine operation. These items are the solid propellant gas generators (SPGG), the SPGG initiators, the squibless igniters, and the conax valves. See H-1 Engine for details on these items.

S-IB STAGE PROPELLANT DISPERSION SYSTEM.

The S-IB Stage Propellant Dispersion System (PDS) will sever each of the nine propellant tanks and disperse the propellants if flight termination becomes necessary. Two exploding bridgewire (EBW) firing units, a safety and arming device (S&A), two EBW detonators, Primacord, and flexible linear-shaped charges (FLSC) make up the PDS (figure 4-72). Primacord assemblies interconnect the FLSC to detonators in the S&A device. The EBW firing units interface with the PDS detonators and the secure range safety command system. If flight termination becomes necessary, the range safety command system (Section I) will provide signals to arm (charge the EBW high voltage storage capacitor) and trigger the firing units, which deliver high-energy electrical pulses to the EBW detonators in the S&A device (figure 4-73). Explosive leads in the S&A device rotor propogate the detonator explosion to the Primacord and subsequently to the FLSC assemblies. The FLSC assemblies rupture the propellant tanks allowing the propellants to disperse radially from the stage and burn rather than explode. The burning propellants result in only a fractional amount of the theoretical yield if the vehicle should explode. The reliability of Saturn propellant dispersion systems was demonstrated during the flights of SA-2 and SA-3. After S-IB stage engine cutoff a destruct command destroyed the vehicle to release water ballast contained in the dummy upper stage (Project Highwater).

EBW Firing Unit.

Exploding bridgewire firing units, used extensively in the Saturn IB launch vehicle, provide electrical pulses to ignite the ordnance. Redundant EBW firing units in each system ensure the desired function. The solid-state electronic EBW firing units generate a high voltage, high energy, short duration pulse to fire EBW detonators or initiators. Each firing unit consists of an input filter, a charging circuit, a storage unit, a trigger circuit, and an electronic switch. Two inputs are necessary to the firing unit function. The first input, the application of 28 Vdc, passes through the input filter, which functions as a noise suppressor, to the charging circuit. In the charging circuit, an oscillator, a step-up transformer, and a silicon controlled rectifier increase the voltage to 2300 Vdc and couple the charge to the capacitor storage unit. A voltage divider network provides a 0- to 4.6-Vdc input to the telemetry system proportional to the 0- to 2300-Vdc charge on the storage unit. The second input applies 28 Vdc to the trigger circuit. The trigger disables the charging circuit and generates a signal voltage for the electronic switch. This permits the storage unit to discharge through the switch and the detonator or initiator. A minimum of 1.5 sec must elapse between inputs.

The S-IB propellant dispersion panel provide control and monitoring of the firing units during checkout and prelaunch operations. Power switches control power application for ground power operations and internal power operations. POWER ON and ON INTERNAL POWER indicators monitor the condition of the firing units. A FIRING UNIT VOLTAGE meter provides a readout of the firing unit charge voltage during checkout. A 0- to 4.6-Vdc signal, proportional to the 0- to 2300-Vdc charge, drives the meters. During flight the charge monitor signal will be telemetered to ground receiving stations and recorded.

During checkout operations, the output pulses from the firing units are checked by connecting the output cables to pulse sensors. The firing units are then charged and triggered by the secure range safety command system. The pulse sensors receive the firing unit output and provide a 28-Vdc output to the FIRED indicators on the S-IB propellant dispersion panel. After checkout, the pulse sensors are removed from the vehicle and the firing unit output cables are attached to the EBW detonators in the S&A device. See figure 4-74 for firing unit characteristics.

EBW Detonator.

EBW detonators provide the electrical-to-ordnance interface for explosive fuse assemblies. The detonators function on the EBW principle: when a high voltage, high energy pulse is applied to a small diameter, low resistance wire element, the element explodes, rapidly releasing a large amount of energy adequate to detonate

4-59

ENGINE COMPARTMENT CONDITIONING

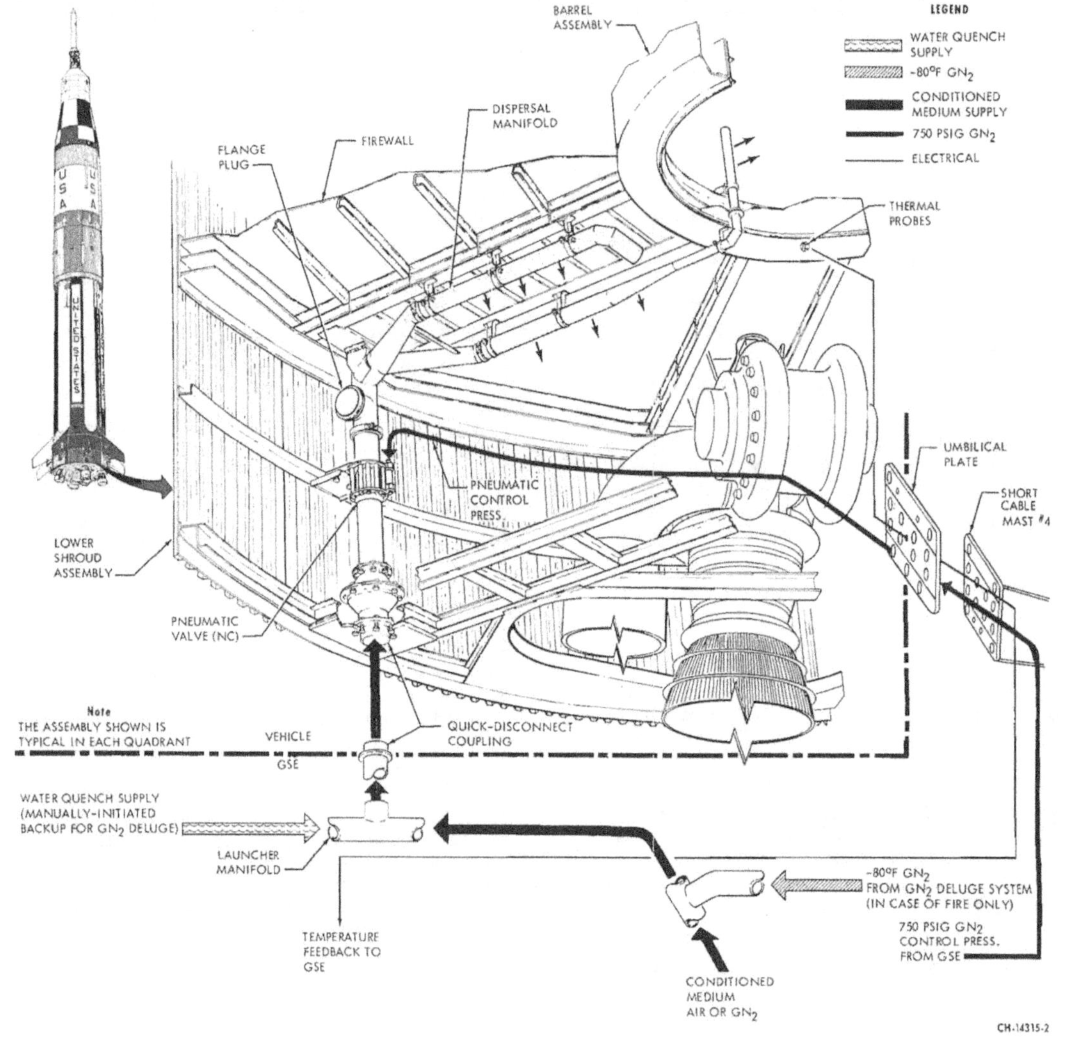

Figure 4-69

an explosive charge. The detonator consists of an acceleration charge, a main charge, a bridgewire, two electrical connector pins, and insulators—all enclosed in a steel case. The case has threads on each end that permit installation into the safety and arming device and provide for attachment of the EBW firing unit cable to the detonator. The detonator is not sensitive to static discharges or RF energy, and the explosives are relatively insensitive to heat. A spark gap in one electrical connector pin precludes accidental dudding of the detonator through inadvertent application of ground power. The gap, created by a thin mica spacer that separates the two metal sections of the pin, has a breakdown voltage of 600 to 1200 Vdc. The spacer is square and creates a gap in the four areas where it does not cover the face of the cylindrical pin sections.

Alumina sleeves and insulators secure the electrical connector pins in the detonator. The bridgewire connects between electrical pins at the base of the acceleration charge. The main charge is located in the end of the detonator just forward of the acceleration charge. The main charge is located in the end of the detonator just forward of the acceleration charge. Closure disc paper separates the charges until detonation. The 2300-Vdc pulse from the EBW firing unit melts the fine bridgewire. The magnetic field created by the current causes the melting wire to form a series of lobes. When the lobes become unconnected spheres of molten metal, arcing occurs between the spheres. Very high internal pressure within the molten spheres explosively propels hot particles into the 1.04 ± 0.03 gr, Class 2 PETN, acceleration charge. The energy release and shock

Section IV S-IB Stage

ENGINE COMPARTMENT CONDITIONING REQUIREMENTS

FLOW MEDIUM	FLOW (lbs/min)	COMPARTMENT INLET CONDITIONS			COMPARTMENT TEMP RANGE (DEGREES F)	PROBE DATA	
		PRESSURE (In of H_2O)	TEMP (DEGREES F)	HUMIDITY (Gr/lb of air)		LOCATION	SETTING (DEGREES F)
AIR	135-170	42	40-150	0-37	50±10	STAGE	50±10
GN_2	300	42	65-95	0-37	40-75	STAGE	50±10
GN_2 DELUGE	420	84	-80	0-37	N/A	STAGE	N/A
WATER QUENCH	8000 gpm	125 psig (SUPPLY)	AMBIENT	N/A	N/A	STAGE	N/A
N/A - NOT APPLICABLE							

Figure 4-70

of the exploding bridgewire is sufficient to initiate PETN detonation. The acceleration charge then detonates the main charge consisting of 1.40±0.25 gr of Class 4 PETN.

Safety and Arming Device.
The safety and arming (S&A) device, an electro-mechanical device that arms the PDS, fulfills the requirements stated in AFETRM 127-1 for arming a flight termination system. The specific requirements are:

a. The unit must complete and interrupt the explosive train by remote control.

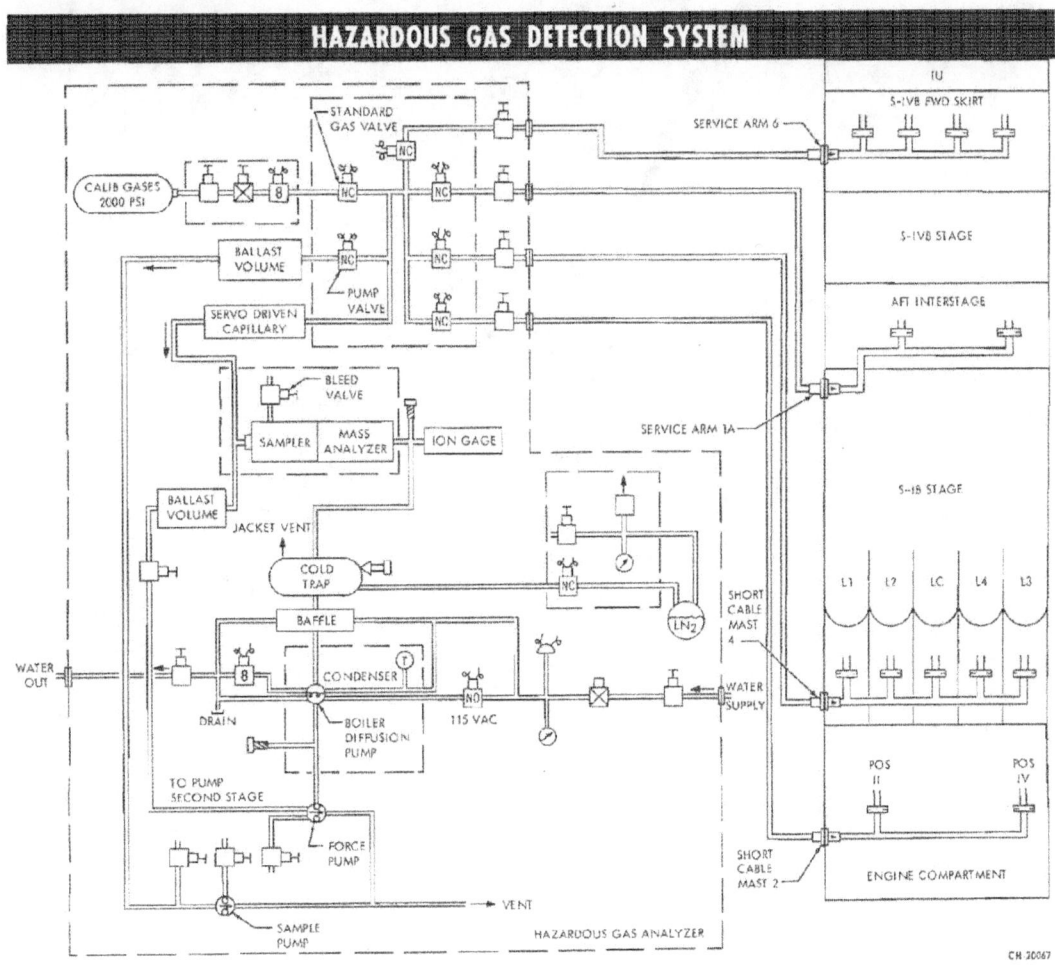

Figure 4-71

4-61

Section IV S-IB Stage

S-IB STAGE PROPELLANT DISPERSION SYSTEM

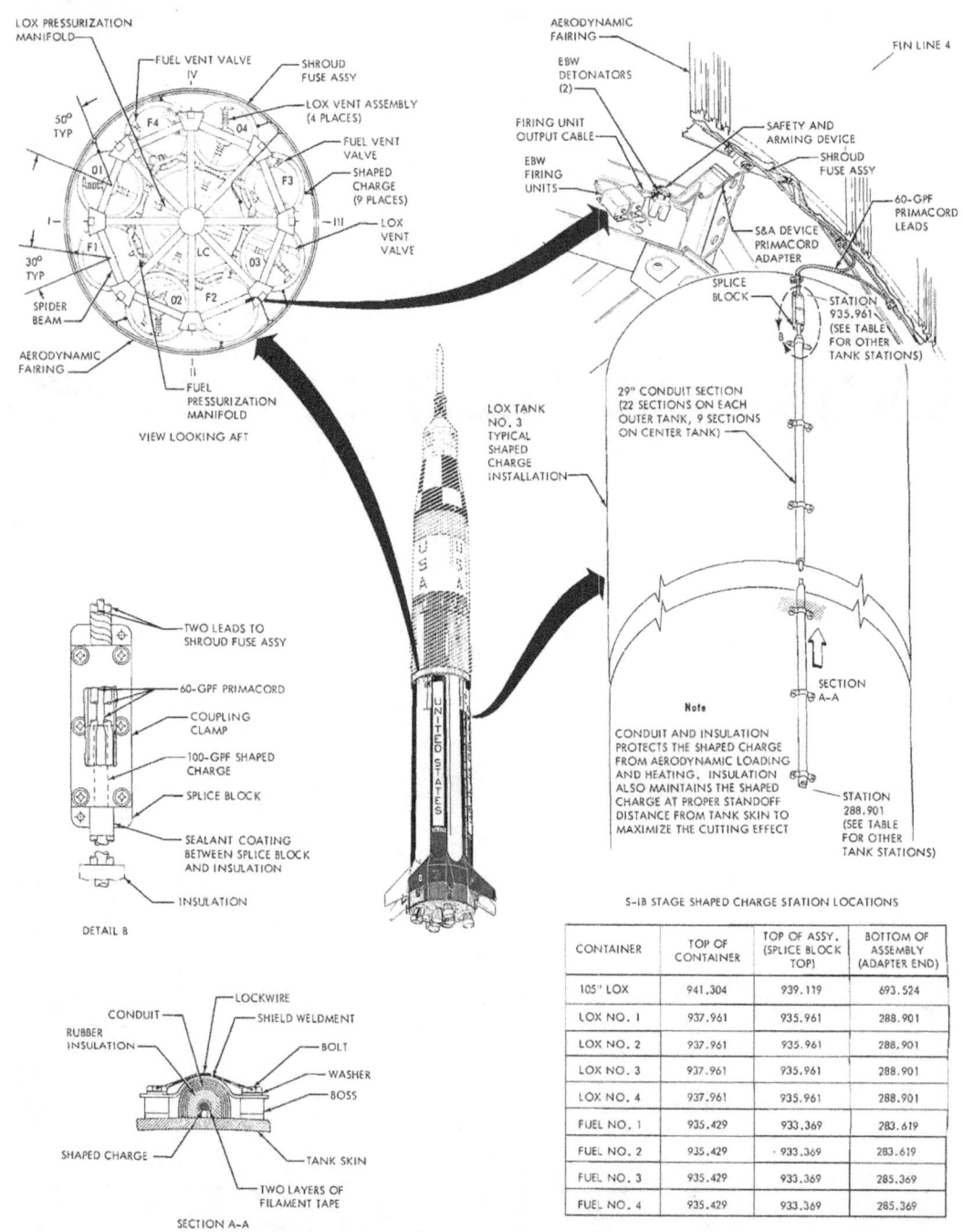

Figure 4-72

Section IV S-IB Stage

b. The unit must provide indications to remote monitoring equipment whether its position is safe or armed.

c. Visual position indication and manual operation of the unit must be possible.

d. Must have separate connections for the arming and firing circuits.

The S&A device consists basically of a Ledex 95-deg rotary solenoid assembly, a metal rotor shaft with two 2.5-gr PETN explosive inserts, and position-sensing and command switches that operate from the rotor shaft cam. On electrical command from the ground system just prior to automatic countdown, the solenoid assembly rotates the shaft containing the two explosive inserts 90 deg. This aligns the inserts between the EBW detonators and the Primacord adapters to form the initial part of the explosive train.

Each power application to the solenoid assembly moves the rotor 90 deg in one direction, thus rotating the explosive inserts in and out of alignment with the explosive train. A spring-loaded detent locks the rotor shaft in position; a spring and ratchet arrangement returns the solenoid to the original position. When the S&A device

EBW FIRING UNIT CHARACTERISTICS	
PARAMETER	VALUE
INPUT POWER	24 TO 32 Vdc
PEAK CURRENT INPUT	2.0 A
STEADY STATE AVERAGE CURRENT INPUT	250 mA MAX
OUTPUT VOLTAGE	2,300 ± 100 Vdc
CHARGE TIME	1.5 sec MAX
TRIGGER INPUT VOLTAGE	24 TO 32 Vdc
TRIGGER CKT SENSITIVITY	a. 8 Vdc OR LESS WILL NOT ACTUATE TRIGGER CIRCUITRY
	b. A 50 Vdc TRANSIENT PULSE OF 50 µSEC DURATION WILL NOT ACTUATE THE TRIGGER CIRCUITRY
TRIGGER CKT RESPONSE	4 ± 1 mS
TRIGGER PEAK CURRENT	250 mA
TRIGGER STEADY STATE AVERAGE CURRENT	75 mA MAX
OPERATING TEMP	-65°F TO +200°F
COMPONENT LIFE	CONTINUOUSLY 'ON,' THE FIRING UNIT WILL PERFORM 1000 OPERATIONS, RECEIVING A TRIGGER AT 15 MIN INTERVALS.

Figure 4-74

is in the safe position, the explosive inserts are 90 deg out of alignment with the EBW detonators and Primacord adapters; the rotor forms a metal barrier to prevent propagation to the explosive train should the detonators inadvertently fire. As an additional safety precaution to prevent damage to the explosive train, vent holes are drilled through the S&A device body, in line with the safe position of the explosive inserts, to vent detonation gases overboard. A debris retainer installed over the vent holes will prevent damage to surrounding equipment should the detonators fire. The vent holes have no function when the S&A device is

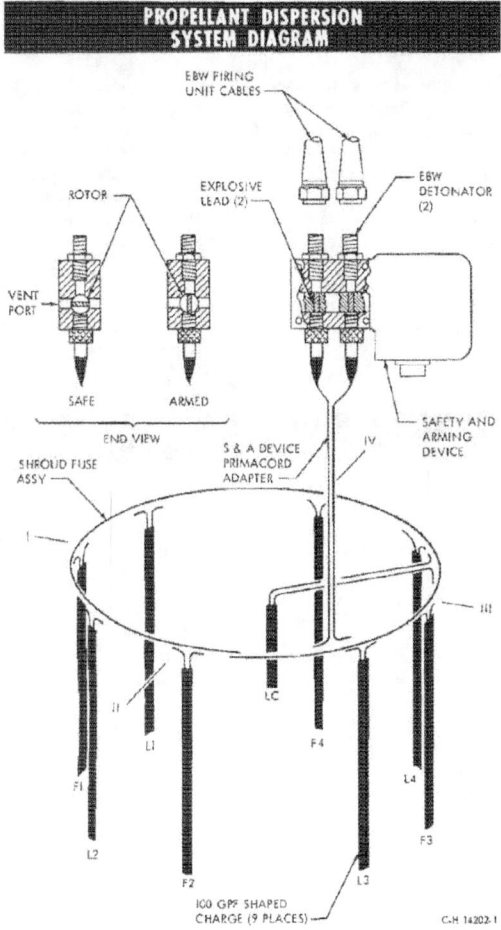

Figure 4-73

SAFE AND ARM DEVICE CHARACTERISTICS	
ITEM	DESCRIPTION
EXPLOSIVE INSERTS OPERATIONAL TEMP RESPONSE TIME	TWO 2.5 gr PETN LEADS -65°F TO +165°F ROTOR SHAFT WILL ROTATE 90 DEG UNDER NO LOAD IN LESS THAN 50 MSEC WITH 28 ± 4 Vdc APPLIED TO SOL
IMPACT SENSITIVITY	THE S&A DEVICE HAS BEEN QUALIFIED THROUGH 8-FOOT DROP TESTS AND SHOCK TESTS OF 35 g MAGNITUDE, THREE SHOCKS IN EACH OF THREE AXES, SHOCK WAVES WERE 1/2 SINE WAVES, 8 MSEC DURATION

Figure 4-75

armed. The S&A device electrical circuitry provides monitoring and actuation functions for the device, but contains no provisions for initiating the explosive train. The position of the device safe or armed can be monitored remotely (by electrical signals from internal sensing switches prior to liftoff only) on the S-IB and S-IVB propellant dispersion panel indicators. A safe/arm switch on the panels permits manual control of the S&A device position for checkout and arms the S&A device approximately 4 min before liftoff. A clear polystyrene plug in the end of the S&A device (at the end of the rotor shaft) provides a means of locally monitoring the shaft position. The plug is removable, providing access to manually reset the S&A device to the safe position, if necessary. See figure 4-75 for S&A device characteristics.

PDS Ordnance.

In addition to the EBW firing units, EBW detonators, and S&A device, the S-IB stage PDS has an adapter fuse assembly, a shroud fuse assembly, and nine FLSC assemblies. The adapter fuse assembly consists of two parallel 61-in. lengths of 60-gpf PETN Primacord with end fittings adhesively bonded to the Primacord on one end of the assembly. The end fittings connect the fuse assembly to the S&A device and contain 6-gr PETN booster charges to ensure propagation of the detonation across the mechanical connection to the Primacord. The adapter fuse assembly extends to the shroud fuse assembly and connects to the shroud fuse assembly by an overlapping splice. In this connection, the ends of the Primacord are placed parallel to the shroud fuse assembly, and both are taped together. See figure 4-76 for PETN characteristics.

The shroud fuse assembly installed on the aft interstage aerodynamic fairing encircles the forward end of the S-IB stage, serving as the main explosive train to which the nine FLSC assemblies connect. The fuse assembly is a 835-in., 60-gpf length of Primacord wrapped with aluminum tape. Quick-release clamps secure the fuse assembly to the aft interstage aerodynamic fairing extension.

An FLSC assembly consisting of a lead sheathed 100-gpf PETN shaped charge, a splice block, and three pieces of 60-gpf Primacord is mounted on each propellant tank. The FLSC assemblies extend the entire length of the four fuel and four outer lox tanks. The center lox tank FLSC assembly extends approximately half way down the tank. The FLSC assemblies are inserted into conduits attached to the tank skin. The conduits provide protection against aerodynamic loading and maintain the correct orientation of the shaped charge to the tank skin. The splice block on each FLSC assembly attaches the Primacord leads to the FLSC. Inside the splice block, one end of the 5-in. Primacord length butts against the end of the FLSC. The two longer Primacord lengths, that connect the FLSC assembly to the shroud fuse assembly, overlap the short Primacord to FLSC splice. An adhesive, which fills the cavity around the Primacord and FLSC in the splice block, and a cover fastened to the splice block secure the Primacord-to-FLSC connection. The parallel Primacord leads extending from the FLSC assemblies to the shroud fuse assembly are 30 in. long for the fuel tanks, 50 in. long for the outer lox tanks, and 150 in. long for the center lox tank. All the Primacord leads attach to the shroud fuse assembly by overlapping splices.

POWER TRANSFER SAFETY SWITCHES.

Two squib-actuated switch assemblies, installed on Power Distributor 12A25 in unit 12 of the S-IB stage, actuate at liftoff to assure application of internal power to stage electrical systems during flight. The one-shot switches parallel the contacts of the command power transfer relays and provide relay contact chatter compensation by maintaining power application to the energizing coils of the power transfer internal relays. Each switch assembly consists of an electrical connector, an adapter, wiring, and a squib-switch. Potting compound encapsulates the squib-switch and wiring leaving only the switch terminals exposed for test purposes. A plastic tube

PETN CHARACTERISTICS

DESCRIPTION	WHITE CRYSTALS
CRYSTAL DENSITY	1.765
MOLECULAR WEIGHT	316.55
CHEMICAL FORMULA	$C(CH_2NO_3)_4$
MELTING POINT	280°F TO 285°F, DECOMPOSES RAPIDLY ABOVE 410°F
DETONATION RATE	27,232 FT/SEC AT 1.70 GRAM/CC (1-INCH DIA SAMPLE)
HEAT OF COMBUSTION	1974 CALORIES/GRAM
FIRE HAZARD	MODERATE, BY SPONTANEOUS CHEMICAL REACTION
IMPACT SENSITIVITY	4.4 LBM BUREAU OF MINES (20 MG SAMPLE) 6.6 IN. PICATINNY ARSENAL (16 MG SAMPLE) 6 INCHES.
ICC CLASSIFICATION	CLASS A
SYNONYM	PENTAERYTHRITOL TETRANITRATE
SPECIFICATION	MIL-P-387

Figure 4-76

encloses the assembly to protect the exposed terminals. The switch assembly has three normally open contacts and one normally closed contact. Firing current is applied to the squib bridgewire through the normally closed contact, which opens when the switch actuates. This action removes power from the squib circuit to prevent possible power drain after the switch actuates. Switch actuation is accomplished by gas pressure generated by the squib. The ICC does not require special handling of the switch assembly because of the small squib charge (approximately 100 mg). The squib has been used extensively in aerospace programs without any known ruptures of the case. The squibs have been qualified per Picatinny Arsenal X PA PD 2145 and per Naval Ordnance Lab. OS 10077, OS 10076, NOTS XS417. See figure 4-77 for safety switches characteristics.

SAFETY SWITCHES CHARACTERISTICS

ITEM	DESCRIPTION
GAS GENERATING CHARGE	
CHEMICAL COMPOUND	LEAD MONONITRORESORCINATE (90%) AND POTASSIUM CHLORATE (10%)
QUANTITY	120 MG. (MAX)
BRIDGEWIRE RESISTANCE	1.8 ± 0.2 OHMS (SWITCH UNASSEMBLED), 1.3 ± 0.3 OHMS (SWITCH ASSEMBLED, WITH SQUIB LEADS SHORTENED).
FUNCTIONING TIME	10 MSEC (MAX)
MAX NO-FIRE CURRENT	0.10 A
MIN ALL-FIRE CURRENT	1.0 A
CLOSED CONTACT CAPACITY	12 A FOR 6 HOURS; 200 A FOR 100 MSEC (SWITCHES WITHOUT HEAT SINK, AT +160°F AND ALL CONTACTS IN SERIES).
OPERATING TEMPERATURE	-65 TO +160°F
DROP TEST	SQUIB SWITCH SUCCESSFULLY COMPLETED 40-FOOT FREE FALL DROP TEST ONTO A STEEL PLATE (207 BRINELL HARDNESS) HAVING A 3-INCH THICKNESS PER MIL-STD-352.

Figure 4-77

SECTION V
S-IVB STAGE

TABLE OF CONTENTS

Introduction	5-1
Structure	5-1
Propulsion	5-8
Pneumatic Control System	5-34
Flight Control	5-36
Electrical	5-56
Instrumentation Systems	5-61
Environmental Conditioning	5-64
Ordnance	5-64

INTRODUCTION.

The second stage, S-IVB, provides thrust from just after first stage burnout and separation, which is approximately 2 min 30 sec after liftoff; thrust is continued until orbital velocity is achieved approximately 10 min after liftoff. During first stage powered flight, the S-IVB stage along with the aft interstage (figure 5-1) provides the load-bearing structure between the S-IB stage and the IU. The propellant tank assembly, the forward and aft skirts, and the thrust structure form the basic stage structure. One J-2 engine mounted on the thrust structure propels the vehicle during second stage flight with 225,000 lbf thrust. The propellant tank assembly is composed of lox and LH_2 tanks separated by a common bulkhead. Most of the stage electrical equipment is mounted around the inner surfaces of the forward and aft skirts (figures 5-2 and 5-3). Each skirt also has an umbilical interface plate for prelaunch ground connection to the various stage systems. Two auxiliary propulsion system modules, mounted on the aft skirt exterior, provide pitch, yaw and roll attitude control during orbital coast periods and roll control during S-IVB burn. The engine-mounted hydraulic system is used for J-2 engine pitch and yaw gimballing to provide attitude control during S-IVB powered flight. The S-IB and S-IVB stages separate at the S-IVB aft skirt and aft interstage joint. The three ullage rockets and four retromotors, which aid this separation, mount respectively on the aft skirt and the aft interstage exteriors. Plumbing and wiring that must pass outside the S-IVB stage are covered by the main tunnel, the auxiliary tunnel, and various fairings on the aft skirt. Figure 5-4 lists a summary of basic stage and stage systems data.

STRUCTURE.

Figure 5-5 shows the primary, load-carrying structural assemblies of the S-IVB stage consisting of the propellant tank, forward and aft skirts, engine thrust structure, and aft interstage in addition to the non-primary structural elements consisting of the main tunnel, auxiliary tunnel, and aerodynamic fairing segments. The primary structural purpose of the stage is to transfer loads imposed by the spacecraft and instrument unit to the S-IB stage as well as maintaining structural integrity for the additional loads generated by the S-IVB stage itself. Local shell structure must also have the ability to support the loads caused by mounting of equipment, protuberances, and externally located systems. The required ultimate safety factor of the stage of 1.4 has been verified by analysis

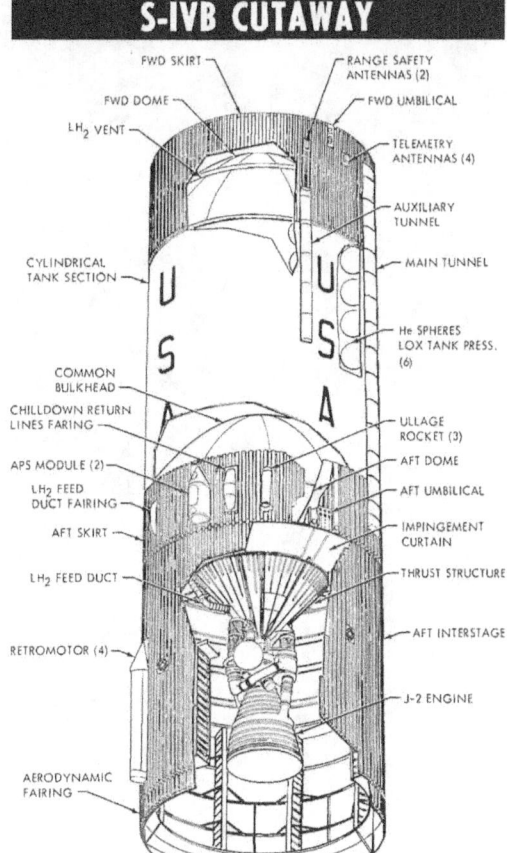

Figure 5-1

and tests. All possible critical shell load conditions including acceptance firing, prelaunch, boost, stage powered flight, and orbital conditions have been investigated. This investigation has established that the structural shell is, in general, critical for the boost condition at the time of maximum αq. The design requirement for each structural assembly, the validity of the design requirement, and the ability of the structural assembly to meet the design requirement have been established. The measured weight of the S-IVB-206 stage (including the S-IVB flight interstage) is 27,771 lbm and it measures 260 in. in diameter by 700 in. in length.

The S-IVB structural material selections were predicated on high strength and cryogenic capabilities, advances in welding and bonding techniques, and a history of previous and similar applied mechanics. The material stock used were casting, sheets, plates, extrusions, forging, and honeycomb. The joining methods were welding, riveting, lockbolting, bonding, and bolt nut/insert removeable

Section V S-IVB Stage

FORWARD SKIRT ELECTRICAL EQUIPMENT INSTALLATION

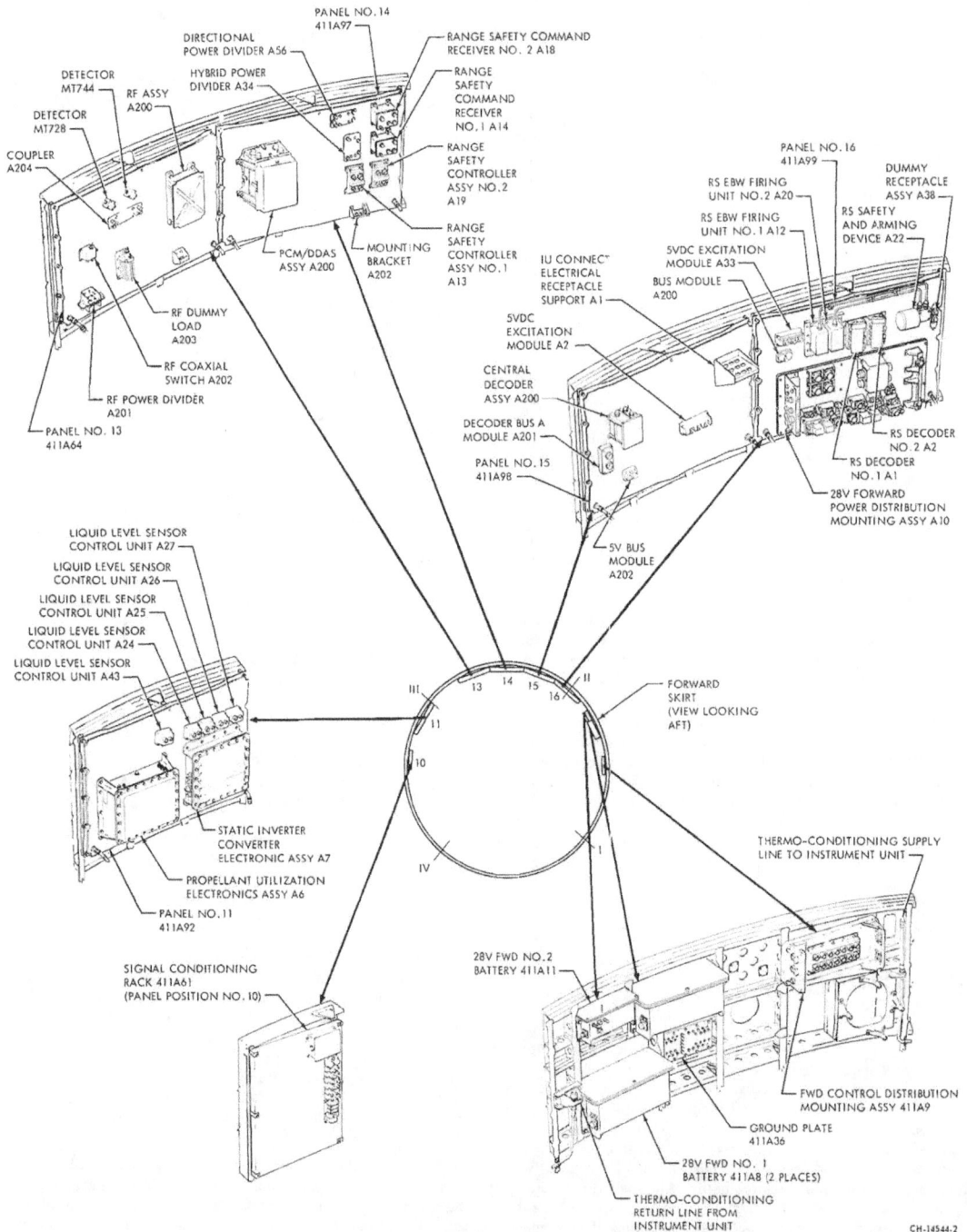

Figure 5-2

Section V S-IVB Stage

AFT SKIRT ELECTRICAL EQUIPMENT INSTALLATION

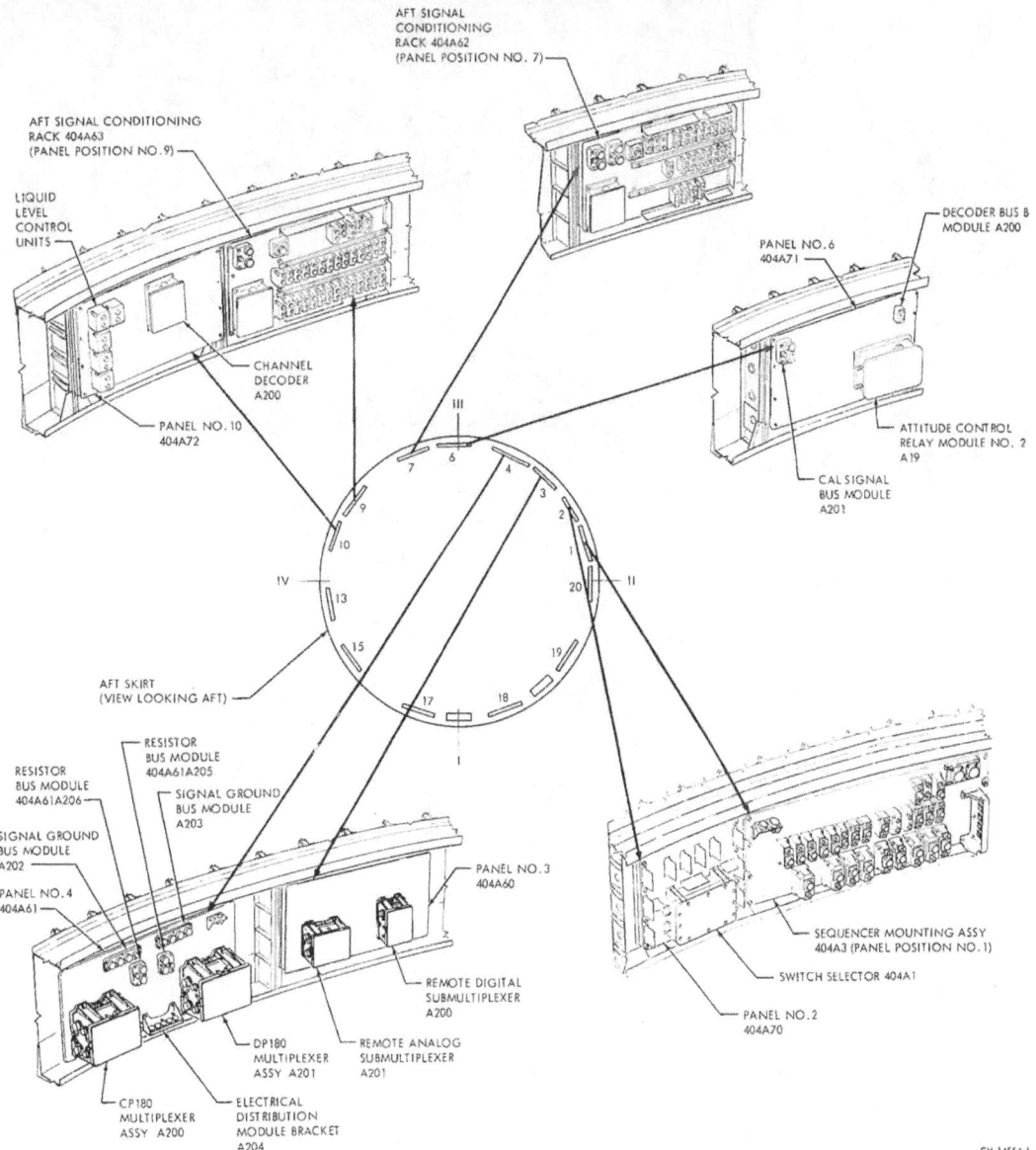

Figure 5-3 (Sheet 1 of 2)

attachments. Aluminum alloys (2014-0 heat treated to 2014-T6, 2014-T651, and 2014-T652) were used in fabricating the welded propellant tank. The skirts, thrust structure, and aft interstage assemblies are skin-stringer-ring frame fabrication. Aluminum alloys (7075-0 heat treated to 7075-T6, 7075-T6 and A356) were used and joined by mechanical fasteners. The common bulkhead is a honeycomb composite structure constructed of welded spherical formed 2014-T6 aluminum alloy face assemblies bonded to heat resistant phenolic honeycomb core material.

PROPELLANT TANK STRUCTURE.

The propellant tank structure shown in figure 5-6 consists of a 268 in. long cylindrical section constructed by butt welding longi-

5-3

Section V S-IVB Stage

AFT SKIRT ELECTRICAL EQUIPMENT INSTALLATION

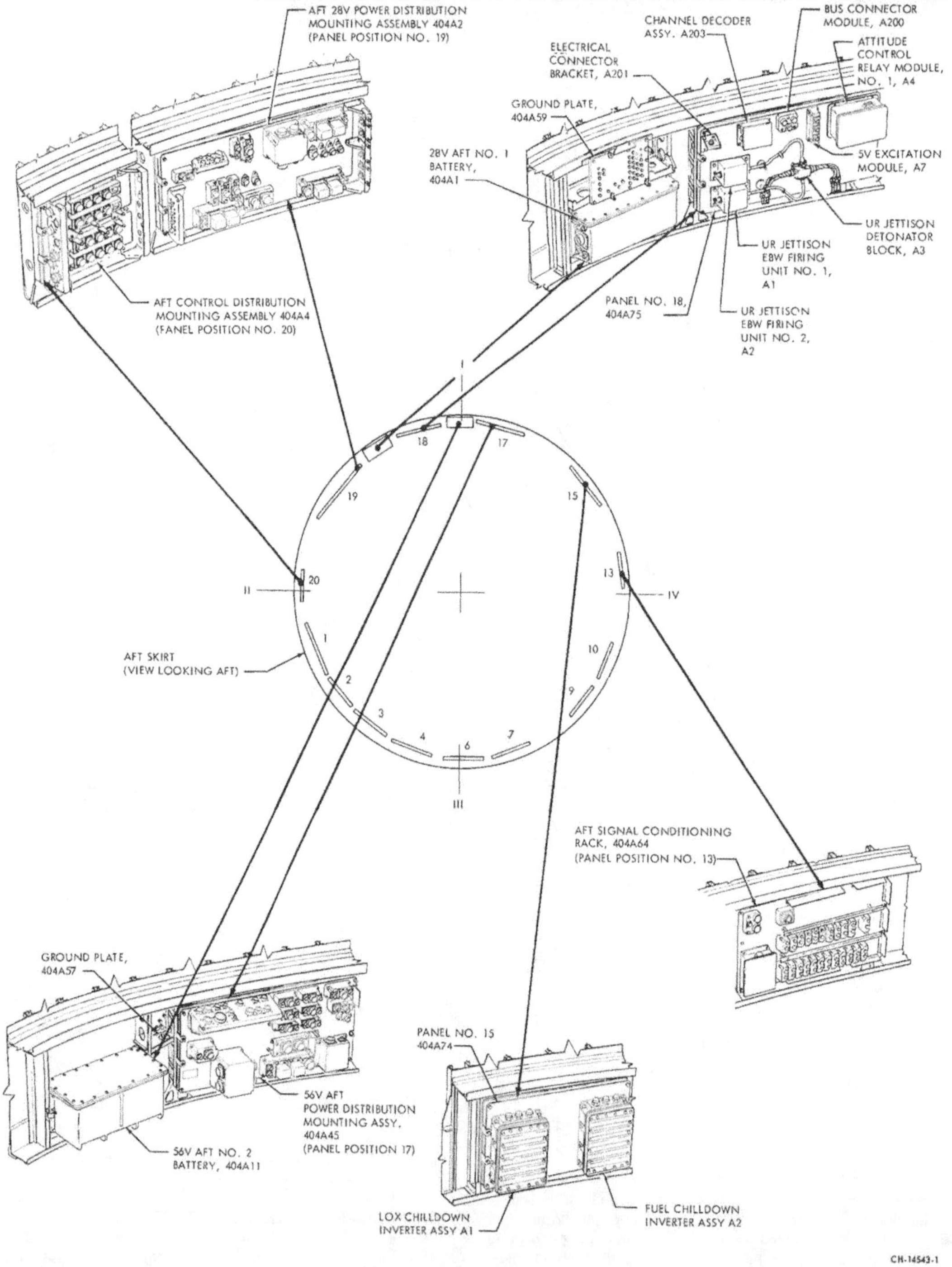

Figure 5-3 (Sheet 2 of 2)

S-IVB STAGE DATA SUMMARY

DIMENSIONS		**ULLAGE ROCKETS**	
LENGTH	59.1 FT	NUMBER OF ENGINES	3
DIAMETER	21.7 FT	THRUST (PER ENGINE)	3,460 LB_f @ 1,000,000 FT
MASS		BURNTIME	3.9 SEC
DRY STAGE	22,150 LB_m	PROPELLANT	SOLID
LOADED STAGE	256,800 LB_m	LOCATION	120 DEG INTERVALS AROUND S-IVB AFT SKIRT. ENGINES CANTED OUTWARD 35 DEG. JETTISONED 15 SEC AFTER STAGE SEPARATION.
AT ORBITAL INJECTION	28,200 LB_m		
S-IB/S-IVB INTERSTAGE	5,661 LB_m		
PROPELLANT LOAD	232,200 LB_m	**PRESSURIZATION SYSTEM**	
ROCKET ENGINES		OXIDIZER CONTAINER	HELIUM
J-2 ENGINES		FUEL CONTAINER	INITIAL: HELIUM FROM GROUND SOURCE, GH_2 FROM J-2 ENGINE DURING S-IVB BURN
BURNTIME	440 SEC		
THRUST	225,000 LB_f @ 200,000 FT	OXIDIZER PRESSURE	
		PREFLIGHT	37 TO 40 psia
PROPELLANT	LOX AND LH_2	INFLIGHT	37 TO 40 psia
MIXTURE RATIO	5.5:1 (MAX), 4.8:1 (MIN)	FUEL PRESSURE	
EXPANSION RATIO	27:1	PREFLIGHT	31 TO 34 psia
OXIDIZER NPSH	35 PSIA MIN	INFLIGHT	26.5 TO 29.5 psia
FUEL NPSH	27 PSIA MIN	**ENVIRONMENTAL CONTROL SYSTEM**	
GAS TURBINE PROPELLANT	LOX AND LH_2	PREFLIGHT AIR CONDITIONING	AFT COMPARTMENT AND FORWARD SKIRT
FUEL TURBINE SPEED	27,000 RPM	PREFLIGHT GN_2 PURGE	AFT COMPARTMENT AND FORWARD SKIRT
OXIDIZER TURBINE SPEED	3,600 RPM	FLIGHT	UNIT CONDITIONING SYSTEM
HYDRAULIC SYSTEM		**ASTRIONICS SYSTEMS**	
ACTUATORS	HYDRAULIC (TWO PER ENGINE)	GUIDANCE	PATH ADAPTIVE GUIDANCE MODE THRU THE IU DURING S-IVB BURN
GIMBAL ANGLE	±7 DEG SQUARE PATTERN	TELEMETRY LINK	258.5 MHz
GIMBAL RATE	8 DEG/SEC EACH PLANE	ELECTRICAL	BATTERIES: 28 Vdc (3 ZINC SILVER-OXIDE) 56 Vdc (1 ZINC SILVER-OXIDE) STATIC INVERTER-CONVERTER: 28 Vdc TO 115 Vac, 400 Hz SINGLE PHASE, 25 Vpp, 400 Hz SQUARE WAVE AND 117, 21, AND 44.2 Vdc. CHILLDOWN INVERTER (2): 56 Vdc TO 56 Vac, 3 PHASE, 400 Hz QUASI-SQUARE WAVE 5-VOLT EXCITATION MODULE: 28 Vdc TO 5 Vdc, −20 Vdc AND 10 Vpp, 2000 Hz SQUARE WAVE 20-VOLT EXCITATION MODULE: 28 Vdc TO 20 Vdc
GIMBAL ACCELERATION	171.5 DEG/SEC^2		
APS			
NUMBER OF ENGINES	6 (3 PER MODULE)		
THRUST (PER ENGINE)	150 LB_f VACUUM		
PROPELLANTS	HYPERGOLIC (MMH AND N_2O_4)		
PROPELLANT LOAD	62 LB_m PER MODULE		
LOCATION	AFT SKIRT AT FIN POSITIONS I AND III	**RANGE SAFETY SYSTEM**	PARALLEL ELECTRONICS, REDUNDANT ORDNANCE
RETROMOTORS			
NUMBER OF ENGINES	4		
THRUST (PER ENGINE)	36,720 LB_f @ 200,000 FT	**Note**	
BURNTIME	1.52 SEC	ALL MASSES ARE APPROXIMATE	
PROPELLANT	SOLID, TP-E-8035 (THIOKOL)		
LOCATION	90 DEG INTERVALS		

Figure 5-4

tudinally the seven cylindrical section segments. The cylindrical section segments are made from 0.750-in. plates, which are first mechanically milled to make the waffle pattern, and then brake-formed to a 130-in. radius. The waffle pattern consists of pockets 9.5 in. on center that are oriented ±45 deg with respect to the vehicle longitudinal axis. These pockets are surrounded by ribs that are 0.627 in. high and 0.144 in. wide. The skin or web thickness at the bottom of the pocket is 0.123 in., and the weld land areas at the edges of these segments are 0.252 in. thick. The forward dome is fabricated from nine "pie-shaped" gore segments welded together and to the jamb manhole, which provides access to the LH_2 tank. These segments are made of 0.150-in. thick aluminum alloy sheet stock, which is formed to a 130-in. radius then chemically milled to a skin thickness of 0.060 in., with a thickness building up to 0.118 in. at the weld-land areas. The increased weld-land thickness is for the purpose of maintaining the stress levels below the yield point of the material in the "as welded" state.

The aft dome, like the forward dome, is fabricated from nine "pie-shaped" gore segments. These segments are made from 0.280-in. thick sheet stock and are chemically milled after forming. The weld-land areas, including the rings around the dome where the common bulkhead and the thrust structure attach, are milled to 0.191 in. The liquid hydrogen portion of the dome (the area forward of the common bulkhead joint) is chemically milled into a waffle pattern with a web thickness of 0.082 in. The rib height is a minimum of 0.245 in. The 0.625-in. wide ribs are located approximately 5.5 in. on center at the maximum, and are oriented at 0 deg and 90 deg with respect to the vehicle longitudinal axis. The web area is milled to a thickness of 0.086 and 0.092 in. for the areas forward and aft of the thrust structure joint, respectively. Access to the liquid oxygen tank is through the sump jamb in the bottom of the dome.

Many honeycomb composite structures are used in the vehicle, the most unique is the common bulkhead that provides both the structural and thermal separation between the LH_2 and the lox tanks. The common bulkhead forward and aft face sheets are similar in construction to the other domes, i.e., from nine chemically milled, welded pie segments. The circumferential edges of the faces are butt-welded to "Y-shaped" extruded rings, which attach the bulkhead to the aft dome. The two face sheets are separated by and adhesively bonded to reinforced plastic honeycomb core material. The extruded Y-rings are joined together by a common seal weld and are attached to the aft dome by mechanical fasteners and by the LH_2 tank and lox tank seal welds.

As stated, all propellant weld areas are thicker than the basic membranes to appropriately reduce stresses transferred through the welds and to allow for lower material allowables due to the welding process. All welds are inspected by use of dye penetrant and X-ray techniques. The final quality control operation on the propellant tank consists of a hydrostatic proof test after which both weld inspection and tank leak checks are performed. One significant problem involving the propellant tank jamb welds was encountered. After hydrostatic proof testing of the S-IVB-502 tank, a few small cracks were discovered in the jamb weld of the lox tank. It was established that these areas had been repaired during manufacture. As a result of this conclusion, all vehicles already manufactured were inspected in the jamb areas of both the lox and LH_2 tank domes. Small cracks were found in S-IVB-201, -202, and the bulkhead test specimen. It was decided, even though all stages had not developed cracks and had successfully passed hydrostatic proof testing, that it would be advisable to reinforce the jamb weld areas on all stages. This was accomplished in two phases. The reinforcement of the earlier stages including S-IVB-206 was accomplished through the addition of internal and external doublers over the jamb weld area of both the lox and LH_2 tank domes. The doublers were attached with mechanical fasteners and also adhesively bonded. This jamb weld reinforcement is shown in a detail view in figure 5-6. Two 5.5-ft domes, including a jamb weld and its reinforcement, were made to the lox dome and to the LH_2 dome configurations. The two domes were then tested to planned destruction to determine their structural capability. Both analysis and test results established that the structural capability was well in excess of the design requirements. In addition, leakage checks were performed to provide assurance that manufacturing capability was adequate to provide a leak-free assembly. The second phase of the reinforcement program consisted of redesign of the domes and utilizing thicker weld lands to eliminate the need for the doublers. The redesign was first accomplished on the S-IVB-209 LH_2 dome and the S-IVB-210 lox dome.

FORWARD SKIRT, AFT SKIRT, AND AFT INTERSTAGE.

These three assemblies, similar in configuration, are cylinders constructed of 7075-T6 aluminum alloy skin, external stringers, and internal frames. Their diameter is 260 in., but their lengths vary. The forward skirt has 108 stringers while the aft skirt and aft interstage each have 112 stringers. The forward skirt assembly extends forward 122 in. from the forward skirt attaching ring of the propellant tank structure. The forward skirt provides an attachment plane for the instrument unit and mounting provisions for electronic equipment on thermo-conditioning panels. After vehicle assembly on the launch pad, the instrument unit access door provides entry to the forward skirt interior. The aft skirt assembly extends aft 85.5 in. from the aft skirt attaching ring of the propellant

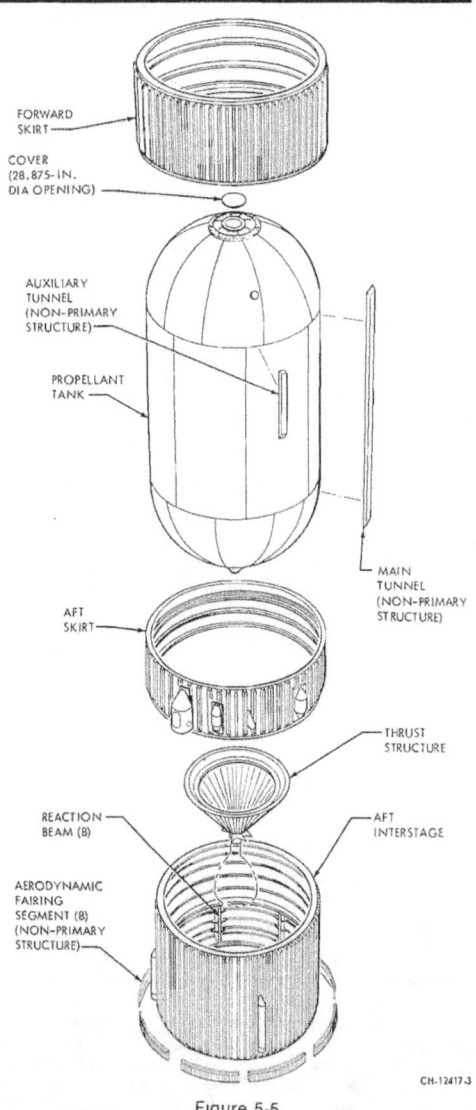

Figure 5-5

Section V S-IVB Stage

S-IVB PROPELLANT TANK STRUCTURE

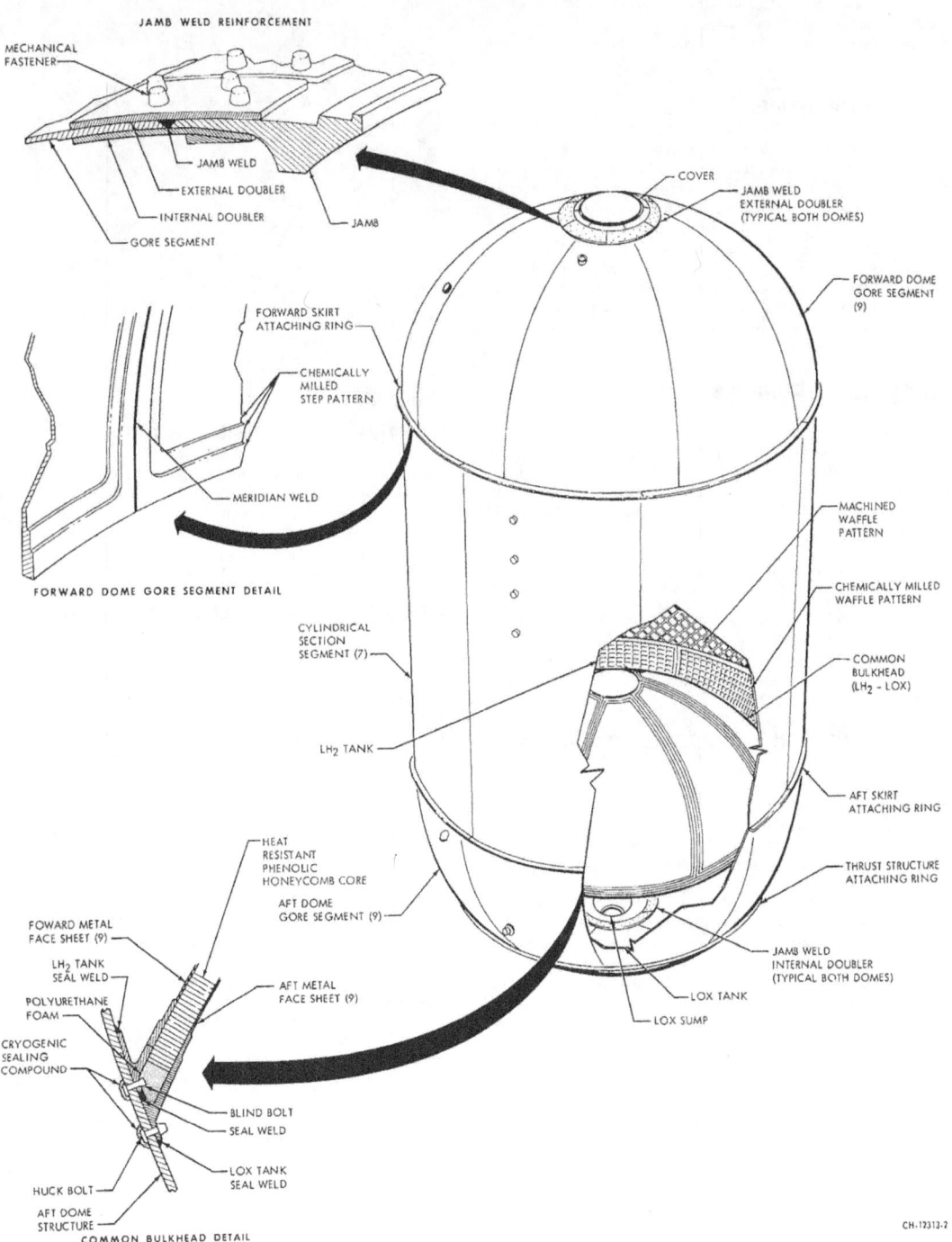

Figure 5-6

Section V S-IVB Stage

tank structure to the aft interstage. Ullage rockets and the auxiliary propulsion system mount on the aft skirt exterior; electrical equipment panels mount on the skirt interior. The aft interstage assembly is 224.5 in. in length and provides the interface between the S-IVB stage and the S-IB stage. The aft interstage has eight vertical internal reaction beams that translate the loads to the S-IB 220-in. diameter structure points. Four retromotors are mounted externally. The separation plane for the S-IVB stage is the aft end of the aft skirt.

THRUST STRUCTURE.

Figure 5-7 shows the truncated-cone-shaped thrust structure. The configuration of the thrust structure is similar to the skirt assemblies in that 7075-T6 aluminum alloy skin, external stringers, and internal frames are utilized. The structure measures 83 in. high, by 168 in. at the base. The J-2 engine is attached at the small diameter end of the thrust structure through the use of an A-356 aluminum casting. The forward end of the thrust structure is attached to the aft dome of the propellant tank assembly. Electro-mechanical and mechanical systems equipment mounts on the inside and outside surfaces of the thrust structure. Two doors provide access to the thrust structure interior.

LONGITUDINAL TUNNELS.

The longitudinal tunnels (main tunnel and auxiliary tunnel) house wiring, pressurization lines, and propellant dispersion system shaped charges. The tunnel covers are made of 7075-T6 aluminum alloy skin stiffened by internal ribs. These structures do not transmit primary shell loads but act only as a fairing reacting to local aerodynamic loading.

AERODYNAMIC FAIRING.

The aerodynamic fairing on the aft end of the interstage is a short cylinder, 260 in. in diameter, made from eight 7075-T6 corrugated aluminum alloy skin panels. This structure does not transmit primary shell loads but must maintain structural integrity when loaded by local aerodynamic pressure.

PROPELLANT TANK CRYOGENIC INSULATION.

Another design requirement for the propellant tank and common bulkhead is that they provide insulation for the LH_2 tank and in the case of the common bulkhead that it prevent lox freezing due to LH_2 temperatures during a ground hold. Figure 5-8 shows the methods and materials used to provide the necessary insulation for the LH_2 tank. Cross-sections show insulation details of the forward dome, aft dome, forward dome access door, aft dome and common bulkhead area, and cylindrical section and aft dome area. Generally, the insulating tile (a polyurethane foam and fiberglass composite) bonds to the tank skin with an epoxy base adhesive system. In certain areas shown, balsa pads are used in conjunction with tile to line the tank walls with insulating materials. An insulating gap filler material consisting of a glass fiber and polyurethane adhesive composite is used to fill gaps between tiles and around fasteners. Liners consisting of woven glass impregnated with polyurethane adhesive are bonded to the tile and balsa insulating material. In critical areas around brackets and joints the liner is covered by doublers of the same material.

ABLATIVE INSULATION.

Figure 5-9 shows the ablative coating material and the patterns and depth to which it is applied to the external skin of the stage.

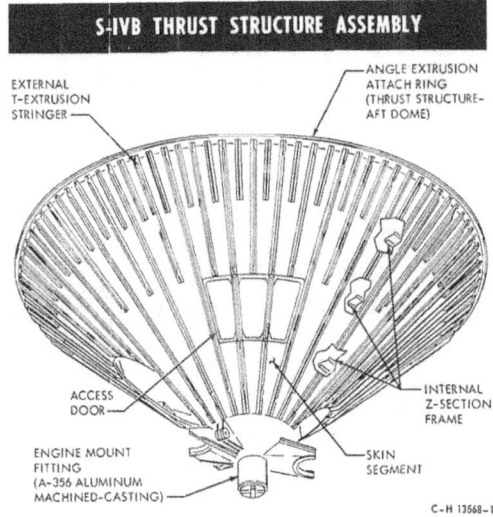

Figure 5-7

FORWARD DOME DEBRIS SHROUD.

A debris shroud surrounds the forward dome of the propellant tank 40.125 in. above the equatorial plane of the dome. Eight nylon cloth shroud segments 9 in. wide by 89 in. long are stitched together with dacron threads and reinforced at their joint with eight nylon cloth splice segments 4 in. by 9 in. The shroud segments are attached to the dome with Velcro tape adhesively bonded to the dome.

RETROROCKET IMPINGEMENT CURTAIN.

A retrorocket impingement curtain is used to shield, at stage separation, the area between the aft dome and aft skirt. The curtain installation spans the area between the aft end of the aft skirt and the engine thrust structure junction with the aft dome. The basic material used in the construction of the curtain is glass cloth. Tape is used to seal openings in the curtain around boots, slots in aluminum, and other openings.

PROPULSION.

Prime thrust for the S-IVB stage is provided by the main propulsion system, which consists of a bipropellant J-2 rocket engine, fuel system, oxidizer system, tank pressurization systems, engine chilldown systems, and a propellant utilization (PU) system. The J-2 engine, burning lox and LH_2, provides the thrust during the second boost phase of flight to inject the S-IVB stage and payload into orbit. In addition to supplying prime thrust, the J-2 engine provides thrust-vector steering (pitch and yaw) for inflight course correction during powered flight. Command signals from the IU guidance and control system effect the flight steering by hydraulically gimballing the engine up to ± 7 deg from the stage longitudinal axis. The PU system controls the mass of propellants loaded into the stage by providing inputs to the propellant loading system.

The stage propellant tanks consist of a cylindrical section enclosed by hemispherical bulkheads with an internal common bulkhead dividing the structure into fuel and lox tanks. The tank capacity is designed to satisfy an engine mixture ratio of 5 to 1 (oxidizer

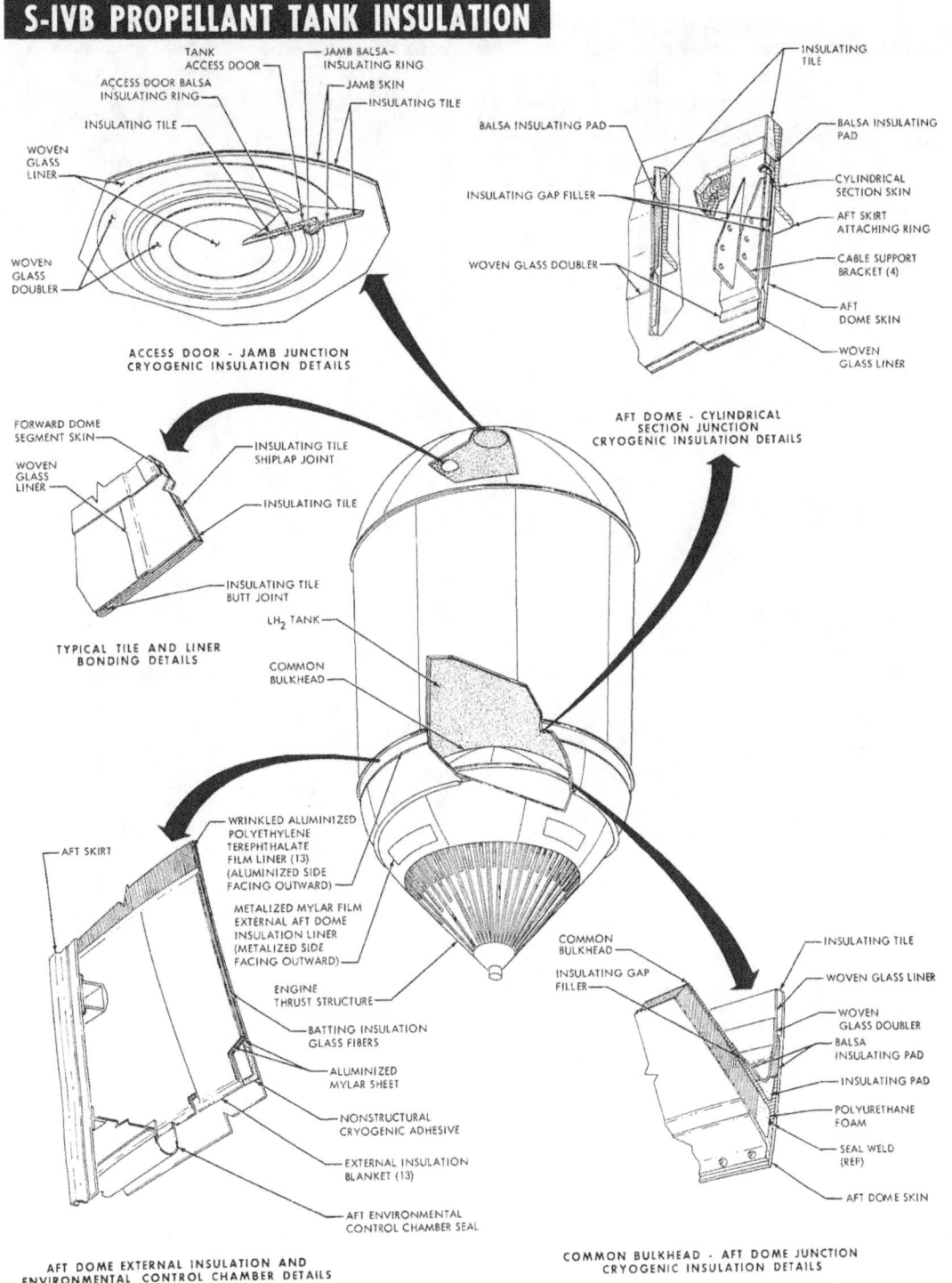

Figure 5-8

Section V S-IVB Stage

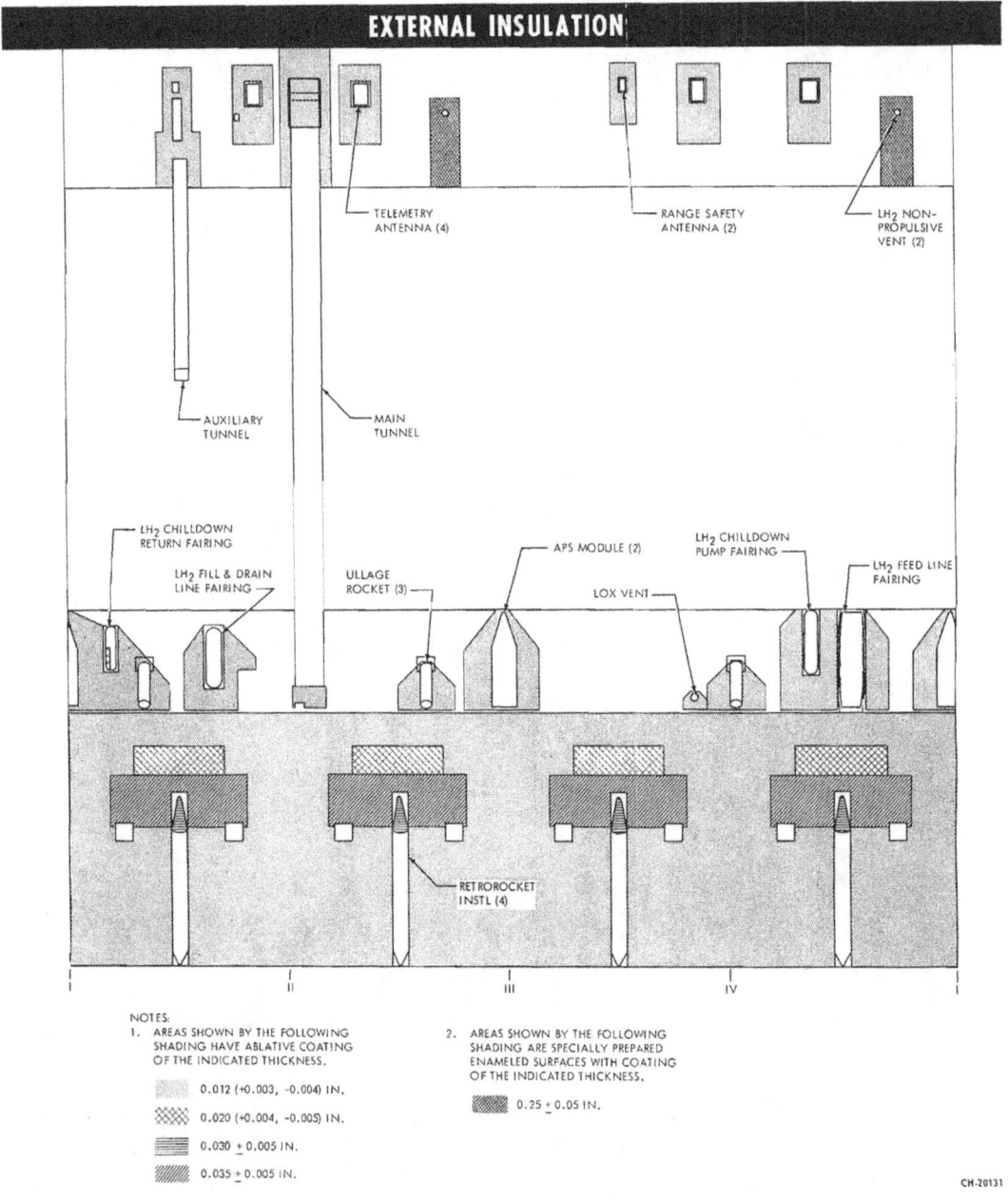

Figure 5-9

to fuel by weight). The fill systems are sized to minimize countdown time and also to be compatible with the other systems loading. Lox design flow is 1000 gpm; LH_2 design flow is 3000 gpm. Initial fill rates are slower to accomplish tank chilldown and to prevent ullage pressure collapse. Final fill rates are also slower to provide tank "topping" during the terminal count.

The propellant tank vent systems were designed to protect the tank structure under all conditions of propellant tank loading, stage power flight, and orbital venting. During loading, tank pressures are maintained well below the normal tank prepressurization levels. The negative pressure differential across the common bulkhead between the fuel and oxidizer tanks is the limiting factor. This

Section V S-IVB Stage

J-2 ENGINE

1. LOX INLET DUCT
2. THRUST CHAMBER INJECTOR
3. LOX TURBOPUMP DISCHARGE LINE
4. PURGE CONTROL VALVE
5. MAIN OXIDIZER VALVE
6. LOX BLEED VALVE
7. SEQUENCE VALVE (LOX)
8. LOX BOOTSTRAP LINE
9. FUEL INLET DUCT
10. ASI FUEL LINE
11. PNEUMATIC CONTROL PACKAGE
12. GAS GENERATOR CONTROL VALVE
13. FUEL BLEED VALVE
14. GAS GENERATOR COMBUSTOR
15. FUEL BOOTSTRAP LINE
16. FAST SHUTDOWN VALVE
17. ELECTRICAL CONTROL PACKAGE
18. FUEL TURBOPUMP
19. FUEL TURBOPUMP TURBINE
20. MAIN FUEL DUCT
21. LOX TURBINE BYPASS DUCT

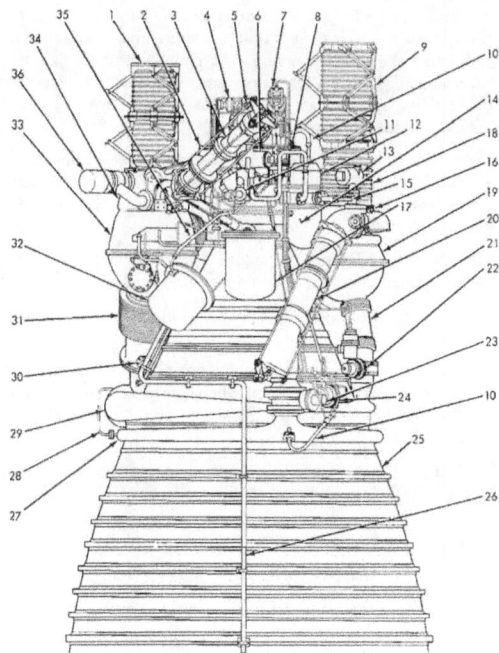

22. LOX TURBINE BYPASS VALVE
23. MAIN FUEL VALVE
24. SEQUENCE VALVE (FUEL)
25. THRUST CHAMBER
26. LOX PUMP PRIMARY SEAL DRAIN
27. FUEL INLET MANIFOLD
28. START TANK RECHARGE LINE (BLOCKED)
29. EXHAUST MANIFOLD
30. HEAT EXCHANGER COLD HELIUM INLET
31. HEAT EXCHANGER
32. PRIMARY INSTRUMENTATION PACKAGE
33. LOX TURBOPUMP TURBINE
34. LOX TURBINE
35. ACCUMULATOR SUPPLY LINE
36. MIXTURE RATIO CONTROL VALVE
37. INTEGRAL START TANK
38. TANK SUPPORT AND FILL VALVE PACKAGE
39. FUEL PUMP DRAIN LINE
40. LOX PUMP DRAIN LINE
41. START TANK DISCHARGE HOSE
42. START TANK DISCHARGE VALVE
43. AUXILIARY INSTRUMENTATION PACKAGE

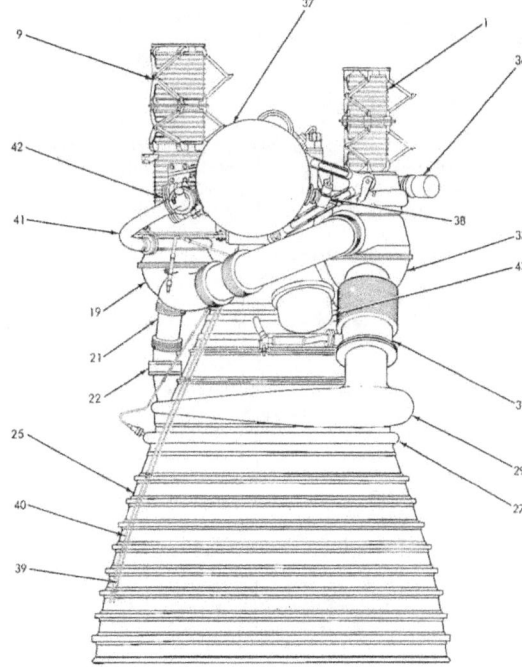

Figure 5-10

Section V S-IVB Stage

ENGINE J-2033 PERFORMANCE TAG VALUES		
PARAMETER	UNIT OF MEAS	NOMINAL TAG VALUES
THRUST	LB	229,294
SPECIFIC IMPULSE	SEC	423.4
CHAMBER PRESS	PSIA	770.6
LOX FLOWRATE	LB/SEC	458.78
FUEL FLOWRATE	LB/SEC	82.78
MIXTURE RATIO	O/F	5.542
GAS GEN LOX FLOWRATE	LB/SEC	3.46
GAS GEN FUEL FLOWRATE	LB/SEC	3.55
LOX PUMP SPEED	RPM	8,682
FUEL PUMP SPEED	RPM	26,494

Figure 5-11

differential pressure is never allowed to exceed 3.0 psid for normal controlled operation.

All components in the propulsion system are designed to function in an explosive atmosphere without providing a source of ignition or electromagnetic interference. Potential leak sources have been lowered and component weight reduced by the modular concept, in which components such as pressure regulators, check valves and solenoid valves are packaged in a single body, thus eliminating many connections.

A temperature conditioning system is provided in both the LH_2 and lox feed systems to provide temperature stabilization at the engine pump inlets to meet minimum net positive suction head requirements.

J-2 ENGINE SYSTEM.

The 225,000 lbf thrust, high performance lox and LH_2 J-2 engine (figure 5-10) powers the S-IVB stage/IU/payload during the second boost phase of powered flight. The J-2 engine accelerates the payload to orbital velocity during its 7 min 30 sec (approximately) burn. Lox, LH_2, and helium constitute the only fluids used on the engine, because the extremely low operating temperature of most of the engine components prevents the use of lubricants or other fluids. The engine features a single tubular-wall bell-shaped thrust chamber, and two independently driven turbopumps. A single gas generator, powered by lox and LH_2 bled off the main LH_2 and lox turbopump discharge lines, drives both the lox and LH_2 turbopumps in series. A mixture ratio control value achieves two mixture ratio levels by passing lox from the discharge side of the turbopump to the inlet. A pneumatic control system provides regulated helium for engine valve operations from a helium tank mounted on the engine. An electrical control system, containing solid-state logic elements, sequences the engine start and shutdown operations. The S-IVB stage electrical power system supplies electrical power for engine operation. An engine-mounted heat exchanger, located in the lox turbopump turbine exhaust duct, heats helium from the S-IVB stage cold helium bottles for pressurization of the fuel tank. A bleed line from the thrust chamber fuel manifold taps off GH_2 for LH_2 tank pressurization. The J-2 engine consists of the following systems: propellant feed, pneumatic-electrical control, gas generator and exhaust, thrust chamber and gimbal, and flight instrumentation. See figures 5-11 and 5-12 for J-2 engine performance values.

Characteristics.

J-2 engine, serial number J2046, will fly on the SL-2 mission. The engine measures 80.75 in. in diameter and 133 in. in length. A gimbal assembly and two hydraulic actuators attach the engine to the S-IVB stage thrust structure. The gimbal assembly permits the entire engine to gimbal for thrust vector control. Engine weight with accessories is 3536 lbm dry and 3697 lbm wet. At burnout the engine weighs 3665 lbm.

Acceleration and Velocity. Engine gimbal angular acceleration shall not exceed 80 rev/sec^2. During the boost phase of flight, before engine ignition, the engine shall not be subjected to accelerations greater than presented in figure 5-13.

Flight Load Limits. During the engine operation phase of flight, the following load conditions should not be exceeded.

a. The longitudinal and lateral inertia load limits as shown in figure 5-14.

b. Aerodynamic moments during gimbal operation shall not exceed the following and shall be such that when combined with all other loads, other maximum limitation specifications in this manual shall not be exceeded.

1. No. 1 or no. 2 gimbal axis (see figure 5-15); aerodynamic moments of ±110,000 in.-lb.

2. X or Z engine axis (see figure 5-15); aerodynamic moments of ±210,000 in.-lb.

c. Maximum aerodynamic pressure on the thrust chamber shall not exceed 500 lb/ft^2.

Gimballing and Vehicle Loads. The designed limit of forward acceleration with respect to engine gimbal angle is shown in figure 5-16. This limit occurs simultaneously with the lateral acceleration relationship to forward acceleration as shown in figure 5-14.

AVERAGE J-2 ENGINE PERFORMANCE VALUES				
ENGINE PARAMETER	UNIT OF MEAS	[1] MEAN VALUE	[1] STD DEV(%) ENG-TO-ENG	[1] STD DEV(%) RUN-TO-RUN
THRUST				
(1) ALTITUDE	lb	225,000 [2]	--	--
(2) SEA LEVEL	lb	156,400	--	--
SPECIFIC IMPULSE				
(1) ALTITUDE	sec	423.8	0.18	0.16
(2) SEA LEVEL	sec	293.81	--	--
MIXTURE RATIO	O/F	5.50 [2]	--	--
RATED DURATION	sec	500	--	--
LOX FLOWRATE (PUMP INLET)	lb/sec	449.3	0.18	0.16
FUEL FLOWRATE (PUMP INLET)	lb/sec	81.68	0.18	0.16
CHAMBER PRESS. (INJECTOR END)	psia	762.6	0.85	0.21
CHAMBER PRESS. (NOZZLE - STAGNATION)	psia	702.2	0.85	0.21
AREA EXPANSION RATIO		27.12:1	0.23	--

[1] BASED ON ROCKETDYNE ACCEPTANCE TEST DATA

[2] RATED CONDITIONS

Figure 5-12

Section V S-IVB Stage

General Purge Requirements.

The engine systems receive purges prior to loading propellants or preconditioning the thrust chamber, to clear the systems of moisture and/or gases, which would solidify or otherwise prove hazardous when the engine hardware is chilled.

Turbopump and Gas Generator Purge Requirements. Fifteen minutes prior to dropping propellants into the engine ducts the turbopump and gas generator receive a 6-scfm (nominal flowrate) helium purge with a temperature range of 50 to 200° F and an 82- to 125-psia pressure range. Helium purges the fuel pump seal cavity, the fuel turbopump turbine seal cavity, the gas generator fuel cavity, and the lox turbopump turbine seal cavity. All four purge lines contain check valves to prevent reverse flow of propellants into the purge manifold. All except the fuel seal cavity purge line contain an orifice downstream of the check valves. The pneumatic control system supplies the helium purge. See Pneumatic Control System for additional information.

Oxidizer Dome, Gas Generator Oxidizer Injector, and Oxidizer Intermediate Seal Purge. Approximately 15 min before dropping propellants into the engine ducts, the oxidizer dome and GG oxidizer injector receive a 230-scfm (nominal flowrate) helium purge. An electrical command from the component test helium control solenoid switch initiates the purge by opening the helium control valve on the pneumatic control package. The control package regulates helium from the integral start tank to 400 ±25 psig. Helium enters the oxidizer dome through an orifice and a check valve in the downstream side of the main lox valve housing and escapes through the thrust chamber injector. The GG oxidizer injector purge enters the gas generator through a check valve and escapes through the turbine exhaust system. The oxidizer pump intermediate seal purge is operative any time the pneumatic control package is activated. The purge flowrate is 2600 to 2700 scim Helium purges the turbopump area between the pump housing and the turbine inlet manifold. The three purges continue for 15 min and are terminated just before lox loading begins approximately 5 hr before liftoff. During flight, the J-2 engine start signal initiates the three purges prior to lox entry into the lox dome and gas generator. The oxidizer dome purge and GG oxidizer injector purge continue for just over 1 sec when the mainstage control valve

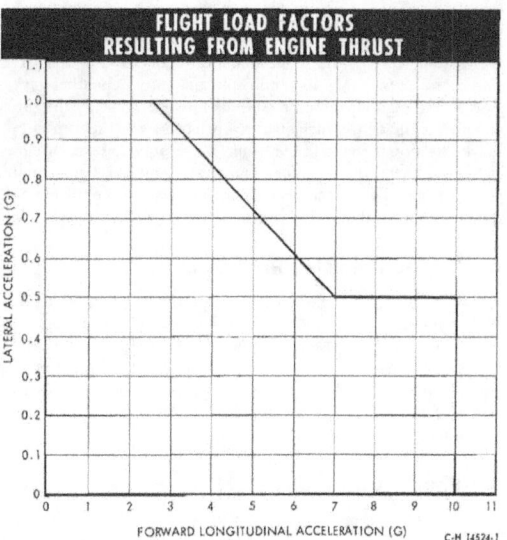

Figure 5-14

terminates the purges by closing the purge control valve. The oxidizer intermediate seal purge continues throughout J-2 engine operation. At J-2 engine cutoff, the mainstage control valve deenergizes and permits the purge control valve to reinstate the oxidizer dome and GG oxidizer injector purges, which terminate 1 sec after the cutoff signal when the helium control valve closes and shuts down the pneumatic control package operation. Oxidizer intermediate seal purge ends at helium control valve closure.

Thrust Chamber Jacket Purge and Chilldown. The thrust chamber jacket purge begins approximately 15 min before liftoff. Helium at 50 psig and ambient temperature enters the engine fuel inlet manifold and flows through the thrust chamber jacket cooling tubes escaping through the thrust chamber injector. S-IVB pneumatic console 433 supplies the purge through the aft umbilical panel at 0.01 lbm/min. The thrust chamber purge valve switch opens the supply valve. Open and closed indicators on the engine panel monitor the supply valve position. This purge lasts for 5 min and

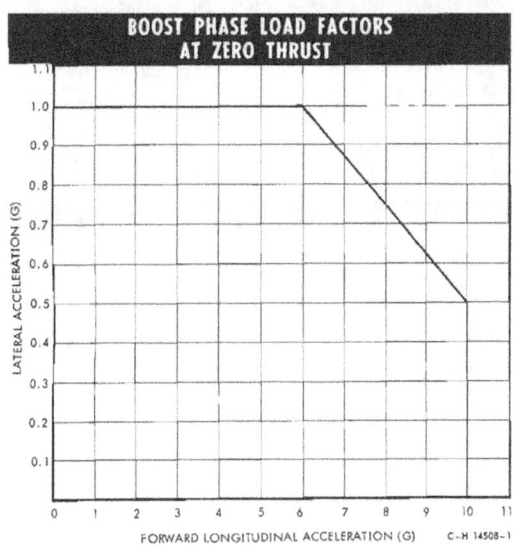

Figure 5-13

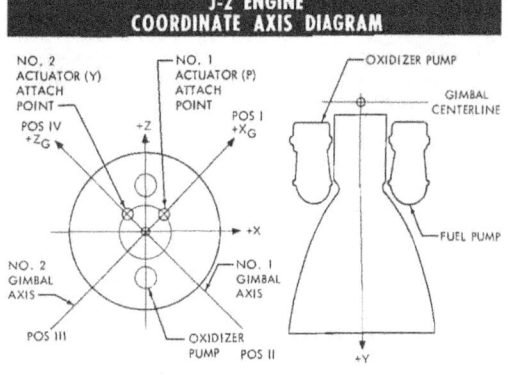

Figure 5-15

5-13

Section V S-IVB Stage

then the thrust chamber chilldown operation begins. Cold helium at approximately -320° F (min) and 1000 psig (max) enters the S-IVB stage through the same umbilical connection as the jacket purge and flows at 15 lbm/min into the thrust chamber jacket. The S-IVB engine panel also controls this operation by the thrust chamber chilldown switch, which opens the cold helium supply valve. Position switches in the supply valve provide open or closed signals to the thrust chamber chilldown indicator lights on the engine panel. Thrust chamber chilldown continues until the S-IB ignition command at T-3 sec terminates the chilldown operation.

J-2 Engine Predicted Performance.

Predicted J-2 engine ignition will occur about 2 min 30 sec into flight with a guidance cutoff occurring at about 10 min into flight. If however, the guidance cutoff does not occur at that time, the LH_2 depletion cutoff sensors in the LH_2 tank are armed to prevent propellant starvation. Thrust decay causing TOPS to deactivate will also initiate cutoff. See figure 5-17 for predicted J-2 engine performance values.

J-2 Engine Operation.

Approximately 2 min 30 sec into flight, the S-IVB stage switch selector issues the engine start signal to the electrical control package on the J-2 engine (figure 5-18). The electrical control package performs all the sequencing functions for proper engine operations and requires only dc power and start and cutoff commands for operations. The engine start signal and an engine ready signal initiate engine operation. An engine ready circuit monitors the conditions and events that are necessary for engine start and issues an engine ready signal when all pre-start conditions have been met. Those prerequisites are: oxidizer turbine bypass valve open; connectors installed; helium control valve, ignition phase control valve, start tank discharge valve control valve, ASI system, mainstage control valve, and GG spark system all deenergized; and absence of an engine running signal. The LVDC also issues an engine ready bypass signal immediately prior to engine start. The start command energizes spark exciters in the sequence controller that provide electrical energy to spark plugs in the gas generator and the augmented spark igniter system. A start tank discharge delay timer (0.640 ± .030 sec) starts and the helium control valve and ignition phase control valve energize simultaneously. 3000-psig helium from the helium tank, in the integral start tank, flows through the helium control valve into the pneumatic control package. Roughing, primary, and control regulators reduce the pressure to 400 ± 25 psig for valve control. Helium from the primary regulator outlet flows to continuously purge the lox turbine intermediate seal during J-2 engine operations. Regulated helium from the pneumatic control package flows into the jacket of the primary flight instrumentation package, which serves as an accumulator for the control pressure system. In the event of supply pressure failure a check valve in the regulator outlet will prevent control pressure loss and the accumulator would then supply a large enough volume of helium to effect safe shutdown of the engine. The helium also closes the fuel and lox bleed valves, and flows through the purge control valve to purge the lox dome and GG oxidizer injector. Helium flows through the normally open port of the mainstage control valve to the main lox valve closing port, to the opening ports of the purge control valve, and the lox turbine bypass valve. Helium flows through the energized ignition phase control valve (normally closed port) to open the main fuel valve and augmented sparkigniter (ASI) oxidizer valve, and to the sequence valves on the main fuel and lox valves. LH_2, under tank pressure, flows through the turbopump and main fuel valve into the thrust chamber fuel manifold. Bootstrap fuel tapped off the fuel manifold flows to the ASI assembly. Lox, under tank pressure, flows through the lox turbopump and ASI valve to the ASI assembly where the

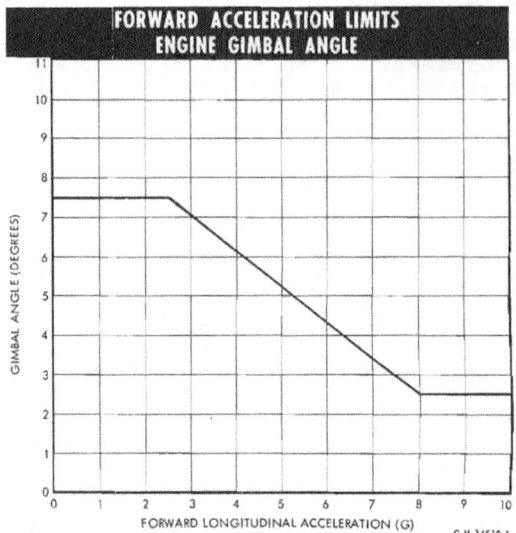

Figure 5-16

sparkplugs ignite the propellant. Fuel enters the fuel manifold and flows through the down tubes to the return manifold and then through the up tubes to the injector where it emerges in a gaseous state. The fuel cools the thrust chamber as it flows through the jacket to the combustion chamber. Lox enters the oxidizer manifold in the injector and flows through equally distributed nozzles where it mixes with fuel and burns. The MFV sequence valve opens when the MFV opens to 90 percent permitting helium flow to the closed start tank discharge valve (STDV) control valve. The start tank discharge timer expires after approximately 1.0 sec, and an ignition phase timer (0.450 ± 0.030 sec) starts. The STDV control valve energizes, admitting helium to the STDV opening port. GH_2 from the integral start tank flows through the series turbine drive system accelerating the fuel and oxidizer turbopumps to the required levels to permit power buildup of the gas generator and to deliver propellant for ASI ignition. The normally open lox turbine bypass valve bypasses a percentage of the GH_2 from the oxidizer turbine during start and later controls the relationship of fuel and oxidizer turbine speed by an orifice in the valve gate.

Expiration of the ignition phase timer initiates: (1) command for the sparks deenergized timer (3.30±0.20 sec), (2) closing signal

J-2 ENGINE PREDICTED PREFORMANCE VALUES		
PARAMETER	PRIOR TO MR SHIFT	AFTER MR SHIFT
AVERAGE THRUST (LB)	230,073	193,655
AVERAGE CHAMBER INJECTOR PRESS. (psia)	774.2	659.2
AVERAGE EFFECT. THRUST COEF.	1.74	1.72
AVERAGE SPECIFIC IMPULSE (sec)	424.2	428.2
AVERAGE MIXTURE RATIO (O/F)	5.509	4.786
AVERAGE FUEL FLOWRATE (lb/sec)	83.324	78.162
AVERAGE LOX FLOWRATE (lb/sec)	459.045	374.068
MR SHIFT TIME ▷	328.1	
ENGINE CUTOFF ▷		448.1

▷ TIME FROM S-IB OBCO (sec)

Figure 5-17

Section V S-IVB Stage

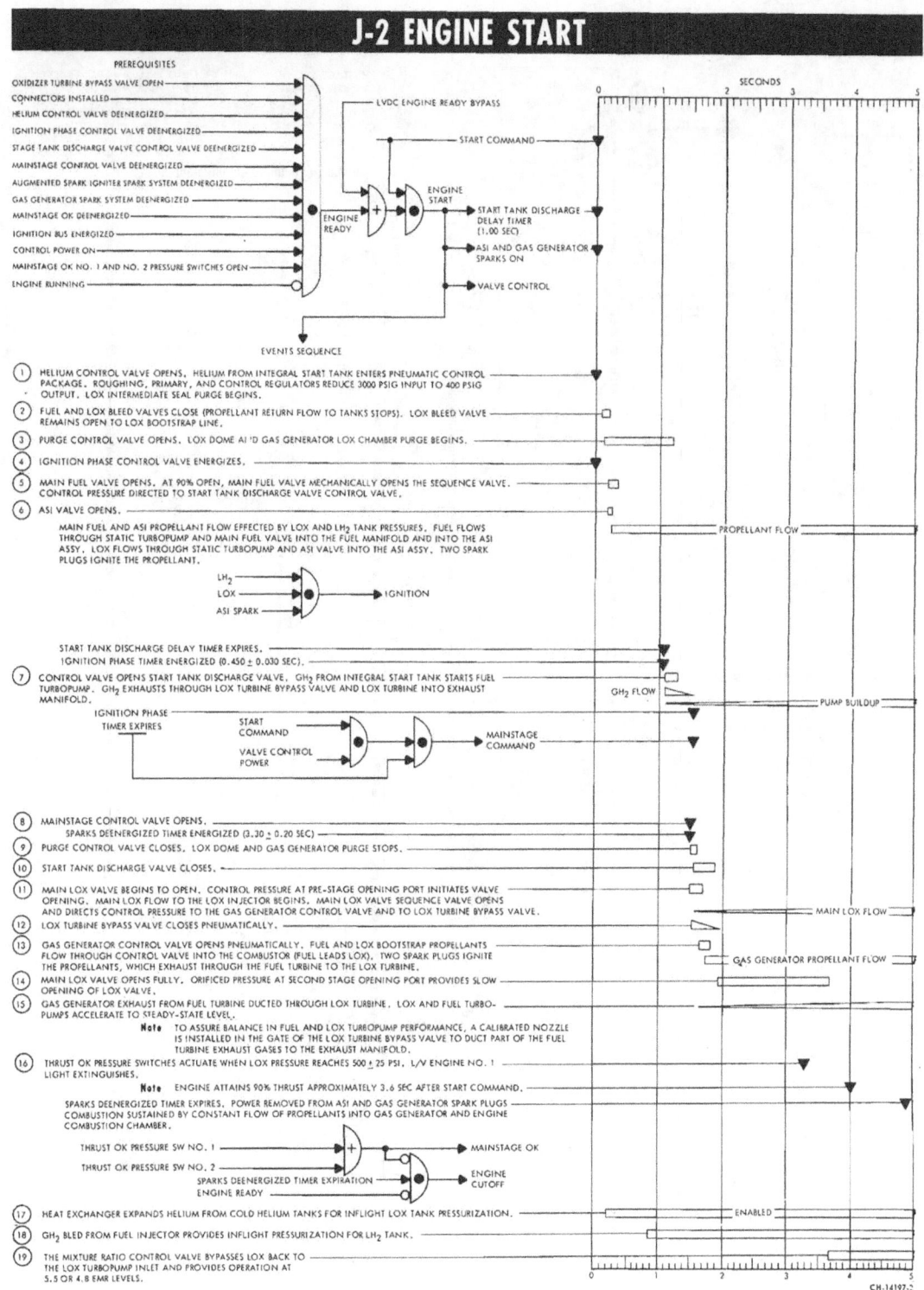

Figure 5-18 (Sheet 1 of 2)

Section V S-IVB Stage

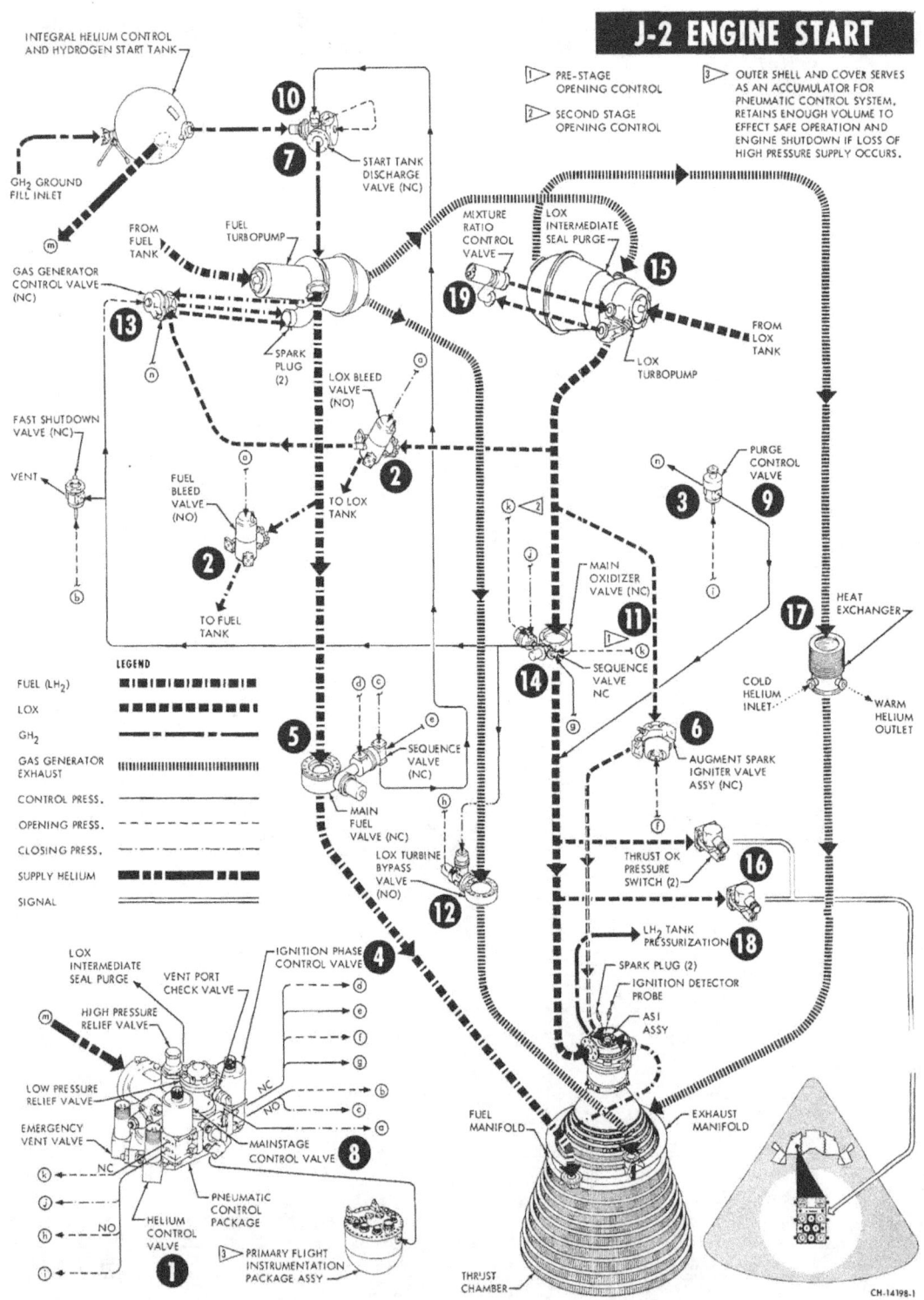

Figure 5-18 (Sheet 2 of 2)

for the start tank discharge valve, and (3) mainstage command. Energizing the mainstage control valve vents pressure from the main oxidizer valve (MOV) closing port, closes the purge control valve terminating the lox dome and GG oxidizer injector purges, vents the lox turbine bypass valve opening pressure, and applies opening pressure to the first- and second-stage opening port of the MOV. (An orifice in the second-stage opening line, series orifices in the actuator closing line, an orifice in the actuator, and an orificed check valve provide a controlled ramp opening of the MOV.) The MOV sequence valve opens with the MOV and supplies helium pressure through orifices to close the lox turbine bypass valve and to open the gas generator control valve. Two spark plugs in the gas generator ignite the lox and LH_2 admitted by the GG control valve. Hot-gas products of combustion pass through the fuel turbine and through the exhaust duct to the lox turbine accelerating the turbopump causing increased propellant flow. The turbine exhaust gases exit through a heat exchanger and into an engine exhaust manifold. The propellant enters the thrust chamber where main propellant ignition occurs. As turbopumps accelerate to operational speeds, oxidizer injection pressure increases actuating two thrust OK pressure switches. Either of the two TOPS, which actuate at 500 ± 30 psi, will issue a mainstage OK signal that must be present before the sparks deenergized timer expires or the sequencer will automatically issue a cutoff command. Expiration of the sparks deenergized timer (3.30 ± 0.20 sec) also turns off the ASI and GG spark exciters. The mainstage OK signal extinguishes the L/V ENGINE 1 light on the MDC, indicating that the engine has attained 90-percent thrust. During engine operation, GH_2 tapped off the fuel injection manifold maintains LH_2 tank pressure. Turbine exhaust gases flowing through the heat exchanger, heats helium from the cold helium bottles for lox tank inflight pressurization. A two-position mixture ratio control valve (MRCV) bypasses lox from the oxidizer pump outlet to the pump inlet and allows the engine to operate at either of two fixed mixture ratios. The MRCV is commanded to the low EMR position prior to engine start and to the high position after 90% thrust is achieved.

J-2 Engine Cutoff.

Guidance cutoff of the J-2 engine occurs at about 10 min into flight. This command will be issued through the switch selector to the sequence controller in the J-2 engine electrical control package. The sequence controller simultaneously deenergizes the mainstage control and ignition phase control valves and energizes the helium control deenergize timer (figure 5-19). Opening control pressure vents from both the first- and second-stage main oxidizer valve opening actuators through the normally closed port of the mainstage control valve while opening control pressure from the augmented spark igniter oxidizer valve and the main fuel valve vents through the normally closed port of the ignition phase control valve. Helium control system pressure is routed from the normally open port of the mainstage control valve to the closing actuator of the main oxidizer valve and to the opening control port of the purge control valve. The purge control valve opens allowing helium control pressure to flow to a check valve in the oxidizer dome purge line and to a check valve in the gas generator oxidizer purge line. Both of these purges will flow when thrust chamber and gas generator chamber pressures decay below the level of control system

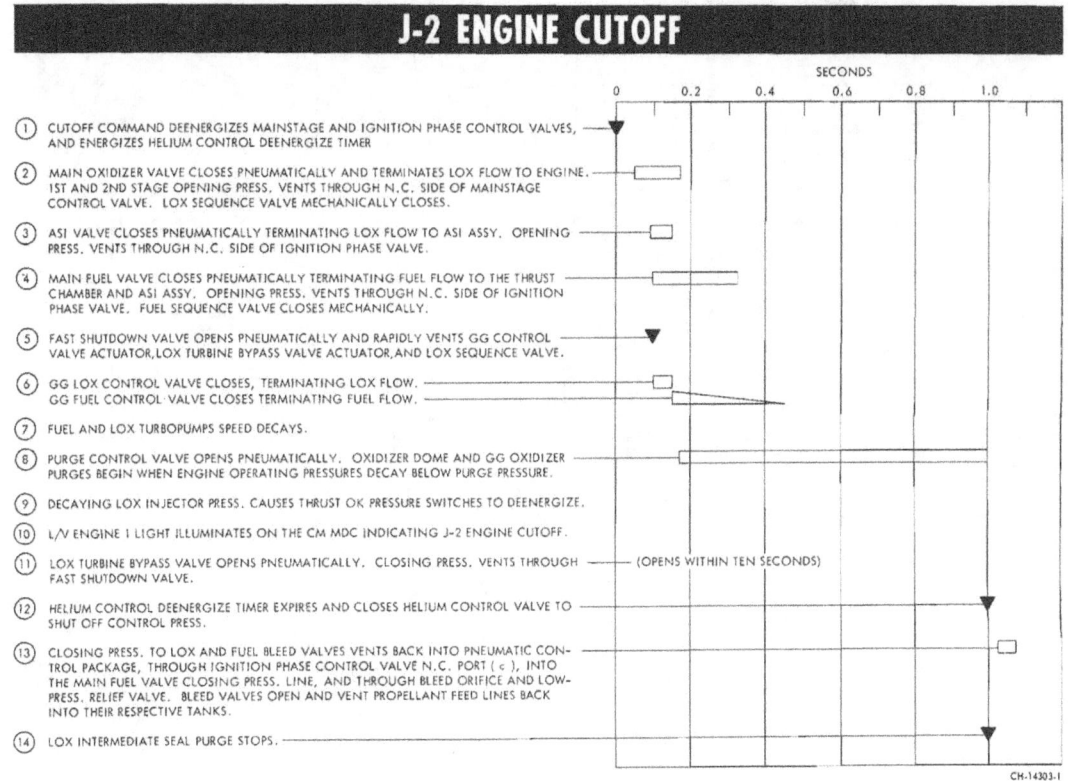

Figure 5-19 (Sheet 1 of 2)

Section V S-IVB Stage

J-2 ENGINE CUTOFF

Figure 5-19 (Sheet 2 of 2)

pressure. The normally open port of the ignition phase control valve routes helium to the closing actuator of the augmented spark igniter oxidizer valve and the main fuel valve and to the opening control port of the pressure-actuated fast shutdown valve which actuates, allowing the gas generator control valve opening control pressure to vent rapidly. All valves, except the augmented spark igniter oxidizer valve and the oxidizer turbine bypass valve, are spring-loaded to the closed position and start to close as soon as opening pressure is vented. Combustion pressure in the gas generator assists the spring in closing of the gas generator control valve. The oxidizer turbine bypass valve closing control pressure also vents through an orifice and the pressure-actuated fast shutdown valve. The oxidizer turbine bypass valve is spring-loaded to the normally open position and starts to open as closing pressure is vented. The normally open port of the mainstage control valve also supplies opening control pressure to the oxidizer turbine bypass valve to assist the spring in opening the valve.

Decaying lox injector pressure causes the thrust OK pressure switches to deactuate. L/V ENGINE 1 light on the main display console in the command module illuminates at TOPS deactuation providing a visual indication of J-2 engine thrust decay. The light remains on until CSM/launch vehicle separation. Expiration of the helium control deenergize timer causes the helium control valve to deenergize, closing the valve and venting control system pressure through the oxidizer dome and gas generator oxidizer purges. As the control system pressure is vented, the normally closed purge control valve actuates closed and the purges stop. Pressure is now locked up in the system between the check valve in the pneumatic control package, through the normally open port of the mainstage and ignition phase control valves, the pneumatic accumulator, and to the propellant bleed valve control ports holding the bleed valves closed. The pressure in this system is bled off through an accumulator bleed orifice located in the line between the closing actuator of the main fuel valve and the normally open port of ignition phase control valve. As this pressure decays, the propellant bleed valves open by spring pressure and the cutoff sequence is complete.

During the transient period, from the cutoff signal to zero thrust, the engine will consume approximately seven gal of lox and 47 gal of LH_2. Thrust will decay to 5% within 0.4 sec after receipt of the cutoff signal. LH_2 and lox turbopump speeds will decay to zero rpm in 10 and 4 sec, respectively.

Thrust Chamber and Gimbal System.

The thrust chamber and gimbal system consists of a thrust chamber, where propellants burn to create thrust; the augmented spark igniter, which ignites the propellants; and the gimbal assembly, which allows the thrust chamber to correct pitch and yaw attitude errors.

Thrust Chamber. The thrust chamber, consisting of a body and an injector, receives liquid propellants under turbopump pressure, converts them to a gaseous state, mixes them, and burns them imparting a high velocity to the expelled combustion gases producing thrust for vehicle propulsion.

Thrust Chamber Injector. The concentric-orificed, porous-faced thrust chamber injector (figure 5-20) atomizes and mixes propellants to produce the most efficient combustion. Oxidizer ports are electrically discharge machined to form part of the injector. Threaded fuel nozzles install over the oxidizer ports to form the concentric orifices. The injector face, formed from a sintered metallic material, is welded at its outside and inside edges to the injector body. Each fuel nozzle is swaged to the injector face. An oxidizer inlet elbow, integral with the dome and injector assembly, admits lox from the turbopump and injects it through the oxidizer ports into the thrust chamber combustion area. Fuel enters the injector from the upper fuel manifold and flows through orifices concentric with oxidizer orifices. Approximately 3 to 4 percent of the fuel flows through the sintered injector face to cool the injector. Combustion zone pressure acts upon the injector face area producing the thrust force, which is transmitted through the gimbal to the vehicle structure.

Thrust Chamber Body. The tubular-walled, bell-shaped thrust chamber, consists of a cylindrical section where combustion occurs, a narrowing throat section, and an expansion section. The thrust chamber body is constructed of longitudinal stainless steel tubes brazed together with bands around the tubes for external stiffening. A fuel inlet manifold on the engine bell admits fuel to 180 down tubes that carry the fuel to the return manifold at the base of the thrust chamber. From the return manifold, 360 tubes carry fuel to the thrust chamber injector. Fuel flowing through the thrust chamber jacket cools the thrust chamber and at the same time absorbs heat converting from LH_2 to GH_2 for injection into the combustion chamber. Propellants burn in the combustion chamber creating large volumes of gases that are forced to exit the combustion chamber through the narrow throat area and out the drainage nozzle producing thrust to propel the vehicle.

Gimbal Assembly. The gimbal is a universal joint consisting of a spherical, socket-type bearing with a teflon-fiberglass composition coating to provide a dry low-friction bearing surface. The gimbal assembly, installed on the thrust chamber dome, attaches the engine to the thrust structure and transmits engine thrust to the vehicle. A boot with a bellows configuration made of silicone-impregnated fiberglass material protects the gimbal assembly from dust, water, and other foreign matter without interferring with gimballing. Two hydraulic actuators, attached to the engine and thrust structure, provide the force to gimbal the engine ±7 deg for thrust vector control. See Flight Control for information on the hydraulic system.

Augmented Spark Igniter Assembly. The augmented spark igniter assembly is chamber mounted in the thrust chamber injector. It consists of a fuel and lox manifold, an injector assembly, and two

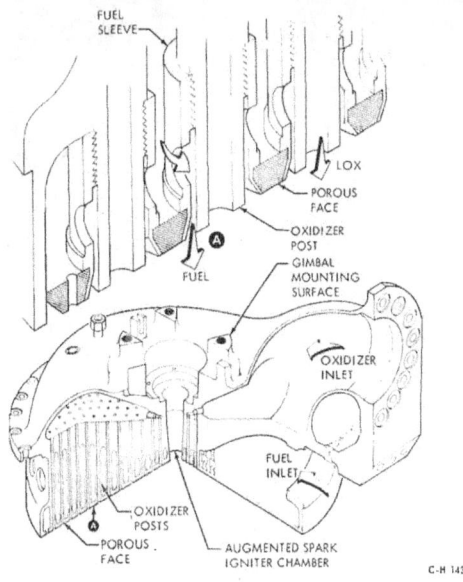

Figure 5-20

Section V S-IVB Stage

J-2 ENGINE FUEL TURBOPUMP

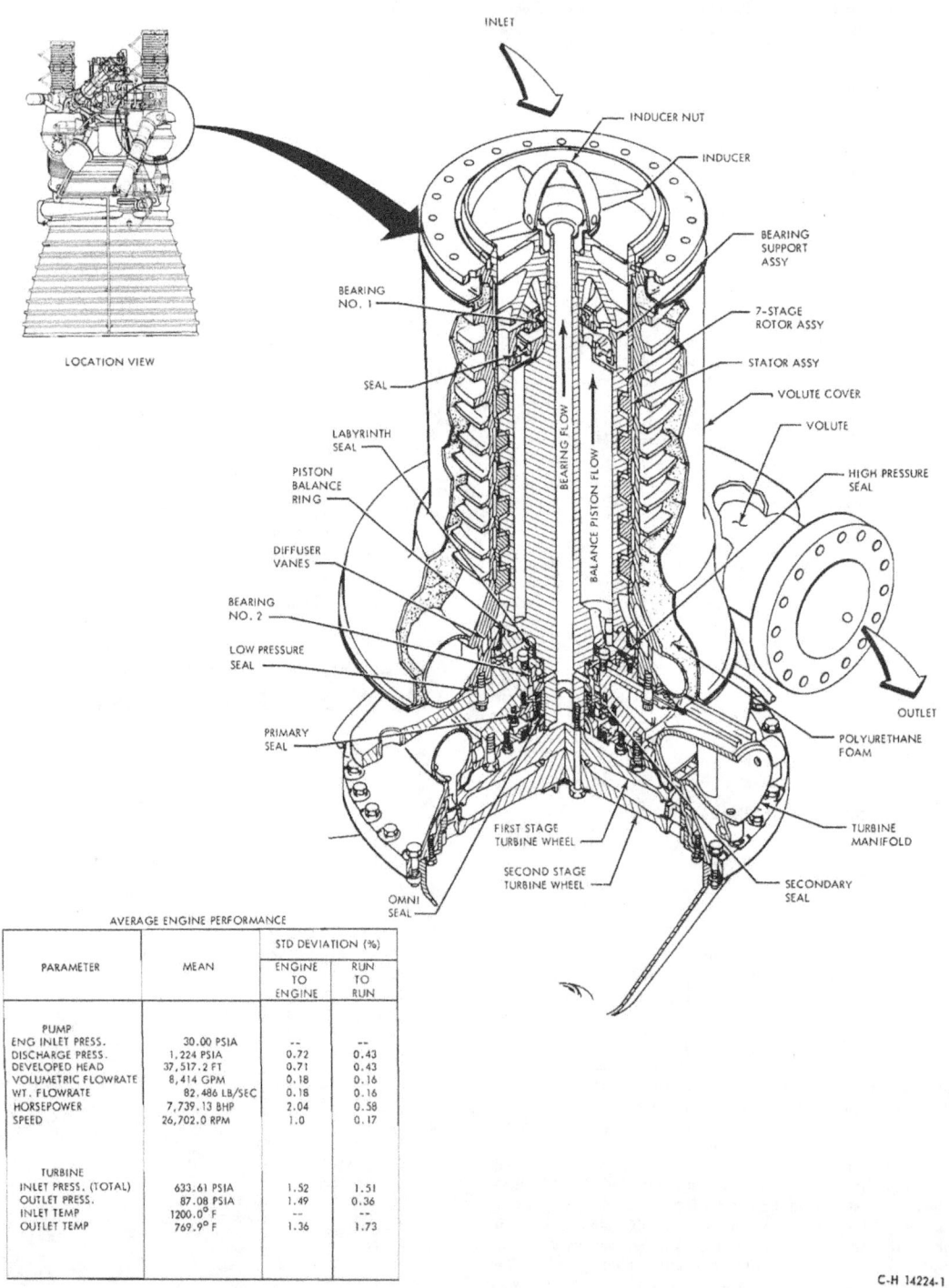

Figure 5-21

PARAMETER	MEAN	STD DEVIATION (%)	
		ENGINE TO ENGINE	RUN TO RUN
PUMP			
ENG INLET PRESS.	30.00 PSIA	--	--
DISCHARGE PRESS.	1,224 PSIA	0.72	0.43
DEVELOPED HEAD	37,517.2 FT	0.71	0.43
VOLUMETRIC FLOWRATE	8,414 GPM	0.18	0.16
WT. FLOWRATE	82.486 LB/SEC	0.18	0.16
HORSEPOWER	7,739.13 BHP	2.04	0.58
SPEED	26,702.0 RPM	1.0	0.17
TURBINE			
INLET PRESS. (TOTAL)	633.61 PSIA	1.52	1.51
OUTLET PRESS.	87.08 PSIA	1.49	0.36
INLET TEMP	1200.0° F	--	--
OUTLET TEMP	769.9° F	1.36	1.73

AVERAGE ENGINE PERFORMANCE

spark igniters and cable assemblies. The ASI assembly receives the initial flow of lox and LH$_2$ and ignites the propellants by discharging electrical energy through the two spark plugs. At engine start, spark exciters in the electrical control package transform 28 Vdc into 27,000-V ($\pm 3,000$) pulses that discharge across the spark plug gap at 40 sparks per sec (min). Hermetically sealed transmission cabling and connections are pressurized with GN$_2$ to ensure operation at high altitudes.

Propellant Feed System.

The propellant feed system consists of a fuel turbopump, oxidizer turbopump, main lox valve, main fuel valve, augmented spark igniter oxidizer valve, mixture ratio control valve, propellant bleed valves, and propellant feed ducts. This system transfers and controls propellant flow from the stage tanks to the thrust chamber and the gas generator.

Fuel Turbopump. The fuel turbopump, a turbine driven, axial-flow pumping unit consisting of an inducer, a seven-stage rotor, and a stator assembly, increases the pressure and flowrate of LH$_2$ entering the thrust chamber. See figure 5-21. The pump is self-lubricated and self-balanced with LH$_2$. The high speed, two-stage turbine, driven by hot gas from the gas generator, drives the inducer and the one-piece seven-stage rotor assembly. Gas enters the turbine inlet manifold and passes through nozzles where it is expanded and directed at high velocity through the first-stage turbine wheel and then through stator blades that redirect the gases through the second-stage turbine wheel. Exhaust ducts direct the gases to the oxidizer turbopump turbine and also to the oxidizer turbine bypass valve. Three dynamic seals in series prevent LH$_2$ and turbine gases from mixing. The inducer increases the LH$_2$ inlet pressure at the pump. Each stage of the seven-stage rotor contributes to the buildup of pressure which forces the fuel through diffuser vanes and outlet volute into the high pressure duct. A self-compensating balance piston absorbs axial thrust loads developed by the rotor. LH$_2$ lubricates and cools the two ball bearings on the rotor shaft. A magnetic pickup senses pump speed as 12 equally spaced slots in the rotor assembly interrupt the magnetic field. Temperature measurements of the fuel turbine inlet and the fuel pump discharge, a fuel pump discharge pressure measurement, and the pump speed measurements are telemetered to ground receiving stations for recording and postflight evaluation. See J-2 Engine Measuring.

Oxidizer Turbopump. The single-stage, direct-turbine-drive centrifugal pump delivers lox to the thrust chamber at increased pressure and flowrates necessary for satisfactory J-2 engine operations. See figure 5-22. The pump is self-lubricated and self-cooled with lox. A high speed, two-stage turbine, driven by exhaust gases from the fuel turbine, provides power for lox turbopump operation. One static and two dynamic seals in series prevent lox and turbine gases from mixing. Turbine exhaust gases enter the oxidizer turbopump turbine manifold and flow through nozzles and the first-stage turbine wheel. Stator blades redirect the gases, which pass through the second-stage turbine wheel and then exhaust through turbine exhaust ducting, a heat exchanger, and through the thrust chamber. A pump shaft transmits the turbine power to the inducer and impeller. Lox enters the turbopump through the inducer, which increases pump inlet pressure, and flows through the impeller into the outlet volute. Passages from the outlet volute permit lox flow to the two pump shaft ball bearings for lubrication. A screen filters lox entering the bearing area. Seven measurements of turbopump operation are taken during flight. The turbine inlet and outlet temperatures and pressures, the pump discharge temperature and pressures, and the pump speed are telemetered to ground receiving stations and recorded for postflight evaluation. See J-2 Engine Measuring. A magnetic pickup located behind the turbine manifold senses pump speed as 12 equally spaced slots in the rotor assembly rotate through the magnetic field. An accessory drive adapter on the oxidizer turbopump drives the hydraulic pump for the two engine actuators. The drive was designed for 30 hp extraction at mainstage speeds. Engine performance balance is based on extraction of 15 hp. The 15 hp value was selected based on nominal power requirements to gimbal the engine at model specification limits.

Main Oxidizer Valve. The main oxidizer valve (MOV) is a butterfly-type valve, spring-loaded to the closed position, pneumatically operated to the open position, and pneumatically assisted to the closed position. The MOV, installed between the high pressure duct and the thrust chamber injector oxidizer inlet, controls the lox flow into the thrust chamber. Pneumatic pressure from the normally open port of the mainstage control valve plus spring pressure maintains the valve closed. During the engine start sequence, the mainstage control valve opens and applies pressure to the first and second stage opening control ports on the MOV actuator. Pressure acting against a small surface in the first stage actuator plus helium flowing through an orifice and acting on a large surface piston, which actuates the valve gate, provides a ramp opening of the valve. Pressure exhausting from the closing actuator exits through an orifice in the actuator housing and through an orifice-check valve to help accomplish the ramp opening. The valve opens in two stages. First stage actuator pressure will start opening the valve 50 ± 20 msec after the mainstage command to the mainstage control valve. Within 50 ± 25 msec the valve will open approximately 14 ± 2 deg. After 610 ± 70 msec the second stage pressure begins to open the MOV, and after 1825 ± 75 msec the valve will be fully open. A sequence valve, installed on the MOV and actuated when the MOV opens, permits helium flow to the gas generator control valve and to the closing port of the lox turbine bypass valve. A spring closes the sequence valve when the opening pressure is vented. A position switch assembly operated by the gate shaft provides signals to the ESE concerning the gate positions. The switch contains a potentiometer that senses valve position and provides a corresponding input to the telemetry system. Open and closed indicator switches operate the MAIN OXIDIZER VALVE OPEN and MAIN OXIDIZER VALVE CLOSED indicators on the S-IVB engine panel during checkout and prelaunch operations. During flight, signals from the MOV open position indicator switch will be telemetered to ground stations and monitored on the engine panel. The closed position indicator and valve position potentiometer signals will be telemetered to a ground receiving station and recorded for postflight evaluation. See J-2 Engine Measuring for additional information. A 5-Vdc reference voltage is supplied to the position potentiometer and 28 Vdc is supplied to the indicator switches.

Main Fuel Valve. The main fuel valve is a butterfly-type valve, spring-loaded to the closed position, pneumatically operated to the open position, and pneumatically assisted to the closed position. The MFV, installed between the fuel turbopump high pressure discharge duct and the fuel inlet manifold, controls LH$_2$ flow to the thrust chamber and to the ASI assembly. The ignition phase control valve on the pneumatic control package opens the MFV during engine start. When pneumatic pressure has opened the valve to 90 percent, the actuator mechanically opens a sequence control valve mounted on the MFV acutator. The sequence valve permits helium flow also from the ignition phase valve to the start tank discharge valve control valve. The MFV and sequence valve arrangement assures that the MFV will be open before the STDV opens to deliver GH$_2$ to start the fuel turbopump. A position switch assembly operated by the gate shaft provides analog position signals of the gate and discrete signals corresponding to the open or closed position of the gate. MAIN FUEL VALVE OPENED and MAIN FUEL VALVE CLOSED indicators on the S-IVB engine panel

Section V S-IVB Stage

J-2 ENGINE LOX TURBOPUMP

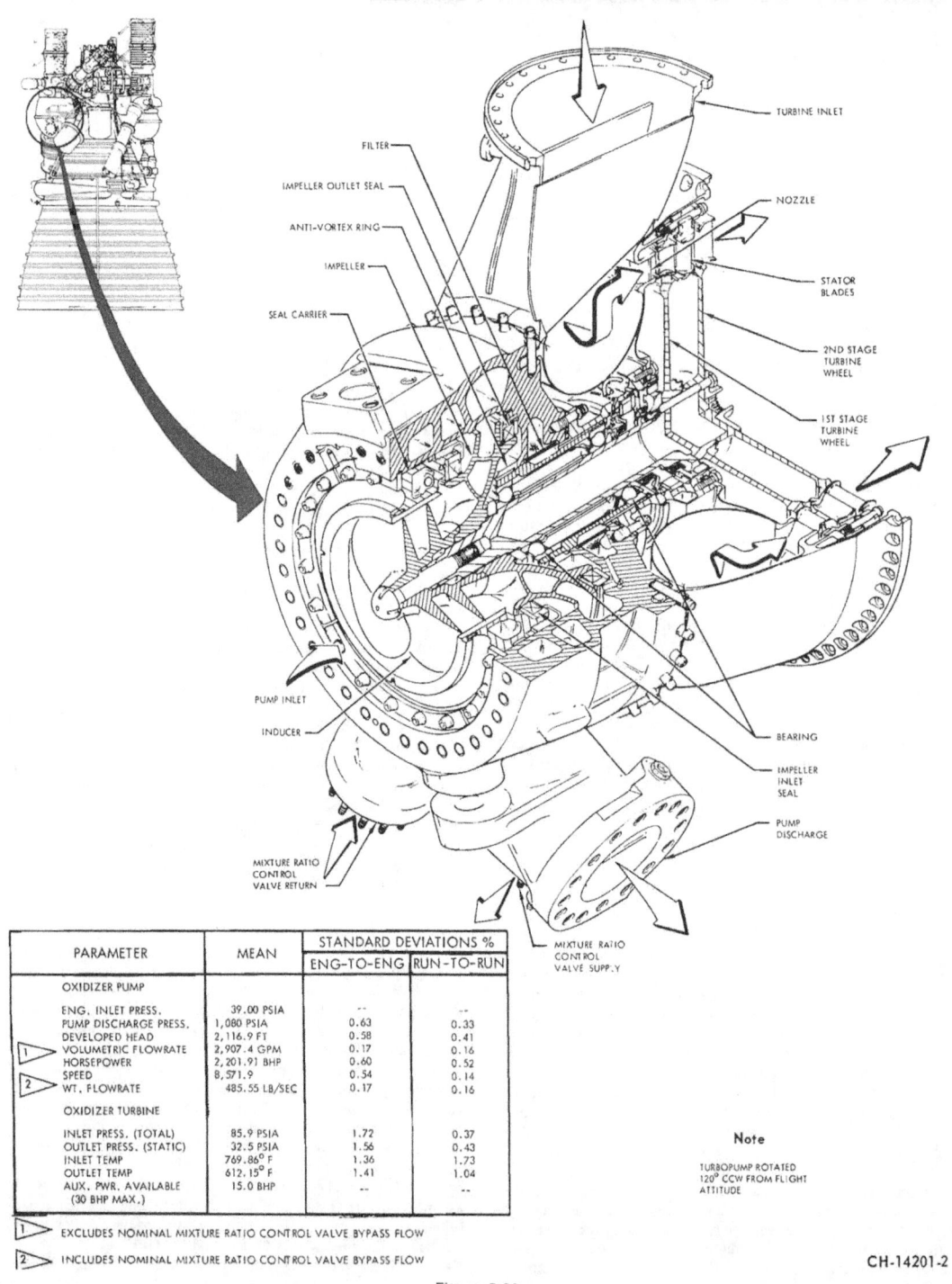

PARAMETER	MEAN	STANDARD DEVIATIONS %	
		ENG-TO-ENG	RUN-TO-RUN
OXIDIZER PUMP			
ENG. INLET PRESS.	39.00 PSIA	--	--
PUMP DISCHARGE PRESS.	1,080 PSIA	0.63	0.33
DEVELOPED HEAD	2,116.9 FT	0.58	0.41
[1] VOLUMETRIC FLOWRATE	2,907.4 GPM	0.17	0.16
HORSEPOWER	2,201.91 BHP	0.60	0.52
SPEED	8,571.9	0.54	0.14
[2] WT. FLOWRATE	485.55 LB/SEC	0.17	0.16
OXIDIZER TURBINE			
INLET PRESS. (TOTAL)	85.9 PSIA	1.72	0.37
OUTLET PRESS. (STATIC)	32.5 PSIA	1.56	0.43
INLET TEMP	769.86° F	1.36	1.73
OUTLET TEMP	612.15° F	1.41	1.04
AUX. PWR. AVAILABLE (30 BHP MAX.)	15.0 BHP	--	--

[1] EXCLUDES NOMINAL MIXTURE RATIO CONTROL VALVE BYPASS FLOW
[2] INCLUDES NOMINAL MIXTURE RATIO CONTROL VALVE BYPASS FLOW

Note

TURBOPUMP ROTATED 120° CCW FROM FLIGHT ATTITUDE

CH-14201-2

Figure 5-22

monitor the valve positions through hardwire connections during checkout and prelaunch operations. During flight, signals from the MFV opened switch are telemetered to the ground, and, in addition to being recorded, they are monitored by the MAIN FUEL VALVE OPENED indicator on the S-IVB engine panel. The MFV closed signal and the continuous position signals are telemetered to the ground and recorded for postflight evaluation. A 5-Vdc reference voltage input to the position potentiometer and 28 Vdc to the indicator switches is stage supplied. Based on Rocketdyne test results, the MFV will begin opening 60 ± 30 msec after the control signal to the ignition phase valve. The valve requires an additional 110 ± 50 msec to fully open. At engine cutoff, the MFV will begin closing 90 ± 25 msec after the cutoff signal issuance, and will be completely closed 225 ± 25 msec later.

Augmented Spark Igniter Oxidizer Valve. The normally closed, pneumatically operated, poppet-type augmented spark igniter oxidizer valve controls lox flow to the ASI assembly during J-2 engine start sequence. The ignition phase control valve pneumatically opens the ASI valve, and lox under tank pressure flows through the valve and into the ASI assembly where it mixes with LH_2. The ASI valve mounts in the MLV gate housing and receives lox through a port just upstream of the MLV gate. A position indicator provides an ASI LOX VALVE OPENED indication to the S-IVB engine panel through hardwire connections during checkout. During flight the indication is telemetered back to ground stations and recorded.

Mixture Ratio Control Valve. The mixture ratio control valve allows the engine to operate at either one of two fixed mixture ratios to achieve maximum vehicle performance. The valve changes mixture ratio by routing a portion of the oxidizer flow from the oxidizer turbopump outlet back to the pump impeller inlet. The valve has an actuator assembly and a gate assembly. The actuator is two-position, electro-pneumatic and is spring-loaded to keep it in the high engine mixture ratio position (valve closed). Pneumatic pressure is directed to the actuator piston by a three-way pneumatic control valve that is energized by a stage signal. The gate assembly consists of a rotating sleeve within a stationary outer sleeve. Each sleeve has three elongated holes; by rotating the inner sleeve (valve gate) the holes are alined or misalined, to control the amount of oxidizer flow through the valve. The valve position indicator is mounted on the valve shaft and consists of a rotary-motion, variable resistor and open and close position switches.

The mixture ratio control valve has two distinct stops, to allow engine operation at engine mixture ratios of either 5.5:1 or 4.8:1 (lox to LH_2 by weight). Pneumatic pressure is supplied to the valve from the engine pneumatic system when the engine helium control valve is energized. At a preselected time during engine operation, a control signal, supplied by the stage, energizes the solenoid control valve. Energizing the solenoid control valve allows pneumatic pressure to enter the valve and apply force to the actuator piston, to overcome the spring tension and move the piston in the direction to rotate the gate to the low engine mixture ratio position (valve open). Opening the valve results in a reduced oxidizer flow to the thrust chamber. If either the pneumatic pressure or the electrical command is lost, the valve will move to the high engine mixture ratio position (valve closed). The position indicator arm rotates with the gate shaft, to remotely indicate valve position.

Propellant Bleed Valves. A propellant bleed valve in the lox system and one in the LH_2 system bleeds trapped gases in the systems back to their respective tanks. During chilldown operations, the chilldown pumps circulate fuel and lox through the respective systems. Propellants flow through the inlet ducts and turbopumps and return to the stage tanks through the bleed valves. The valves are poppet-type, spring-loaded to the open position and pressure actuated to the closed position. The oxidizer bleed valve is mounted on the lox bootstrap line located on the oxidizer high pressure duct just upstream from the MOV. The lox bleed valve has an inlet port, two outlet ports, and an actuation port. Lox flows through the valve to a return line to the lox tank and through the lox bootstrap line to the gas generator. When actuated, the lox tank return line port closes and the bootstrap port remains open. The fuel bleed valve is mounted on the LH_2 bootstrap line at the fuel turbopump outlet. This valve has an actuator port, an LH_2 inlet port, and one LH_2 outlet port for fuel return to the LH_2 tank. At engine start, pressure from the pneumatic control package actuates both bleed valves terminating propellant flow back to the tanks. A position indicator in each valve feeds a 28-Vdc BLEED VALVE CLOSED signal to the respective indicators on the S-IVB engine panel during checkout operation. In flight, these signals are telemetered to ground stations as events of valve actuation and are recorded for evaluation.

Propellant Inlet Ducts. The fuel and lox inlet ducts convey the propellants from the stage tanks to the fuel and lox turbopumps. The ducts employ flexible bellows sections to permit freedom of movement during engine gimballing. Bipod clevis assemblies stabilize the bellows convolutions allowing maximum engine gimballing without collapsing the bellows. A vacuum jacket insulates the fuel inlet duct to reduce boiloff of LH_2.

Gas Generator and Exhaust System.

The gas generator and exhaust system consists of a gas generator, which supplies the hot gases to drive the turbopumps; turbine exhaust ducts, which transfer the exhaust gases from the fuel turbine to oxidizer turbine and to the thrust chamber exhaust manifold; the heat exchanger, which expands cold helium for lox tank inflight pressurization; and the turbine bypass valve, which allows a portion of the fuel turbine exhaust gases to bypass the oxidizer turbopump turbine.

Gas Generator Assembly. The gas generator, which produces the hot gases to drive the oxidizer and fuel turbines, consists of a combustor containing two spark plugs, a control valve containing oxidizer and fuel poppets, and an injector assembly. When engine start is initiated, spark exciters in the electrical control package are energized providing energy to the spark plugs in the gas generator combustor. Propellants flow through the open poppets of the control valve to the injector assembly and into the combustor where they are mixed and burned, resulting in hot gases that pass through the combustor outlet and are directed to the fuel turbine and then to the oxidizer turbine. See figure 5-23 for gas generator characteristics.

Gas Generator Control Valve. The gas generator control valve is a pneumatically operated, spring-loaded to the closed position, poppet valve. The oxidizer and fuel poppets are mechanically linked by an actuator. The purpose of the gas generator control valve is to control the flow of propellants through the gas generator injector. When the mainstage signal is received, pneumatic pressure is applied against the gas generator control valve actuator assembly which moves the piston and opens the fuel poppet. During the fuel poppet opening, an actuator contacts the piston that opens the oxidizer poppet. LH_2 and lox from the bootstrap lines flow through the control valve into the combustion chamber. Orifices in the bootstrap lines control the propellant flowrate to the gas

GAS GENERATOR CHARACTERISTICS

PARAMETER	MEAN	STD DEV (%) ENG-TO-ENG	STD DEV (%) RUN-TO-RUN
CHAMBER PRESS. (INJECTOR END)	654.7 psia	1.46	0.51
OXIDIZER FLOWRATE	3.4 lb/sec	1.08	0.51
FUEL FLOWRATE	3.62 lb/sec	1.08	0.51
OUTLET TEMP	1,200°F [1]	--	--

[1] RATED CONDITION

Figure 5-23

generator. A line from the gas generator housing to the sequence valve vent port on the MOV provides a vent to equalize pressure in the housing to prevent premature opening of the fuel control poppet. A position indicator assembly consisting of switches and a potentiometer provides valve position signals through DDAS to the S-IVB engine panel. The position switches provide signals to the gas generator valves OPENED and CLOSED indicators during checkout and prelaunch operations. During flight, the signals from the switches and potentiometer are telemetered to ground stations and recorded. During engine start sequence, opening of the gas generator control valve is monitored on the S-IVB engine panel as is event measurement (VK117). The potentiometer indicates the control valve position from the closed position through the fully open position (0 to 100 percent). See J-2 Engine Measuring. At engine cutoff, the fast fill valve vents the pneumatic pressure from the control valve, and a spring returns the valve to the closed position.

Gas Generator Injector Assmebly. The gas generator injector assembly consists of a circular metal plate containing a normally closed, spring-loaded oxidizer poppet valve and injector, centered within a fuel injector ring. The purpose of the gas generator injector assembly is to distribute propellants into the gas generator combustor. The injector is welded to the gas generator combustor, and the oxidizer poppet and injector is threaded into the gas generator injector assembly. During operation, fuel enters the injector assembly fuel inlet, fills a manifold in the top of the combustor, and flows through drilled passages in the fuel injector ring. Oxidizer pressure displaces the oxidizer poppet valve and allows oxidizer flow through the injector to impinge on the fuel flowing through the fuel injector ring.

Gas Generator Combustor. The gas generator combustor is a cylindrical chamber in which the propellants are mixed and burned. Two spark plugs initiate combustion. The inlet port mates with the gas generator injector assembly and the outlet port and short duct section is welded to the fuel turbine manifold. Propellants entering the combustor are ignited by the spark plugs; combustion hot gases pass through the combustor outlet into the fuel turbine manifold.

Exhaust Ducting. The exhaust ducting and turbine exhaust hoods are welded sheet metal construction. Dual (Naflex) seals are used in flanges at all component connections. The ducting conducts the fuel turbopump turbine exhaust gases to the lox turbopump turbine and subsequently through the heat exchanger and into the thrust chamber exhaust manifold. A second duct from the fuel turbopump turbine directs exhaust gases through the oxidizer turbine bypass valve into the thrust chamber exhaust manifold.

Heat Exchanger. The shell-assembly heat exchanger consists of a duct, bellows, flanges, and coils. It mounts in the exhaust duct between the oxidizer turbine exhaust and the thrust chamber exhaust manifold. During flight, cold helium flows through one of the four coils, which are heated by the flow of exhaust gases through the exhaust duct, and expands then returns to the lox tank as ullage pressurant. The remaining three coils are blanked-off.

Oxidizer Turbine Bypass Valve. The oxidizer turbine bypass valve is a normally open, spring-loaded gate valve mounted in the oxidizer bypass duct. The valve gate is equipped with a nozzle whose size is determined during engine calibration. The purpose of the valve is to prevent an overspeed condition of the oxidizer turbopump and to act as a calibration device for the turbopump performance balance. When the fuel turbopump turbine starts to spin, the exhaust gas in the turbine exhaust duct passes through a duct to the oxidizer turbopump turbine. A percentage of the gas volume bypasses the oxidizer turbine through the open oxidizer turbine bypass valve and vents through the thrust chamber. During engine transit into mainstage, pneumatic pressure, directed to the closing port of the oxidizer turbine bypass valve, closes the valve to divert the turbine exhaust gases, except for a volume of gas which passes through the valve gate nozzle, through the oxidizer turbopump turbine. During engine shutdown, the ignition phase control valve deenergizes and vents the closing pressure from the oxidizer turbine bypass valve. The normally open port of the mainstage control valve supplies pressure to the valve opening control port to assist the spring in opening the valve. A potentiometer and position switches monitor the bypass valve gate position and provide signals through the DDAS to the ESE. The position switches provide inputs to illuminate the LOX TURBINE BYPASS CLOSED, LOX TURBINE BYPASS OPENED, and LOX TURBINE BYPASS OPEN indicators on the S-IVB engine panel. During flight these indications are telemetered back to the ESE as events when the bypass valve closes and opens during the J-2 engine start and cutoff operations. These signals are also monitored on the events panel. The potentiometer provides an analog signal of the valves' position from fully opened to fully closed (0 to 100 percent). See J-2 Measuring for additional information.

Control System.

The control system includes the pneumatic control package, which controls helium flow to the various valves in the engine system; the electrical control package, which controls electrical signals in the system, and electrical harnesses.

Pneumatic Control Package. The pneumatic control package is a combination of two regulators, two relief valves, an actuator assembly, a filter unit, and four solenoid valves. The purpose of the pneumatic control package is to control the flow of helium to the engine control system and to supply opening and closing control pressure for all pneumatically operated valves. The control package also supplies helium for the oxidizer dome, gas generator oxidizer injector, and oxidizer intermediate seal purges. Helium from the integral start tank flows into the pneumatic control package through a filter and a roughing regulator, which reduces the helium pressure to 450 psi. A spring-loaded, ball-type relief valve that relieves at 3800 psi and reseats at 3500 psi, will bleed off excessive pressure at the control package inlet. At engine start command, the helium control valve opens and loads the primary regulator dome with 450 psi. The control regulator samples the output of the primary regulator and controls primary regulator positions to maintain system pneumatic control pressure at 400 psi (approximately). Helium exits the pneumatic control package through a check valve to the mainstage control valve, the ignition phase control valve, the pressure accumulator, and to the engine purge lines. Maximum flowrate through the regulators is 1300 scfm. A low pressure relief valve prevents over-pressurization of the control system. Cracking pressure is 497 psig, and reseat pressure is 420 psig. The helium control and helium tank emergency vent valves are three-way, electrically operated solenoid valves. The mainstage, ignition phase, and STDV control valves are four-way, electrically operated solenoid valves with opening and closing functions arranged so that one is venting while the other is pressurizing. The STDV control valve is not part of the pneumatic control package but is mounted on and controls the start tank discharge valve. The ignition phase control valve and the helium control valve are energized by the J-2 engine start command. An STDV timer controls the STDV control valve, and the mainstage control valve is energized by an ignition phase timer during the engine start sequence. During checkout and prelaunch operations, commands to energize the solenoids are monitored on the S-IVB engine panel through the DDAS. Components test switches on the S-IVB engine panel permit manual actuation of all four solenoids as necessary during checkout. In flight, commands that energize the valves are telemetered back to ground receiving stations and recorded as event measurements in addition to being displayed on the engine panel.

Fast Shutdown Valve. The fast shutdown valve vents the gas generator control valve permitting it to close and vent the turbine bypass valve permitting it to open during engine shutdown. Pneumatic pressure from the normally open port of the ignition phase control valve opens the normally closed two-position poppet valve. The valve is mounted on the main fuel duct at the fuel turbopump outlet.

Purge Control Valve. The purge control valve is used to control the thrust chamber oxidizer dome and gas generator purges. The valve is actuated by pneumatic pressure entering the control port, acting upon the piston to overcome the force of the piston spring, and moving the piston to open position. When the valve is not actuated, the force of the spring maintains the valve in closed position and the outlet side of the valve is vented through the vent port.

Pneumatic Accumulator. The pneumatic accumulator is an integral part of the primary instrumentation package, which is enclosed within an outer shell and cover. The volume between the primary instrumentation package and the outer shell and cover serves as the pneumatic accumulator. The purpose of the pneumatic accumulator is to provide the necessary gas volume for the safe operation and shutdown of the engine in the event of a loss of high pressure pneumatic supply.

Electrical Control System. The electrical control system consists of a sequence controller to properly sequence engine start and cutoff, and a spark ignition system to establish ignition in the gas generator and in the augmented spark igniter chamber. The purpose of the electrical control system is to control engine operation by means of electrical signals and to supply power to establish ignition. The electrical control package (figure 5-24) is a sealed, dome-shaped, pressurized control assembly, containing spark exciters and sequence controller circuitry. At engine start or cutoff, the sequence controller performs the necessary sequencing and timing functions required to properly operate the engine system. The electrical control package circuitry will automatically reset for restart capability. The sequence controller, mounted in the electrical control package, is composed of solid-state switching elements, which perform the necessary logic functions to properly sequence, time, and monitor the engine system. The system is completely self-contained and requires only dc power and external engine start and cutoff signals for operation. Additional signals are provided to the stage to allow monitoring of the engine condition at significant points of engine operation. This system also has the capability to actuate individual components through properly designed checkout equipment.

Thrust OK Pressure Switches. Mainstage thrust OK pressure switches consist essentially of an inlet port, a checkout port, 2 diaphragms, toggle blades, a toggle spring, a housing, and an electrical switch and connector. Pressure entering the inlet port

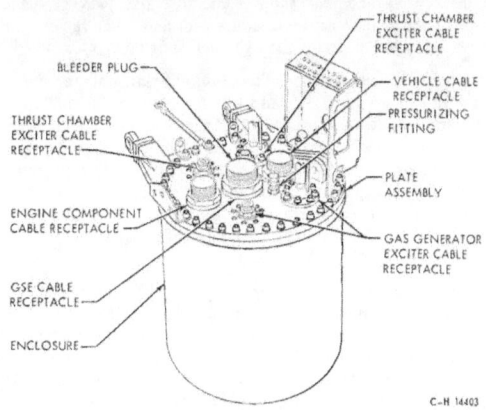

Figure 5-24

acts upon a diaphragm linked to the electrical switch through the toggle spring and toggle blades. As engine oxidizer injector pressure increases, electrical continuity is switched from normally closed contact to normally open contact and an electrical circuit is completed for producing a mainstage OK signal. If oxidizer injector pressure deteriorates, the pressure switch deactuates, breaking the contact, interrupting the mainstage OK signal, and re-establishing a mainstage OK depressurized signal. Proper operation of the switch may be verified by applying pressure to the checkout port which is independent of the inlet pressure port. The two switches are mounted opposite each other on the injector at the lox inlet. A sensing line attached to each pressure switch and the customer connect panel (engine/stage interface) permits remote checkout. Signals from the pressure switches are monitored on the S-IVB engine panel. During flight, actuation and deactuation signals from the switches are telemetered to ground receiving stations as event measurements. The MAINSTAGE SW NO. 1 PRESS and MAINSTAGE SW NO. 2 PRESS indicators on the engine panel and events panel illuminate when lox injector pressure is above 500 ± 330 psi. The MAINSTAGE NO. 1 DEPRESSURIZED and MAINSTAGE NO. 2 DEPRESSURIZED indicators illuminate when lox injector pressure is below the 375-psig minimum actuating pressure of the thrust OK pressure switches.

Flexible Armored Harness. The flexible armored harness consists of Teflon-insulated wires that terminate in modified RD (MS R series) connectors. The conductor wires are wrapped in a layer of Mylar tape, sleeved inn a silicone rubber tube, and sheathed in two layers of nickel-plated copper wire braid. The silicone rubber tube is for thermal protection and the wire braid is for protection against abrasion and radio frequency interference. Mylar tape is used only to facilitate installation of the silicone rubber tube. After installing the wire braid sheaths, exposed ends of braid are soldered to unify the strands. The wires are soldered or brazed to pins of the connectors sealed in the connector housing by neoprene rubber grommets. To smooth the contour of the harness, potting compound is applied to each Y-joint and connector prior to installing the wire braid sheaths. A compound of polyurethane is overmoulded at the Y-joints and connectors to secure and cover the braid pigtails. When the harness is installed on the engine, a thermoprotective boot is installed over each connector.

Start System.

The start system is comprised of an integral helium and hydrogen start tank, which contains hydrogen (GH_2) and helium gases for starting and operating the engine; a start tank discharge valve, which contains the GH_2 in the tank until engine start; a helium fill-check valve and tank support and fill valve package, which supply gases to the start tank; and a vent and relief valve, which relieves pressure or drains the hydrogen start tank.

Integral Helium and Hydrogen Start Tank. The integral tank consists of a 4.2 ft^3 sphere for GH_2 and a 0.58 ft^3 sphere inside for helium. GN_2 stored under 1250 psig and 200° F provides the energy source for starting the engine while helium stored under 3000 psig supplies the pneumatic requirements for engine control system operation. Since a restart capability is not required for the SL-2 mission, a start tank inflight refill line has been blanked off. Insulation covering the start tank prevents excessive internal pressure buildup from effects of external temperatures.

During prelaunch operations the integral start tank is purged with helium, and then pre-chilled, and filled with cold GH_2 and helium. Replenish supply is maintained to the helium tank and to the GH_2 start tank until T-3 sec. GH_2 enters the start tank through the tank support and fill package mounted on the start tank. GH_2 from the ground source, or from the engines having repressurization capability, flows through poppet check valves into the tank. The recharge fill port contains a filter to remove contaminants from GH_2 entering the start tank from the thrust chamber.

Helium enters the helium tank through the helium cover and fill check valve. The cover contains a tank support for mounting the start tank, an outlet line to the pneumatic control package, and a mount for temperature transducer VXC7-401. The poppet-type fill check valve prevents pressure loss at umbilical disconnect.

Start Tank Discharge Valve. A pneumatically controlled, spring-closed STDV contains the GH_2 in the start tank until needed for engine start. A control valve directs pneumatic pressure from the main-fuel-valve sequence valve to open and close the STDV. Until the start tank discharge delay timer expires during the engine start sequence, the STDV control valve maintains pressure to the closing side of the STDV actuator. When the timer expires, the control valve vents the closing actuator and pressurizes the opening actuator. GH_2 then flows through the STDV to the fuel turbopump turbine. A gate-type check valve on the STDV outlet prevents gas-generator combustion products from entering the STDV and contaminating the STDV poppet. Open and closed position switches and a potentiometer monitor the STDV position. During checkout operations, the STDV position (START TANK DISCHARGE OPEN or START TAANK DISCHARGE CLOSED) is monitored on the S-IVB engine panel. During flight the STDV positions are telemetered back to KSC and recorded. The STDV control valve opening command and the STDV OPEN signals are monitored in real time on the S-IVB engine panel. See J-2 Engine Monitoring.

Start Tank Vent and Relief Valve. The STDV vent and relief valve controls GH_2 flow from the start tank. During servicing operations, control pressure from the S-IVB pneumatic control system opens the vent and relief valve to permit GH_2 flow through the start tank for chilling. The GH_2 exits the start tank through the vent and relief valve and flows through umbilical lines to the facility burn pond. Removal of the control pressure permits the STDV to fill with GH_2. During flight the valve relieves excessive pressure in the start tank. Cracking pressure is 1395 psi minimum and reseat pressure is 1335 psi.

Start Tank Emergency Vent Valve. The start tank emergency vent valve provides a redundant means of venting pressurized gas from the start tank. The valve is actuated, in an emergency, from a ground source. The emergency vent valve is a solenoid-operated, two-way, spring-loaded to the closed position, poppet-type valve mounted on the start tank support-and-fill valve. The valve spring holds the poppet on the seat, against start tank pressure, when

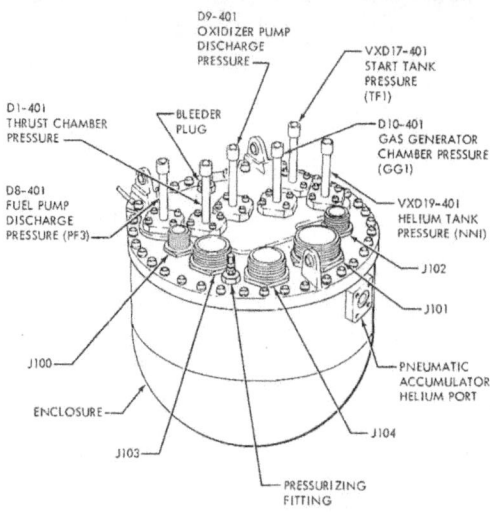

Figure 5-25

the solenoid is deenergized. When the solenoid is energized, the armature moves to overcome the spring pressure, allowing the flexure to unseat the poppet and vent start tank pressure through the valve outlet. The outlet is connected to a line that is teed to the fuel turbopump primary seal drain line.

J-2 Engine Measuring.

The J-2 engine measuring system, consisting of a primary instrumentation package (figure 5-25), an auxiliary instrumentation package (figure 5-26), and transducers, monitor 36 conditions in

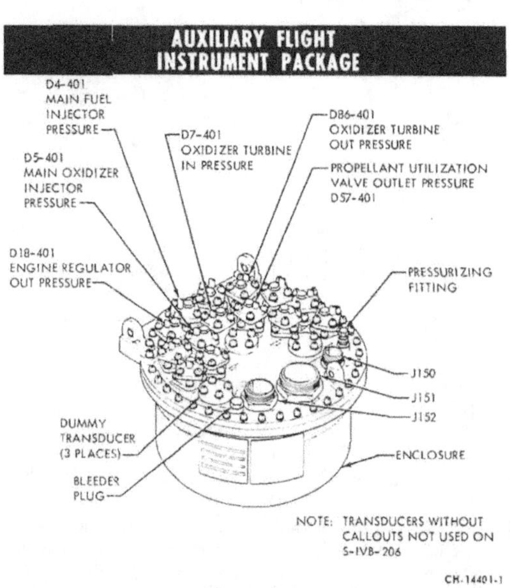

Figure 5-26

6 parameters and 27 event signals generated by valve positioning and sequencer commands. The primary instrumentation package provides instrumentation for monitoring critical engine parameters during static testing and vehicle flight. The auxiliary instrumentation package monitors non-critical engine parameters. System design allows for substitution or deletion of auxiliary package functions without affecting primary instrumentation function. Temperature sensors located in the area of temperature samplings, pressure transducers located in the instrumentation packages with sensing lines to the sampling points, flowmeters located in the lines, and position indicator portions of the valve assemblies acquire the measurement data and supply it to the S-IVB stage telemetry system. The S-IVB stage telemetry system telemeters the measurement information to ground receiving stations for recording.

Some measurements are monitored in real time by indicators and gages on panels, others are displayed on analog recorders. See figure 5-27 for J-2 engine measurement summary. Measurements are taken in the following parameters: temperature, C; pressure, D; flow, F; position, G; events, K; voltage, M; and RPM, T. Two-thousand-ohm potentiometers and/or position switches provide position indicator signals for the main oxidizer, main fuel, gas generator control, oxidizer turbine bypass, start tank discharge, mixture ratio control, augmented spark igniter oxidizer, and propellant bleed valves. Both turbopumps are equipped with magnetic pickups to measure turbopump speed and to provide a turbine overspeed cutoff signal for static testing. The magnetic pickups are utilized to provide turbopump speed measurements for the instrumentation system. The output of the magnetic transducers is designed for generation of a 1- to 3-volt pulse suitable for direct telemetry. The fuel turbopump rotor is fabricated from K-monel, which does not exhibit magnetic qualities until chilled to -300° F. Therefore, checkout of the measurement by spinning the turbopump is not feasible at ambient temperatures. Electrical checkout can be accomplished, however, by applying a voltage to the checkout coil and inducing a voltage in the signal coil. This check may be made at either ambient or cryogenic temperatures. Flowmeters provided within the high pressure propellant discharge ducts measure main fuel and main oxidizer flowrates. The basic element of the flowmeter is a helical-vaned rotor, which is turned by propellant flow to measure flow velocity. The flow diameter is closely controlled to permit accurate determination of the volumetric flowrate. Within the fuel flowmeter is a four-vane rotor that produces four electrical impulses per revolution and turns approximately 3600 rpm at nominal flow. The oxidizer flowmeter includes a six-vane rotor producing six electrical impulses per revolution and turns approximately 2400 rpm at nominal flow. The output of the magnetic transducers is designed for generation of a 1- to 3-volt pulse suitable for direct telemetry. Electrical checkout of the flowmeter can be accomplished by supplying a voltage to the checkout coil and inducing a signal in the measurement coil. The 12 event (K) measurements are discrete measurements that indicate when the various engine functions occur. These measurements are telemetered to the ground receiving stations and recorded. See figure 5-28. The measurements with a 'V' prefix are monitored in real-time on the LCC-SIVB engine panel. Mission control center monitors flight control measurements in real-time. These displays assist the flight controller in making decisions affecting the mission and safety of the crew.

Acceptance Firing.

The S-IVB-206 stage was installed in the Sacramento test center beta complex test stand III on June 30, 1966 for acceptance testing. Prefiring checks on the propulsion system included a manual stage and GSE control checkout, a system leak check, an automatic checkout, and a final leak check procedure. Acceptance firing of the stage was conducted on August 19, 1966. After the propulsion

J-2 ENGINE ANALOG MEASURMENTS

NUMBER	NAME	RANGE
A12-403	ACCELERATION, GIMBAL BLOCK	0 TO 5 G
C1-401	TEMP, FUEL TURBINE INLET	460 TO 2260°R
C2-401	TEMP, OXID TURBINE INLET	460 TO 1660°R
VXC6-401	TEMP, GH$_2$ START BOTTLE	110 TO 560°R
VXC7-401	TEMP, ENGINE CONTL He	110 TO 560°R
XC11-401	TEMP, ELECT CONTL PKG	160 TO 660°R
XC12-401	TEMP, GG FUEL BLD VLV	35 TO 85°R
C133-401	TEMP, OXID PUMP DISCHARGE	160 TO 210°R
C134-401	TEMP, FUEL PUMP DISCHARGE	35 TO 60°R
C197-401	TEMP, PRI INSTR PKG	160 TO 660°R
C198-401	TEMP, AUX INSTR PKG	160 TO 660°R
VXC199-401	TEMP, THRUST CHAMBER JACKET	35 TO 560°R
XC200-401	TEMP, FUEL INJECTION	35 TO 560°R
C215-401	TEMP, OXID TURBINE OUTLET	440 TO 1460°R
D1-401	PRESS, THRUST CHAMBER	0 TO 1000 PSIA
D4-401	PRESS, MAIN FUEL INJECTOR	0 TO 1000 PSIA
D5-401	PRESS, MAIN OXID INJECTOR	0 TO 1000 PSIA
D7-401	PRESS, OXID TURB INLET	0 TO 200 PSIA
D8-401	PRESS, FUEL PUMP DISCHARGE	0 TO 1500 PSIA
D9-401	PRESS, OXID PUMP DISCHARGE	0 TO 1500 PSIA
D10-401	PRESS, GG CHAMBER	0 TO 1000 PSIA
*VXD17-401	PRESS, GH$_2$ START BOTTLE	0 TO 1500 PSIA
*D18-401	PRESS, ENG REG OUTLET	0 TO 750 PSIA
*VXD19-401	PRESS, ENG CONTL He SPHERE	0 TO 3500 PSIA
D57-401	PRESS, PU VALVE OUTLET	0 TO 500 PSIA
D86-401	PRESS, OXID TURBINE OUTLET	0 TO 100 PSIA
XD241-401	PRESS, GH$_2$ START BOTTLE, BACKUP	0 TO 1500 PSIA
XD242-401	PRESS, ENG CONTL He SPHERE, BACKUP	0 TO 3500 PSIA
D266-401	PRESS, THRUST CHAMBER OSCILL	-5 TO +5 PSIA
F1-401	FLOWRATE, OXID	0 TO 3000 GPM
F2-401	FLOWRATE, FUEL	0 TO 9000 GPM
*G3-401	POSITION, MAIN OXID VALVE	0 TO 100%
*G4-401	POSITION, MAIN FUEL VALVE	0 TO 100%
G5-401	POSITION, GAS GEN VALVE	0 TO 100%
*G8-401	POSITION, OXID TURBINE BYPASS VALVE	0 TO 100%
G9-401	POSITION, GH$_2$ START VALVE	0 TO 100%
*VXG17-401	POSITION, MIXTURE RATIO CONTROL VALVE	0 TO 65 DEG
M6-401	VOLTAGE, ENGINE CONTROL BUS	0 TO 30V
M7-401	VOLTAGE, ENGINE IGNITION BUS	0 TO 30V
T1-401	SPEED, OXID PUMP	0 TO 12 K RPM
T2-401	SPEED, FUEL PUMP	0 TO 30 K RPM

MEASUREMENT NUMBER PREFIXES INDICATE THE FOLLOWING:
* FLIGHT CONTROL; V-ESE DISPLAY; X-AUXILIARY DISPLAY

Figure 5-27

Section V S-IVB Stage

| J-2 ENGINE EVENT MEASUREMENTS ||
NUMBER	NAME
VK5-401	MAINSTAGE CONTL SOL ON
VK6-401	IGNITION PHASE CONTL SOL ON
K7-401	HELIUM CONTL SOL ON
VXK8-401	IGNITION DETECTED
VK10-401	TC SPARK SYS ON
VK11-401	GG SPARK SYS ON
*K12-401	ENGINE READY SIGNAL
*VXK13-401	CUTOFF SIGNAL
*VK14-401	MAINSTAGE OK PRESS SW 1
K20-401	ASI LOX VALVE OPEN
VK95-401	TC INJECTOR TEMP OK
VK96-401	STDV CONTL SOL ON
K116-401	GG CONTL VALVE CLOSED
VK117-401	GG CONTL VALVE OPEN
VK118-401	MAIN FUEL VALVE OPEN
K119-401	MAIN FUEL VALVE CLOSED
VK120-401	MAIN OXID VALVE OPEN
K121-401	MAIN OXID VALVE CLOSED
VK122-401	STDV OPEN
K123-401	STDV CLOSED
VK124-401	OXID TURBINE BYPASS VALVE OPEN
VK125-401	OXID TURBINE BYPASS VALVE CLOSED
*K126-401	OXID BLEED VALVE CLOSED
*K127-401	FUEL BLEED VALVE CLOSED
*VK157-401	MAINSTAGE OK PRESS SW 2
K158-401	MAINSTAGE OK PRESS SW 1 DEPRESSURIZED
K159-401	MAINSTAGE OK PRESS SW 2 DEPRESSURIZED

MEASUREMENT NUMBER PREFIXES INDICATE THE FOLLOWING:
*FLIGHT CONTROL; V-ESE DISPLAY; X-AUXILIARY DISPLAY

CH-14400-1

Figure 5-28

system achieved 436.1 sec of mainstage operation, a computer controlled program issued the cutoff command. Major engine events of the acceptance test firing included:

a. Normal engine start sequence.

b. Propellant utilization system activation 6 sec after engine start command.

c. Engine side load restrainer links were released approximately 25 sec after mainstage control.

d. An automatically controlled gimbal program was initiated after restrainer link release.

e. After 300 sec of mainstage control, the propellant utilization valve repositioned from the lox-rich approximate mixture ratio of 5.5 to 1.0 to the nominal position for reference mixture ratio of 4.7 to 1.0 for duration of firing.

f. After 436.1 sec of mainstage operation, automatic cutoff was initiated at 2216 lbm of lox, or 596 lbm of LH_2. Depletion sensors armed for 3-percent residual mass served as backup for the cutoff.

g. Post-firing inspection revealed that metal chips were present in the turbine exhaust system and a damaged turbine seal was discovered. The lox turbopump was replaced and a 70 sec verification firing at 5.5 EMR was conducted on September 14, 1966.

Postfiring propulsion system checks included test equipment removal, leak check of the system, and system automatic checks. Inspection checks reverified engine alignment and structural integrity after the static acceptance firing test.

PROPELLANT SYSTEMS.

The S-IVB propellant tanks were sized to accommodate the Saturn V mission requirements. The propellant masses loaded provide hydrogen and oxygen for J-2 engine operation, boiloff, LH_2 tank pressurization, and usable and unusable residuals. The propellant tank fill systems were designed to minimize countdown and prelaunch time and to be compatible with the loading times scheduled for the other stages of the Saturn V vehicle. The fill systems were sized to flow 1000 gpm of liquid oxygen and 3000 gpm of liquid hydrogen. Initial fill rates are slower to accomplish tank chilldown and to prevent ullage pressure collapse. Final fill rates are also slower to provide greater precision in obtaining the desired 100% load.

The propellant tank vent systems were designed to protect the tank structure under all conditions of propellant tank loading, stage powered flight and orbital venting. During loading, tank pressures are maintained well below the normal tank prepressurization levels. The negative pressure differential across the common bulkhead between the fuel and oxidizer tanks is the limiting factor. This differential pressure is never allowed to exceed 3.0 psi for normal controlled operation. Vent outlets are so located and directed as to prevent disturbing moments on the stage during venting in flight.

Propellant Characteristics.

Figure 5-29 lists the physical and chemical properties of LH_2, and figure 5-30 shows the LH_2 vapor pressure curve. See the S-IB stage lox system description in Section IV for lox characteristics.

Fuel Tank.

The LH_2 is stored in an insulated tank with a capacity of approximately 44,300 lbm at a temperature of -423° F. An ullage volume of approximately 400 ft^3 is maintained at this load level; however, for Saturn IB earth orbit missions a full load of LH_2 is not carried. The approximate LH_2 load for S-IVB-206 is 38,000 lbm, which includes quantities required for programmed mixture ratio operation (5.5 to 1 for approximately 325 sec) and unusable propellants. The LH_2 tank is pressurized between 28 and 31 psia. Pressure is provided by a ground supply of helium for prepressure and is maintained by a hydrogen bleed from the engine during burn. Tank venting and relief is accomplished through parallel valves installed in a top-mounted vent system which exits through nonpropulsive vent outlets in the forward skirt. An anti-vortex screen is installed over the engine feed duct inlet.

Fuel Loading.

Chilldown. The automatic fuel loading sequence is started 4 hr and 19 min before liftoff by manually depressing the fill push button on the LH_2 control panel in the storage facility. The chilldown sequence begins and the following five operations occur simultaneously.

a. The vehicle LH_2 vent-and-relief valve opens (figures 5-31 and 5-32). With the directional control valve now in the ground position, the vent gases flow through the umbilical to the facility burn pond.

b. The facility line chilldown valve, fill valve (on LH_2 control sled),

Section V S-IVB Stage

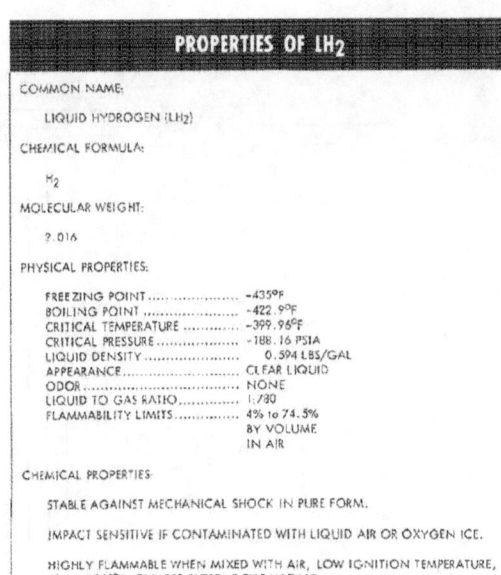

Figure 5-29

and LH_2 debris valve (on service arm 6) open; and the vehicle fill and drain valve opens establishing a flow path from the storage facility LH_2 tank to the vehicle fuel tank.

c. The gas heat exchanger inlet valve (on the umbilical tower) opens and internal level sensors become active filling the heat exchanger with LH_2 to a controlled level. See Lox Pressurant Loading for gas heat exchanger function.

d. A 4-min facility timer starts.

e. The helium nozzle purge starts. (Purge of the LH_2 fill and drain disconnect).

f. At the end of four min, the facility storage tank pressurization valve opens allowing LH_2 to flow through a vaporizer producing GH_2 that flows into the storage tank ullage space providing pressure to force out the LH_2 for vehicle loading.

As pressure builds in the LH_2 storage tank, and at the end of four min, LH_2 begins flowing through the transfer line increasing to 500 gpm. This flow chills the transfer system including the vehicle LH_2 tank.

Initial Fill. When the facility 4-min timer expires, the facility fill valve partially closes, the line fill valve opens, and the line chilldown valve closes starting the initial fill sequence. LH_2 flows into the vehicle tank at 500 gpm until the PU (propellant utilization) mass probe signals the LH_2 tanking computer that LH_2 has reached the 5-percent level. This requires about 7 min.

Main Fill. When LH_2 reaches the 5-percent level, the tanking computer commands the facility fill valve to fully open increasing the flow rate to 3000 gpm. When either the PU mass probe or the fast fill sensor indicate the LH_2 has reached the 96-percent level (after approximately 24 min), the tanking computer starts the slow fill sequence.

Slow Fill. The facility fill valve partially closes and the replenish valve (on LH_2 control sled) opens establishing a flow of 500 lbm/min until the LH_2 level reaches 100-percent. This requires approximately 9 min after which the tanking computer starts the replenish sequence.

Replenish. The facility fill valve closes and LH_2 flow through the replenish valve maintains the tank level at 100 percent. Should boiloff exceed the replenish valve capacity and drop the level below 90 percent, the tanking computer will partially open the facility fill valve until the level again reaches 100 percent.

Umbilical Drain. Replenish ends and umbilical drain begins at start of automatic sequence. Then the vehicle LH_2 fill and drain valve closes, the umbilical line drain valve opens, and helium purges the umbilical line. At liftoff the service arm debris valve closes.

Fuel Tank Pressurization.

Prepressurization. The prepressurize-for-launch command 1 min 37 sec before liftoff starts 800-psig ground helium flow through the stage pressurization system (figures 5-33 and 5-34) into the tank ullage space to assure the necessary head at the engine pump at ignition. When the tank ullage reaches 31 psia, a pressure switch actuates, sending a signal to close the ground prepressurization supply valve. If the pressure drops to 28 psia, the switch will deactuate opening the supply valve replenishing the ullage gas. After liftoff no pressurization supply is required until engine ignition after which the flight pressurization system is enabled. Overpressurization during any period, though not expected, would relieve through mechanical operation of the vent and relief valve or latching vent and relief valve in the 31 to 34 psia band. During flight with the directional control valve in the flight position, vent gases pass through two 4-in. ducts that exhaust through the skirt directly opposite each other for total thrust cancellation.

Flight Pressurization. GH_2 bled from the J-2 engine flows to the pressurization control module, which has three orificed flow paths and into the ullage space of the fuel tank. One path is always open, but normally-open solenoid valves control the other two. A switch selector command (channel 68) 2.6 sec after J-2 ignition closes one solenoid valve and enables the circuit so the flight control

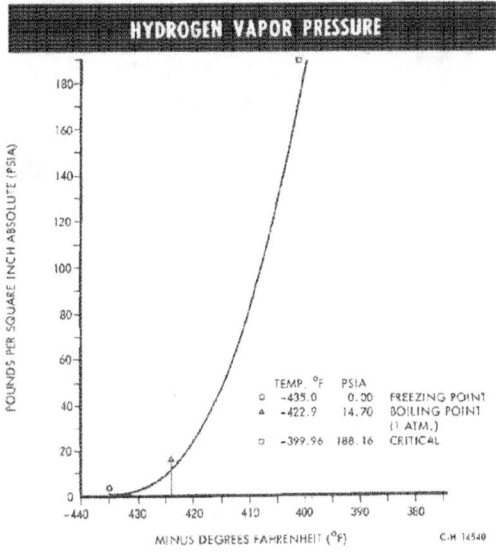

Figure 5-30

5-29

Section V S-IVB Stage

S-IVB FUEL SYSTEM

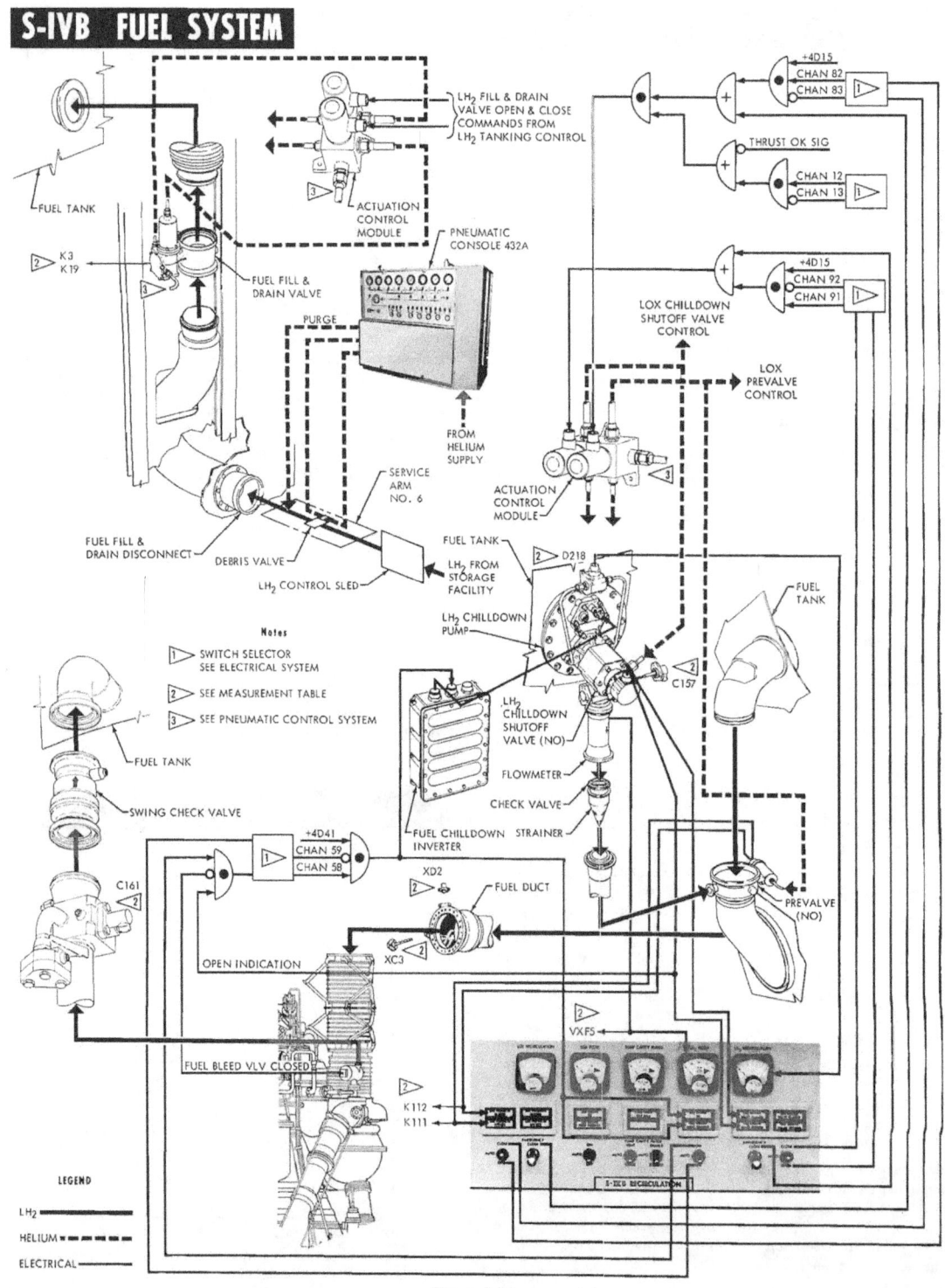

Figure 5-31

pressure switch can control the other valve. When the tank pressure is above 28 psia, the pressure switch actuates holding the valve closed restricting the GH_2 flow to a single path. When the tank pressure drops below 28 psia, the switch deactuates allowing the valve to return to the normally-open position providing two GH_2 flow paths. After 5 min 0.2 sec, another command (channel 69) disables the control circuits opening both controlled paths and allowing maximum GH_2 flow during the last portion of J-2 engine burn (about 2 min 15 sec) when the liquid level reaches a point that bulk temperature stratification requires increased ullage pressure to meet engine inlet requirements.

LH_2 Chilldown.

During fuel loading, the 10-in. prevalve remains open (its normal position) providing a partial chilldown of the feed system. At 10 min before liftoff, the LH_2 chilldown pump is started. An interlock is provided to ensure that the vehicle LH_2 tank is at least 10 percent full prior to energizing the pump. The prevalve is closed 5 sec after the pump starts. The LH_2 is now routed from the S-IVB stage tank, through the pump, chilldown shutoff valve, line strainer, and J-2 engine feed duct, fuel turbopump and fuel bleed valve, and returned to the LH_2 tank through the return line and check valve. The pump delivers 135 gpm with a differential pressure of 6 to 9 psi. This recirculation is required to maintain a liquid phase in the turbopump to ensure proper fuel quality for engine start and to prevent turbine overspeed and possible disintegration at engine start. The LH_2 chilldown pump operates continuously until 0.6 sec prior to J-2 engine start. At 4.5 sec before J-2 start command, a switch selector command (channel 83) opens the LH_2 prevalve. Since the chilldown pump is still operating, the LH_2 is diverted from the normal circulation path and flows back through the LH_2 prevalve and into the fuel tank. The backflow of LH_2 clears any vapor entrapment from the fuel tank outlet area.

Fuel Measurements.

Figure 5-35 lists the fuel system flight measurements and indicates the information that is monitored in real time. Measurements VXD177 and VXD178 are also displayed in analog form on the command module main display console providing fuel tank pressure data for the crew.

Lox Tank.

The lox is stored in a tank formed by the aft dome and common bulkhead with a capacity of approximately 194,000 lbm of lox allowing for approximately 70 ft^3 of ullage. The lox tank is loaded to capacity for Saturn IB missions. The tank ullage pressure is maintained between 37.0 and 40.8 psia during boost and engine operation using GHe. Two parallel valves provide venting and overpressure relief through the aft skirt. An anti-vortex screen is installed over the engine feed duct inlet.

Lox Loading.

The following steps describe the automatic lox loading operation as controlled and monitored in the LCC. See figures 5-36 and 5-37.

Chilldown. When the S-IB lox level reaches 65%, the S-IVB main line chilldown begins by opening the S-IVB replenish valve on the lox control sled, the S-IVB lox vent valve, the lox debris valve, and the S-IVB lox fill and drain valve. When these valves are open, main storage tank pressure causes lox to flow slowly through the S-IVB main line causing chilldown. When the S-IB lox level reaches 97%, S-IVB slow fill starts and the tank is filled to 5 percent at a maximum flowrate of 5225 lbm/min.

Main Fill. When the level reaches 5 percent (sensed by the PU lox mass probe), the flowrate is increased to a maximum of 10,494 lb/min. After approximately 22 min, the level reaches 94 percent and slow fill begins.

Slow Fill. When the level reaches 94 percent, the flowrate is reduced to a maximum of 3146 lbm/min. In approximately 1 min 30 sec, 99 percent of flight mass is loaded.

Replenish. When the level reaches 99 percent, the tanking computer enables the lox replenish flow control valve. The computer controls the replenish flowrate to match the boiloff rate. The maximum replenish flowrate is 575 lbm/min. The replenish sequence ends at the start of the automatic sequence. At that time the lox fill and drain valve closes, the lox replenish flow control valve closes, and the umbilical line vent valve and the umbilical purge valve open. The lox debris valve closes at liftoff.

Lox Tank Prepressurization.

The automatic lox tank prepressurization operation begins 2 min 47 sec before liftoff. At this automatic command the lox tank vent-and-relief-valve closes (figures 5-38 and 5-39). The two normally-closed helium supply valves are opened allowing helium to flow at a rate of 20 lbm/min maximum, at -360° F and 2000 psig for 15 sec. The flow is through the aft umbilical into the plenum chamber, through the normally open J-2 engine heat exchanger bypass valve, and the pressure regulating orifices, and into the lox tank. When the lox tank pressure reaches 40 psia, the prepressurization and flight control pressure switch controlling the helium supply shutoff valve actuates closing the valves. If the pressure drops below 37 psia, the switch deactuates allowing more helium flow into the tank. The helium ground supply valve remains active until S-IB ignition command (3 sec before liftoff) when it closes automatically and the supply line vent valve opens, allowing the

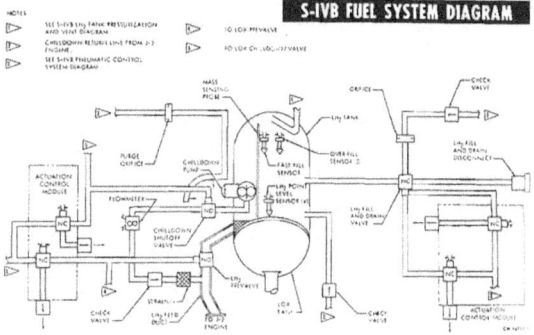

Figure 5-32

Section V S-IVB Stage

S-IVB FUEL TANK PRESSURIZATION AND VENT SYSTEM

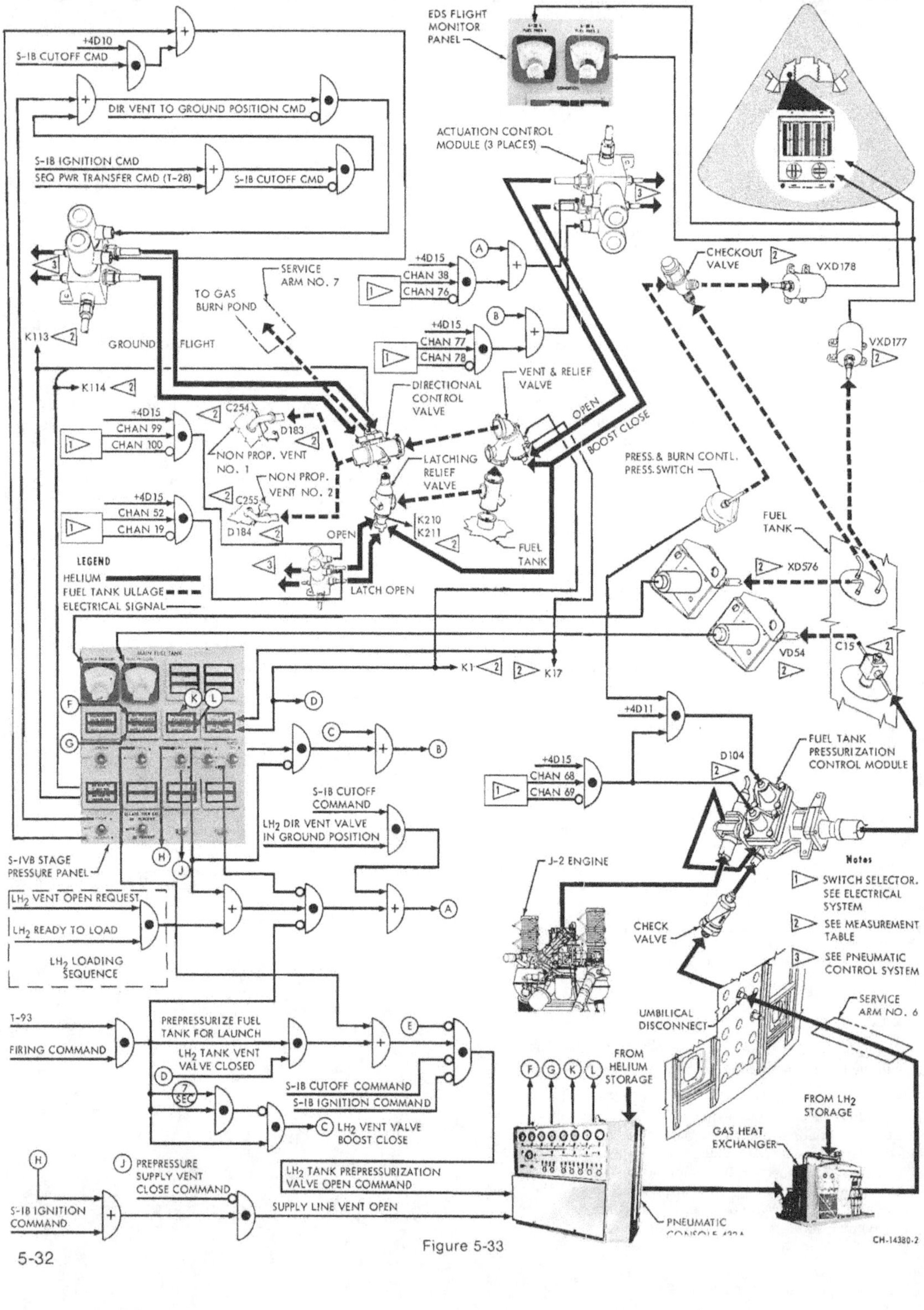

Figure 5-33

helium supply line to become inerted from the stage supply check valve to the closed helium supply valve.

Lox Pressurant Loading.

Helium is used as the pressurant in the lox tank during flight. The helium is stored in the vehicle in six spheres manifolded together in the fuel tank. The filling of these spheres is a ground controlled three step operation. The first transfer is at a pressure of 950 ± 50 psig, ambient temperature, and a flowrate of 5.1 lbm/min maximum. The minimum time required to make this transfer of 14 lbm of helium is 4 min. The helium pressure is regulated as it flows through pneumatic console 432A the gas heat exchanger, and pneumatic console 433A and then flows through service arm 6 into the aft umbilical of the stage and into the spheres.

Prior to the lox loading, the second step of helium transfer starts. The pressure is increased to 1450 ± 50 psig, still at ambient temperature, and the flowrate is 7.6 lbm/min maximum. During this step, 7 lbm of helium is transferred in 2 min 24 sec. The third step begins after the helium spheres become immersed in LH_2 approximately 4 hr before liftoff. The gas heat exchanger is filled with LH_2 and the helium is chilled to -410° F as it passes through. The cold helium pressure is increased to 3100 ± 100 psig and loaded at a flowrate of 32.5 lbm/min maximum, until a total mass of helium reaches 246 lbm. Continuous pressurization is available to the spheres until 3 sec before launch. The cold helium dump module contains a solenoid vent valve that will vent the spheres upon command from the LCC S-IVB stage pressure panel. This module also contains a relief valve that guards against over pressurization by cracking at 3500 psig. It reseats at 3200 psig.

Lox Tank Flight Pressurization.

During S-IB powered flight, lox ullage makeup cycles are initiated. The cold helium shutoff valves are opened 2.5 sec prior to S-IVB engine start and remain open throughout burn. Cold helium flows from the storage spheres through the control module, where it is filtered and regulated to 385 (+28-32) psia, and into the compressed gas tank plenum. The medium pressure switch senses plenum pressure and backs up the regulator by closing the two shutoff valves if plenum pressure reaches 465 psia. When the pressure decreases to 350 psia, the switch deactuates allowing the shutoff valves to open again. From the plenum, helium flows to the lox tank ullage space through three paths: (1) through an orifice and directly into the tank, (2) through the heat exchanger coil and an orifice, and (3) through the heat exchanger coil, and the normally-open heat exchanger bypass valve.

The helium that flows through the heat exchanger is warmed and expanded. At engine start plus 23.4 sec, the flight control pressure switch control of the heat exchanger bypass valve is enabled. The

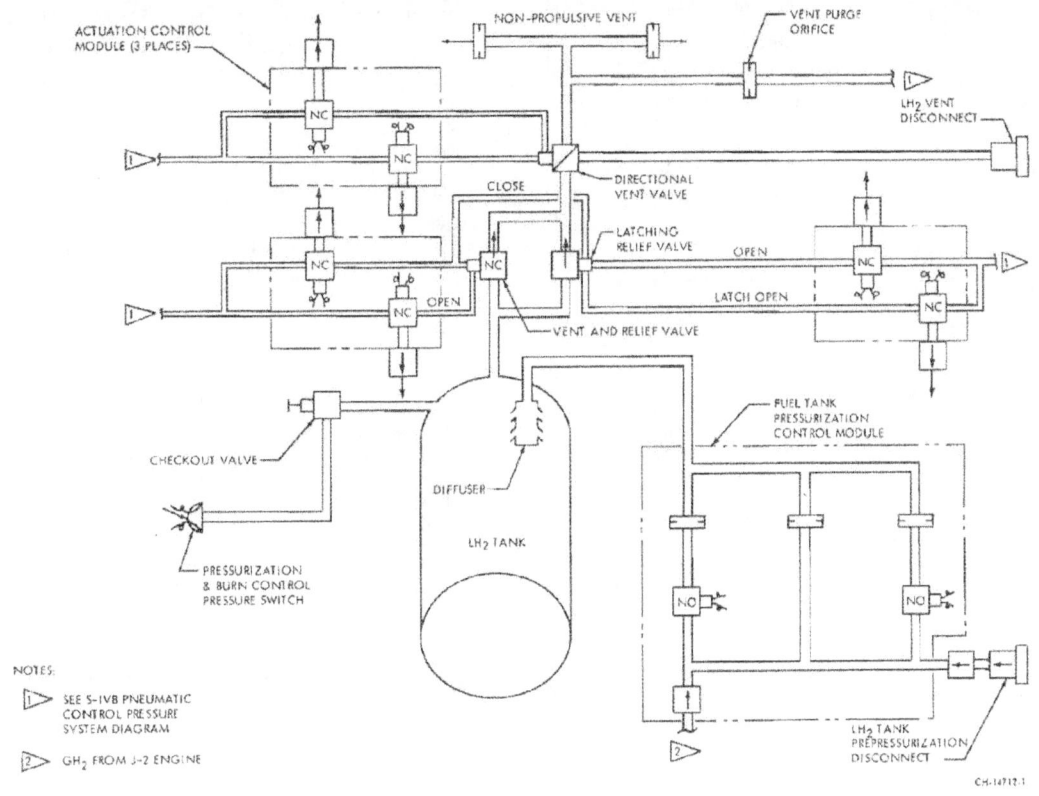

Figure 5-34

Section V S-IVB Stage

\multicolumn{3}{c}{LIQUID HYDROGEN SYSTEM MEASURMENTS}		
NUMBER	NAME	RANGE
*XC3-403	TEMP, FUEL PUMP INLET	35 TO 47°R
C15-410	TEMP, FUEL TANK GH_2 INLET	50 TO 300°R
C52-408	TEMP, FUEL TANK POS. 1	35 TO 47°R
C157-404	TEMP, FUEL CIRC PUMP OUTLET	35 TO 50°R
C161-424	TEMP, LH_2 CIRC RET LINE TANK INLET	35 TO 50°R
C254-409	TEMP, LH_2 TANK NON-PROP VENT 1	25 TO 260°R
C255-409	TEMP, LH_2 TANK NON-PROP VENT 2	25 TO 260°R
*XD2-403	PRESS, FUEL PUMP INLET	0 TO 60 psia
VD54-410	PRESS, FUEL TANK INLET	0 TO 100 psia
D104-403	PRESS, LH_2 PRESS MODULE INLET	0 TO 1000 psia
*VXD177-408	PRESS, FUEL TANK ULLAGE EDS 1	0 TO 50 psia
*VXD178-408	PRESS, FUEL TANK ULLAGE EDS 2	0 TO 50 psia
*D183-409	PRESS, LH_2 TANK NON-PROP VENT 1	0 TO 50 psia
*D184-409	PRESS, LH_2 TANK NON-PROP VENT 2	0 TO 50 psia
*D218-403	PRESS, LH_2 CHILLDOWN PUMP DIFF	-30 TO 30 psid
*VXF5-404	FLOWRATE, FUEL CIRCULATION PUMP	0 TO 160 gpm
*K1-410	EVENT, FUEL TANK VENT VALVE CLOSED	
K3-427	EVENT, FUEL FILL VALVE CLOSED	
*K17-410	EVENT, FUEL TANK VENT VALVE 1 OPEN	
K19-403	EVENT, FUEL FILL VALVE OPEN	
*K111-404	EVENT, FUEL PREVALVE OPEN	
*K112-404	EVENT, FUEL PREVALVE CLOSED	
K113-411	EVENT, LH_2 TANK VENT VALVE C CLOSED	
K114-411	EVENT, LH_2 TANK VENT VALVE D CLOSED	
*K210-410	EVENT, LH_2 LATCH RELIEF VALVE CLOSED	
*K211-410	EVENT, LH_2 LATCH RELIEF VALVE OPEN	
L1-408	LEVEL, FUEL TANK POS 1	0 TO 5 vdc
L2-408	LEVEL, FUEL TANK POS 2	0 TO 5 vdc

MEASUREMENT NUMBER PREFIXES INDICATE THE FOLLOWING:
*FLIGHT CONTROL; V-ESE DISPLAY; X-AUXILIARY DISPLAY   CH-20125

Figure 5-35

pressure switch opens and closes the valve as it senses ullage pressure and regulates the quantity of helium that flows into the tank. A switch selector command (channel 79) one sec after J-2 cutoff opens the circuit and closes the shutoff valves.

Lox Chilldown.

During lox loading, the 10-in. prevalve in the engine feed duct remains open (its normal position) providing a partial chilldown of the feed system. Chilldown is required to properly condition the lox turbopump and feed duct to meet turbopump inlet requirements at engine ignition. A ground command starts the lox chilldown pump 10 min before liftoff, and it runs continuously until 0.4 sec before J-2 ignition command. The prevalve closes 5 sec after the pump starts. The pump circulates lox through the normally open lox chilldown shutoff valve, the lox feed duct, the lox turbopump, the high pressure propellant duct, the lox bleed valve, and the return check valve back into the tank. The pump delivers 38 gpm at 15 psia. The prevalve opens on a switch selector command (channel 83) 4.5 sec before J-2 start command allowing reverse lox flow through the feed duct removing trapped bubbles through the duct and anti-vortex screen.

Lox Measurements.

Figure 5-40 lists the lox system flight measurements and indicates the information that is monitored in real time. Measurements VD179 and VD180 are also displayed in analog form on the command module main display console providing lox tank pressure data for the crew.

Propellant Utilization.

The PU (propellant utilization) system (figure 5-41) is provided for mass indications during loading to within ±1.0 percent accuracy, and inflight propellant mass history to the telemetry system to an accuracy of ±1.0 percent of total propellant mass. The PU electronic assembly, located in the forward skirt, is pressurized and mounted on an environmental panel. Inputs to the PU electronic assembly are from mass sensor probes, one located in each propellant tank. The probes are cylindrical capacitors, varying in capacitance linearly with the liquid mass in the tanks. Each mass sensor forms one leg of a servo-balanced capacitance bridge. The output signal of the bridge drives a servo-motor (pot positioner) which positions a potentiometer wiper to yield a signal that is supplied to the ground loading computer during propellant loading. During flight, the same signals are telemetered (figure 5-42) to provide a flight history of propellant masses.

SA-206 S-IVB ORBITAL SAFING.

The propellant tanks will be vented immediately after engine cutoff to ensure that no stage disturbances occur during CSM separation, which will occur approximately six min after S-IVB cutoff. Subsequently, the following safing operations will be accomplished.

a. The lox tank will be vented through the lox NPV system.

b. The LH_2 tank will be vented through the LH_2 NPV system.

c. The cold helium bottle residuals will be dumped through the lox NPV system.

d. The stage pneumatic bottle residuals will be dumped through the engine pump purge control module.

e. The engine start bottle will not be refilled during engine powered flight so it will not have to be dumped.

f. The engine control bottle residuals will be dumped through the engine purge system.

Rapid elimination of these residuals will preclude high tank pressures and result in a completely safe vehicle for orbital coast.

Propellant Tank Orbital Venting.

Orbital venting starts after J-2 engine cutoff. Figures 5-43 and 5-44 illustrate the sequence of events during this period. Switch selector commands are sent 0.2 and 0.4 sec after the J-2 cutoff command to open the LH_2 and lox valves, respectively. The lox tank vents for about 3 min 20 sec while the LH_2 tank vents for about 5 min 20 sec. The LH_2 and lox tank vent systems are non-propulsive. The non-propulsive vent valves are latched open at 15 min after J-2 engine cutoff to allow continuous tank venting, which precludes any long term tank pressure buildup.

PNEUMATIC CONTROL SYSTEM.

Pneumatic control of all propulsion system components (figures 5-45 and 5-46) requiring command actuation was chosen to provide the capability of rapid response and high force where needed with minimum weight of hardware required for the power source and

Section V S-IVB Stage

S-IVB LOX SYSTEM

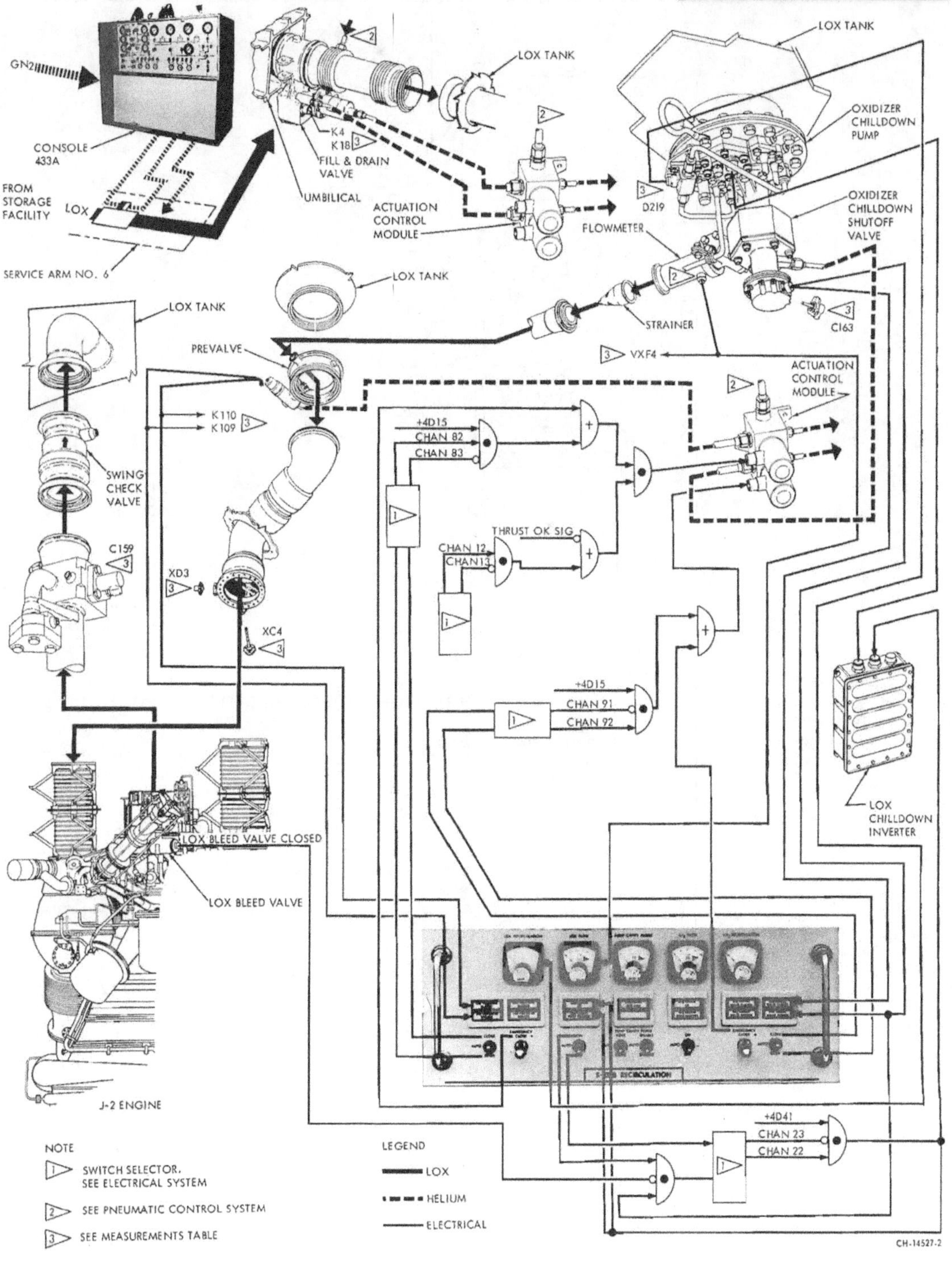

Figure 5-36

Section V S-IVB Stage

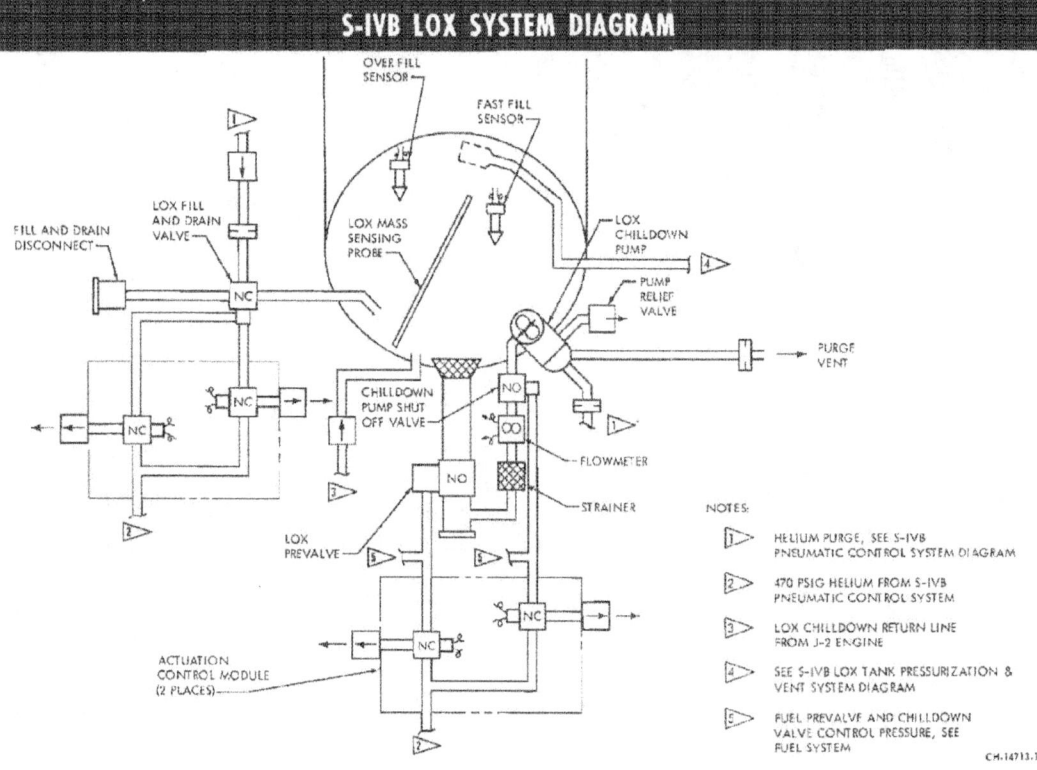

Figure 5-37

control components. A storage pressure of 3000 psi was chosen to be compatible with other vehicle requirements and to take advantage of components previously developed for this pressure. The pneumatic control system provides supply pressure for stage pneumatically operated valves including the J-2 engine start tank vent. Pneumatic power for all the other engine valves is supplied by the engine pneumatic supply. A pneumatic regulator assembly regulates filtered ambient helium flowing from the ambient storage sphere at 3000±100 psia and 70° F. The module regulates pressure down to 470±12 psig for operation of the LH_2 directional control valve, the lox and LH_2 vent valves, the lox and LH_2 fill and drain valves, the J-2 engine GH_2 start system vent-relief valve, the lox and LH_2 prevalves, the lox and LH_2 chilldown shutoff valves, lox NPV valve, and LH_2 latching relief valve, and for purge of the J-2 engine LH_2 and lox turbopump turbines and the lox chilldown pump housing. Several other components are purged during preflight operations; for a list of purges and flowrates, see figure 5-47.

PREPRESSURIZATION OPERATION.

Prepressurization is accomplished by opening the helium dome supply valve on console 432A and by actuating the switch on the helium control panel located in the LCC. See figure 5-48 for pneumatic control system measurements and ground displays. The storage sphere is pressurized to 1000 psig and the regulator maintains a plenum chamber pressure of 470 psig downstream of the storage sphere. The ground supply helium source through pneumatic console 432A is maintained preventing depletion of the control pressure.

PREFLIGHT PRESSURIZATION OPERATION.

The preflight pressurization begins at approximately 6 hr 37 min before liftoff. A regulator dome pressure supply solenoid valve in console 432A is opened and helium at 3000 psig flows to the control helium sphere. The control helium storage sphere is made of titanium and will withstand a minimum proof pressure of 4800 psig in a temperature range of -40 to 210° F without structural failure. The ground supply pressure is maintained until 3 sec before liftoff at which time the supply is terminated by ignition command. A check valve maintains subsystem pressure for inflight operations.

FLIGHT CONTROL.

Two modes of flight control, burn and coast, maintain attitude control of the S-IVB/IU/payload vehicle configuration from S-IB/S-IVB separation until shortly after CSM separation from the launch vehicle. During the burn mode, attitude and steering adjustments are made by hydraulically gimballing the J-2 engine ±7 deg maximum in the pitch and yaw planes. Auxilary propulsion system (APS) modules on the S-IVB stage aft skirt provide vehicle roll control during the burn mode by thrusting tangentially to the vehicle. During the coast mode, which begins at J-2 engine cutoff, the APS modules provide pitch, yaw, and roll attitude control. All attitude and steering commands originate in the IU guidance, navigation, and control system. As computations are made the flight control computer issues the commands that gimbal the J-2 engine

S-IVB LOX TANK PRESSURIZATION AND VENT SYSTEM

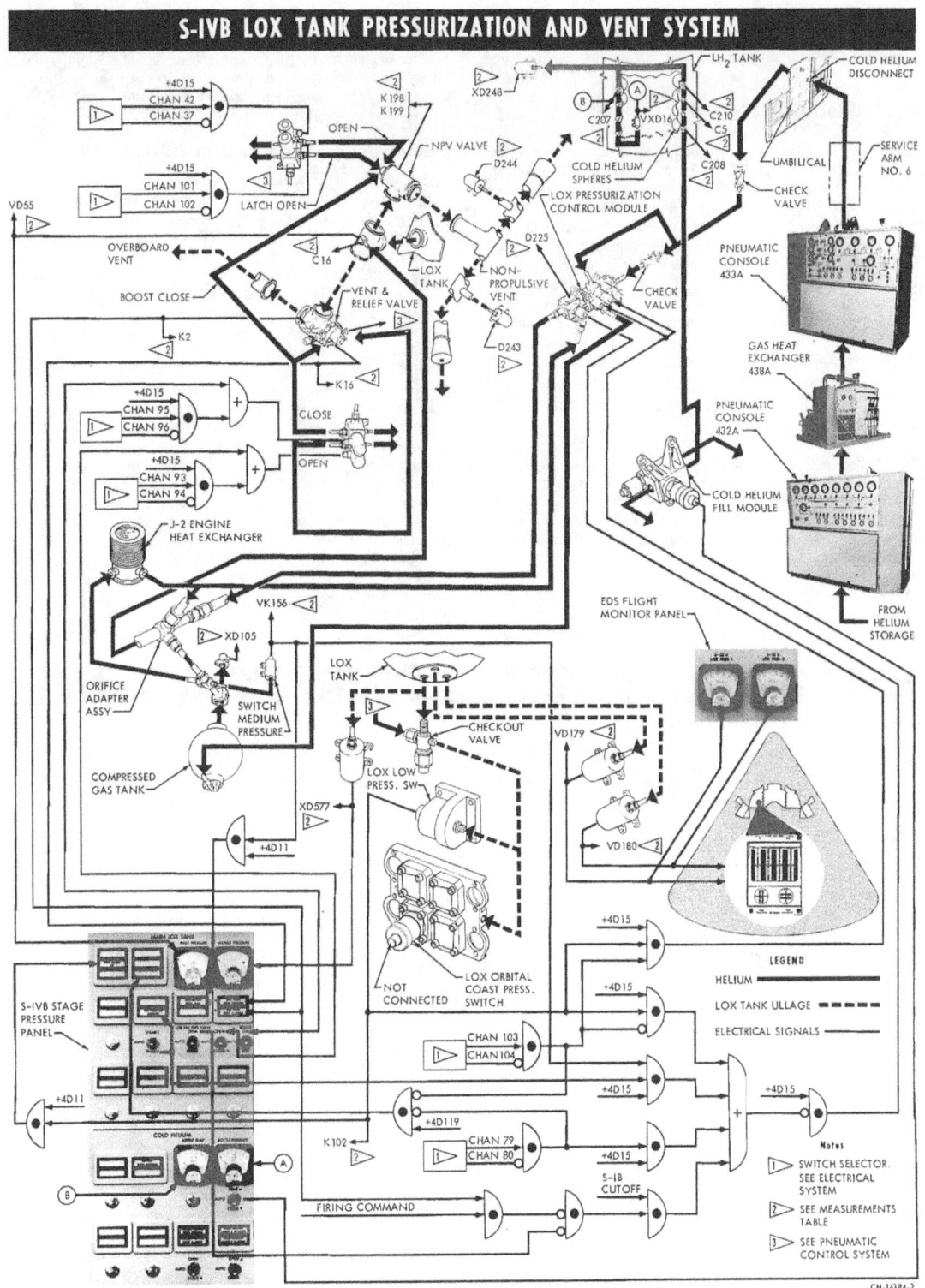

Figure 5-38

Section V S-IVB Stage

and/or fire the APS engines. The control system for pitch and yaw during the burn mode is essentially a linear, continuous type whereas the APS during burn and coast is a nonlinear 'bang-bang' system.

HYDRAULIC SYSTEM.

An independent closed-loop hydraulic system (figure 5-49) on the S-IVB stage gimbals the J-2 engine during firing and non-firing operations. A main hydraulic pump driven by the lox turbopump turbine, a compressed air tank, an auxiliary motor-driven hydraulic pump, an accumulator-reservoir, two servoactuator assemblies, and interconnecting tube and hose assemblies comprise the hydraulic system. The servoactuators, connected between the J-2 engine and the thrust structure (one in the pitch plane and one in the yaw plane), provide the forces necessary to gimbal the engine as commanded by the flight control computer. The actuators can operate individually or together, either in phase or out of phase, to accomplish gimballing. The double-acting actuators are capable of delivering 42,000 lbf at a pressure of 3650 psi in either the extend or retract direction.

Design Philosophy.

The philosophy of the design was to keep the system simple without compromising reliability. The system is very similar to the hydraulic system that was flown on the S-IV stages of the Block II Saturn vehicles. The system features modular design to minimize the number of external lines and reduce leakage paths. The auxiliary motor-driven hydraulic pump, which is plumbed into the system, provides an onboard checkout capability and provides a degree of redundancy by operating in parallel with the engine-driven pump during J-2 engine operation. A mechanical feedback loop and a dynamic pressure feedback loop between the actuator and servovalve eliminate the necessity for an electrical feedback system. Thermal conditioning features, such as a thermal isolator mount for the engine-driven pump and circulation of hydraulic fluid through the auxiliary pump motor jacket to distribute heat to the hydraulic system components, prevent the system from freezing during a ground hold.

Main Hydraulic Pump, Engine-Driven.

In the single stage, yoke type, variable displacement, pressure compensated, axial piston, S-IVB main pump (figure 5-50), pumping is accomplished by the reciprocation of nine pistons mounted parallel to the cylinder block axis within the cylinder block. The piston rods attach to the input drive shaft with ball sockets. The cylinder block rotates synchronously with the input shaft, both joined with a double universal joint. The rotating cylinder block mounts in a yoke, which pivots to control the angular relations between the cylinder block and drive shaft axes. When the axis of the cylinder block aligns with the axis of the drive shaft (zero angle), the pumping elements are in a zero-displacement or no-flow position. As the yoke moves to increase the angle between the shaft and the cylinder block, the pistons reciprocate over an increasing stroke length, thereby increasing the displacement and the output flow of the pump. A fixed valve plate, attached to the shaft on which the cylinder block rotates, contains two ports, an inlet port connecting the cylinder suction with pump inlet and an outlet port connecting cylinder discharge with pump outlet. As shown in the main pump elements schematic, pistons at position A are completely extended into the cylinder bores and pistons at position B are completely withdrawn from the cylinder bores. As the cylinder block rotates against the valve plate, each piston begins its suction stroke at A, drawing fluid into the cylinder bore through the valve plate inlet port. As the pistons pass position B, the discharge or pressure stroke begins, forcing fluid from the cylinder bores through

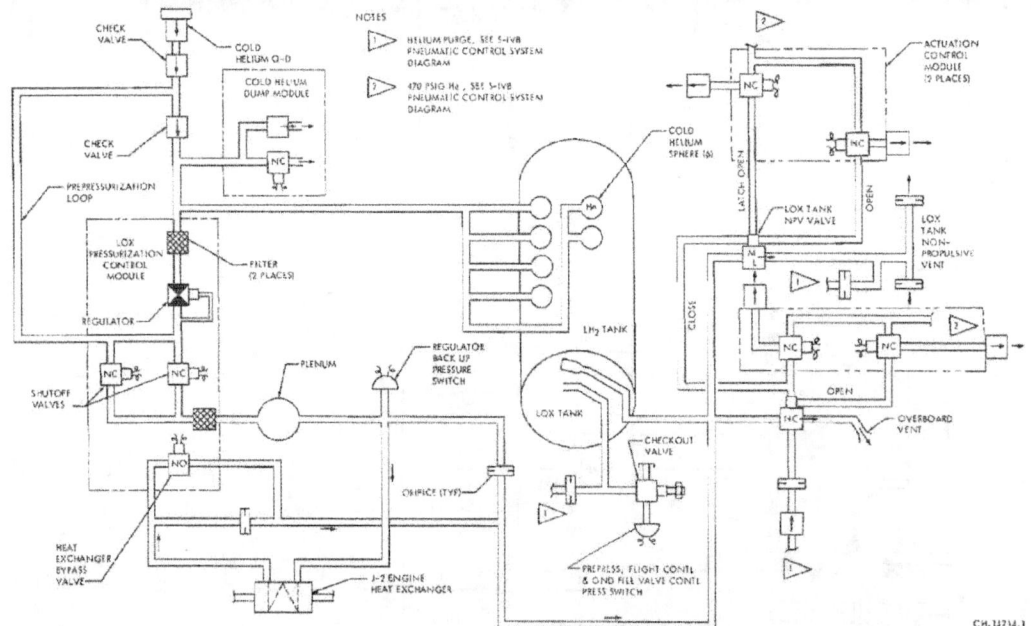

Figure 5-39

the valve plate outlet port; this continues until the piston reaches position A where the cycle begins all over again.

Operation of the yoke type, variable displacement mechanism is shown in the compensator schematic (figure 5-50). The pump yoke is mounted on bearings, permitting the yoke to pivot and vary the angle between the cylinder block axis and input shaft from 0 deg, zero displacement or no-flow to 30 deg, maximum displacement or full-flow. The yoke is positioned by an actuating cylinder mechanically coupled to an arm on the yoke. The yoke spring retains the actuating cylinder in the maximum displacement position under no-load conditions. During operations, this spring force is opposed by control pressure fluid delivered by the compensator valve.

The compensator valve senses discharge pressure. When pump outlet pressure exceeds 3650 psig, the hydraulic force on the compensator valve spool overcomes the preset compensator valve spring force and displaces the spool downward, delivering high pressure discharge fluid to the yoke actuating cylinder. This flow is proportional to the compensator valve opening and to the excess discharge pressure above 3650 psig. The yoke actuating cylinder reduces the yoke angle until the flow is sufficient to maintain 3650 psig. The compensator valve spool then centers to lock the yoke in the new position.

A thermal isolator attaches the main pump to an accessory pad on the lox turbine gas collector dome. The isolator protects the pump, pump oil seal, and hydraulic fluid from temperature extremes effected by gases in the turbine that change suddenly from -300 to +900° F during J-2 engine start. A dual element hot-gas shaft seal prevents impingement of hot gases on the hydraulic seal. Any leakage past the seals is vented overboard. A crown-spline quill shaft extending from the turbine shaft drives the main pump.

Compressed Gas Tank.

The compressed gas tank stores dry air to maintain an atmospheric condition in the auxiliary pump motor case and to ensure proper heat transfer from the motor to the hydraulic fluid circulating through the motor case. The pressurized environment within the motor also prevents excessive or rapid brush wear. A quick-disconnect fitting on the tank accommodates filling with air and a dial-indicator type pressure gage provides local monitoring of tank pressure. The tank is constructed of forged titanium and has a proof pressure of 1200 ± 25 psig.

Auxiliary Hydraulic Pump, Motor-Driven.

The single stage, electrically driven, variable delivery, fixed angle, constant-displacement pump (figure 5-51) supplies operating pressure, hydraulic fluid, for preflight engine gimballing checkout, null positioning during the boost phase, and emergency back-up during S-IVB powered flight. As shown in the auxiliary pump schematic, view one, there is the piston and driver assembly, cylinder block, and the valve plate. The pistons at position A are completely extended into the cylinder bores and pistons at position B are completely withdrawn from the cylinder bores. The pump accomplishes pumping with the nine pistons attached by rods to the drive plate at a fixed angle to the cylinder block axis. In view two, the piston, the driver assembly, and the cylinder block are shown connected. They rotate synchronously causing the pistons to reciprocate within their cylinder bores. The valve plate does not rotate with the cylinder block. The pump output, hydraulic fluid flow to and from the cylinder bores, is controlled by the position of the valve plate. In view two (the maximum delivery position) as the assembly rotates, fluid enters the cylinder bores during rotation from A to B (suction stroke) and discharges during rotation from B to A (discharge stroke). The valve plate inlet port connects each cylinder to the pump inlet during the entire suction stroke, and the valve plate outlet port connects each cylinder with the pump outlet during the entire discharge stroke, thus the pump delivers fluid.

During operation, control pressure fluid metered by a compensator valve opposes the compensator spring (as shown in the compensator valve schematic in figure 5-51). The compensator valve is sensitive to discharge pressure and varies inversely with the flow. When the pump discharge pressure exceeds 3650 psig, the compensator valve spool displaces, delivering high pressure fluid to the valve plate vane. The hydraulic force acting on the valve plate vane

LOX SYSTEM MEASURMENTS

NUMBER	NAME	RANGE
XC4-403	TEMP, OXID PUMP INLET	160 TO 170°R
VXC5-405	TEMP, COLD He SPHERE NO. 3 GAS	25 TO 80°R
C16-424	TEMP, OXID TANK He INLET	175 TO 560°R
C40-406	TEMP, OXID TANK POS 1	160 TO 173°R
*C159-424	TEMP, LOX CIRC RET LINE TK INLET	160 TO 320°R
C163-424	TEMP, OXID CIRC PUMP OUTLET	163 TO 200°R
C207-425	TEMP, COLD He SPHERE NO. 5 GAS	25 TO 80°R
C208-405	TEMP, COLD He SPHERE NO. 1 GAS	25 TO 80°R
C210-405	TEMP, COLD He SPHERE NO. 4 GAS	25 TO 560°R
C2030-404	TEMP, LOX NPV NOZZLE NO. 1	100 TO 300°R
C2031-404	TEMP, LOX NPV NOZZLE NO. 2	100 TO 300°R
*XD3-403	PRESS, OXID PUMP INLET	0 TO 60 psia
*VXD16-425	PRESS, COLD He SPHERE	0 TO 3500 psia
VD55-424	PRESS, OXID TANK INLET	0 TO 300 psia
VXD105-403	PRESS, LOX TANK PRESS MOD He GAS	0 TO 500 psia
*VXD179-406	PRESS, OXID TK ULLAGE EDS 1	0 TO 50 psia
*VXD180-406	PRESS, OXID TK ULLAGE EDS 2	0 TO 50 psia
D219-403	PRESS, LOX CHILLDOWN PUMP DIFF	-30 TO 30 psid
D243-404	PRESS, LOX NPV NOZZLE NO. 1	0 TO 50 psia
D244-404	PRESS, LOX NPV NOZZLE NO. 2	0 TO 50 psia
XD248-425	PRESS, COLD He SPHERES	0 TO 3500 psia
D265-403	PRESS, LOX PUMP INLET, CL COUPLED	0 TO 60 psia
VXF4-424	FLOWRATE, OXID CIRCULATION PUMP	0 TO 50 gpm
*K2-424	EVENT, OXID TANK VENT VLV CLOSED	
K4-404	EVENT, OXID FILL VLV CLOSED	
K16-404	EVENT, OXID TANK VENT VLV 1 OPEN	
K18-404	EVENT, OXID FILL VLV OPEN	
K102-404	EVENT, LOX PREPRESS FLIGHT SW ENABLED	
K109-403	EVENT, OXID PREVALVE OPEN	
*K110-403	EVENT, OXID PREVALVE CLOSED	
VK156-404	EVENT, LOX TANK REG BACKUP PRESS ENABLED	
K198-424	EVENT, LOX NPV VALVE OPENED	
K199-414	EVENT, LOX NPV VALVE CLOSED	
L4-406	LEVEL, LOX TANK POS 1	0 TO 5 V
L5-406	LEVEL, LOX TANK POS 2	0 TO 5 V

MEASUREMENT NUMBER PREFIXES INDICATE THE FOLLOWING:
* FLIGHT CONTROL; V-ESE DISPLAY; X-AUXILIARY DISPLAY

Figure 5-40

Section V S-IVB Stage

PROPELLANT UTILIZATION SYSTEM

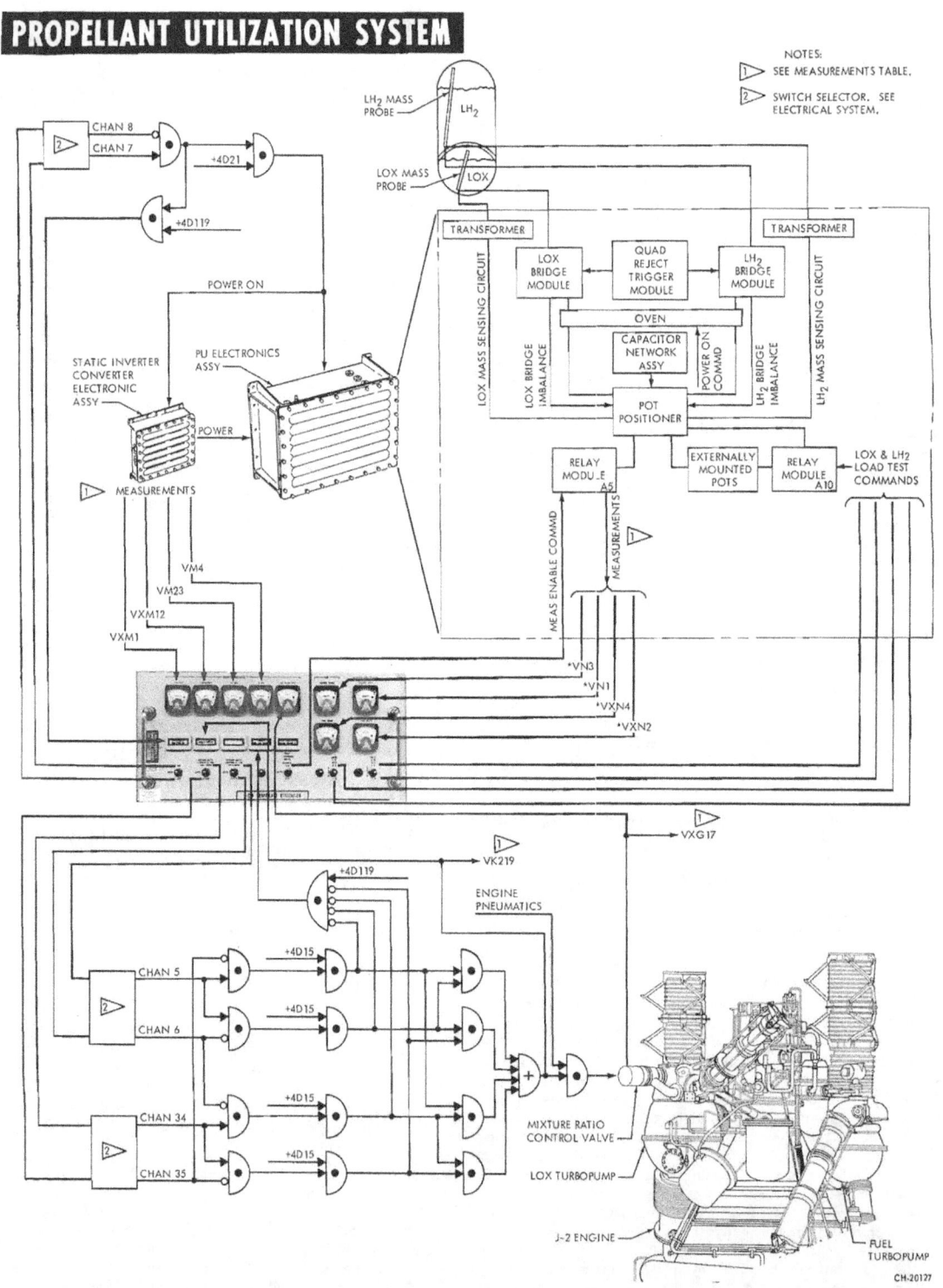

Figure 5-41

overcomes the compensator spring force and causes the valve plate to rotate until pump output is just sufficient to maintain 3650 psig. In view three (the reduced delivery position) although the same volume is displaced by the pistons, the valve plate ports a portion of the discharge stroke (B to D) to the pump inlet and a portion of the suction stroke (A to C) to the pump discharge. If the pump outlet was completely restricted, the control fluid would rotate the valve plate until (A–B) is perpendicular to (C–D).

In view four, the zero delivery position, each cylinder is open to pump inlet and outlet during its entire suction and discharge stroke, thus pump delivery is zero.

A regulator in the auxiliary pump manifold regulates the 475 psig air from the compressed air tank to 15 ± 5 psi for motor cavity pressurization. The air serves as a heat transfer medium for thermal conditioning of the hydraulic fluid circulating through the motor jacket, and provides proper atmospheric conditions for the motor operation.

The design safety factors for the auxiliary pump motor ensure capable performance after operating at 125-percent speed at maximum rated discharge pressure and capable performance with the motor temperature of 150° F at start of cycle and 275° F at the end of a 5-min cycle.

Accumulator-Reservoir.

The hydraulic system accumulator-reservoir assembly (figure 5-52) stores low pressure hydraulic fluid and provides pump ripple suppression, pressure surge damping, and a source of high pressure fluid to supplement pump outputs during S-IB/S-IVB separation transients. The accumulator-reservoir assembly consists of a manifold assembly, a floating piston, a gas-loaded accumulator, a bootstrap piston, reservoir, and reservoir piston. The manifold assembly contains relief valves, bleed valves, quick-disconnects, a system filter, and instrumentation. The accumulator is precharged with GN_2 at 2350 ± 50 psig and 70° F. A dial indicator pressure gage on the accumulator provides local monitoring of the precharge pressure. During pump operations fluid at 3650 ± 50 psig enters the accumulator. Accumulator filling continues until the floating piston compresses the gas charge equal to the pressure of the fluid.

The high pressure fluid acts against the reservoir piston to maintain low pressure in the reservoir. During non-operating periods of the pumps, the precharge pressure forces the bootstrap piston against the reservoir piston to keep the reservoir pressurized to 170 psi. The low pressure fluid in the reservoir prevents cavitation in the pumps during start and run operations.

A high pressure, full flow, cartridge type filter element filters the

PU SYSTEM MEASUREMENTS

NUMBER	NAME	RANGE
*VXG17-401	POSITION, MRC VALVE	0 TO 65°
VK219-404	MRC VALVE OPEN ON	
VXM1-411	VOLT, STATIC INVERTER-CONVERTER	90 TO 135 V
VM4-411	VOLT, STATIC INVERTER-CONVERTER, 5 VDC	4.5 TO 5.5 V
VXM12-411	FREQUENCY, STATIC INVERTER-CONVERTER	390 TO 410 Hz
VM23-411	VOLT, STATIC INVERTER-CONVERTER, 21 VDC	20 TO 23.5 V
*VN1-411	PU SYS LH_2 COARSE MASS, VOLT	0 TO 5 V
*VXN2-411	PU SYS LH_2 FINE MASS, VOLT	0 TO 5 V
*VN3-411	PU SYS LOX COARSE MASS, VOLT	0 TO 5 V
*VXN4-411	PU SYS LOX FINE MASS, VOLT	0 TO 5 V

MEASUREMENT NUMBER PREFIXES INDICATE THE FOLLOWING:
*FLIGHT CONTROL; V-ESE DISPLAY; X-AUXILIARY DISPLAY

Figure 5-42

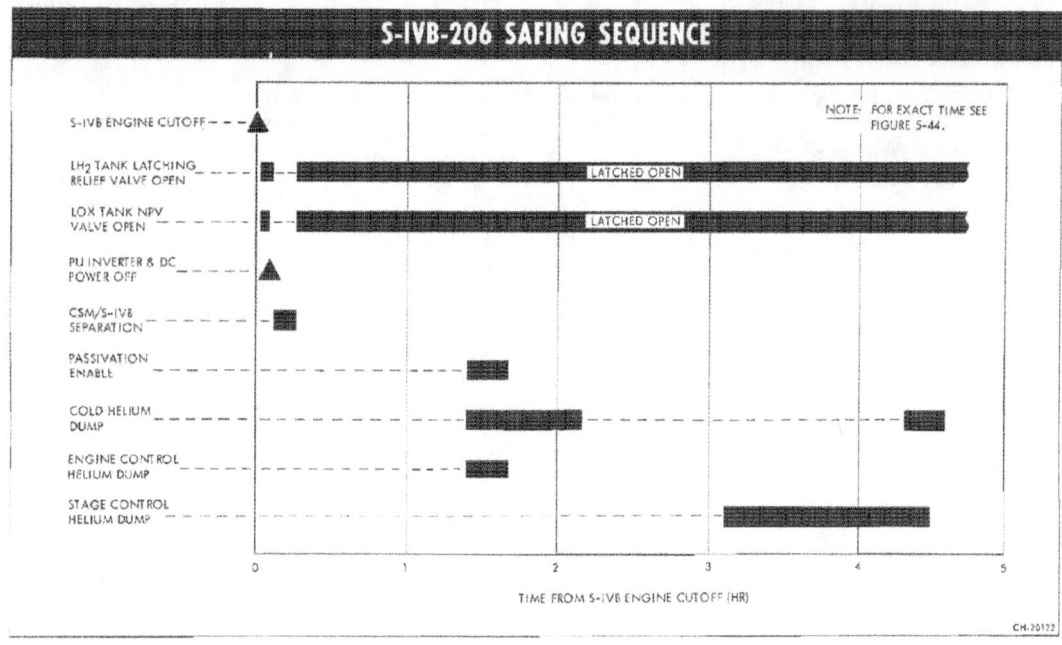

Figure 5-43

Section V S-IVB Stage

SEQUENCE OF EVENTS FOLLOWING J-2 CUTOFF (S-IVB-206 ORBITAL SAFING)

COMMAND	CHANNEL	TIME (T_4 + HR: MIN:SEC)
S-IVB ENGINE CUTOFF NO. 1 ON		
(START OF TIME BASE 4)	12	0:00:0.1
S-IVB ENGINE CUTOFF NO. 2 ON	48	0:00:0.2
LOX TANK NPV VALVE OPEN ON	42	0:00:0.3
LH$_2$ TANK LATCHING RELIEF VALVE OPEN ON	99	0:00:0.4
PREVALVES CLOSE ON	82	0:00:0.5
CHILLDOWN SHUTOFF VALVES CLOSE	91	0:00:0.5
LOX TANK PRESSURIZATION SHUTOFF VALVES CLOSE	79	0:00:0.8
LOX TANK FLIGHT PRESSURIZATION SYSTEM OFF	104	0:00:1.0
PROPELLANT DEPLETION CUTOFF DISARM	98	0:00:1.8
MIXTURE RATIO CONTROL VALVE CLOSE	6	0:00:2.2
MIXTURE RATIO CONTROL VALVE BACKUP CLOSE	35	0:00:2.4
FLIGHT CONTROL COMPUTER S-IVB BURN MODE OFF "A"	IU/12	0:00:3.5
FLIGHT CONTROL COMPUTER S-IVB BURN MODE OFF "B"	IU/3	0:00:3.7
AUX. HYDRAULIC PUMP FLIGHT MODE OFF	29	0:00:3.9
S/C CONTROL OF SATURN ENABLE	IU/18	0:00:5.0
S-IVB ENGINE EDS CUTOFF DISABLE	IU/3	0:00:10.0
LOX TANK NPV VALVE OPEN OFF	37	0:03:20.0
LOX TANK VENT AND NPV VALVES BOOST CLOSE ON	95	0:03:23.0
LOX TANK VENT AND NPV VALVES BOOST CLOSE OFF	96	0:03:25.0
P.U. INVERTER AND DC POWER OFF	8	0:04:00.0
LH$_2$ TANK LATCHING RELIEF VALVE OPEN OFF	100	0:05:20.0
LH$_2$ TANK VENT AND L/R VALVES BOOST CLOSE ON	77	0:05:23.0
LH$_2$ TANK VENT AND L/R VALVES BOOST CLOSE OFF	78	0:05:25.0
LH$_2$ TANK LATCHING RELIEF VALVE OPEN ON	99	0:15:00.0
LOX TANK NPV VALVE OPEN ON	42	0:15:0.2
LH$_2$ TANK LATCHING RELIEF VALVE LATCH ON	52	0:15:2.0
LOX TANK NPV VALVE LATCH OPEN ON	101	0:15:2.2
LH$_2$ TANK LATCHING RELIEF VALVE OPEN OFF	100	0:15:3.0
LOX TANK NPV VALVE OPEN OFF	37	0:15:3.2
LH$_2$ TANK LATCHING RELIEF VALVE LATCH OFF	19	0:15:4.0
LOX TANK NPV VALVE LATCH OPEN OFF	102	0:15:4.2
PASSIVATION ENABLE	85	1:23:10.2
LOX TANK PRESSURIZATION SHUTOFF VALVES OPEN	80	1:23:20.0
ENGINE He CONTROL VALVE OPEN ON	109	1:23:20.2
ENGINE He CONTROL VALVE OPEN OFF	110	1:40:00.0
PASSIVATION DISABLE	86	1:40:0.1
LOX TANK PRESSURIZATION SHUTOFF VALVES CLOSE	79	2:10:00.0
ENGINE PUMP PURGE CONTROL VALVE ENABLE ON	24	3:06:40.0
LOX TANK PRESSURIZATION SHUTOFF VALVES OPEN	80	4:20:00.0
ENGINE PUMP PURGE CONTROL VALVE ENABLE OFF	25	4:30:00.0
LOX TANK PRESSURIZATION SHUTOFF VALVES CLOSE	79	4:36:40.0

Figure 5-44

fluid entering the accumulator reservoir from the pumps. The 15-micron filter has an operating pressure of 3700 psig and a differential pressure of 15 psi maximum at 12 gpm and 100 ± 3° F.

A high pressure relief valve prevents fluid over-pressurization in the accumulator by relieving excessively pressurized fluid into the reservoir. The valve cracks at 3760 psi, is fully open at 4250 psi, and reseats at 3620 psi. The valve has a proof pressure of 5600 psi.

Two balanced, low pressure relief valves connected in series protect the low pressure reservoir and fluid return side of the system. Both valves have a cracking pressure of 275 psig at -70 to +110° F, a relief pressure of 290 psig, and a reseat pressure of 225 psig minimum at 120° F. Instrumentation transducers are covered in Hydraulic System Measurements.

Hydraulic Servoactuators.

The S-IVB actuator (figure 5-53) consists of a tailstock, a body-cylinder member, a piston, and a front bearing member. The actuator attaches to the thrust structure with a spherical bearing housed in the tailstock.

The two self-contained, equal area, double acting actuators convert hydraulic power into a mechanical output that controls engine movement and vehicle attitude in two perpendicular planes (pitch and yaw). The actuator body flange mounts to the tailstock. The forged aluminum body houses the pressure and return ports, the electrical connectors, the prefiltration valve, the servofilter, the cylinder bypass valve, bleed valve, and the servovalve piston. Servovalve and telemetry potentiometers are mounted to the actuator body.

The servovalve is a four-way, two-stage, flow control valve with a dynamic pressure feedback loop. The valve consists of a polarized, dry coil, electrical torque motor and two stages for hydraulic amplification. The polarizing magnetic flux is generated by two permanent magnets mounted between the upper and lower pole pieces. The motor armature extends into the air gaps of the magnetic flux, supported in this position by the flexible tube member. Two motor coils surround the armature, one located on each side of the flexure tube member. The flexure tube member also acts as a seal between the electromagnetic and hydraulic sections of the servovalve.

The flapper of the first hydraulic amplifier stage is rigidly attached to the midpoint of the armature. The flapper extends through the flexure tube member and passes between two nozzles in the control pressure loop and two nozzles in the dynamic feedback loop, creating two variable orifices between the nozzle tips and the flapper. Filtered fluid from the pressure source is supplied to the variable orifices in the control pressure loop through two fixed upstream orifices. Pressure from a point between the fixed and variable orifices is transmitted to the ends of the spool in the second hydraulic amplifier stage.

When a signal is applied to the motor windings, the flapper moves toward one of the nozzles (which nozzle depends upon the polarity of the signal). The flapper movement toward one nozzle tends to restrict fluid flow through that nozzle which results in a pressure build-up that acts against the spool. The opposite nozzle has less restriction and a greater fluid flow results in a lower pressure on the opposite end of the spool. This pressure differential causes the spool to reposition. In doing so, one side of the spool aligns high pressure supply with one of the actuator fluid loops while the other side of the spool ports fluid from the actuator to the low pressure return passages.

The second stage spool is a conventional four-way, sliding design

Section V S-IVB Stage

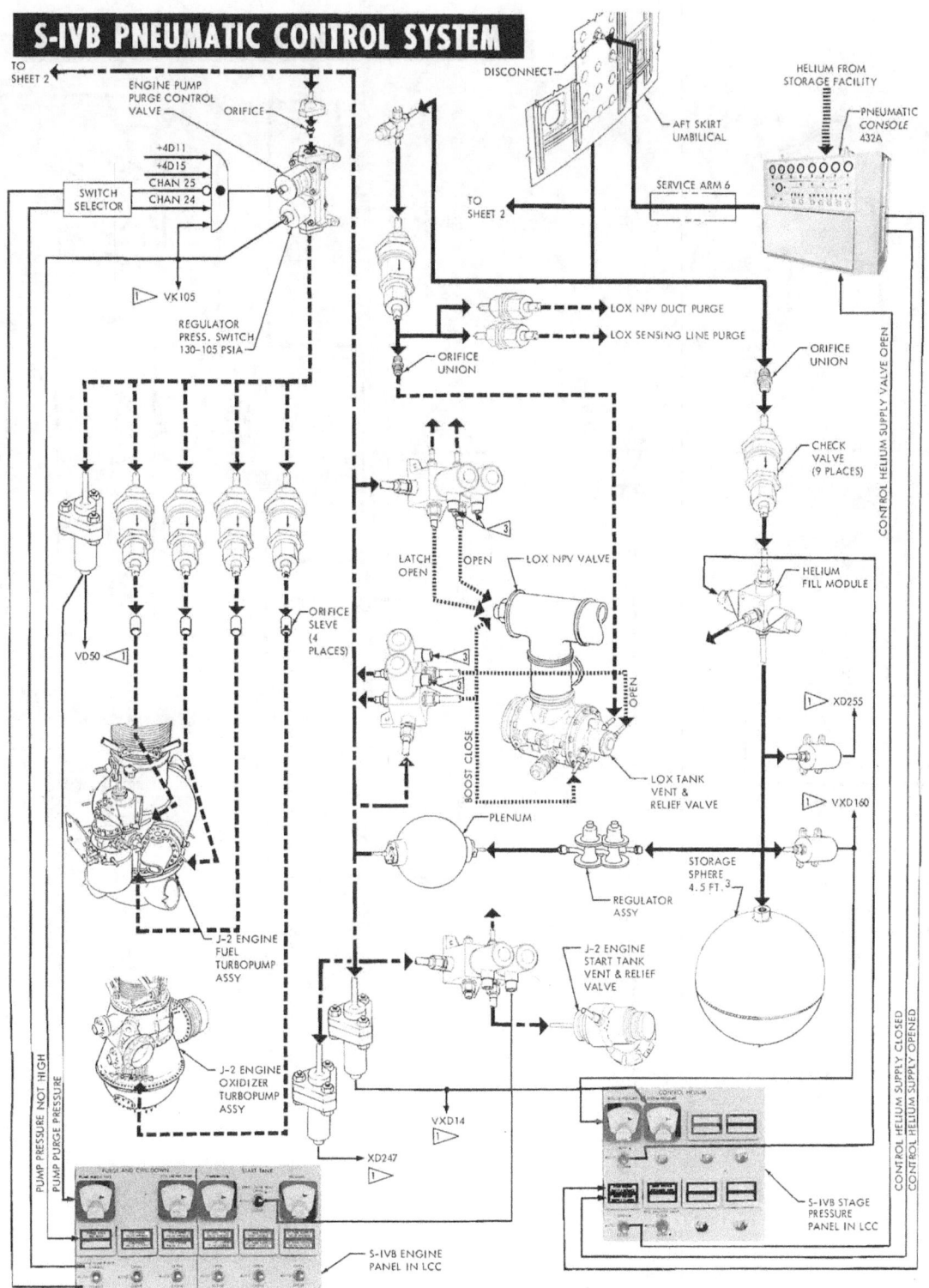

Figure 5-45 (Sheet 1 of 2)

Section V S-IVB Stage

S-IVB PNEUMATIC CONTROL SYSTEM

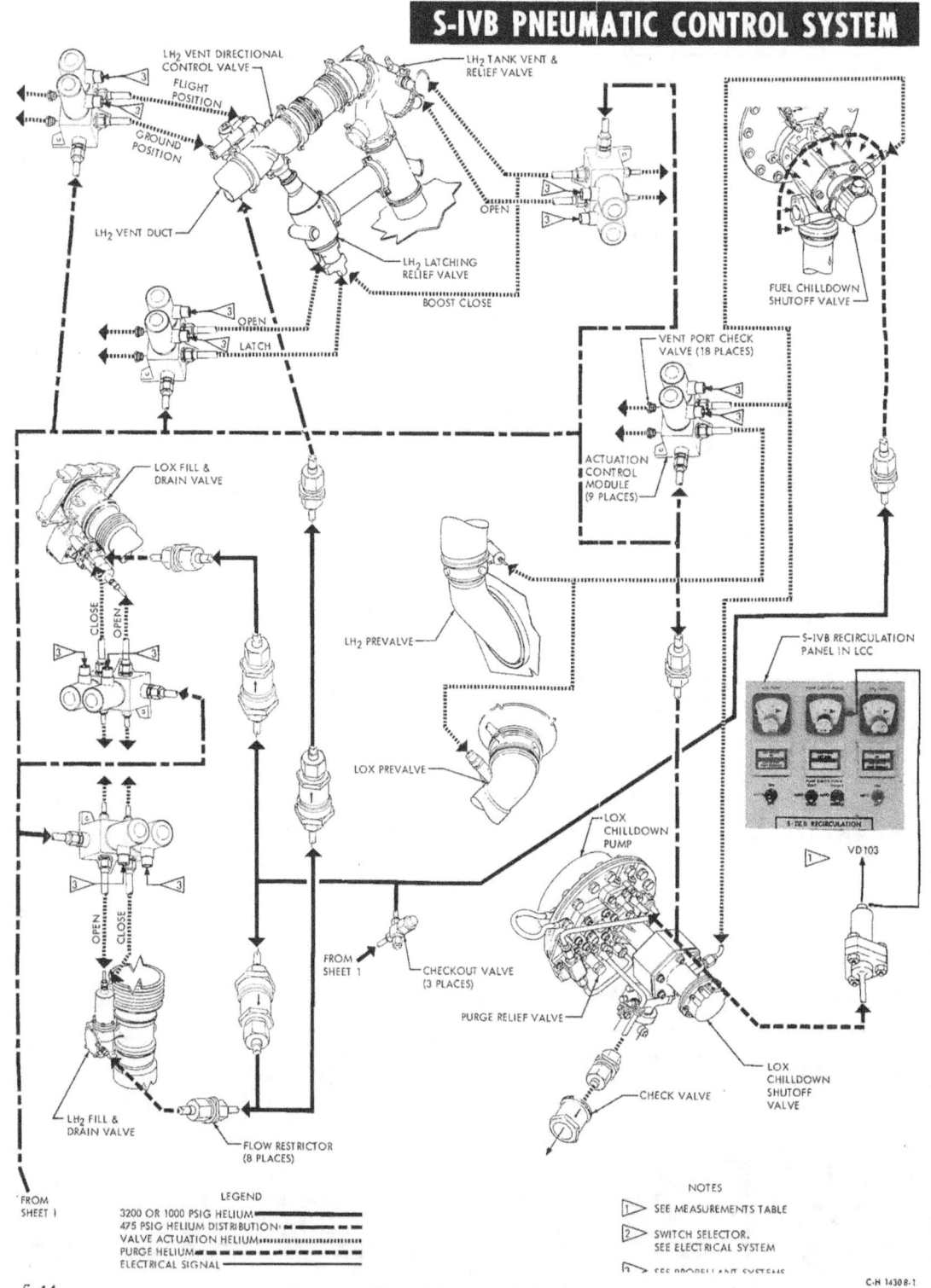

Figure 5-45 (Sheet 2 of 2)

Section V S-IVB Stage

S-IVB PNEUMATIC CONTROL SYSTEM DIAGRAM

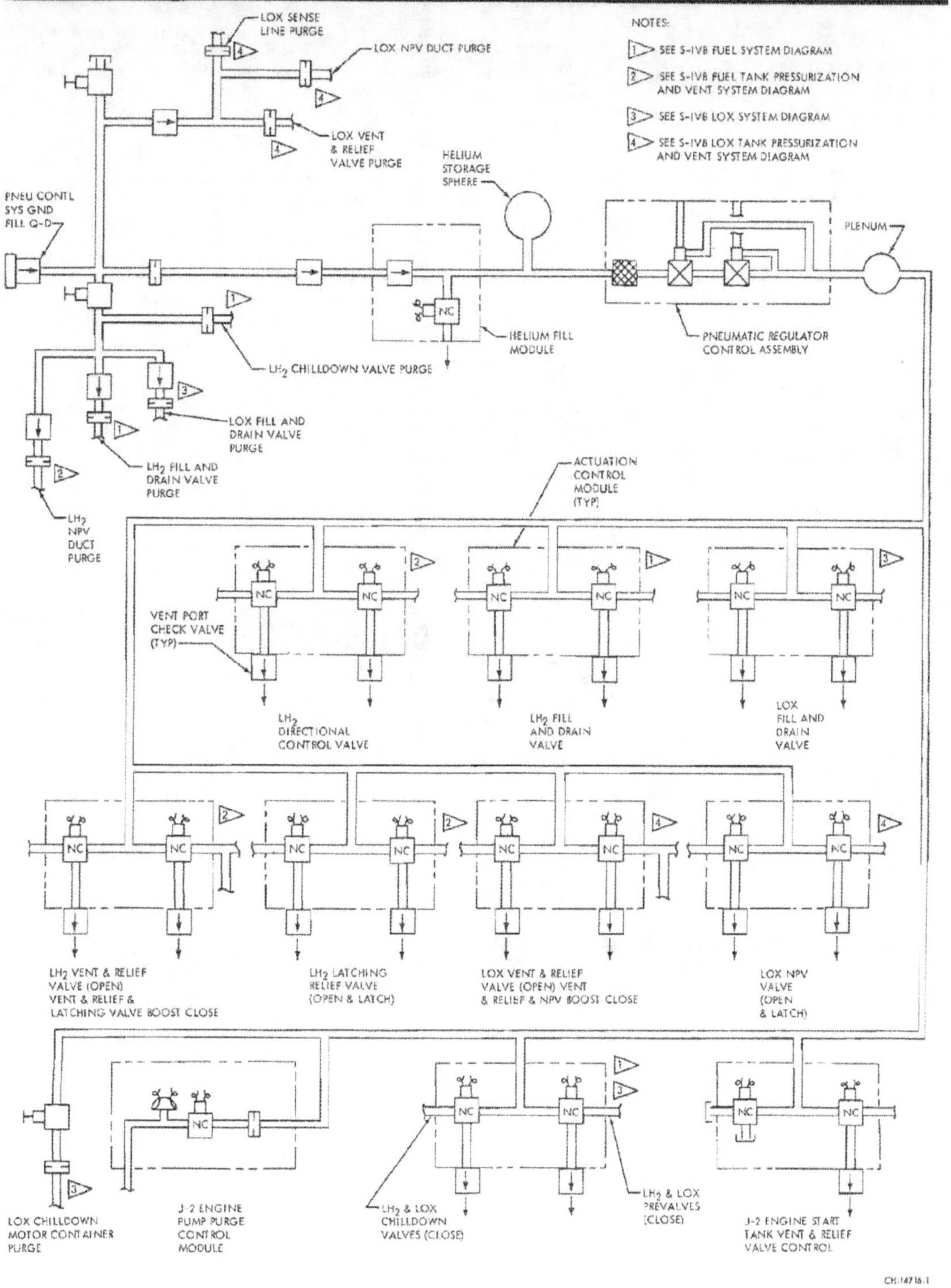

NOTES:
[1] SEE S-IVB FUEL SYSTEM DIAGRAM
[2] SEE S-IVB FUEL TANK PRESSURIZATION AND VENT SYSTEM DIAGRAM
[3] SEE S-IVB LOX SYSTEM DIAGRAM
[4] SEE S-IVB LOX TANK PRESSURIZATION AND VENT SYSTEM DIAGRAM

Figure 5-46

Section V S-IVB Stage

PURGE FLOW RATES

PURGED COMPONENT	FLOW RATE
LH₂ VENT DUCT	1728 SCIM
LOX NPV DUCT	1728 SCIM
LOX FILL AND DRAIN VALVE	15 SCIM
LH₂ FILL AND DRAIN VALVE	15 SCIM
FUEL CHILLDOWN SHUTOFF VALVE	24,200 SCIM
LOX VENT AND RELIEF VALVE	112,300 SCIM
LOX CHILLDOWN PUMP	600 SCIM
LOX ULLAGE SENSING LINE	1728 SCIM
ENGINE TURBOPUMPS AND GAS GENERATOR	16,400 SCIM

CH-14511-1

Figure 5-47

in which output flow from the valve at a fixed pressure drop is proportional to spool displacement from the null position. A cantilever feedback spring is fixed to the armature and extends through the flapper to engage a slot at the center of the spool. Displacement of the spool deflects the feedback spring, creating a restoring torque on the armature. The output flow from the valve is ported to either side of actuator piston, causing the piston to extend or retract and moving the required load.

A major feature of this servovalve is its dynamic pressure beedback (DPF) loop. This loop dampens gain peaking under dynamic conditions, providing the necessary stability. Under static conditions the feedback loop is ineffective in eliminating positional errors due to static loads. The feedback loop consists of a spring-centered, DPF piston connected to a second pair of nozzles located on either side of the flapper. Each chamber at the end of the DPF piston is open to an actuator chamber; therefore, differential pressures across the actuator, act on the ends of the DPF piston, causing the piston to seek a new equilibrium position and to displace a quantity of hydraulic fluid in the process. The fluid flows through one of the DPF nozzles and impinges against the flapper, providing a feedback force proportional to actuator forces. (This feedback condition occurs only under dynamic loads; under static loads the DPF piston ahcieves the new equilibrium position as forces on the centering spring equalize, ceasing feedback fluid flow.)

A mechanical feedback device, which consists of a lanyard connected between the torque motor armature and roller-ramp assembly helps maintain the servoactuator assembly in a null position

PNEUMATIC CONTROL SYSTEM MEASURMENTS

NUMBER	NAME	RANGE
*VXD14-403	PRESS, CONTROL He REG DISCHARGE	0 TO 650 psia
*VD50-403	PRESS, ENGINE PUMP PURGE REG	0 TO 150 psia
VD103-403	PRESS, He PRESS TO LOX MOT CNTR	0 TO 150 psia
*VXD160-403	PRESS, AMBIENT He SPHERE	0 TO 3500 psia
*XD247-403	PRESS, CONT He REG DISCHARGE	0 TO 650 psia
*XD255-403	PRESS, AMBIENT He SPHERE	0 TO 3500 psia
VK105-404	EVENT, PUMP PURGE REG BACKUP DEENERGIZED	

MEASUREMENT NUMBER PREFIXES INDICATE THE FOLLOWING:
*FLIGHT CONTROL; V-ESE DISPLAY; X-AUXILIARY DISPLAY

CH-20128

Figure 5-48

by keeping the flapper centered in the servovalve. A mechanical adjustment on the armature permits alignment of the servoactuator null position.

A cartridge type, full flow, vented seal, stainless steel mesh filter unit filters all operating fluid before it reaches the servovalve. Fluid passages are so designed that direct impingement of fluid upon the filter element is avoided. The element is rated at 5 microns nominal, 15 microns absolute, both at 1.0 gpm flow.

A manually operated, prefiltration valve bypasses all actuator components by connecting the supply pressure port with the return pressure port. The prefiltration valve permits flushing of the hydraulic system without exposing sensitive components of the actuator to the contaminated fluid. The servovalve (bypass valve) is a manually operated, normally closed valve that interconnects the actuator ports, allowing manual movement of the actuator. A midstroke locking device provides a means for actuator installation without accidental actuator piston movement and prevents random engine movement during servicing, shipping, and storing.

There are several bleed valves in the system. Bleed valves are connected to the cylinder ports and to the servovalve operating pressure fluid inlet passage. The cylinder bleed valves allow gas or hydraulic fluid to vent to atmosphere, also the servovalve bleed valve allows samples of filtered operating fluid to be taken.

Hydraulic System Servicing.

The hydraulic system servicing consists of assuring the system is charged with approximately 12 lbm of MIL-H-5606 hydraulic fluid, and the accumulator is charged with GN_2 to the equivalent of 2350 ± 50 psig at $70°$ F. The total quantity of the accumulator GN_2 will be 1.9 ± 0.1 lbm. The auxiliary pump air supply tank

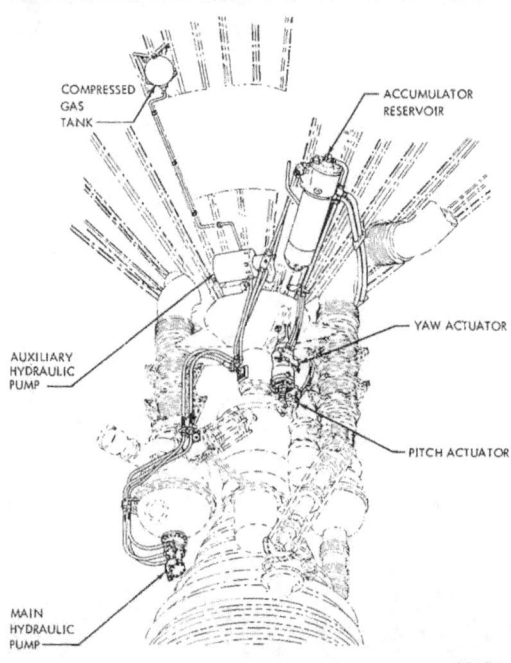

Figure 5-49

Section V S-IVB Stage

ENGINE DRIVEN HYDRAULIC PUMP

LOCATION	J-2 ENGINE, OXIDIZER TURBINE EXHAUST DUCT
SERVICE	HYDRAULIC FLUID MIL-H-5606

OPERATING PRESSURE
  OUTLET PRESS.
  INLET PRESS.
  CASE PRESS.
  INLET & CASE PRESS.
	
	3650 psig MAX.
	190 psig MAX.
	190 psig MAX.
	320 psig MAX. - SERVICING (NON-OPERATING)

PROOF PRESSURE
  OUTLET PRESS.
  INLET PRESS.
  CASE PRESS.
	
	5475 psig MIN.
	480 psig MIN.
	480 psig MIN.
	THE PUMPS SHALL MEET ALL REQUIREMENTS AFTER SUBJECTION TO THE ABOVE PRESSURES FOR 3 MINUTES AT 275°F.

BURST PRESSURE
  OUTLET PRESS.
  INLET PRESS.
  CASE PRESS.
	
	9125 psig MIN.
	800 psig MIN.
	800 psig MIN.
	THE PUMPS SHALL MEET ALL REQUIREMENTS AFTER SUBJECTION TO THE ABOVE PRESSURE FOR 3 MINUTES AT 275°F WITH NO RUPTURE OR STRUCTURE FAILURE.

OVERSPEED	THE PUMPS SHALL MEET ALL REQUIREMENTS WITHOUT DAMAGE AFTER OPERATING AT 10,000 rpm's FOR 30 MINUTES.

TEMPERATURE
  AMBIENT
  FLUID
	
	-35 TO +275°F
	-35 TO +275°F

RATED FLOW	7 gpm AT THESE OPERATING CONDITIONS:
	  SPEED MAX. 7000 rpm
	  OUTLET PRESS. MIN. 3550 psig
	  INLET PRESS. MAX. 150 psig

STARTING TORQUE	THE MAXIMUM ALLOWABLE TORQUE TO ACCELERATE THE PUMP FROM STATIC CONDITION TO 9000 rpm LINEARLY IN 1.5 sec. WITH STEADILY APPLIED 3550 psig OUTLET PRESSURE AND 150 psig INLET PRESSURE SHALL NOT EXCEED 220 in.-lbs. AFTER 8 hrs. SOAK PERIOD AT 0°F.
RESPONSE	THE PUMP FLOW COMPENSATOR SHALL RESPOND WITHIN 50 ms a) FROM 80% FULL FLOW TO 5% AND b) FROM 5% FULL FLOW TO 80% FULL FLOW.

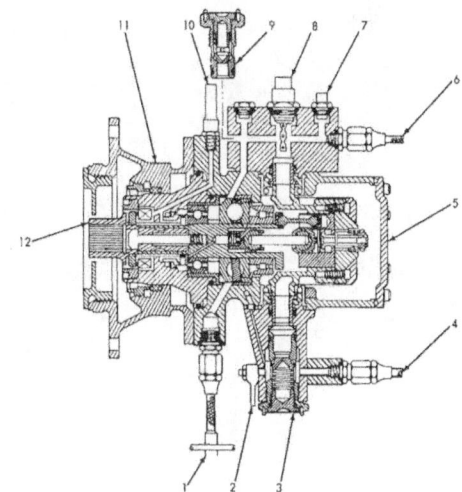

1. THERMOCONDITIONING FLUID TO AUXILIARY PUMP
2. VALVE, HYDRAULIC BLEEDER
3. VALVE, CHECK, HIGH PRESSURE
4. FLUID DISCHARGE
5. PUMP, HYDRAULIC, ENGINE DRIVEN
6. FLUID RETURN
7. SWITCH, THERMAL, MAIN PUMP INLET
8. TRANSDUCER, TEMPERATURE, MAIN PUMP INLET
9. VALVE, CHECK, HIGH PRESSURE
10. HYDRAULIC PUMP, PIPE PLUG ASSEMBLY
11. THERMAL ISOLATOR
12. DRIVE SHAFT

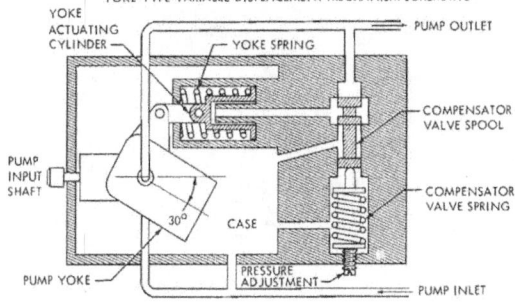

Figure 5-50

receives an initial charge equivalent to 475 psig at 70° F prior to propellant loading operations.

Hydraulic System Operation.

The S-IVB stage hydraulic system performance is monitored at the S-IVB hydraulics panel during countdown and launch operations. Switches and indicators on the hydraulics panel provide control and monitoring of system pressures, temperature, and hydraulic oil level. During propellant loading, the auxiliary hydraulic pump is cycled on from the hydraulic panel using the AUX HYDRAULIC PUMP POWER switch when the MAIN PUMP INLET OIL TEMP indicates −10° F minimum and remains on until the temperature indicates approximately 80° F. The presence of the auxiliary pump on command illuminates the AUX HYDRAULIC PUMP POWER ON light on the panel. The auxiliary pump is programmed to operate continuously through the S-IVB switch selector in the flight mode from approximately T-11 min prior to launch, throughout the boost phase and S-IVB engine operation. In the event the reservoir oil level reaches approximately 10%, the RESERVOIR OIL LEVEL LOW talkback light will illuminate. In the event the inlet oil temperature should reach −15° F, the MAIN PUMP INLET TEMP LOW talkback light will illuminate.

During normal system operation, low pressure fluid from the accumulator-reservoir flows through the main engine-driven pump

Section V S-IVB Stage

AUXILIARY MOTOR-DRIVEN HYDRAULIC PUMP

LOCATION	J-2 ENGINE THRUST STRUCTURE
SERVICE	ELECTRICAL & HYDRAULIC FLUID MIL-H-5606

OPERATING PRESSURE
  OUTLET PRESS.
  INLET PRESS.
  CASE PRESS.
	
	3650 psig MAX.
	45 psig MIN.
	45 psig

PROOF PRESSURE
  OUTLET PRESS.
  INLET PRESS.
  CASE PRESS.
	AT 275°F
	5475 psig MIN.
	480 psig MIN.
	480 psig MIN.

BURST PRESSURE
  OUTLET PRESS.
  INLET PRESS.
  CASE PRESS.
	at 275°F
	9125 psig MIN.
	800 psig MIN.
	800 psig MIN.

TEMPERATURE
  AMBIENT
  FLUID
	
	-30 TO +275°F
	0 TO +275°F

ELECTRICAL MOTOR
  TYPE
  VOLTAGE
  STARTING CURRENT
  RUNNING CURRENT
	
	CONTINUOUS DUTY SHUTFIELD
	50 TO 60 vdc
	250 A
	75 A

RATED FLOW	1.5 gpm
RESPONSE	100 ms UNDER NORMAL OPERATING CONDITIONS.

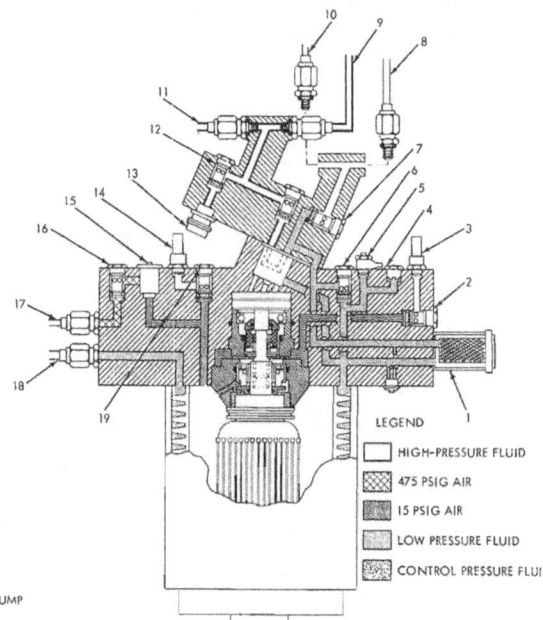

LEGEND
- HIGH-PRESSURE FLUID
- 475 PSIG AIR
- 15 PSIG AIR
- LOW PRESSURE FLUID
- CONTROL PRESSURE FLUID

1. FILTER
2. VALVE, RELIEF
3. HYDRAULIC PUMP PIPE PLUG ASSEMBLY
4. DETECTOR, CHIP
5. VALVE, HYDRAULIC BLEEDER
6. VALVE, RELIEF
7. VALVE, CHECK
8. RETURN FROM ACCUMULATOR
9. DISCHARGE TO ACCUMULATOR
10. RETURN TO MAIN PUMP
11. DISCHARGE FROM MAIN PUMP
12. VALVE, CHECK
13. COUPLING HALF, QUICK DISCONNECT
14. HYDRAULIC PUMP PIPE PLUG ASSEMBLY
15. REGULATOR, AIR PRESSURE
16. VALVE, CHECK
17. AIR FROM DRY AIR BOTTLE
18. THERMOCONDITIONING FLUID FROM MAIN PUMP
19. VALVE, RELIEF

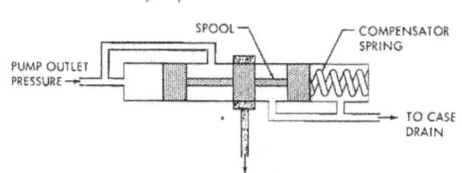

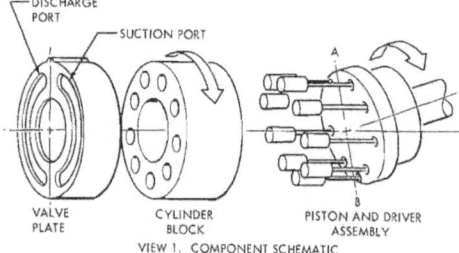

VIEW 1. COMPONENT SCHEMATIC

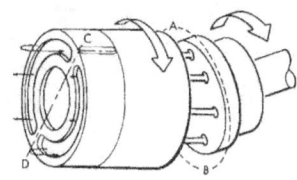

VIEW 3. REDUCED DELIVERY POSITION

VIEW 2. MAXIMUM DELIVERY POSITION

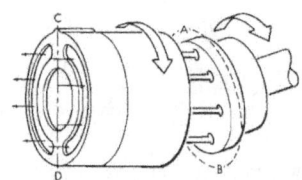

VIEW 4. ZERO DELIVERY POSITION

Figure 5-51

Section V S-IVB Stage

ACCUMULATOR-RESERVOIR

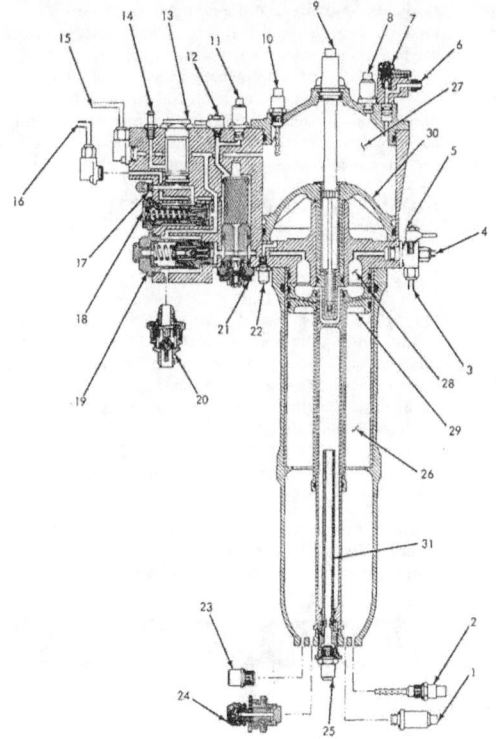

LOCATION	J-2 ENGINE THRUST STRUCTURE
SERVICE	HYDRAULIC FLUID MIL-H-5606
TEMPERATURE	-12 TO +275°F
PRESSURE ACCUMULATOR RESERVOIR LOW HIGH	2350 ± 50 psig GN$_2$ 170 psig 3650 psig

1. TRANSDUCER, PRESSURE, GN$_2$ ACCUMULATOR
2. TRANSDUCER TEMPERATURE, GN$_2$ ACCUMULATOR
3. DISCHARGE TO YAW ACTUATOR
4. DISCHARGE TO PITCH ACTUATOR
5. VALVE, HYDRAULIC BLEEDER
6. RETURN FROM BOTH ACTUATORS
7. VALVE, HYDRAULIC BLEEDER
8. TRANSDUCER, PRESSURE, RESERVOIR OIL
9. POTENTIOMETER, PISTON POSITION
10. TRANSDUCER, TEMPERATURE, RESERVOIR OIL
11. TRANSDUCER, PRESSURE, HYDRAULIC SYSTEM
12. VALVE, HYDRAULIC BLEEDER
13. VALVE, LOW PRESSURE, BALANCED
14. FITTING, HOSE ADAPTER
15. LOW-PRESSURE FLUID RETURN TO PUMP
16. HIGH-PRESSURE FLUID FROM PUMP
17. COUPLING HALF, QUICK DISCONNECT
18. VALVE, RELIEF, LOW PRESSURE BALANCED
19. VALVE, RELIEF, HIGH PRESSURE
20. COUPLING, HALF, QUICK DISCONNECT
21. FILTER ELEMENT, MAIN SYSTEM
22. VALVE, RELIEF, VENT, LOW PRESSURE
23. GAGE PRESSURE, DIAL INDICATOR
24. VALVE, AIR, HIGH PRESSURE, CHARGING
25. VALVE, RELIEF, VENT, LOW PRESSURE
26. ACCUMULATOR
27. LOW-PRESSURE RESERVOIR
28. HIGH-PRESSURE RESERVOIR
29. FLOATING PISTON
30. RESERVOIR PISTON
31. BOOTSTRAP PISTON

Figure 5-52

case and enters the auxiliary pump motor jacket for thermal conditioning. The fluid then flows through a filter and enters the auxiliary hydraulic pump. A relief valve will permit the fluid to bypass the filter if the differential pressure exceeds 90 psid. A check valve in the low pressure auxiliary pump inlet will permit return flow of main pump case fluid if the auxiliary pump should fail. The pump increases the fluid pressure to 3,650 psig and delivers the fluid through a pump discharge check valve to the accumulator reservoir.

The hydraulic fluid flows through the main system filter element into the accumulator chamber where the hydraulic volume increases until a fluid-to-gas pressure balance equivalent to the discharge pressure is obtained. The fluid exits the accumulator and flows to the pitch and yaw servoactuators. The fluid flows through an actuator filter into the servovalve. Inside the servovalve, the fluid is filtered again before it enters the critical control passages. With the servovalve commanded to the null position, the fluid flows through the variable orifices into the low pressure return passages, which port the fluid back to the accumulator-reservoir. Flowrate in the null mode of operation is 0.4–0.8 gpm, which is the total of both servovalve leakage rates.

With the system pressurized to full operating pressure by the auxiliary pump at J-2 engine start, the main pump compensator moves the main pump yoke toward the no-flow position, thereby reducing starting torque requirements upon the turbine-driven quill shaft. As the pump accelerates, the compensator increases the pump-yoke angle which controls the output. When the pump reaches full speed, its discharge flow adds to the auxiliary pump flow and the hydraulic system is in a state of readiness. Check valves in the discharge lines of the pumps prevent pressure reversal whenever the system is pressurized with the pumps inactive, and prevent interaction between the pumps during their operation.

Hydraulic System Fluid Requirements Summary.

Hydraulic Fluid. The hydraulic fluid shall meet the requirements of MIL-H-5606A. The properties of the petroleum base fluid are: fluid temperature range of -65° F to +275° F for closed system; pour point of less than -90° F; flash point of +230° F.

The solid contaminant particles content shall not exceed the allowable number of particles per 100 ml sample of fluid: 1340 particles 0-25 microns in size; 210 particles, 25-50 microns in size; 28 particles, 50-100 microns in size; 3 particles, including fibers, over 100 microns in size.

Foaming characteristics of the hydraulic fluid when tested at 75° shall be: stable foaming tendency and complete foam collapse after a 5-minute blowing period followed by a 10-minute settling period. A ring of small bubbles around the edge of the graduate constitutes

Section V S-IVB Stage

complete collapse. A random sample of filled unit containers and a sample of shipping containers fully prepared for delivery shall be selected from each lot of the hydraulic fluid.

Air. Air used for filling the auxiliary hydraulic motor compressed gas tank conforms to the following specification: In a 30-ft^3 volume of air the maximum allowable number of particles 30 to 100 microns in size is 25 and no particles over 100 microns in size are allowed; the total allowable hydrocarbon content is 25 ppm expressed by equivalent weight of carbon; the maximum allowable moisture content is 25 ppm by volume (equivalent to a dew point of -64° F).

Gaseous Nitrogen. GN_2 used for the accumulator-reservoir precharge conforms to MSFC-SPEC-234. The purity of GN_2 is 99.98 percent (min) nitrogen by volume.

Oxygen content is 150 ppm (max) by volume, hydrocarbon content (expressed as methane) is 15 ppm (max) by volume, and moisture content is 11.5 ppm (max) by volume at standard conditions.

Interconnection Plumbing.

Both high and low pressure flexible hoses are constructed of polytertrafluoroethylene (teflon) lined hose reinforced with steel wire braid. The flex hoses are normally straight (not precurved) and use straight fittings.

The hydraulic tube assemblies are constructed of steel tubing corrosion-resistant (CRES) 304, aerospace vehicle hydraulic system 1/8 hard condition.

Critical Components.

All the hydraulic system components are flight critical. They are: servoactuators; main hydraulic pump, engine-driven; auxiliary hydraulic pump, motor-driven; and accumulator-reservoir.

Hydraulic System Measurements.

Nine measurements (figure 5-54) are utilized to monitor system readiness condition and operation. These measurements are telemetered through the S-IVB stage telemetry system to ground receiving stations where they are monitored in real-time in addition to being recorded. Seven meters on the S-IVB hydraulics panel provide monitoring for each measurement. Pitch and yaw actuator-position measurements are monitored on the S-IVB engine deflection panel meters. Flight control measurements in figure 5-54 are relayed to mission control center for real-time monitoring.

AUXILIARY PROPULSION SYSTEM.

The auxiliary propulsion system (APS) provides attitude control for the S-IVB stage/IU and payload during both powered and coast phases of flight. During the powered phase, the APS provides only roll control; gimballing the J-2 engine provides yaw and pitch control. During the coast phase, the APS provides pitch, yaw, and roll control (figure 5-55). The APS configuration (figure 5-56) consists of two aerodynamically shaped modules, each 80 in. high, installed on the aft skirt 180 deg apart at positions I and III. Each module contains three liquid-propellant rocket engines, a fuel and oxidizer storage and supply system, a high pressure helium storage and regulator system, and control components for preflight servicing and inflight operation.

The APS uses nitrogen tetroxide (N_2O_4) as oxidizer, and monomethylhydrazine (MMH) as the fuel. The fuel and oxidizer are hypergolic; therefore, an ignition system is not required. Attitude

HYDRAULIC ACTUATOR ASSEMBLY

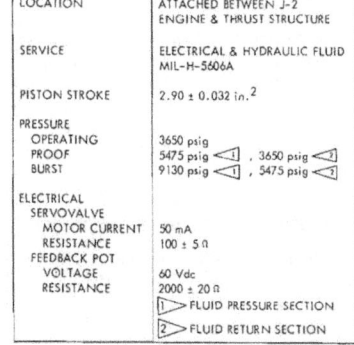
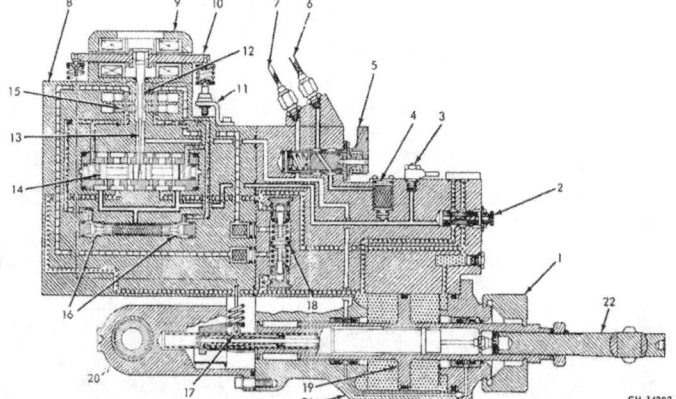

Figure 5-53

Section V S-IVB Stage

corrections are made by firing the 150-lbf engines, individually or in any combination, in short bursts of approximately 65 msec minimum duration. Commands from the flight control computer through the attitude control relay modules actuate fuel and oxidizer solenoid-valve clusters that admit propellant to the engine combustion chambers. Helium pressure exerted against stainless steel bellows assemblies, which contain the fuel and oxidizer, forces propellant into the engine. For a detail description of the engine actuation control, see Section VI (Navigation, Guidance, and Control).

Attitude Control System Concept.

A detachable modular subsystem concept for the attitude control system was selected for the S-IVB stage to maintain an independent propulsion subsystem that facilitated development, qualification checkout, manufacturing, and replacement if required. Also, the modular concept is appropriate for isolating the thermal control problems associated with space envrionments and the cryogenic space vehicle systems. The extended orbital operation of the S-IVB stage requires thermal protection of the hypergolic propellant systems to prevent feeezing and overheating. Compact system modules, with short propellant lines and buried engine installations greatly simplify the temperature control problem by permitting the use of a passive protection. Conventional insulation and control of the surface emissivities are the only thermal protection required.

The attitude control system module was developed and qualified as a system at the Sacramento test center. Firing tests verified the interaction performance and operational characteristics of the pressurization, positive expulsion propellant, and engine systems. Also, verification of the checkout and operational procedures was accomplished. The verification of thermal capabilities was accomplished in the 39-ft diameter space chamber at the McDonnell Douglas space systems center. In addition, the modular concept allowed a complete system vibration test to verify the installations under this environment.

The module was designed to operate with a 3000-psi helium supply and a 196-psia propellant expulsion system. This allows the engine to operate with a 100-psia chamber pressure at 150 lbf. The 3000-psi helium pressurization system was chosen to be consistent with the requirements for the main stage, which also utilizes helium at 3000 psi. The requirement for 196 psia was based on an optimization weight study which provides the minimum weight tank system, and still provided adequate pressure for operating the engines at a 100-psia chamber pressure. A chamber pressure of 100 psia was considered to be consistent with the requirements of a 150-lbf engine for a satisfactory engine envelope and performance. The thrust requirement was established based on the maximum thrust required to control the S-IVB stage.

Hypergolic bi-propellants are utilized since they provide the advantages of storability and spontaneous ignition; hypergolic ignition allows a more simple engine design with a fast response rate. The quantity of propellant required was established as 61 lbm (nitrogen tetroxide and monomethylhydrazine) at an oxidizer to fuel ratio of 1.60. This quantity was based on a system study of the mission requirements for attitude control and vehicle stabilization requirements, with consideration of engine performance. The requirements also include a need for high-response-rate engine valves to provide a minimum impulse bit. It was determined that a total impulse of 7.5 ± 0.75 lb-sec is satisfactory without significant degradation of engine performance. This assumes that all propellant valves on each engine will operate at a maximum pulse rate of 10 starts/sec. Each engine has a maximum design life of 10,000 starts with a maximum pulse rate of 10 starts/sec.

APS Major Subsystems.

Each of the two modules is comprised of three major subsystems:

S-IVB HYDRAULIC SYSTEM MEASURMENTS

NUMBER	NAME	RANGE
*VXC50-401	TEMP, HYD PUMP INLET OIL	400 TO 785°R
*VXC51-403	TEMP, RESERVOIR OIL	400 TO 785°R
VC138-403	TEMP, ACCUMULATOR GN_2	400 TO 785°R
*VXD41-403	PRESS, HYDRAULIC SYSTEM	1500 TO 4500 PSIA
*VXD42-403	PRESS, RESERVOIR OIL	0 TO 400 PSIA
*VXD43-403	PRESS, GN_2 ACCUMULATOR	1500 TO 4000 PSIA
*VXG1-403	POSITION, ACTUATOR PISTON POT, PITCH	7.5 TO -7.5 DEG
*VXG2-403	POSITION, ACTUATOR PISTON POT, YAW	-7.5 TO 7.5 DEG
*VXL7-403	LEVEL, RESERVOIR OIL	0 TO 100%

MEASUREMENT NUMBER PREFIXES INDICATE THE FOLLOWING:
*FLIGHT CONTROL; V-ESE DISPLAY; X-AUXILIARY DISPLAY

Figure 5-54

helium pressurization, positive expulsion propellant feed, and attitude control engines. The helium pressurization subsystem supplies and regulates the flow of pressurized helium to the positive expulsion subsystem (propellant tankage). The positive expulsion subsystem provides propellant (fuel and oxidizer), on demand, to the engines under zero and random gravity environment. The attitude control engines provide the necessary thrust for attitude control and maneuvering of the stage during various phases of flight. Instrumentation on the module consists of the various pressure and temperature readings for performance evaluation. A description of each of these subsystems and their regulated components is presented in the following paragrpahs.

Helium Pressurization Subsystem. The pressurization subsystem consists of a fill disconnect, two check valves in series, a helium storage tank, and a pressure regulator. Downstream from the regulator, the system splits into fuel and oxidizer pressurant branches. Each branch includes a quad-check value and a low pressure module. During prelaunch operations, the 268 cu in helium storage tank is pressurized through the fill disconnect and two check valves to 3100 ± 100 psig. The helium tank incoprorates a temperature probe for monitoring tank temperature. Tank pressure is measured at the inlet to the helium pressure regulator.

Helium stored in the tank at 3100 ± 100 psig is supplied to the helium pressure regulator assembly. The helium gas entering the

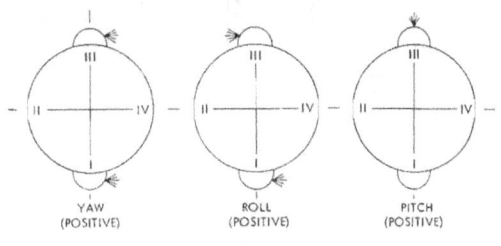

Figure 5-55

Section V S-IVB Stage

AUXILIARY PROPULSION SYSTEM

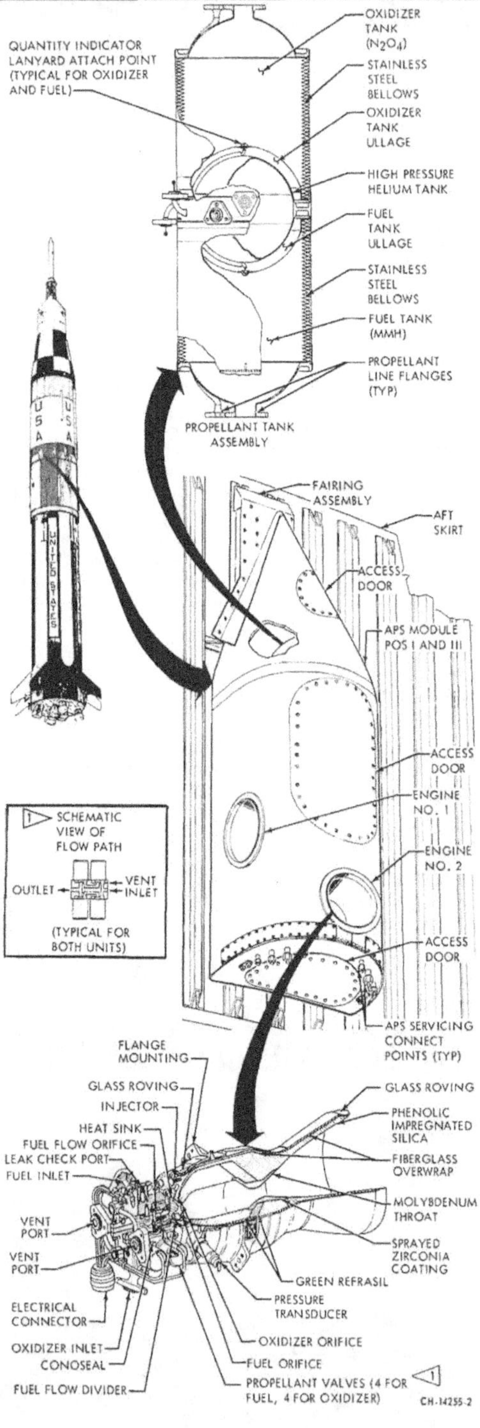

Figure 5-56

assembly first passes through a filter and then through two regulators in series, both of which sense downstream pressure. During normal operation, the primary regulator maintains regulated pressure at 196 (+3, −6) psig. Should the primary regulator fail open, the secondary regulator would maintain the output at 200 (+3, −6) psig.

Ambient pressure sensing ports, provided on both regulators, furnish the necessary ambient pressure references and provide a checkout capability for enabling the regulated outlet pressure settings to be checked. Regulator performance is evaluated by pressure transducers installed immediately upstream and downstream from the regulator. Regulated helium is fed through the quad check valves to the ullage area of the fuel and oxidizer tanks.

Two sets of quadruple check valves are employed in the helium pressurization subsystem; one set in the fuel tank pressurization line and the other set in the oxidizer tank pressurization line. These check valves are a redundant safety feature used to prevent contact of fuel and oxidizer vapors in the pressurization subsystem should both positive expulsion bellows develop a leak. Each set of check valves consists of four check valves contained in one body connected in a series-parallel arrangement. Failure of a check valve set requires open failure of two check valves in series or closed failure of two check valves in parallel.

The low pressure helium modules provide capabilities for ground venting of propellant tank ullage as well as a means of establishing pneumatic control of the expansion rate of the expulsion bellows during loading and checkout. These modules also provide ground and flight relief capability. The relief value cracking pressure is 325 to 350 psia; it reseats at 225 psia minimum. No command venting capability during flight is provided.

Positive Expulsion Propellant Feed Subsystem. The positive expulsion propellant feed subsystem provides hypergolic propellant transfer to the engines under zero and random gravity conditions. This subsystem consists of an integral propellant tank assembly containing two metal bellows for propellant expulsion, two propellant control modules with propellant filters and auxiliary ports for servicing operations and propellant manifolds for distribution of propellants to the engines.

The propellant tank (figure 5-56) is an integral assembly consisting of cylindrical compartments separated from each other by a helium storage sphere. Each compartment contains a positive expulsion bellows, with one bellows being used for fuel and the other for oxidizer. The bellows are constructed of welded stainless steel convolutions incorporating a hemispherical movable end. A corrugated liner is installed between the tank walls and the bellows assembly. This liner serves as a guide for axial movement of the bellows and as a bellows vibration damper.

The bellows can be moved to either of its extreme positions (collapsed or extended) through the application of a small differential pressure across the bellows movable end. Pressurized helium, when applied to the bellows movable end, compresses the bellows and expels propellant on demand from the end of the tank. When the bellows is completely collapsed, the movable hemispherical end comes to rest against the mating tank bottom. This close stacking of the mating parts permits an expulsion efficiency of 98 percent. The bellows assemblies have a minimum service life requirement of 200 cycles. One cycle consists of extension from the stacked hieght (collapsed) to the fully expanded position and returned to the stacked height.

A bellows position indicator connected externally to each propellant tank, consists of a potentiometer connected to the bellows by means of a flexible cable attached to a spring wound drum. The potentiometer is connected to the drum and provides an indication of bellows position. The position indicator assembly contains a microswitch which is actuated at 95 percent of the fully extended bellows position (100 percent fill level). These microswitches perform no flight function and can be employed during servicing to indicate that the tank has been adequately loaded.

The propellant control modules filter propellants and provide access into the propellant subsystem for servicing. The module contains a propellant transfer valve, a recirculation valve, facility lines purge check valves, and a system filter. The propellant transfer valve is a direct operating, normally closed solenoid valve. This valve is employed to fill and drain the attitude control system module.

The propellant recirculation valve is a direct acting, normally closed solenoid valve with two independent poppets and seats. The two-poppet design isolates the engine recirculation line from the tank recirculation line and all propellant flowing to the engine passes through the filter. The filter has a 10-micron nominal and 25-micron absolute rating.

The propellant manifold delivers propellant to the engine from the propellant control module and provides a secondary path of recirculation of propellant to the control module.

Engine and propellant control modules, which attach to the manifold, use Marman conoseal flanges. Bellows are employed in the lines to provide manifold flexibility.

Attitude Control Engines. Three 150-lbf TRW, Inc. engines (figure 5-56) are employed in each attitude control system module. The combustion chamber is lined with ablatively refrasil material. The engines have quadruple propellant or injector valves for redundancy.

The thrust chamber is integrally fabricated and composed of three major elements: the combustion chamber, the nozzle throat section, and the nozzle expansion cone.

The engine has an expansion ratio of 33.9 to 1. The injector consists of twelve pairs of unlike-on-unlike doublets arranged to minimize hot spots in the combustion chamber. The valve side of the injector is filled with a silver braze heat sink which reduces injector operating temperature.

The engine was qualified for a total pulse operation of 300 sec. During the 300-sec life requirement, the external wall temperature does not exceed 600° F ..nd the maximum valve body external temperature does not exceed 165° F. The maximum expected duty cycle requirements on the Saturn S-IB/S-IVB is approximately 90 sec. The rocket engines have a reliability design goal of 0.9992 at a 90-percent confidence level while operating.

The engine propellant valve assembly consists of eight normally closed, quad redundant propellant valves (four oxidizer and four fuel), arranged in two series parallel arrangements. Dual failure within the manifold fuel or oxidizer arrangement is required to cause failure of the rocket engine assembly. An assembly closed failure prevents any engine operation while an assembly open failure results in continuous flow and loss of all propellant. Assembly valve failure cannot occur unless two valves fail open in series or two valves fail closed in parallel.

The injector valves provide positive on-off control of propellant flow upon command from an external power source. Four valves, integral in an assembly, are capable of simultaneous operation and are synchronized to allow flow or terminate flow within 3 msec of each other. The opening time for each valve assembly, defined as the time from initiation of open signal to fully open valve package, does not exceed 23 msec.

Each valve plunger and solenoid consists of a complete assembly that may be removed from the propellant valve assembly to allow thorough valve cleaning and solenoid replacement. Each valve assembly (four fuel and four oxidizer valves) may be removed from the rocket engine. All valve assemblies are interchangeable within their own subsystem; oxidizer with oxidizer and fuel with fuel. All external leak paths, at individual solenoid coils, are sealed by Marman conoseals (excluding welded joints).

The leakage of the rocket engine fluids at each propellant injector valve will not exceed a seepage rate of 0.03 in.^{3}/24 hr during exposure to ambient pressures and any propellant pressures up to 185 psia.

APS Flight Operation.

The system on the S-IVB stage is enabled in flight after the S-IB lower stage retrorockets have been ignited. First command actuation occurs just prior to J-2 engine start to provide roll control of the stage during engine burn. The single J-2 engine performs pitch and yaw control only.

Helium supplied from a 3100-psia tank is regulated to 196 (+3, -6) psig to expel propellants from each propellant tank to the engines. During propellant flow, the liquid passes out of the bellows tank into the engine mainfold for injection into the engines. See figures 5-57 and 5-58.

Commands for control are provided by the instrument unit. Output from a guidance platform indicating measured vehicle attitude is received by the instrument unit and a comparison is made with the desired or programmed attitude. If a deviation exists, the instrument unit provides the required commands via the control relay package to the attitude control system engine injector valves for correction proportional to the magnitude of the deviation. The engines operate in short pulse-type bursts and may operate 65 msec or longer as required.

At J-2 engine cutoff the pitch and yaw control is activated and all control (roll, pitch and yaw) remains active throughout the coast phase.

Attitude Control System Performance.

The maximum impulse usage expected for SA-206 up to and including CSM separation is 12.5 lbm of oxidizer and 7.9 lbm of fuel per module. This includes S-IVB burn roll control, insertion transients, maneuver to local horizontal and hold, and CSM separation transients. This consumption is 33% of the propellants loaded. The total capacities are 39.4 lbm of oxidizer and 23.6 lbm of fuel per module. The propellants remaining after CSM separation will be used for passivation and the M415 experiment.

APS Servicing.

The North American Rockwell services supply the MMH and N_2O_4 to the APS modules through the McDonnell Douglas models 472 and 473 propellant loading carts, respectively. Two 0-100 psig regulator gage assemblies, one fuel and one oxidizer, provide GN_2 for loading, pressurization, and purging of GSE and facility lines.

The model 472 and 473 carts provide control for fuel and oxidizer during propellant recirculate and fill procedures. APS servicing is conducted while the mobile service structure is at the pad. All servicing operations are accomplished through fittings on the aft end of each APS module, except helium system purge and pressurization, which are accomplished through a single coupling on the aft umbilical panel.

Figure 5-59 presents information pertinent to servicing the APS. The APS propellant systems (figure 5-57) are purged with GN_2 prior to propellant loading. This purge clears the propellant fill and return line and the fuel and oxidizer fill modules. After the purge operation, GN_2 supplied through the fuel and oxidizer low

AUXILIARY PROPULSION SYSTEM OPERATION

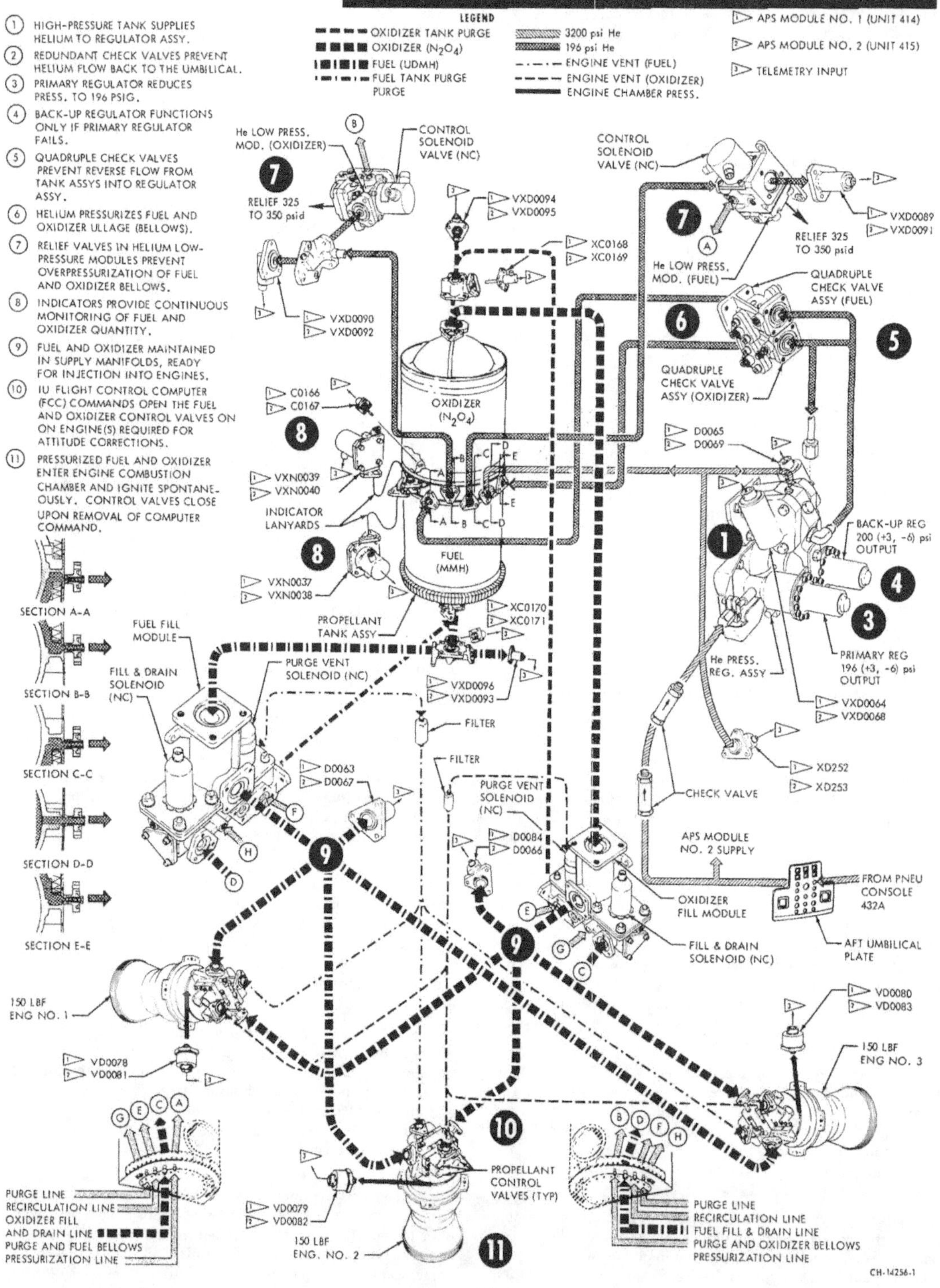

1. HIGH-PRESSURE TANK SUPPLIES HELIUM TO REGULATOR ASSY.
2. REDUNDANT CHECK VALVES PREVENT HELIUM FLOW BACK TO THE UMBILICAL.
3. PRIMARY REGULATOR REDUCES PRESS. TO 196 PSIG.
4. BACK-UP REGULATOR FUNCTIONS ONLY IF PRIMARY REGULATOR FAILS.
5. QUADRUPLE CHECK VALVES PREVENT REVERSE FLOW FROM TANK ASSYS INTO REGULATOR ASSY.
6. HELIUM PRESSURIZES FUEL AND OXIDIZER ULLAGE (BELLOWS).
7. RELIEF VALVES IN HELIUM LOW-PRESSURE MODULES PREVENT OVERPRESSURIZATION OF FUEL AND OXIDIZER BELLOWS.
8. INDICATORS PROVIDE CONTINUOUS MONITORING OF FUEL AND OXIDIZER QUANTITY.
9. FUEL AND OXIDIZER MAINTAINED IN SUPPLY MANIFOLDS, READY FOR INJECTION INTO ENGINES.
10. IU FLIGHT CONTROL COMPUTER (FCC) COMMANDS OPEN THE FUEL AND OXIDIZER CONTROL VALVES ON ON ENGINE(S) REQUIRED FOR ATTITUDE CORRECTIONS.
11. PRESSURIZED FUEL AND OXIDIZER ENTER ENGINE COMBUSTION CHAMBER AND IGNITE SPONTANEOUSLY. CONTROL VALVES CLOSE UPON REMOVAL OF COMPUTER COMMAND.
LEGEND
OXIDIZER TANK PURGE
OXIDIZER (N2O4)
FUEL (UDMH)
FUEL TANK PURGE PURGE
3200 psi He
196 psi He
ENGINE VENT (FUEL)
ENGINE VENT (OXIDIZER)
ENGINE CHAMBER PRESS.
1 APS MODULE NO. 1 (UNIT 414)
2 APS MODULE NO. 2 (UNIT 415)
3 TELEMETRY INPUT
He LOW PRESS. MOD. (OXIDIZER)
CONTROL SOLENOID VALVE (NC)
RELIEF 325 TO 350 psid
VXD0094
VXD0095
XC0168
XC0169
VXD0090
VXD0092
CONTROL SOLENOID VALVE (NC)
VXD0089
VXD0091
RELIEF 325 TO 350 psid
He LOW PRESS. MOD. (FUEL)
QUADRUPLE CHECK VALVE ASSY (FUEL)
QUADRUPLE CHECK VALVE ASSY (OXIDIZER)
C0166
C0167
OXIDIZER (N2O4)
D0065
D0069
VXN0039
VXN0040
INDICATOR LANYARDS
BACK-UP REG 200 (+3, -6) psi OUTPUT
FUEL (MMH)
VXN0037
VXN0038
PROPELLANT TANK ASSY
XC0170
XC0171
SECTION A-A
FUEL FILL MODULE
PURGE VENT SOLENOID (NC)
FILL & DRAIN SOLENOID (NC)
SECTION B-B
VXD0096
VXD0093
FILTER
He PRESS. REG. ASSY
PRIMARY REG 196 (+3, -6) psi OUTPUT
VXD0064
VXD0068
SECTION C-C
FILTER
D0063
D0067
PURGE VENT SOLENOID (NC)
D0084
D0066
XD252
XD253
CHECK VALVE
SECTION D-D
APS MODULE NO. 2 SUPPLY
FROM PNEU CONSOLE 432A
OXIDIZER FILL MODULE
AFT UMBILICAL PLATE
FILL & DRAIN SOLENOID (NC)
SECTION E-E
150 LBF ENG NO. 1
VD0080
VD0083
150 LBF ENG NO. 3
VD0078
VD0081
PROPELLANT CONTROL VALVES (TYP)
PURGE LINE
RECIRCULATION LINE
OXIDIZER FILL AND DRAIN LINE
PURGE AND FUEL BELLOWS PRESSURIZATION LINE
VD0079
VD0082
150 LBF ENG. NO. 2
PURGE LINE
RECIRCULATION LINE
FUEL FILL & DRAIN LINE
PURGE AND OXIDIZER BELLOWS PRESSURIZATION LINE
CH-14256-1

Figure 5-57

Section V S-IVB Stage

AUXILIARY PROPULSION SYSTEM DIAGRAM

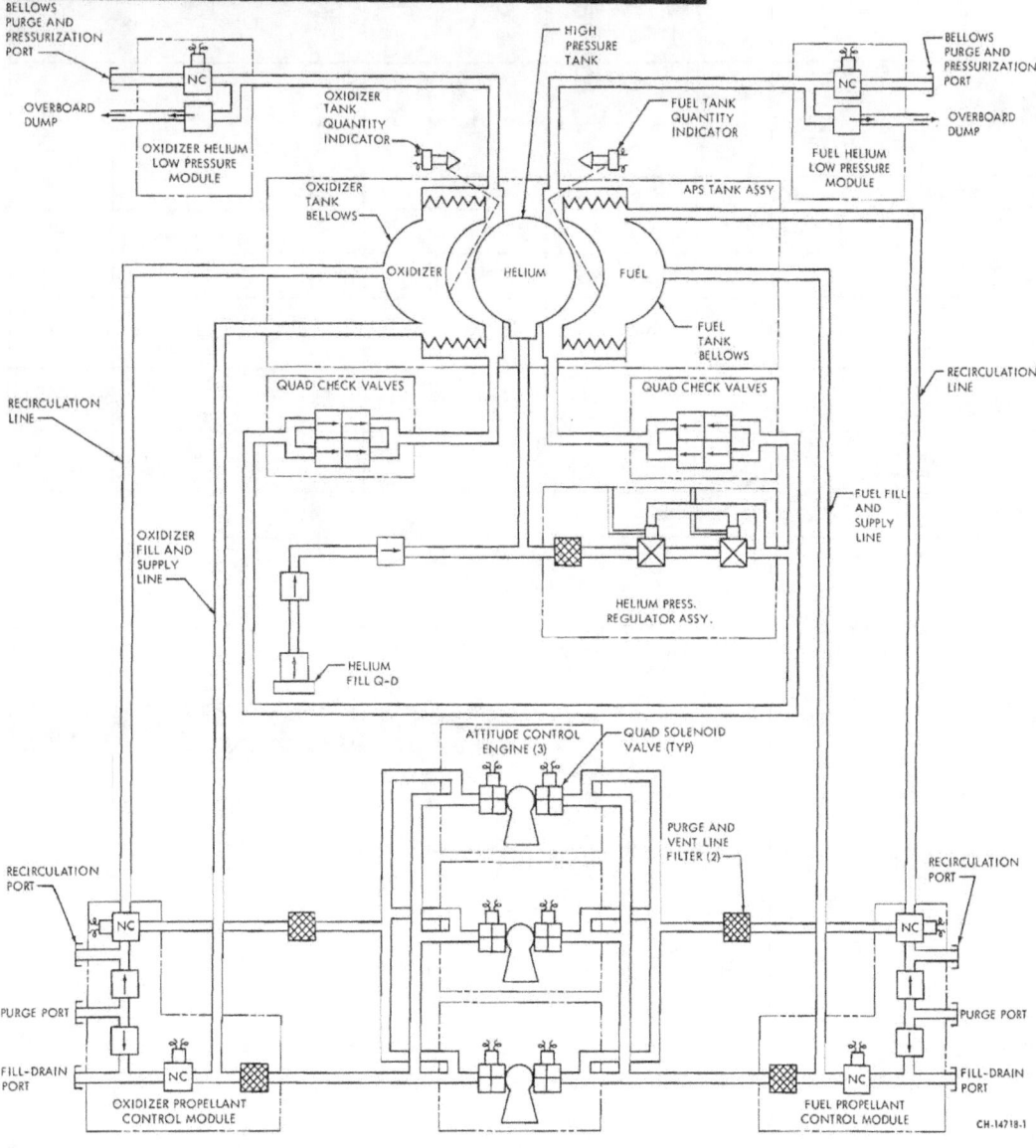

Figure 5-58

pressure modules collapses the bellows to begin the propellant loading operations.

At the initiation of loading, a recirculation flow is first established through each module. Propellants flow through the module propellant lines and out through the recirculation valves. Flow is continued until all trapped gas is eliminated by the recirculation procedure. Once the lines are cleared of trapped gases, the recirculation valve is closed and the bellows is allowed to expand by venting the GN_2 back through the low pressure helium modules allowing the tank to be filled. Bellows position is monitored during loading by use of position potentiometers installed on each tank assembly.

Helium used for propellant expulsion is loaded into the module through a pneumatic service line connected to the stage through the stage umbilicals.

APS Fluids Summary.

Helium. Helium conforming to MSFC-SPEC-364A is used as a pressurant. The purity of the helium shall be not less than 99.95 percent helium by volume when tested in accordance with MSFC requirements. Impurities shall not exceed: oxygen, 90 ppm by volume; hydrogen, no test required; nitrogen, 360 ppm by volume; inert gases, no test required.

5-55

Section V S-IVB Stage

APS FLUIDS FUNCTIONAL REQUIREMENTS

FUNCTIONAL OPERATION	MEDIA	PRESS. (psig)	TEMP. (°F)	FLOW RATE (GPM)	TIME REQUIRED (MIN)	QUANTITY (LB)	REMARKS
FUEL LOADING	MMH	25 TO 35	70 TO 90	0.5 TO 20	7.5	23.6	REQUIREMENTS ARE GIVEN FOR ONE MODULE. LOADING ACCURACY IS DEPENDENT ON MISSION REQUIREMENTS. ACCURACY CAN BE HELD WITHIN $\pm$ 1% OF MASS REQUIRED.
OXIDIZER LOADING	N_2O_4	25 TO 35	70 TO 90	0.5 TO 2.5	7.5	39.4	
FUEL CIRCULATION (PRELOADING)	MMH	25 TO 35	70 TO 90	1.5 TO 2.5	15	109	PRIOR TO FILLING, FUEL AND OXID ARE CIRCULATED THROUGH SYSTEM WITH BELLOWS COLLAPSED.
OXIDIZER CIRCULATION (PRELOADING)	N_2O_4	25 TO 35	70 TO 90	1.5 TO 2.5	15	179	
OXIDIZER AND FUEL LOADING AND TRANSFER LINES PURGE (POSTLOADING)	GN_2	25 TO 35	AMB	1.0 TO 2.0	15		THE PURGE IS TO INERT THE LINES. REQUIREMENTS FOR ONE MODULE, TWO REQUIRED.
OXIDIZER TANK BELLOWS AND FUEL TANK BELLOWS COLLAPSING PRESSURE	GN_2	35 TO 45	AMB	STATIC	N/A	0.015 NOM	REQUIREMENTS FOR ONE MODULE, TWO REQUIRED. NOT A MAIN UMBILICAL.
OXIDIZER AND FUEL ULLAGE PRESSURIZATION	GN_2	50 $\pm$ 5	AMB	STATIC	N/A	N/A	
HELIUM STORAGE SPHERE	HELIUM	3100 $\pm$ 100	AMB	0.75 LB/MIN MAX	65 $\pm$ 5	0.160	

Figure 5-59

Nitrogen. Nitrogen conforming to MSFC-SPEC-234A is used for purging and pressurization during loading. The nitrogen purity shall not be less than 99.983 percent nitrogen by volume when tested in accordance with MSFC requirements. Impurities shall not exceed: oxygen, 150 ppm by volume; carbon dioxide, no test required; inert gases, no test required.

Freon 113 (TF 113). Freon 113 is used to flush the oxidizer system in accordance with MSFC requirements. The product requirement for TF 113 is the same as for Freon 12.

Nitrogen Tetroxide (Inhibited). The nitrogen tetroxide (N_2O_4) per specification MSC-PPD-2B is used as an oxidizer propellant. It contains 0.8 $\pm$ 0.2% by weight nitricoxide as an inhibitor. See figures 5-60 and 5-61 for N_2O_4 properties and vapor pressure.

Monomethylhydrazine. The monomethylhydrazine (MMH) per specification MIL-P-27404 is used as a fuel propellant for the fuel system. See figures 5-62 and 5-63 for MMH properties and vapor pressures.

APS Measuring.

Thirty-eight measurements, nineteen per module, are taken during flight and telemetered to ground receiving stations (figure 5-64). All measurements are recorded for postflight evaluation. The flight control measurements are monitored in real-time at mission control center. The results of the flight control measurements are used in decision making affecting the mission in contingency situations.

ELECTRICAL.

The stage electrical system produces and distributes all ac and dc power for flight functions. Four batteries, two each in the forward and aft skirts, supply 28-Vdc and 56-Vdc primary power. This power buses directly to an associated distribution assembly. From here power routes through networks operational equipment and secondary power sources (to convert and regulate for specialized functions). There are four primary networks—one from each battery. Two networks (+4D21 and +4D31) originate in the forward skirt

dary power sources (to convert, invert and regulate for specialized functions). There are four primary networks—one from each battery. Two networks (+4D21 and +4D31) originate in the forward skirt

PROPERTIES OF N_2O_4

COMMON NAME:

NITROGEN TETROXIDE, NITROGEN PEROXIDE OR LIQUID NITROGEN DIOXIDE.

CHEMICAL FORMULA:

N_2O_4 $2NO_2$

MOLECULAR WEIGHT:

92.016

PHYSICAL PROPERTIES:

FREEZING POINT .... 11.84°F
BOILING POINT .... 70.07°F
CRITICAL TEMPERATURE .... 316.8°F
CRITICAL PRESSURE .... 1469 PSIA
LIQUID DENSITY (77°F) .... 11.94 LBS/GAL
APPEARANCE .... AT ROOM TEMPERATURE A HEAVY BROWN LIQUID. AS TEMPERATURE DECREASES, COLOR BECOMES LIGHTER. FUMES ARE YELLOWISH TO REDDISH BROWN.
ODOR .... CHARACTERISTIC PUNGENT ODOR
MAXIMUM ALLOWABLE
CONCENTRATION .... 2.5 PPM

CHEMICAL PROPERTIES:

A CORROSIVE OXIDIZING AGENT. HYPERGOLIC WITH UDMH, HYDRAZINE, ANILINE, FURFURYL ALCOHOL AND SOME OTHER FUELS.

NOT SENSITIVE TO HEAT, MECHANICAL SHOCK OR DETONATION.

IS NONFLAMMABLE WITH AIR, HOWEVER, IT CAN SUPPORT COMBUSTION WITH COMBUSTIBLE MATERIALS. VERY STABLE AT ROOM TEMPERATURE.

Figure 5-60

Figure 5-61

PROPERTIES OF MMH

COMMON NAME:

MONOMETHYL HYDRAZINE (MMH)

CHEMICAL FORMULA:

CH_3NHNH_2

MOLECULAR WEIGHT:

46.075

PHYSICAL PROPERTIES:

FREEZING POINT	-62.5°F
BOILING POINT	192.5°F
CRITICAL TEMPERATURE	593.6°F
CRITICAL PRESSURE	1195.1 PSIA
LIQUID DENSITY (77°F)	7.29 LBS/GAL
APPEARANCE	CLEAR, OILY WATER WHITE LIQUID
ODOR	AMMONIACAL
MAXIMUM ALLOWABLE CONCENTRATION	0.5 PPM
FLAMMABILITY LIMITS	4 TO 97% IN AIR

CHEMICAL PROPERTIES:

MMH IS NOT SENSITIVE TO IMPACT OR FRICTION, HOWEVER, ITS VAPOR WILL PROPAGATE DETONATION WITHIN THE FLAMMABILITY LIMITS.

EXTREMELY MISCIBLE IN WATER.

WILL REACT WITH CARBON DIOXIDE AND OXYGEN IN AIR.

IRON, COPPER, MOLYBDENUM, AND THEIR OXIDES CAUSE MMH TO DECOMPOSE.

VERY CAUSTIC.

VERY HYGROSCOPIC.

IGNITES SPONTANEOUSLY IF EXPOSED TO A LARGE SURFACE (SUCH AS SATURATED RAGS).

FREEZING HAS NO EFFECTS ON THE CHEMICAL PROPERTIES.

C-H 14534

Figure 5-62

and two others (+4D11 and +4D41) originate in the aft skirt. All networks supply 28 Vdc, except +4D41 which supplies 56 Vdc. Each network is independent of the others. Figure 5-65 shows the distribution network arrangement.

DESIGN GROUND RULES.

The electrical system design is analogous to those of earlier weapon systems (Thor) and space vehicles (S-IV stage). These proven concepts influence the electrical system ground rules.

Power and Signal Distribution.

The stage has two separate primary power distribution networks and source power completely independent from other vehicle stages. Each network has switching provisions to selectively alternate between ground source power (prelaunch) and stage battery power (flight). The stage wiring arrangement limits voltage drop to a 2-V (maximum) level from bus to load. Placing separate networks above and below the propellant tanks reduces cabling and eliminates need for excessive power lines in the connecting cable tunnels (see Structure). Electrical bonding techniques (metal-to-metal) ensure a unipotential structure. Electrical circuits are insulated from their respective cases. These cases bond to the structure for personnel safety and short circuit protection. Low current signals (measurements and telemetry) use separate networks; the cables are shielded and spaced to prevent electrical interference with high current lines.

Packaging and Installation.

Electrical components (or assemblies) are modular wherever possible. This packaging concept increases reliability, speeds part changeout, and reduces the number of cable interconnections. The modules, encapsulated when design permits, bond to open assembly panels for convenient maintenance and replacement. Physical parameters and heat dissipation requirements sometimes prevent encapsulation (epoxy) of equipment; instead, these components are installed in pressurized boxes. Mounting each assembly on its center of gravity minimizes induced vibration. Most electronic equipment is in the forward skirt where it mounts on environmental (cold-plate)

Figure 5-63

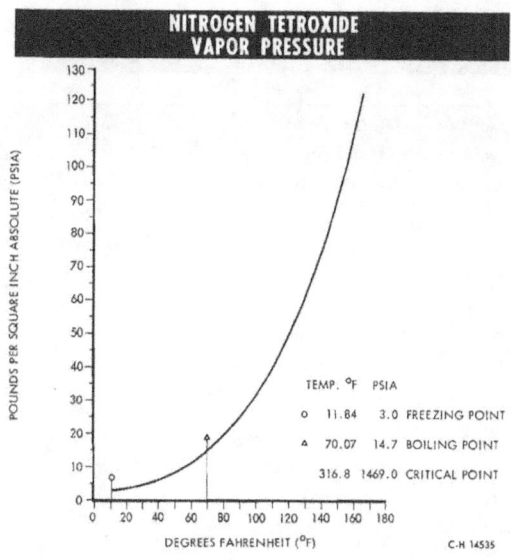

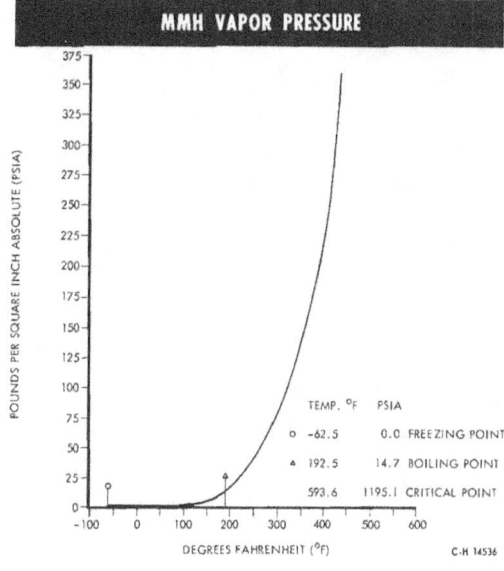

Section V S-IVB Stage

panels; the aft skirt has only ambient panels. Figures 5-2 and 5-3 show the equipment location in both skirt areas.

BATTERIES.

Studies proved that the best type of power source for the stage mission is a battery. Consideration was given to various other sources such as fuel cells, sterling cycle engines, and solar arrays. However, batteries best satisfy the power level requirements for the mission time duration. Weight and voltage regulation characteistics are basic reasons for using primary (one-shot) batteries instead of secondary (rechargeable) batteries. The original system design required only two batteries: one in the forward skirt for noise-free power to the data acquisition, propellant utilization, and range safety systems; another in the aft skirt to provide power switching and sequencing functions. Subsequent requirements for redundant range safety and emergency detection systems forced the addition of a second battery in the forward skirt. High power requirements of the chilldown pump motors and auxiliary hydraulic pump motor forced the addition of a second battery in the aft skirt. A 56-Vdc battery was selected because it drives these motors more efficiently and reduces complexity in the chilldown inverter circuitry.

Physical Construction.

The same manufacturer (Eagle Picher) produces all four stage batteries; in fact, all vehicle batteries. Section IV gives a detailed description of battery construction that is basically true for these batteries. Some features change such as mounting provision, electrical connection, overall size, cell configuration, and rated capacity. Notice that forward battery no. 1 is packaged as two separate units.

Electrical Characteristics.

The four batteries use silver oxide (pos) and zinc (neg) electrodes in a potassium hydroxide electrolytic solution to produce the electromotive force. Forward batteries no. 1 and no. 2 and aft battery no. 1 have output voltages of 28 ± 2 Vdc; aft battery no. 2 has an output of 56 ± 4 Vdc. These outputs hold true when each battery loads to the profile in figure 5-66.

The following list gives the nominal capacity and the predicted usage for each battery:

a. Fwd No. 1 Battery: 300 A-hr capacity, 40 percent usage.

b. Fwd No. 2 Battery: 5 A-hr capacity, 76 percent usage.

c. Aft No. 1 Battery: 58 A-hr capacity, 87 percent usage.

d. Aft No. 2 Battery: 22 A-hr capacity, 51 percent usage.

Internal heaters maintain battery temperature close to the normal operating range of 70 to 90° F. Circuitry inside the battery provides inflight monitoring of battery voltage (bridge network), current (current transformer), and temperature (thermistor). These measurements are listed in figure 5-67 with the ground station that displays them.

CHILLDOWN INVERTERS.

Motor-driven pumps, one each for the lox and fuel systems, circulate the liquid propellants to remove latent heat and stabilize temperature at the J-2 engine turbopump inlets before engine start (see Propulsion). These two pump motors operate on ac voltage from separate power supplies. These secondary power supplies, called chilldown inverters, receive 56 Vdc from aft battery no. 2 (+4D41 bus) and invert it to 3-phase current with 1500 VA capacity. The lox inverter is operable 5 min before liftoff; the fuel inverter powers

S-IVB AUXILIARY PROPULSION SYSTEM MEASUREMENTS

NUMBER	NAME	RANGE
XC166-414	TEMP, He SPHERE MODULE 1	360° TO 660°R
XC167-415	TEMP, He SPHERE MODULE 2	360° TO 660°R
*XC0168-414	TEMP, OXIDIZER TANK OUTLET MODULE 1	460° TO 590°R
*XC0169-415	TEMP, OXIDIZER TANK OUTLET MODULE 2	460° TO 590°R
*XC0170-414	TEMP, FUEL TANK OUTLET MODULE 1	460° TO 590°R
*XC0171-415	TEMP, FUEL TANK OUTLET MODULE 2	460° TO 590°R
*D0063-414	PRESS, FUEL SUPPLY MANIFOLD MODULE 1	0 TO 400 PSIA
*VXD0064-414	PRESS, He REGULATOR INLET MODULE 1	0 TO 3500 PSIA
*D0065-414	PRESS, He REGULATOR OUTLET MODULE 1	0 TO 400 PSIA
*D0066-415	PRESS, OXIDIZER SUPPLY MANIFOLD MODULE 2	0 TO 400 PSIA
*D0067-415	PRESS, FUEL SUPPLY MANIFOLD MODULE 2	0 TO 400 PSIA
*VXD0068-415	PRESS, He REGULATOR INLET MODULE 2	0 TO 3500 PSIA
*D0069-415	PRESS, He REGULATOR OUTLET MODULE 2	0 TO 400 PSIA
*D0084-414	PRESS, OXIDIZER SUPPLY MANIFOLD MODULE 1	0 TO 400 PSIA
*VD0078-414	PRESS, ATTITUDE CONTROL CHAMBER 1-1	0 TO 200 PSIA
*VD0079-414	PRESS, ATTITUDE CONTROL CHAMBER 1-2	0 TO 200 PSIA
*VD0080-414	PRESS, ATTITUDE CONTROL CHAMBER 1-3	0 TO 200 PSIA
*VD0081-415	PRESS, ATTITUDE CONTROL CHAMBER 2-1	0 TO 200 PSIA
*VD0082-415	PRESS, ATTITUDE CONTROL CHAMBER 2-2	0 TO 200 PSIA
*VD0083-415	PRESS, ATTITUDE CONTROL CHAMBER 2-3	0 TO 200 PSIA
VXD0089-414	PRESS, FUEL TANK ULLAGE MODULE 1	0 TO 400 PSIA
VXD0090-414	PRESS, OXIDIZER TANK ULLAGE MODULE 1	0 TO 400 PSIA
VXD0091-415	PRESS, FUEL TANK ULLAGE MODULE 2	0 TO 400 PSIA
VXD0092-415	PRESS, OXIDIZER TANK ULLAGE MODULE 2	0 TO 400 PSIA
VXD0093-414	PRESS, FUEL TANK OUTLET MODULE 2	0 TO 400 PSIA
VXD0094-414	PRESS, OXIDIZER TANK MODULE 1	0 TO 400 PSIA
VXD0095-415	PRESS, OXIDIZER TANK OUTLET MODULE 2	0 TO 400 PSIA
VXD0096-415	PRESS, FUEL TANK OUTLET MODULE 2	0 TO 400 PSIA
*XD252-414	PRESS, He REG IN MODULE 1	0 TO 3500 PSIA
*XD253-415	PRESS, He REG IN MODULE 2	0 TO 3500 PSIA
VK132-404	EVENT, ENGINE 1-1/1-3 FD VLV OPEN	
VK133-404	EVENT, ENGINE 1-2 FD VLV OPEN	
VK134-404	EVENT, ENGINE 2-1/1-3 FD VLV OPEN	
VK135-404	EVENT, ENGINE 2-2 FD VLV OPEN	
*VXN0037-414	QTY, OXIDIZER TANK MODULE 1	0 TO 10 IN.
*VXN0038-415	QTY, OXIDIZER TANK MODULE 2	0 TO 10 IN.
*VXN0039-415	QTY, FUEL TANK MODULE 1	0 TO 10 IN.
*VXN0040-415	QTY, FUEL TANK MODULE 2	0 TO 10 IN.

MEASUREMENT NUMBER PREFIXES INDICATE THE FOLLOWING:
*FLIGHT CONTROL; V-ESE DISPLAY; X-AUXILIARY DISPLAY

Figure 5-64

up 23 sec later. See Sequencing for power off commands in flight. The two inverters are identical units developed for this specific task. Previous converters (using 28 Vdc) were 80 percent heavier, had 80 percent more components, and were 30 percent less efficient.

Section V S-IVB Stage

PRIMARY POWER DISTRIBUTION

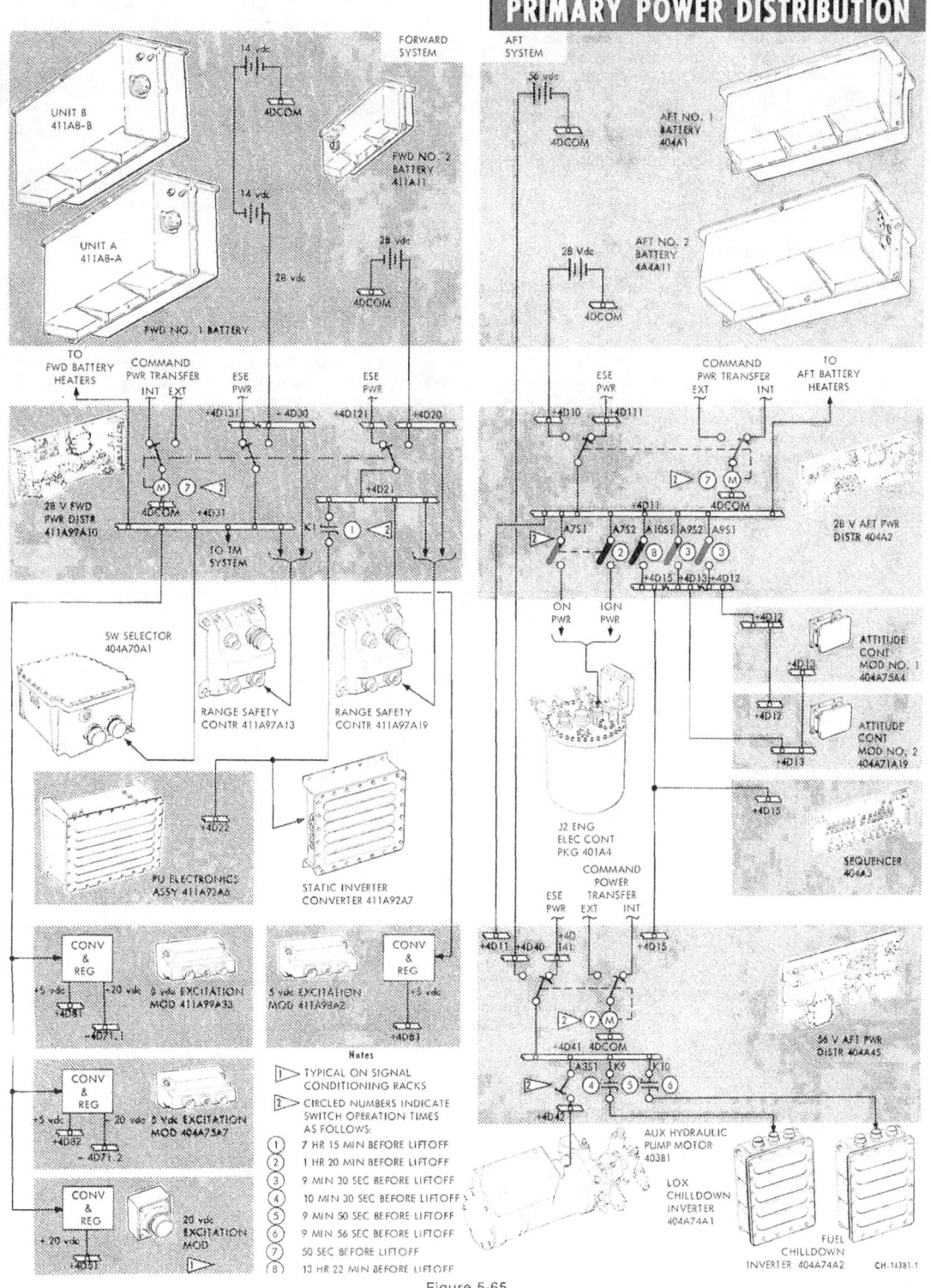

Figure 5-65

Section V S-IVB Stage

PU STATIC INVERTER-CONVERTER.

This unit is a secondary power supply that regulates voltages to the PU electronics assembly (see Propulsion) for various PU functions. The unit is operable 5 hr 45 min before liftoff; a switch selector function switches power off in flight (see Sequencing). Forward battery no. 2 (+4D21 bus) provides 28-Vdc operating power and the unit supplies the following specialized outputs:

a. 115 Vac, 400 Hz drives the bridge rebalancing servo motors.

b. 21 Vdc operates PU bridge networks.

c. 5 Vdc excites PU fine and coarse mass potentiometers.

EXCITATION MODULES.

Two different types of signal excitation modules, 5 Vdc and 20 Vdc, provide excitation for transducers and instrumentation networks. As such, these excitation modules are secondary power supplies for the instrumentation system. Ground rules specify that measurements must be in the 0 to 5 Vdc range when applied to the digital data acquisition system. Since measurement signals originate in various forms, many require 0 to 5 Vdc conditioning. Several types of solid-state conditioning modules perform this function; for example, dc amplifiers, temperature bridge networks, and frequency-to-dc converters. These conditioning modules receive −20 Vdc, +20 Vdc and +28 Vdc operating power. In addition, the measuring networks use a regulated 5-Vdc reference for calibration at ground receiving stations. Since 28 Vdc is available from the primary netowrks, secondary power sources supply the other converted voltages for measurements excitation.

5-Vdc Excitation Module.

This module contains solid-state electronic circuitry that transforms an unregulated 28-Vdc input voltage into accurately regulated output voltages. It consists of two conversion sections: one section supplies up to 400 mA at 5 Vdc and up to 100 mA at −20 Vdc; the other section supplies a 10-Vpp, 2000-Hz square wave and has a capacity of 5 VA. The 5-Vdc output is monitored as a flight measurement (see figure 5-67). Three modules are on the stage; two are in the forward skirt, and one is in the aft skirt (see figures 5-2 and 5-3).

20-Vdc Excitation Module.

This is a conventional dc-to-dc converter with regulation circuitry. It receives unregulated 24 to 30 Vdc and regulates the output to 20 ± 0.04 Vdc. A module mounts on all four signal conditioning racks (one in the forward skirt and three in the aft skirt) to power measuring system components.

DISTRIBUTORS.

There are five distribution assemblies that bus battery power to

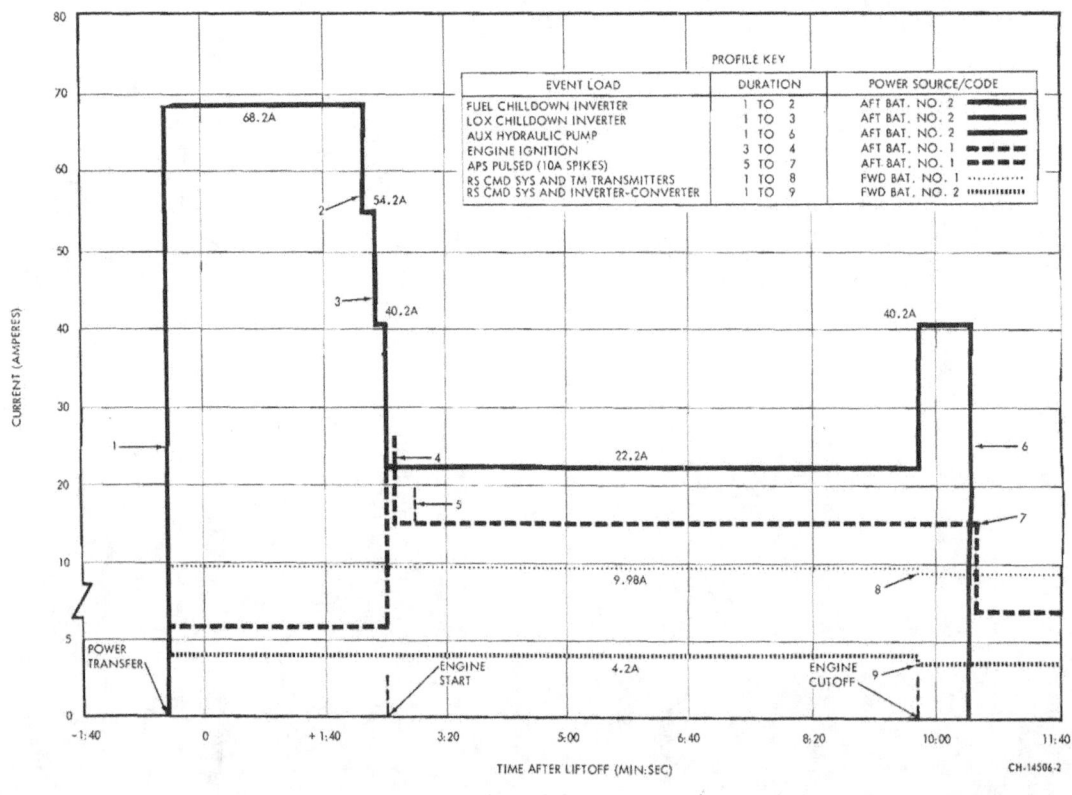

Figure 5-66

Section V S-IVB Stage

ELECTRICAL SYSTEM MEASUREMENTS		
NUMBER	NAME	RANGE
*C102-411	TEMP, FWD BATTERY NO. 1, UNIT 1	460 TO 660°R
*C103-411	TEMP, FWD BATTERY NO. 2	460 TO 660°R
*XC104-404	TEMP, AFT BATTERY NO. 1	460 TO 660°R
*XC105-404	TEMP, AFT BATTERY NO. 2	460 TO 660°R
*C211-411	TEMP, FWD BATTERY NO. 1, UNIT 2	460 TO 660°R
*XM14-404	VOLT, AFT BATTERY NO. 1 OUTPUT	0 TO 40 V
*XM15-404	VOLT, AFT BATTERY NO. 2 OUTPUT	0 TO 80 V
*XM16-411	VOLT, FWD BATTERY NO. 1 OUTPUT	0 TO 40 V
*XM18-411	VOLT, FWD BATTERY NO. 2 OUTPUT	0 TO 40 V
*XM19-411	CURRENT, FWD BATTERY NO. 1 LOAD	0 TO 200 A
*XM20-411	CURRENT, FWD BATTERY NO. 2 LOAD	0 TO 20 A
*XM21-404	CURRENT, AFT BATTERY NO. 1	0 TO 200 A
*XM22-404	CURRENT, AFT BATTERY NO. 2 LOAD	0 TO 200 A
*VM24-411	VOLT, 5-V EXCITATION MOD, FWD 1	4.5 TO 5.5 V
*VM25-404	VOLT, 5-V EXCITATION MOD, AFT	4.5 TO 5.5 V
*VM68-411	VOLT, 5-V EXCITATION MOD, FWD 2	4.5 TO 5.5 V

MEASUREMENT NUMBER PREFIXES INDICATE THE FOLLOWING:
*FLIGHT CONTROL; V-ESE DISPLAY; X-AUXILIARY DISPLAY

Figure 5-67

the forward skirt distributes primary power from both batteries in that area. Since 56-Vdc and 28-Vdc batteries are in the aft skirt, separate distributors accommodate them. Two other distribution assemblies, called control distributors, provide no flight function and are not shown in the distribution diagram; they provide ESE interface via the umbilicals for control and talkback functions during prelaunch operations. Notice that the power distribution assemblies use modular construction as mentioned in the design ground rules. For example, each motorized transfer switch between battery and primary bus is contained in separate modules. An identical module is on each distribution assembly. There are also relay modules, bus modules, and connector modules that house the complete assembly circuitry.

SEQUENCING.

As on all stages, a switch selector is the communications link between the stage and LVDC where the flight sequencing program is stored. Details on switch selector operations are outlined in Section IV, Sequencing. As shown in figure 5-68 the switch selector issues its commands to the sequencer assembly. The sequencer assembly is modular in construction and contains logic circuitry to perform its functions. It has magnetic latching relays, non-latching relays, transistor networks, and diodes that enable or disable circuits, or switch power on-and-off to perform the various switch selector commands. Notice that some commands are repetitious, such as tank venting during the coast mode which is explained in AS-206 S-IVB Orbital Safing subsection.

INSTRUMENTATION SYSTEMS.

The S-IVB stage instrumentation systems consist of measuring subsystems and one PCM/DDAS telemetry link. The major instrumentation equipment for the S-IVB stage is located on the interior of the forward and aft skirt assemblies.

MEASURING SYSTEM.

The S-IVB measuring system consists of various types of transducers, four signal conditioning rack assemblies, and a central decoder assembly and associated channel decoder assemblies. Approximately 252 measurements made on the S-IVB stage of the AS-206 vehicle provide both preflight and inflight data. The measurement coding system employed in the S-IVB stage is the same as that used on the S-IB stage and has been previously described in Section IV of this manual. Those measurements that are unsuitable for use by the telemetry system in original form are modified by signal conditioning modules before being routed to the telemetry multiplexers. These signal conditioning modules mount on four signal conditioning racks; three in the aft skirt and one in the forward skirt. A central decoder assembly and associated channel decoder assemblies translate coded commands to calibrate the measuring system. The remote automatic calibration system (RACS) permits a remote calibration of the measuring system prior to launch. During vehicle checkout, calibration of measurements is accomplished through the RACS and various corrections can be made. Prior to launch, the RACS is operated to determine if the system drifts or deviates from the final adjustments. The data obtained is used to correct the flight data for more accurate measurements.

TELEMETRY SYSTEM.

The S-IVB telemetry system (figure 5-69) is a PCM/DDAS (pulse-code modulated/digital data acquisition system) link (CP1) that transmits real-time checkout data before launch and measuring program information during flight. The system consists of two Model 270 multiplexers, a remote digital submultiplexer, a remote analog submultiplexer, a Model 301 PCM/DDAS assembly, and an RF assembly. Components of the telemetry system are located in the forward skirt, with the exception of the multiplexers that are located in the aft skirt. The use of two time division multiplexers, a digital submultiplexer, and an analog submultiplexer allows increased data transmission capability. The multiplexers sample measurement inputs to produce a train of varying amplitude output pulses. The PAM (pulse-amplitude modulated) data from the multiplexers are converted to digital words in the Model 301 PCM/DDAS assembly, which then encodes the digital words into a data frame. The data frame is transferred to the RF assembly where it modulates the RF carrier. The RF signal is transmitted through two antennas installed on opposite sides of the vehicle by means of a bidirectional coupler and an RF power divider. A coaxial switch transfers the telemetry signal to RF dummy load during checkout. Two power detectors measure the forward and reflected power from the directional coupler. These measurements are telemetered to ground stations and are used to determine the VSWR and the actual RF power transmitted from the vehicle. The PCM/DDAS feeds data through a coaxial cable to the DDAS ground receiving station during prelaunch checkout. With the exception of the remote analog submultiplexer (RASM), the telemetry components in the S-IVB stage are common to the S-IB stage and are discussed in Section IV. The RASM is a PAM analog submultiplexer, which provides the capability to remotely locate any 6 primary channels of a Model 270 time division multiplexer. The RASM has 6 data output channels, each of which has a capacity of 10 submultiplexer channels. Each submultiplexed channel is sampled 12 times per second.

Approximately 178 measurements originating in the S-IVB stage are routed through multiplexer DP1B0 to the IU PCM/DDAS assembly for transmission of these 178 measurements; approximately 78 are also routed through multiplexer CP1B0 for transmis-

tion V S-IVB Stage

ELECTRICAL SEQUENCING

THIS CIRCUITRY IS SIMPLIFIED TO SHOW ONLY THE RELAYS AND DIODES NECESSARY TO EXPLAIN THE RESULTS OF THE SWITCH SELECTOR COMMANDS. THE TABLE LISTS SEQUENCED EVENTS UNTIL CSM SEPARATION. CERTAIN EVENTS CONCERNING TM CALIBRATION AND TANK VENTING CONTINUE UNTIL END OF S-IVB ACTIVE LIFETIME. THIS SEQUENCE DOES NOT INCLUDE STAGE PASSIVATION OPERATIONS (SEE ORBITAL SAFING).

1. S-IB PROP LEVEL SENSOR ACTUATION
2. S-IB O.B. ENG CUTOFF
3. S-IB/S-IVB SEPARATION
4. TURNED ON APPROX 10 MIN 30 SEC BEFORE LIFTOFF ON SW SEL CHAN 28
5. TURNED ON APPROX 9 MIN 30 SEC BEFORE LIFTOFF ON SW SEL CHAN 22
6. TURNED ON APPROX 9 MIN 56 SEC BEFORE LIFTOFF ON SW SEL CHAN 58
7. OPEN APPROX 3 SEC BEFORE LIFTOFF ON SW SEL CHAN 80
8. NOT USED IN FLIGHT
9. TURNED ON APPROX 7 HR 15 MIN BEFORE LIFTOFF ON SW SEL CHAN 7
10. ONLY ONE OF THE THREE ULLAGE ROCKETS IS SHOWN FOR CLARITY. THE OTHER TWO CHARGE AND FIRE THROUGH PARALLEL FUNCTION OF SAME RELAYS
11. THE FLIGHT PROGRAM PLANS FOR A GUIDANCE CUTOFF (30) BASED ON ADEQUATE VELOCITY. COMMAND (30) STARTS T_4 AT 9 MIN 50.1 SEC AFTER LIFTOFF. HOWEVER, PROPELLANT DEPLETION CUTOFF (29) IS ARMED TO PROTECT AGAINST PROPELLANT STARVATION.
12. HELD OPEN UNTIL PHYSICAL SEPARATION OF S-IB/S-IVB STAGES AT APPROX 3.1 SEC AFTER S-IB LEVEL SENSOR ACTUATION
13. OPENED BY MAINSTAGE THRUST OK SIGNAL
14. TURNED ON BY LOX TANK PRESS SOL VLV CLOSE CMD & LOX TANK PRE-PRESS FLT CONTL PRESS SW DEENERGIZED
15. S-IVB ENGINE CUTOFF

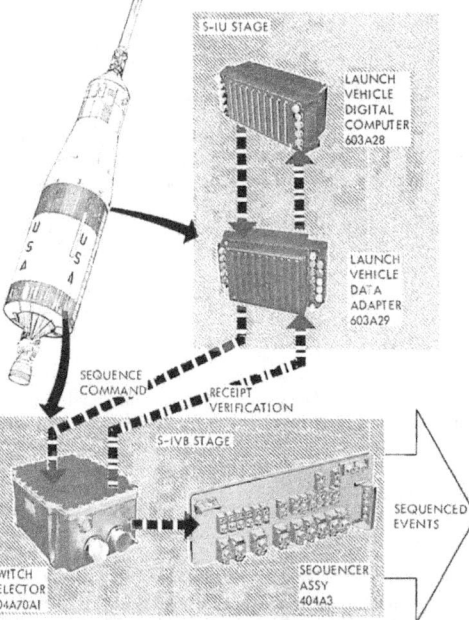

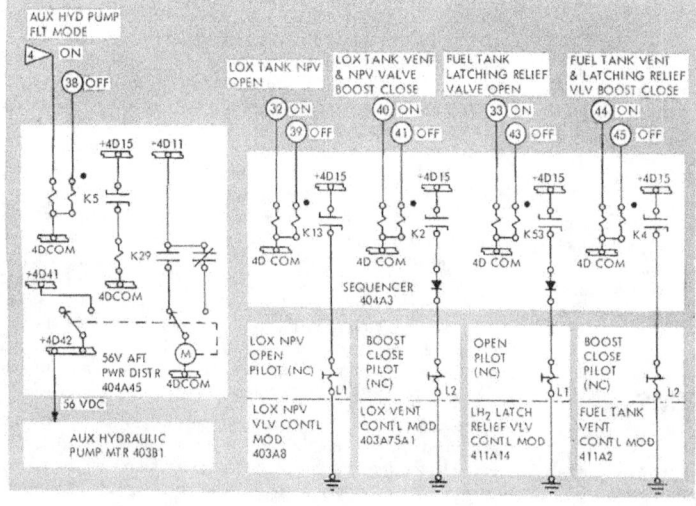

	TIME AFTER LIFTOFF (MIN:SEC)	SWITCH SELECTOR CHAN	TIME BASE (SEC)
1	00:06.0	79	$T_1 + 6.0$
2	02:07.5	62	$T_1 + 127.5$
3	02:08.5	63	$T_1 + 128.5$
	02:14.5	1	$T_2 + 0.0$
4	02:18.1	54	$T_2 + 3.6$
5	02:18.8	83	$T_2 + 4.3$
	02:21.6	2	$T_3 + 0.0$
6	02:21.8	80	$T_3 + 0.2$
7	02:21.9	103	$T_3 + 0.3$
8	02:22.0	13	$T_3 + 0.4$
9	02:22.1	49	$T_3 + 0.5$
10	02:22.4	5	$T_3 + 0.8$
11	02:22.5	34	$T_3 + 0.9$
12	02:22.7	56	$T_3 + 1.1$
	02:22.9	3	$T_3 + 1.3$
13	02:23.5	10	$T_3 + 1.9$
14	02:23.7	59	$T_3 + 2.1$
15	02:23.9	23	$T_3 + 2.3$
16	02:24.3	9	$T_3 + 2.7$
17	02:24.8	27	$T_3 + 3.2$
18	02:25.3	11	$T_3 + 3.7$
19	02:26.9	68	$T_3 + 5.3$
20	02:30.3	6	$T_3 + 8.7$
21	02:30.5	35	$T_3 + 8.9$
22	02:31.8	55	$T_3 + 10.2$
23	02:34.9	57	$T_3 + 13.3$
24	02:35.3	16	$T_3 + 13.7$
25	02:40.9	88	$T_3 + 19.3$
26	02:41.1	73	$T_3 + 19.5$
27	02:45.6	50	$T_3 + 24.0$
28	07:24.5	69	$T_3 + 302.9$
10	07:49.7	5	$T_3 + 328.1$
11	09:01.6	34	$T_3 + 328.3$
29	13:13.1	97	$T_3 + 400$
30	09:49.9	12	$T_4 - 0.2$
31	09:50.0	48	$T_4 - 0.1$
	09:50.1	15	$T_4 + 0.0$
30	09:50.2	12	$T_4 + 0.1$
31	09:50.3	48	$T_4 + 0.2$
32	09:50.4	42	$T_4 + 0.3$
33	09:50.5	99	$T_4 + 0.4$
34	09:50.6	82	$T_4 + 0.5$
35	09:50.7	91	$T_4 + 0.6$
1	09:50.9	79	$T_4 + 0.8$
36	09:51.1	104	$T_4 + 1.0$
37	09:51.9	98	$T_4 + 1.8$
20	09:52.3	6	$T_4 + 2.2$
21	09:52.5	35	$T_4 + 2.4$
38	09:54.0	29	$T_4 + 3.9$
39	13:10.1	37	$T_4 + 200.0$
40	13:15.1	95	$T_4 + 203.0$
41	13:15.1	96	$T_4 + 205.0$
42	13:50.1	8	$T_4 + 240.0$
43	15:10.1	100	$T_4 + 320.0$
44	15:13.1	77	$T_4 + 323.0$
45	15:15.1	78	$T_4 + 325.0$

SHEET 2

CH-20130

Figure 5-68 (Sheet 1 of 2)

Section V S-IVB Stage

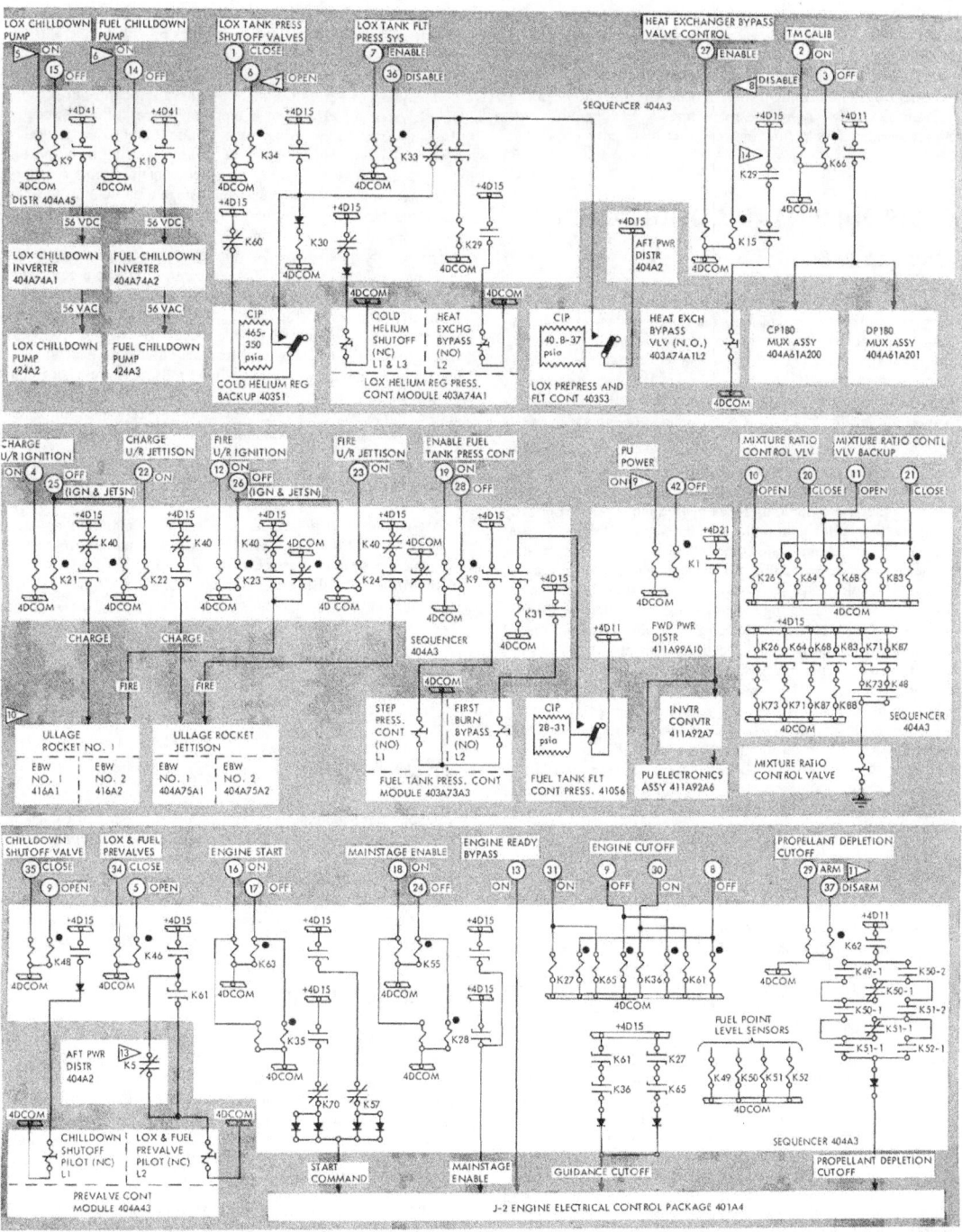

Figure 5-68 (Sheet 2 of 2)

sion from the S-IVB stage. Approximately 73 measurements originating outside the S-IVB are routed through multiplexer CP1A0, located in the IU, to the S-IVB stage PCM/DDAS assembly for transmission. These measurements originate in the spacecraft, IU, and S-IB stage. This configuration is necessary for the measuring program to satisfy battery life requirements, EDS requirements, and requirements for redundancy of other flight critical measurements.

ENVIRONMENTAL CONDITIONING.

FORWARD SKIRT.

The S-IVB stage forward skirt area is conditioned from a ground environmental control system during preflight preparations through a purge manifold located in the instrument unit (see Section VI). The system provides ventilating air to the compartment during the preflight phases of IU checkout. During LH_2 loading operations the system maintains an inert GN_2 atmosphere in the compartment to preclude accumulation of combustibles. No inflight purge is required.

During flight, a cold-plate system provides thermal conditioning for forward skirt equipment. A maximum of 16 cold plates connect to the IU thermal conditioning system through a parallel hookup of supply and return manifolds. SA-206 contains five cold plates and 11 cold plate simulators that maintain normal system flow characteristics through calibrated orifices where panels are not needed. The cold plates serve as mounting surfaces and heat exchangers for heat-generating electronic equipment. A silicate ester (Oronite Flo-Cool 100 dielectric coolant) liquid flows from the IU thermal conditioning system through the cold plate supply manifold, through the series-flow heat exchangers, into the return manifold and back to the IU thermal system for conditioning. A fluid minimum flowrate of 7.4 gpm and maximum temperature of 70° F is required to maintain the cold plates at a maximum design temperature of 80° F for acceptable component heat extraction. Radiation shields are installed over electronic components and attached to the mounting panels to assist in maintaining acceptable cold plate and equipment temperature. The thermal conditioning system is operational during prelaunch electronic component operations and during flight.

AFT SKIRT.

The aft skirt thermal conditioning system provides for thermal conditioning of launch critical electronic equipment, the attitude control system, and the hydraulic accumulator-reservoir while purging the aft skirt and interstage area prior to launch. Design requirements for the aft skirt thermal conditioning and purge system are listed in figure 5-70.

The system (figure 5-71) consists of a main manifold which circles the lox container dome near the aft skirt attach flange, a thrust structure manifold which circles the interior of the thrust cone structure, a duct connecting the two manifolds, and a hydraulic accumulator-reservoir manifold attached to the connecting duct. The system utilizes air as the conditioning medium until 20 min prior to propellant loading. At this time the conditioning medium is switched to GN_2 which provides inert purging of the aft skirt and interstage until liftoff; no conditioning is provided after liftoff. Compartment conditioning requirements are presented in figure 5-72.

The conditioning medium is supplied by the ground ECS and enters the system through service arm no. 6 umbilical ducting and flows into the main manifold. A portion of the medium flows into each of two attitude control modules to maintain the propellant within the proper temperature range. Temperature sensors in the gas discharge of the attitude control modules provide temperature input to the LCC S-IVB ECS panel which controls the conditioning medium inlet temperature. The remaining portion of the conditioning medium flows from the main manifold through orifices to the aft skirt area for electronic equipment temperature control and through a duct to the thrust cone manifold and to the hydraulic accumulator-reservoir shroud. The thrust cone ring manifold distributes the medium to purge the interior of the thrust structure. It is then exhausted into the aft interstage area, joining the flow exhausted from the accumulator-reservoir shroud and flow past the impingement curtain from the aft skirt area. Slots in the aft interstage structure then vent the conditioning medium to the atmosphere. Thermal control of temperature sensitive equipment during flight is achieved by passive methods. Flight environment is maintained by controlling preflight temperatures and by controlling equipment thermal radiation and conduction paths. Low emissivity, goldized Kapton is used to reduce radiative heat loss and maintain equipment within allowable limits. A retrorocket impingement curtain between the aft skirt/aft interstage attach flange and the thrust cone provides protection to electronic components from hot retrorocket exhaust gases during the separation sequence.

HAZARDOUS GAS DETECTION.

A one-quarter-inch aluminum leak detection manifold mounted in the forward skirt has four orificed ports for obtaining atmosphere samples. The manifold connects to a hazardous gas analyzer (HGA) in the ML through the service arm no. 7 umbilical. A one-quarter-inch aluminum leak detection manifold mounted in the aft interstage has two orificed ports for obtaining atmosphere samples. This manifold connects to the HGA through the S-IB stage service arm 1A umbilical. For additional information, see Hazardous Gas Detection System in Section IV.

ORDNANCE.

Ordance components perform staging of the Saturn IB vehicle by severing a tension strap that secures the S-IB and S-IVB stages, by providing the force to decelerate the spent S-IB stage, and by providing a slight acceleration to the S-IVB stage/payload to keep the S-IVB stage propellants seated until J-2 engine ignition. Shaped charges installed on the LH_2 and lox tanks provide the capability to destroy the vehicle by severing the tanks and dispersing the propellants.

STAGE SEPARATION.

S-IB/S-IVB stage separation occurs 1.3 sec after S-IB stage outboard engine cutoff on command by the S-IB stage switch selector. The S-IVB stage separates from the S-IB stage/aft interstage at vehicle station 1186.804 by simultaneous operation of: (1) the stage separation ordnance system, which severs a circumferential tension plate; (2) four retromotors, which decelerate the S-IB stage/aft interstage assembly; and (3) three ullage rockets, which maintain a slight acceleration on the S-IVB stage and payload.

Operation of the stage separation systems begins when the S-IB stage fuel level sensor no. 1 or no. 2 or lox level sensor no. 2 or no. 3 actuates. Sensor actuation applies 28 Vdc to the charge circuits of the separation ordnance system and retromotors ignition system exploding bridgewire (EWB) firing units. At 2.5 sec before separation, an S-IVB stage switch selector command applies 28 Vdc to the charge circuits of the ullage rocket ignition system EWB firing units.

Section V S-IVB Stage

S-IVB TELEMETRY SYSTEM

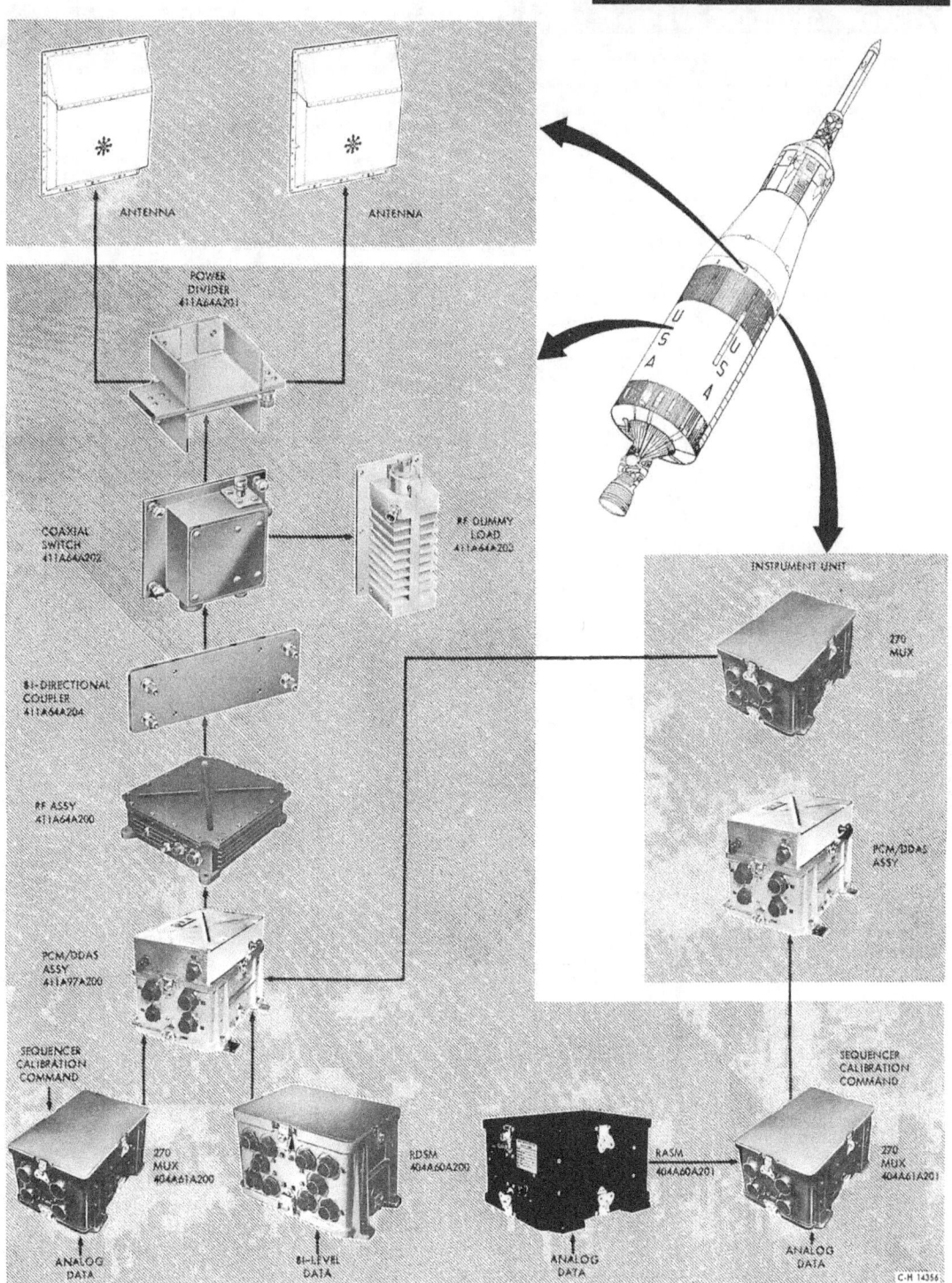

Figure 5-69

Section V S-IVB Stage

DESIGN REQUIREMENTS: AFT SKIRT THERMAL CONDITIONING AND PURGE SYSTEM	
REQUIREMENT	LIMITS/NOTES
SYSTEM LEAKAGE AREA	2.5 in.^{2} Max.
SYSTEM PROOF PRESSURE	0.75 psig
PRELAUNCH PURGE CONTROLS	OBTAIN 4% BY VOLUME O_2 CONTENT IN 30 MINUTES (Max)
PRELAUNCH THERMAL CONTROL	
AFT SKIRT ELECTRONIC EQUIPMENT ENVIRONMENT	0 - 120°F
ACCUMULATOR RESERVOIR	0 - 100°F
ATTITUDE CONTROL MODULE GAS DISCHARGE	65 -90° F
INFLIGHT THERMAL CONTROL	PASSIVE METHODS

Figure 5-70

At 0.1 sec before stage separation command, the S-IVB stage switch selector triggers the EBW firing units that fire the ullage rockets and, at separation, the S-IB stage switch selector triggers the EBW firing units that fire the retromotors and detonate the separation ordnance. Approximately 15 sec after stage separation, the spent ullage rockets and their fairings are jettisoned to reduce stage weight.

Separation Ordnance Subsystem.

The stage separation ordnance system consists of two EBW firing units, two EBW detonators, a detonator block, and a mild detonating fuse (MDF) assembly. See figures 5-73 and 5-74.

The MDF assembly, consisting of two parallel lengths of 10-gpf lead-sheathed PETN enclosed in a polyethylene plastic covering, is installed in a groove around the aft skirt assembly just forward of the separation plane. A tension plate, consisting of eight sections permanently attached to the aft skirt assembly and bolted to the aft interstage assembly, secures both assemblies until stage separation command. The MDF assembly lies just beneath the thinnest part of the tension plate, which is approximately 0.050 in. thick. The explosive force created by either MDF will sever the plate;

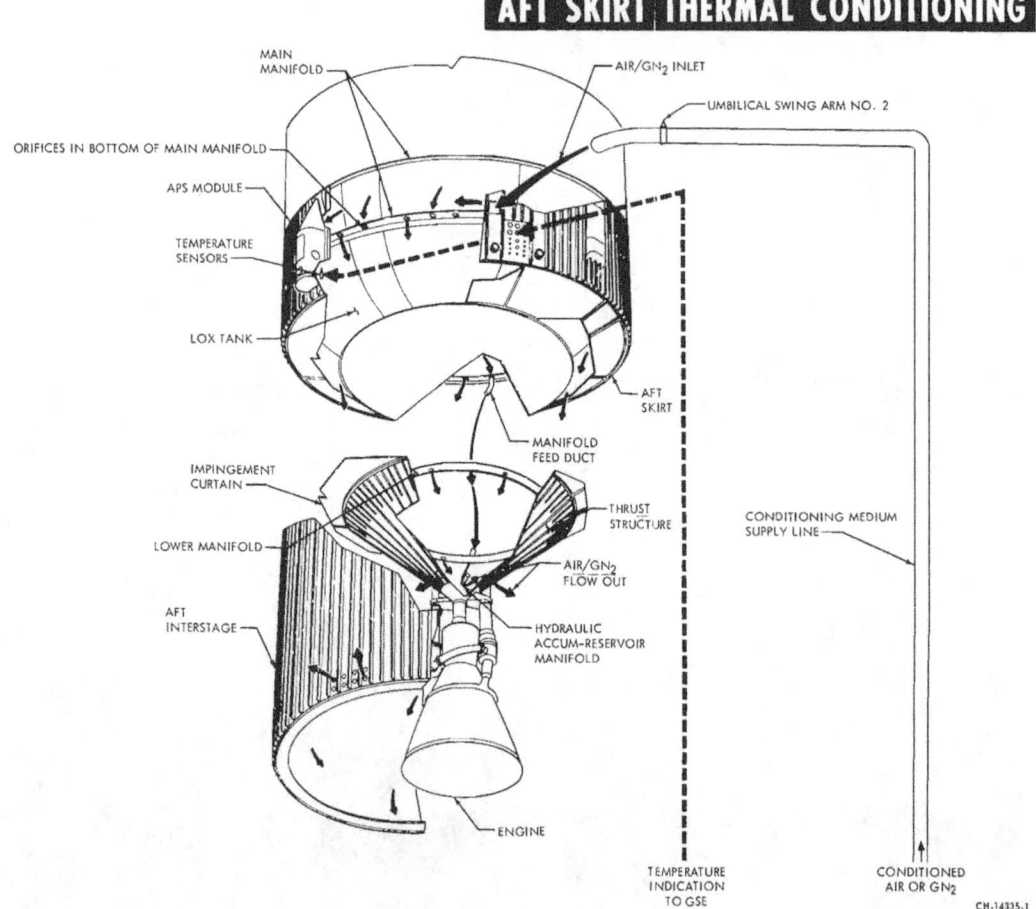

Figure 5-71

Section V S-IVB Stage

AFT SKIRT COMPARTMENT CONDITIONING REQUIREMENTS

FLOW MEDIUM	FLOW (Lbs/Min)	COMPARTMENT INLET CONDITIONS			COMPARTMENT TEMP. RANGE (Degrees F)	PROBE DATA	
		PRESSURE (In. of H_2O)	TEMP. (Degrees F)	HUMIDITY (Gr/Lb of Air)		LOCATION	SETTING (Degrees F)
AIR	225 - 300	10 - 25	75 - 140	0 - 43	N/A	STAGE	65 - 90
GN_2	267 - 300	10 - 25	75 - 140	0 - 1.0	N/A	STAGE	70 - 90

Figure 5-72

however, the redundant fuses ensure severance. After installing the MDF assembly around the aft skirt, each end is secured in a common detonator block, which contains two EBW detonators. Two spacers in the detonator block orient the fuse assembly ends with the detonators. Each detonator fires one end of the MDF assembly, thereby providing redundant ignition. Electrical cables deliver a 2300 ± 100 Vdc pulse from the EBW firing units to the detonators at separation command. The detonators fire the MDF assembly, which propagates the detonation around the stage, severing the tension plate. Five blast deflectors, mounted just forward of the separation plane, protect the APS modules and the ullage rockets from tension plate fragments. The stage separation ordnance ignition components are located in the aft interstage, which remains with the S-IB stage at separation.

The separation ordnance has successfully met all development and qualification objectives. Tests included 71 short segments of the separation joint and three 360-deg, full-scale structural separations, all without failure. The successful tests of the separation system and components included: 60 fuse qualification tests, 10 detonator block tests, six mild detonating fuse installation tests, three full-scale system tests, and five tests with the system functioning under load.

Retromotor Subsystems.

Four solid-propellant retromotors, mounted at 90-deg intervals around the aft interstage assembly (figure 5-75), decelerate the S-IB stage and aft interstage assembly during the stage separation sequence. Nose, center, and aft aerodynamic fairings enshroud each retromotor. The aft and center fairings are permanently installed; however, the nose fairing jettisons when the retromotor fires to expose the motor nozzle. Exhaust gases, acting against the internal surface of the fairing, shear a retaining pin that secures the forward end of the fairing to the aft interstage assembly. The fairing then assumes an aftward rotation, pivoting about a hinge and hook arrangement. After an approximate 70-deg rotation, the fairing separates from a hook on the center fairing and falls away from the vehicle.

Each retromotor nozzle cants 9.5 deg outboard from the motor centerline to direct the exhaust plume away from the S-IVB stage. With an average burn time of 1.52 sec, each retromotor develops a 36,720 lbf thrust at 200,000 ft altitude and 60° F. Thiokol Chemical Corporation, Elkton, Maryland, manufactures the E17029-02 recruit motors used as the S-IB stage/aft interstage assembly retromotors. Each motor weighs 376 lbm, including 267 lbm of propellant having a tapered, internal-burning, 5-point-star configuration.

Independent ignition systems, consisting of two EBW firing units and two EBW initiators, ignite the retromotors. An igniter, which is a part of each retromotor, has two threaded receptacles for initiator installation. A pair of EBW firing units mounts inside the aft interstage assembly at each retromotor position. An electrical cable, which is an integral part of each firing unit, connects the firing unit to its respective initiator. At separation command, each firing unit delivers a 2300 ± 100 Vdc pulse to its initiator. The initiators detonate and fire the igniter, which directs hot particles and gases to the solid propellant surface, thereby igniting the retromotor.

To demonstrate satisfactory firing of the motors, tests of the EBW initiator with the motor igniter were conducted. Eight compatibility tests were run on the retromotors, including six tests with dual ignition and two tests with single ignition; all were successful. The motors were fired under temperature extremes of $-10°$ F to $+155°$ F after being subjected to vibration and temperature cycling tests.

Ullage Rocket Subsystem.

Three solid-propellant ullage rockets, mounted approximately 120 deg apart of the S-IVB stage aft skirt (figure 5-76), induce a slight forward acceleration to the S-IVB stage and payload during stage separation. The motors begin firing 0.1 sec before separation and terminate shortly after J-2 engine ignition. By providing continuous acceleration to the S-IVB stage during stage separation system operations, the S-IVB stage propellant remains properly seated in the bottom of the tanks for J-2 engine start. Aerodynamic fairing assemblies house the ullage rockets and provide for attachment to the aft skirt. The fairing assemblies cant each ullage rocket center line 35 deg outward from the vehicle longitudinal axis, thus directing the exhaust gases away from the S-IVB stage. Each fairing assembly also houses a transducer for chamber pressure measurement and two EBW firing units for ullage rocket ignition.

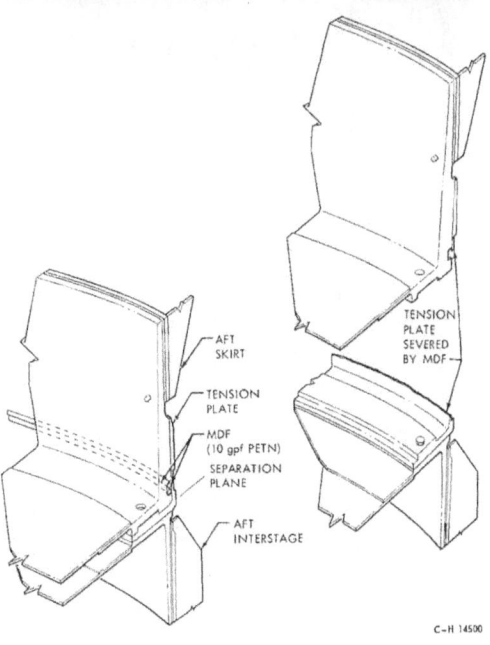

Figure 5-73

Section V S-IVB Stage

SEPARATION SYSTEM

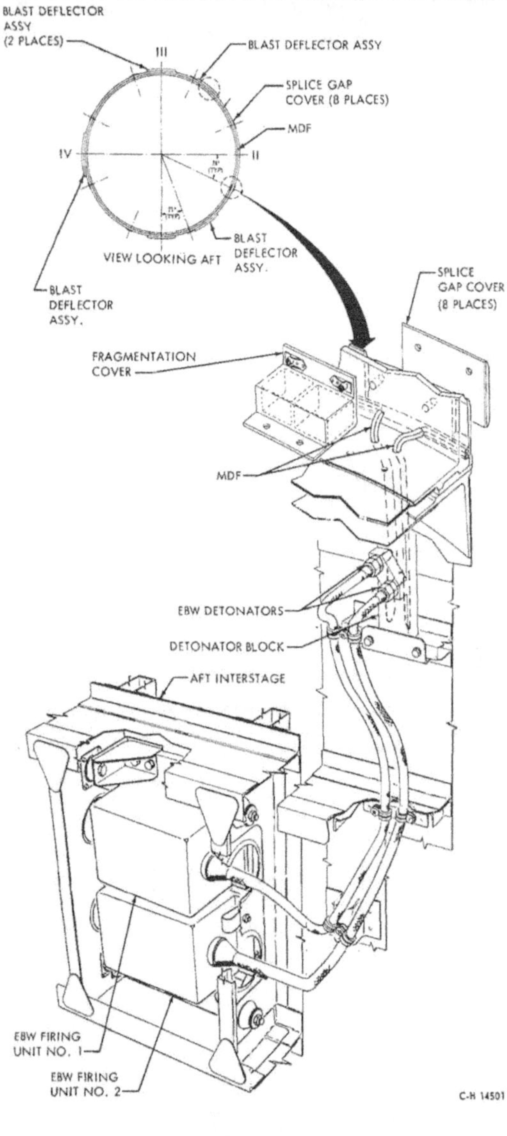

RETROMOTORS AND EBW FIRING UNITS INSTALLATION

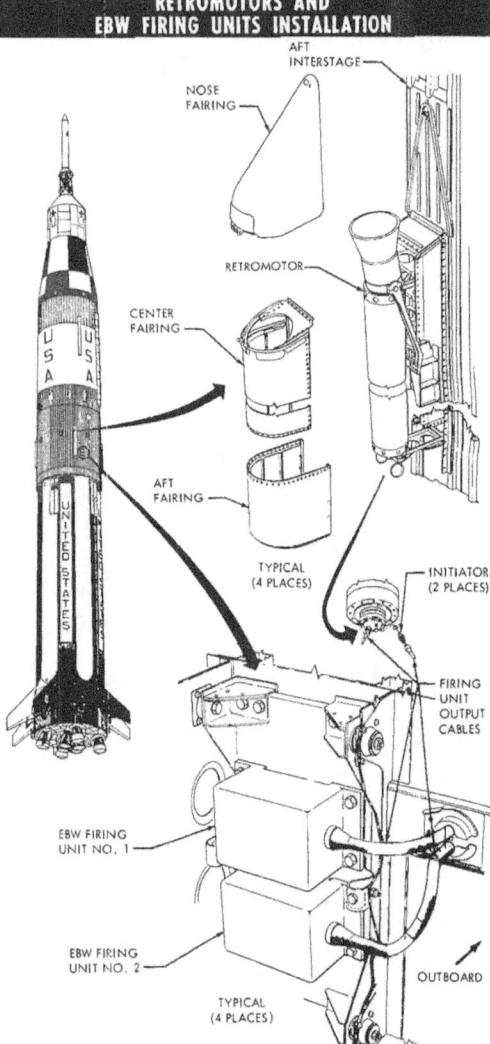

Figure 5-74

Figure 5-75

Thiokol Chemical Corporation manufactures the TX-280-10 rocket motors. The motors are 8.316 in. in diameter and a maximum of 37 in. in length. The internal-burning, 5-point-star-configuration solid propellant develops an average thrust of 3,460 lbf at a temperature of 70° F and at an altitude of 1,000,000 ft. At 70° F the propellant burns for 3.9 sec. An igniter, which is part of the rocket motor and is installed in the forward end of the motor casing, has two receptacles for EBW initiators. Two initiators, the Thiokol TX-346-1 and the Aerojet-General AGX 2008, have been approved for use with the ullage rockets. The output cables from the EBW firing units in each fairing assembly attach to the respective ullage rocket initiators.

The S-IVB stage switch selector issues the ullage rocket ignition command 0.1 sec before stage separation. The command simultaneously triggers the two EBW firing units at each ullage rocket position. Use of redundant EBW firing units and EBW initiators insures ullage rocket ignition. A 2300 ± 100 Vdc pulse from the EBW firing units fires the EBW initiators installed in the motor igniter. The initiators then detonate pellets contained in the igniter. The pellets eject hot particles and gases through perforations in the igniter case to the solid-propellant surface, igniting the propellant.

Thirteen ullage rocket motors were tested in motor qualification, and all were successful. Eleven were fired utilizing a dual ignition system, and two were fired utilizing a single ignition system. To demonstrate EBW initiator/igniter compatibility, six dual ignition tests and four single ignition tests were conducted successfully. The

Section V S-IVB Stage

test firings were conducted under temperature extremes of −30° F to +145° F after the motors had been subjected to vibration tests, temperature cycling tests, and long duration tests of temperature conditioning.

Ullage Rocket Jettison Subsystem.

Approximately 15 sec after stage separation, the ullage rockets and their fairings are jettisoned to reduce stage weight. To accomplish this operation, the jettison system uses two EBW firing units, two EBW detonators, a detonator block, two confined detonating fuse (CDF) assemblies, six frangible nuts, and three spring-loaded jettison assemblies. See figures 5-77 and 5-78.

The S-IVB stage switch selector issues the ullage rocket jettison command, which triggers the EBW firing units. The 2300 ± 100 Vdc output pulses from the firing units detonate the EBW detonators. The explosion propagates from the detonators through the CDF assembly, simultaneously breaking all six frangible nuts. The spring-loaded jettison assemblies then propel the ullage rockets away from the S-IVB stage.

The EBW firing units, EBW detonators, and detonator block are mounted at panel position 18 in the aft skirt assembly. A tray with quick-release clamps, installed on the aft skirt's inner periphery just forward of the separation plane, secures the CDF assembly

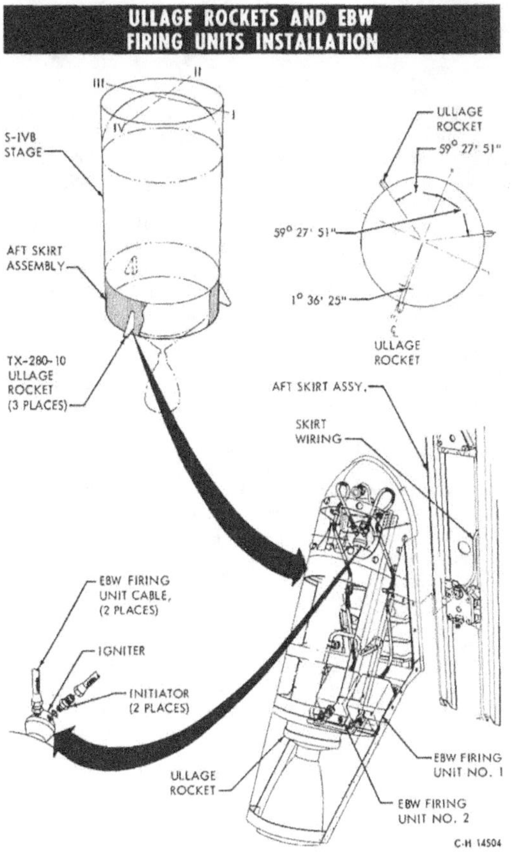

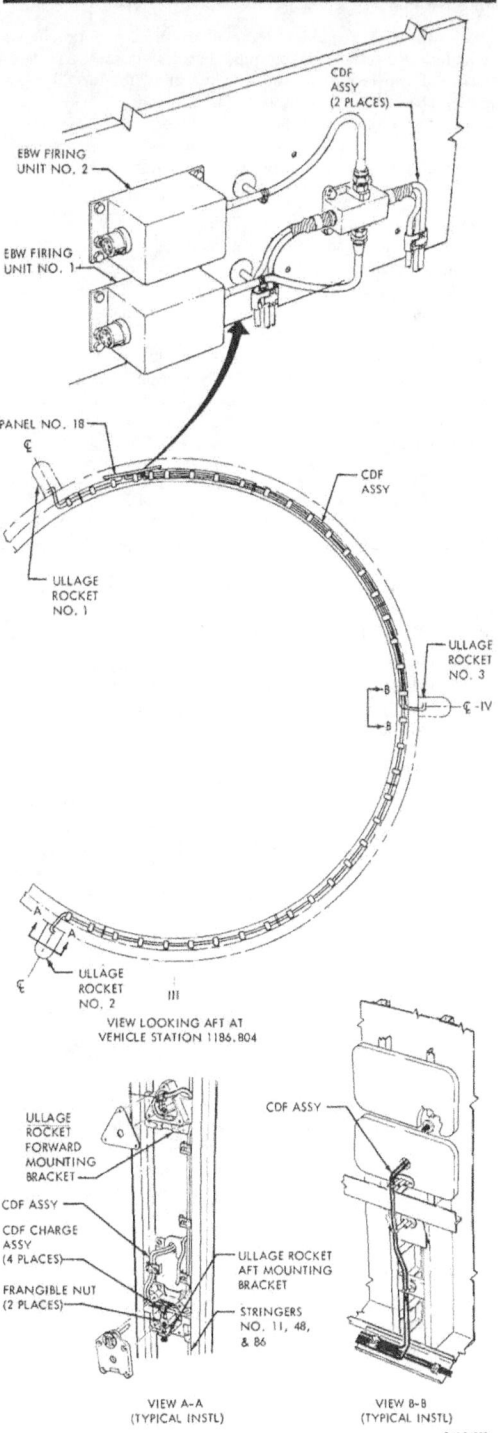

Figure 5-76

Figure 5-77

5-69

Section V S-IVB Stage

leads between the detonator block and each ullage rocket position. A pair of bolts and frangible nuts attach each ullage rocket and fairing assembly to the aft skirt assembly.

The detonator block contains four threaded receptacles for ordnance installation: two receptacles, on opposite sides, for the EBW detonators; and two receptacles, on opposite ends, for the CDF assemblies. The EBW detonators and CDF assemblies form an X-type ordnance connection inside the detonator block to provide system redundancy. The proximity of the explosives within the detonator block ensures propagation of the explosion from either (or both) EBW detonator to both CDF assemblies.

Each CDF assembly has three 2-gpf PETN leads with two charge assemblies on each lead, each charge assembly containing 4.66 gr of PETN. One end of each lead is adhesively bonded in an end fitting, which also contains a 2-gr booster charge. The PETN cores from the three leads butt against the booster charge inside the end fitting. The end fitting serves two purposes: (1) to install the CDF assembly in the detonator block and (2) to propagate the EBW detonator explosion to the PETN core in each lead. Each CDF assembly lead extends to one ullage rocket position where the charge assemblies are installed, one in each frangible nut. Two retaining pins secure the charge assemblies in each nut.

The frangible nuts have a structurally weak plane, which permits the nut to break open when the charge assemblies detonate. One charge assembly will fracture a frangible nut; however, the redundant charges ensure nut fracture. After the frangible nuts release the attachment bolts, a spring-loaded jettison assembly that is permanently attached to the fairing assembly propels the spent ullage rocket and fairing assembly away from the stage. Jettisoning the ullage rockets reduces stage mass by approximately 216 lbm.

Qualification of the explosive release mechanism for ullage rocket jettison has been completed. Fifty successful tests of the detonator block and confined detonating fuse and fittings were completed, and three complete jettison systems were successfully tested.

Flight History.

The stage separation ordnance flown on the three Saturn IB launch vehicles performed well within design limits. Flight separation was completed (S-IVB stage engine clear of the interstage) at approximately 1.5 sec on AS-201, 1.07 sec on AS-202, and 1.04 sec on AS-203 after the separation command. The retromotors performed satisfactorily although AS-201 burn times were longer, and thrusts, impulses, and pressures were slightly lower than predicted.

Retromotor no. 2 on AS-202 flight experienced a short burn time, but separation dynamics were well within tolerance. Retromotors with a larger expansion ratio and a higher thrust rating were flown on the AS-203 flight. The motors were moved 6.75 in. outboard from the S-IB-1 and S-IB-2 retromotor positions. These changes were made to provide a one-retromotor-out capability for S-IB-3 and subsequent vehicles. Retromotor performance on AS-203 flight was very close to predicted performance. The ullage rockets performed satisfactorily and jettisoned properly on all three flights. All rocket motors performed within design limits. During the three flights, J-2 engine thrust increased to the 8000 lbf necessary to keep the propellants seated before ullage rocket thrust decayed below the 8000 lbf minimum.

Firing Units Monitoring.

During checkout operations the EBW firing units in the separation system, retromotor ignition system, ullage rocket ignition system, and ullage rocket jettison system are connected to pulse sensors instead of the EBW initiators or detonators. A simulated fuel or lox level sensor actuation command (simulating propellant level sensor actuation during flight) causes the retromotor and separation system firing units to charge to 2300 Vdc. Meters on the S-IB EBW

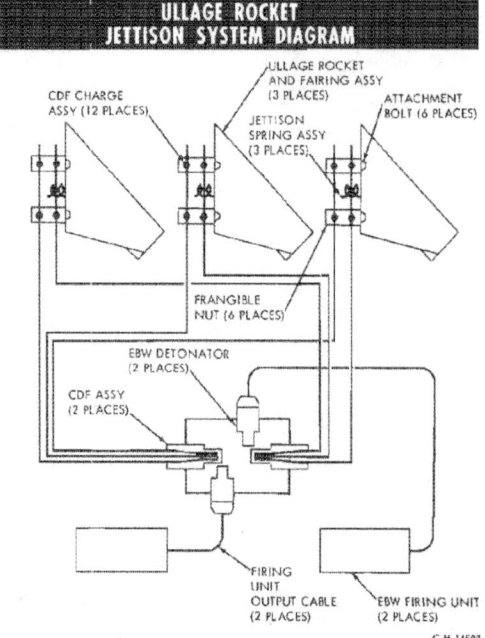

Figure 5-78

ordnance panel provide a readout of the charge voltage of each firing unit, which must be 2300 ± 100 Vdc. A 0 to 4.9 Vdc signal from each firing unit, proportional to the 0 to 2300 Vdc charge, operates the meters. The separation command issued by the switch selector triggers the firing units and the 2300-Vdc pulse discharges into the pulse sensors. Each pulse sensor provides a continuous 28-Vdc signal to the FIRED indicators (one for each firing unit) on the S-IB EBW ordnance panel. After checkout completion the pulse sensors are removed from the vehicle and the firing units output cables are attached to the initiators and detonators. During flight, the charge voltage of the firing units is telemetered back to KSC and monitored on the S-IB EBW ordnance panel meters. The measurement numbers for the retromotor firing units are VM42-400 through VM49-400, and VM68-400 and VM69-400 for the separation system firing units. These measurements will be monitored from liftoff until the S-IB stage impacts in the ocean. Checkout of the ullage rocket ignition and jettison firing units also requires the use of pulse sensors. The S-IVB stage switch selector issues the charge and trigger commands to the firing units. Voltage meters on the S-IVB EBW ordnance panel provide readout of the charge voltage and FIRED indicators illuminate when the pulse sensors receive the 2300-Vdc pulse from the firing units. During flight the firing units charge voltage measurements are telemetered back to KSC and displayed on the S-IVB EBW ordnance panel meters. Measurement numbers for the ullage rocket ignition firing units are VM32-416, VM33-416, VM34-417, VM35-417, VM36-418, and VM37-418. For the ullage rocket jettison firing units, the measurement numbers are VM38-404 and VM39-404. These measurements are monitored from liftoff through S-IVB burn.

PROPELLANT DISPERSION SYSTEM.

The S-IVB stage propellant dispersion system (PDS) severs the cylindrical portion LH$_2$ tank between stations 1276 and 1536, and the bottom of the lox tank at station 1144 if flight termination becomes necessary (figure 5-79). The propellants disperse away

S-IVB STAGE PROPELLANT DISPERSION SYSTEM

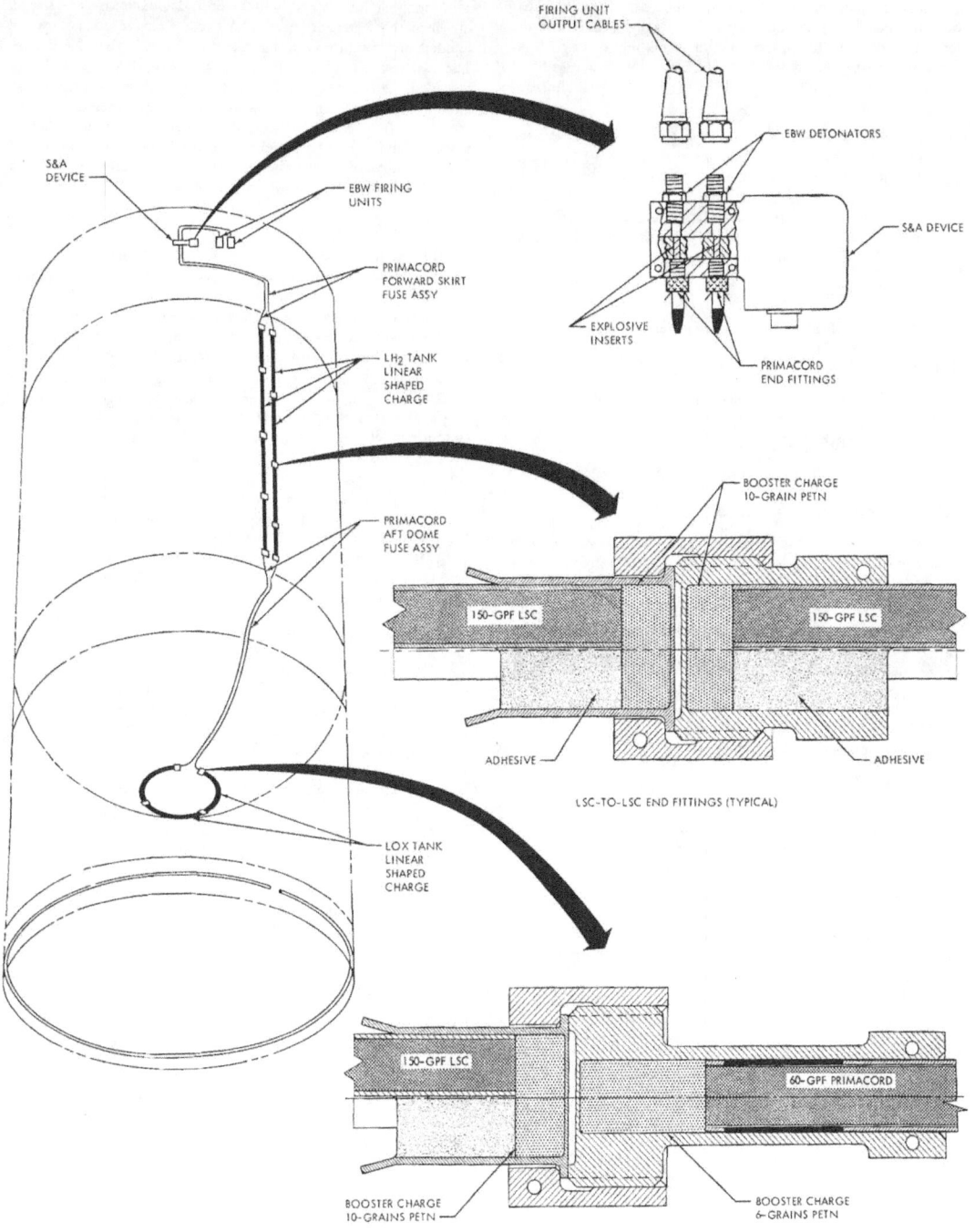

Figure 5-79

from each other, thus reducing the yield of the burning propellants. Two EBW firing units, two EBW detonators, a safety and arming device, a forward skirt fuse assembly, aft dome fuse assembly, LH$_2$ tank linear shaped charge (LSC) assemblies, and lox tank, LSC assemblies make up the redundant S-IVB stage PDS. To fire the PDS ordnance, the EBW firing units deliver a high-voltage high-energy pulse to the EBW detonators. The detonators then propagate the explosion through the two leads in the S&A device and subsequently to the LSC.

Tests that demonstrate the ability of the 150-gpf LSC to cut the LH$_2$ tank and lox tank at cryogenic temperatures have been successfully completed. Eighteen shaped-charge tank-cutting tests, eighteen end-fitting propagation tests, and one safe-arm device installation test were all successful. The continued reliability of the ordnance is assured by a 25 percent lot sampling of the components. A single failure would cause rejection of the lot.

EBW Firing Unit.

See Section IV, S-IB Stage Propellant Dispersion System.

EBW Detonator.

See Section IV, S-IB State Propellant Dispersion System.

Safety and Arming Device.

See Section IV, S-IB Stage Propellant Dispersion System.

Forward Skirt Fuse Assembly.

The forward skirt fuse assembly propagates the detonation from the S&A device to the LH$_2$ tank LSC. Two 60-gpf Primacord leads, each approximately 155 in. long, form a redundant explosive train. Each lead has a fitting on each end to permit connection of the fuse assembly to the S&A device and LH$_2$ tank LSC. A 6-gr PETN booster charge contained in each end fitting assures propagation across the mechanical connections. Each of the two end fittings connecting the fuse assembly to the S&A device consists of a sleeve, in which the Primacord lead and booster charge are adhesively bonded, and a knurled nut that secures the lead in the S&A device. The two end fittings that interface with the LH$_2$ tank LSC are threaded sleeves. The Primacord leads are wrapped together in aluminum tape along their entire length. Quick-release clamps secure the fuse assembly along the power distribution mounting assembly and mounting panel no. 31 inside the forward skirt, and along the forward-skirt tunnel-equipment panel assembly installed on the forward skirt exterior under the tunnel.

LH$_2$ Tank LSC.

The LH$_2$ tank LSC installation consists of eight sections of 150-gpf cyclotrimethylene trinitramine (RDX) enclosed in an aluminum sheath. Each RDX section has bonded end fittings that contain 10-gr booster charges. Redundant explosive trains consisting of four RDX sections connected by the end fittings effect LH$_2$ dispersion by severing the tank skin. The LSC design confines the explosive force to a very narrow path along the tank skin to produce the cutting effect. The use of different LSC lengths staggers the end-fitting connections along the LH$_2$ tank to assure that the proximity of the parallel LSC assemblies will produce a continuous cut between stations 1276 and 1536. If the end fittings were installed side by side, the LSC could fail to completely sever the tank skin underneath the end fittings. Quick-release clamp assemblies, which are adhesively bonded to the LH$_2$ tank (under the tunnel), secure the LSC assemblies against the skin. In addition to severing the LH$_2$ tank skin, the LSC propagates the detonation to the aft dome fuse assembly through end fittings at station 1276. See figure 5-80 for RDX characteristics.

Aft Dome Fuse Assembly.

Two 60-gpf Primacord leads, each approximately 250 in. long, form the redundant explosive train between the LH$_2$ tank LSC and the lox tank LSC. Threaded end fittings containing 6-gr PETN booster charges are adhesively bonded to the ends of the Primacord, providing the means of attachment to the LH$_2$ and lox tank LSC end fittings. Quick-release clamps install the fuse assembly on a panel underneath the tunnel on the aft skirt assembly and along the lox tank aft dome and thrust structure. Under the tunnel, the fuse assembly is wrapped with aluminum tape. Along the aft dome and thrust structure, the fuse assembly has three counterwrapped layers of asbestos tape with an outer layer of glass fiber tape for protection against heat from the J-2 engine and retromotor exhaust plumes. The aft dome fuse assembly interfaces with the lox tank LSC at vehicle station 1144 (approx). See Section IV, Ordnance, for PETN characteristics.

Lox Tank LSC.

The lox tank LSC installation consists of three 150-gpf RDX sections joined together by end fittings. The LSC encircles the base of the lox tank on a 23-in. radius at vehicle station 1144 (approx). The end fittings each contain a 10-gr booster charge to ensure propagation of the detonation across the mechanical connections. Quick-release clamp assemblies that are adhesively bonded to the lox tank aft dome secure the LSC.

RDX CHARACTERISTICS	
DESCRIPTION	WHITE CRYSTALLINE POWDER
CRYSTAL DENSITY	1.816
CHEMICAL FORMULA	$C_3H_6N_6O_6$
MOLECULAR WEIGHT	222.15
MELTING POINT	204°C
DETONATION RATE	8,180 M/SEC AT 1.65 GM/CC (1-INCH DIA SAMPLE)
HEAT OF COMBUSTION	2307 CALORIES/GM
FIRE HAZARD	MODERATE, BY SPONTANEOUS CHEMICAL REACTION
ICC CLASSIFICATION	CLASS A
SYNONYMS	CYCLONITE, HEXAGON
SPECIFICATION	MIL-R-00398
IMPACT SENSITIVITY	GREATER THAN 120 FOOT-POUNDS. 100 GPF LSC TEST (DAC REPORT SM-42678), NO DETONATION OCCURRED WHEN IMPACTED WITH A 10-LB WEIGHT FROM A 12-FOOT HEIGHT.

Figure 5-80

SECTION VI
INSTRUMENT UNIT

TABLE OF CONTENTS

Introduction	6-1
Structure	6-1
Environmental Control System	6-2
Electrical Power Systems	6-10
Emergency Detection System	6-12
Launch Vehicle Navigation, Guidance, and Control	6-13
Measurement and Telemetry	6-24
Command System	6-25
Saturn Tracking Instrumentation	6-25
Ground Support Equipment	6-27
IU/SLA Interface	6-28
Experiments	6-29

INTRODUCTION.

The Instrument Unit (IU) is a cylindrical structure installed on top of the S-IVB stage (figure 6-1). The IU contains the guidance, navigation, and control equipment which will guide the vehicle through its launch trajectory and subsequently to provide attitude control during orbital operations. In addition, it contains telemetry, communications, tracking, and crew safety systems, along with their supporting electrical power and environmental control systems.

This section of the Flight Manual contains a description of the physical characteristics and functional operation for the equipment installed in the IU.

STRUCTURE.

The basic IU structure is a short cylinder fabricated of an aluminum alloy honeycomb sandwich material (see figure 6-2). The top and bottom edges are made from extruded aluminum channels bonded to the honeycomb sandwich. This type of construction was selected for its high strength-to-weight ratio, acoustical insulation, and thermal conductivity properties. The cylinder is manufactured in three 120 degree segments which are joined by splice plates into an integral structure. The three segments are the access door segment, the flight control computer segment, and the ST-124M-3 segment. The access door segment has an umbilical door, as well as an equipment/personnel access door. The access door has the requirement to carry flight loads and still be removable at any time prior to flight.

Attached to the inner surface of the cylinder are cold plates which serve both as mounting structures and thermal conditioning units for the electrical/electronic equipment. Mounting the electrical/electronic equipment around the inner circumference of the IU

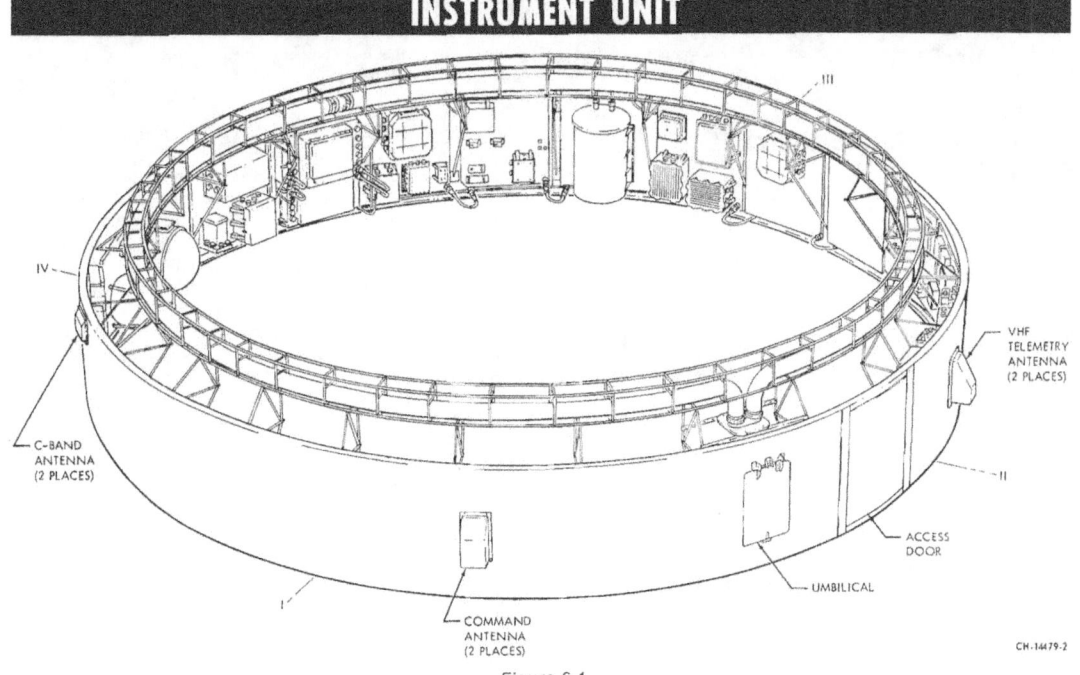

Figure 6-1

Section VI Instrument Unit

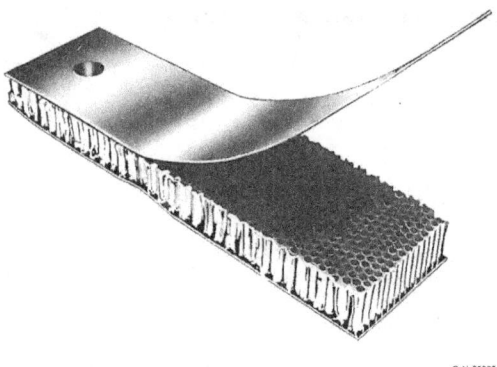

Figure 6-2

leaves the center of the unit open to accommodate the convex upper tank bulkhead of the S-IVB stage.

Cross section "A" of figure 6-3 shows equipment mounting pads bolted and bonded to the honeycomb structure. This method is used when equipment is not mounted on thermal conditioning cold plates. The bolts are inserted through the honeycomb core, and the bolt ends and nuts protrude through the outside surface. Cross section "B" shows a thermal conditioning cold plate mounting panel bolted to brackets which, in turn, are bolted to the honeycomb structure. The bolts extend through the honeycomb core with the bolt heads protruding through the outer surface. Cross section "C" shows the cable tray supports bolted to inserts, which are potted in the honeycomb core at the upper and lower edges of the structure.

Figure 6-4 shows the relative locations of all equipment installed in the IU.

ENVIRONMENTAL CONTROL SYSTEM.

The Environmental Control System (ECS) maintains an acceptable operating environment for the IU and S-IVB forward skirt equipment during preflight and flight operations. The ECS is composed of the following:

a. The thermal conditioning system (TCS) which maintains a circulating coolant at a controlled temperature to the cold plates and certain internally-cooled components mounted on the IU and the S-IVB forward skirt.

b. The preflight purging system which maintains a supply of temperature and pressure related air/GN_2 to the IU/S-IVB equipment area.

c. The gas-bearing supply which furnishes GN_2 to the ST-124M-3 inertial platform gas bearings.

THERMAL CONDITIONING SYSTEM (TCS).

Up to sixteen thermal conditioning panels; (cold plates), each capable of absorbing up to 420 watts of thermal energy, may be located in each of the IU and S-IVB stages. Each cold plate contains tapped bolt inserts in a fixed grid pattern which provide flexibility of component mounting. Temperature control is accomplished by circulation of a coolant fluid, Oronite Flocool 100*, through passages in the cold plates.

A functional flow diagram is shown in figure 6-5. Two heat exchangers are employed in the system. One is used during the preflight mode and employs GSE supplied circulating Oronite as the heat exchanging medium. The other is the flight mode unit, which uses demineralized water and the principle of sublimation to effect heat dissipation.

The manifold, plumbing, and both accumulators are filled during the prelaunch preparations. The accumulators serve as positive pressure reservoirs supplying fluid to their respective systems on a demand basis. There is a flexible diaphragm in each accumulator, backed by regulated low pressure GN_2.

During operation of the TCS, the Oronite coolant is circulated through a closed loop by an electrically driven centrifugal pump (a secondary pump provides redundancy). The supply manifold diverts part of the coolant to the cold plates in the S-IVB stage and the remainder to the cold plates, gas bearing heat exchanger, inertial platform, LVDC/LVDA, and flight control computer in the IU.

During the preflight mode, the sublimator, which functions only in a vacuum environment, is inactive and a solenoid valve blocks the water flow. The preflight heat exchanger transfers heat from the closed loop fluid to GSE fluid.

Approximately 180 seconds after liftoff, the water solenoid valve is opened and the sublimator becomes active. During the period between GSE disconnect (at liftoff) and sublimator activation, the capacity of the pre-cooled system is sufficient to preclude equipment overheating.

The sublimator element is a porous plate. Since the sublimator is not activated until approximately 180 seconds after launch, the ambient temperature and pressure outside the porous plates are rapidly approaching the vacuum conditions of space. Water flows readily into the porous plates and into the pores. The water freezes when it meets the low temperature and pressure of the space environment, and the resulting ice formation blocks the pores (figure 6-6).

As heat is generated by the equipment, the temperature in the Oronite solution rises. This heat is transferred within the sublimator to the demineralized water. This heat is then dissipated to space through the latent heat of sublimation as the ice is transformed to vapor by the space environment. The vapor is vented into the IU compartment. As the heat input decreases, the rate of sublimation is reduced, decreasing the water flow. Thus, the sublimator is a self-regulating system. However, if the coolant temperature falls below the lower limit of approximately 59° F, a switch selector command is issued to close the water solenoid valve causing the cessation of sublimator operation. GN_2 for the Oronite and water accumulators is stored in a 165 cubic inch sphere in the IU at a pressure of 3,000 psig. The sphere is filled prior to liftoff by applying high pressure GN_2 through the umbilical. A solenoid valve controls the flow into the sphere, and a pressure transducer indicates to the GSE when the sphere is pressurized. The GN_2 flow from the sphere is filtered and applied to the accumulators through a pressure regulator, which reduces the 3,000 psig to 16.5 psia. An orifice regulator further reduces the pressure at the water accumulator to approximately 5 psia. The GN_2 flow is then vented within the IU compartment.

PREFLIGHT AIR/GN_2 PURGE SYSTEM.

The preflight air/GN_2 purge system directs ground supplied, temperature and pressure regulated, filtered air or GN_2 to the IU/S-IVB

*Reg. TM, Chevron Chemical Co.

Section VI Instrument Unit

INSTRUMENT UNIT STRUCTURAL DETAILS

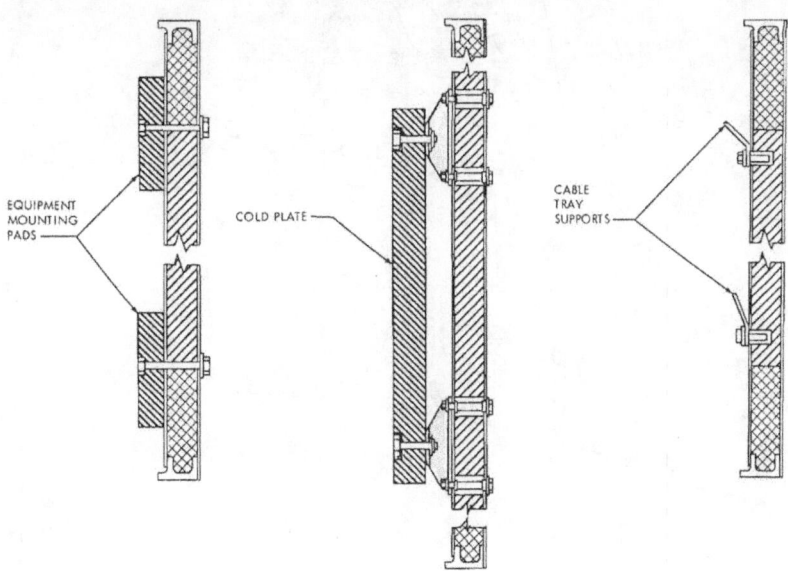

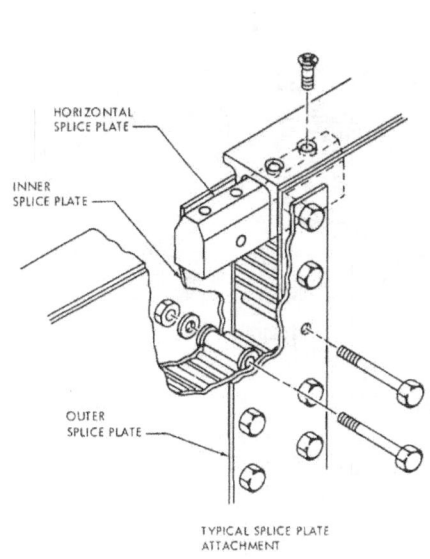

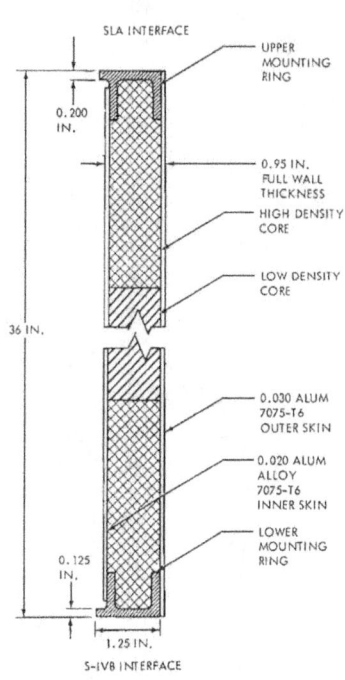

Figure 6-3

Section VI Instrument Unit

IU COMPONENTS

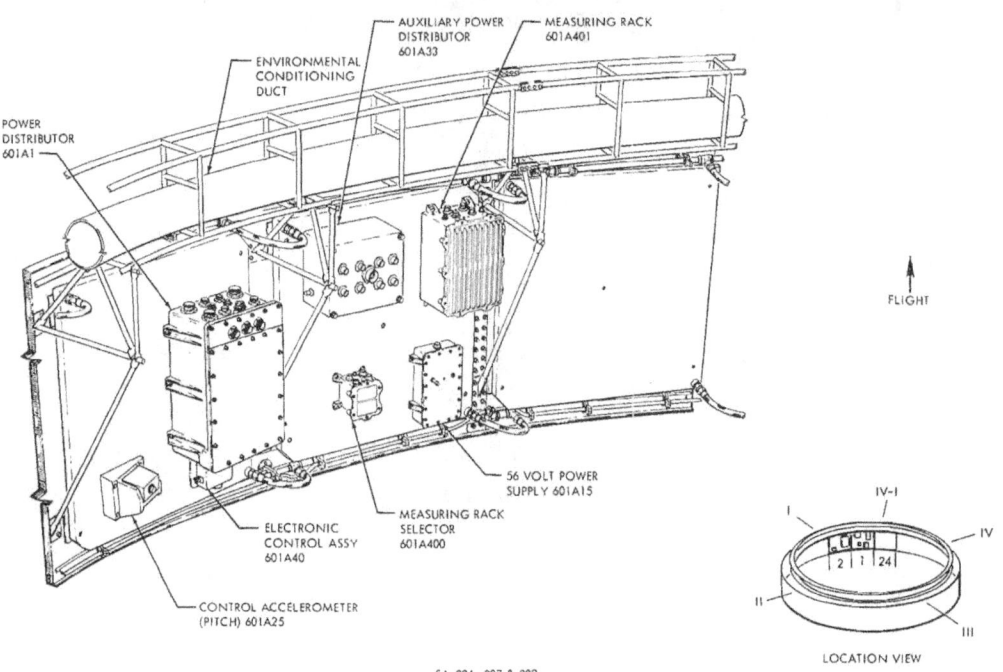

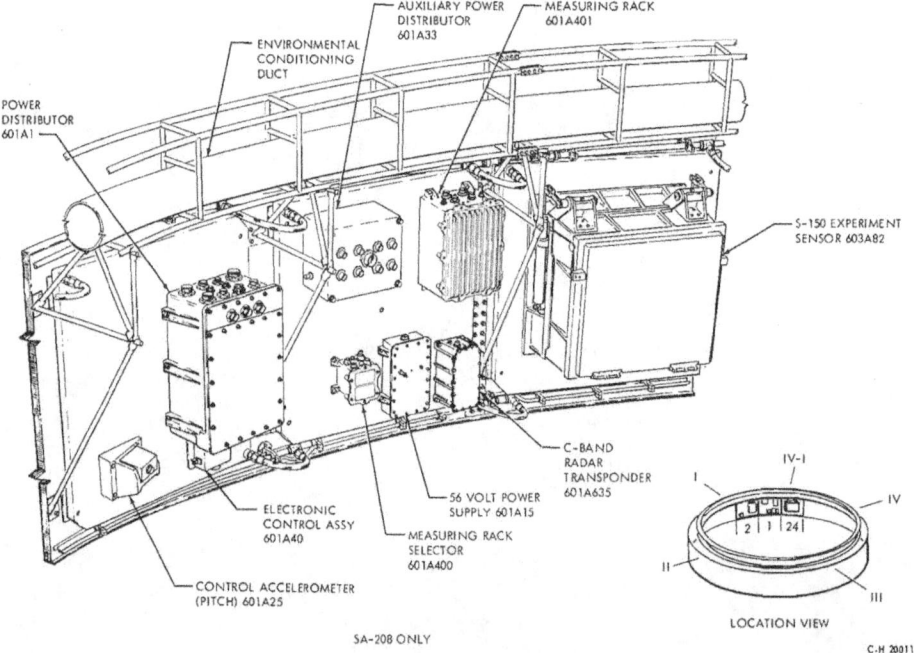

Figure 6-4 (Sheet 1 of 6)

Section VI Instrument Unit

IU COMPONENTS

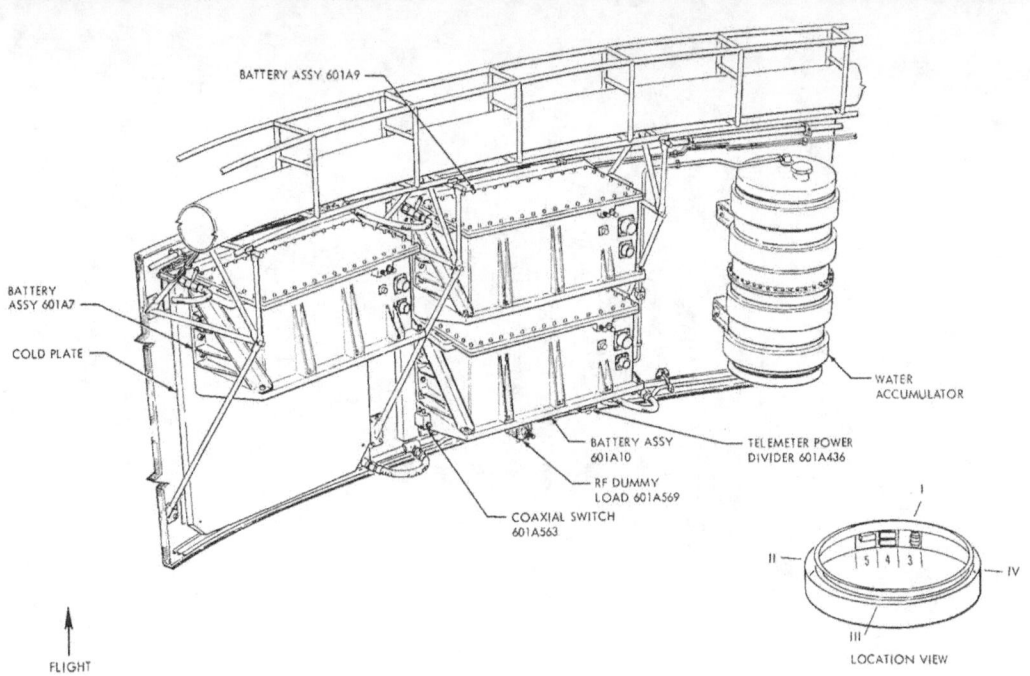

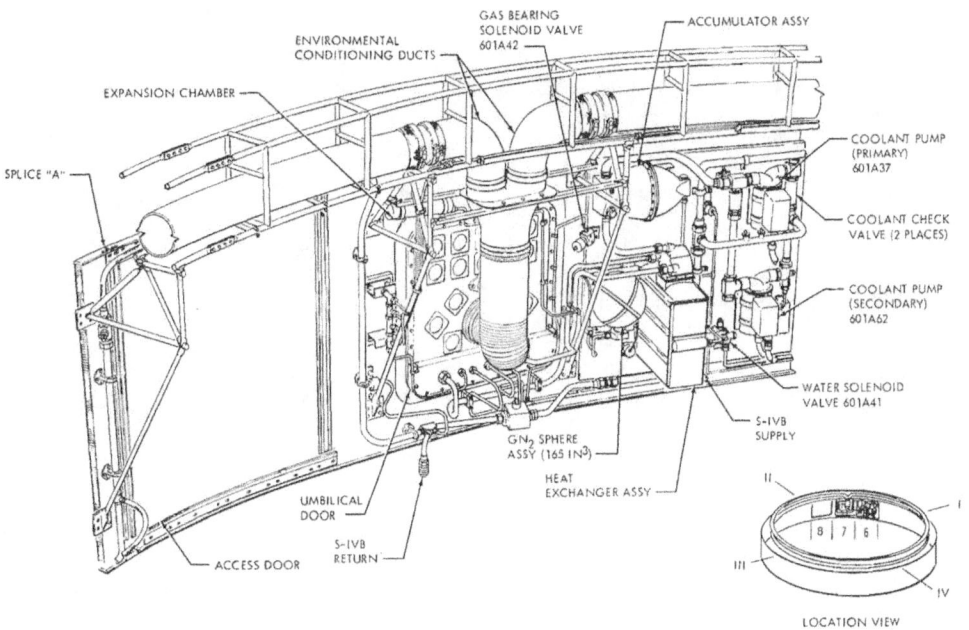

Figure 6-4 (Sheet 2 of 6)

Section VI Instrument Unit

IU COMPONENTS

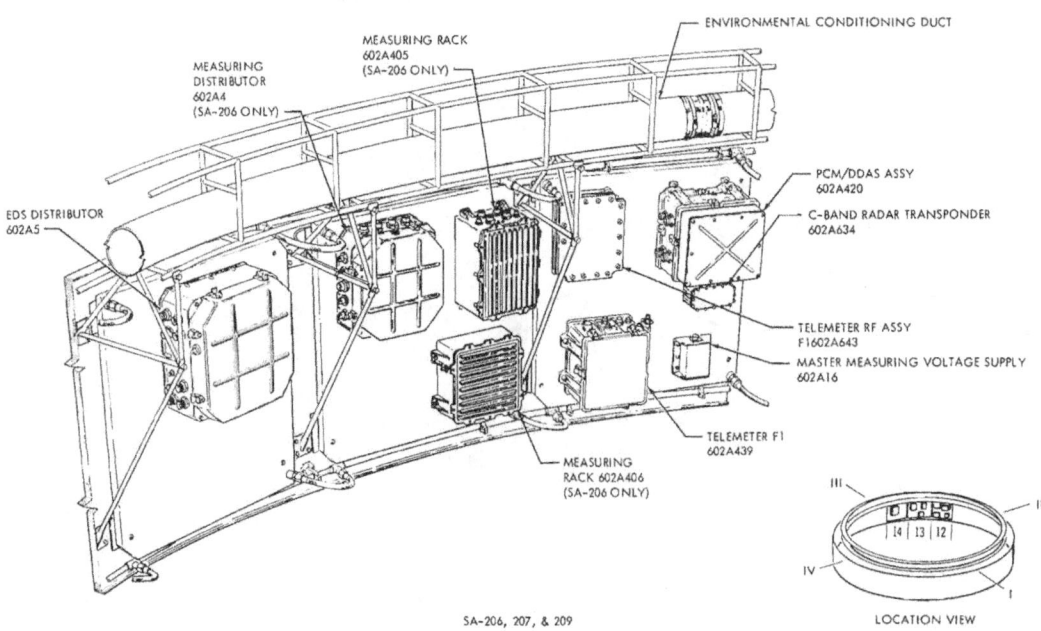

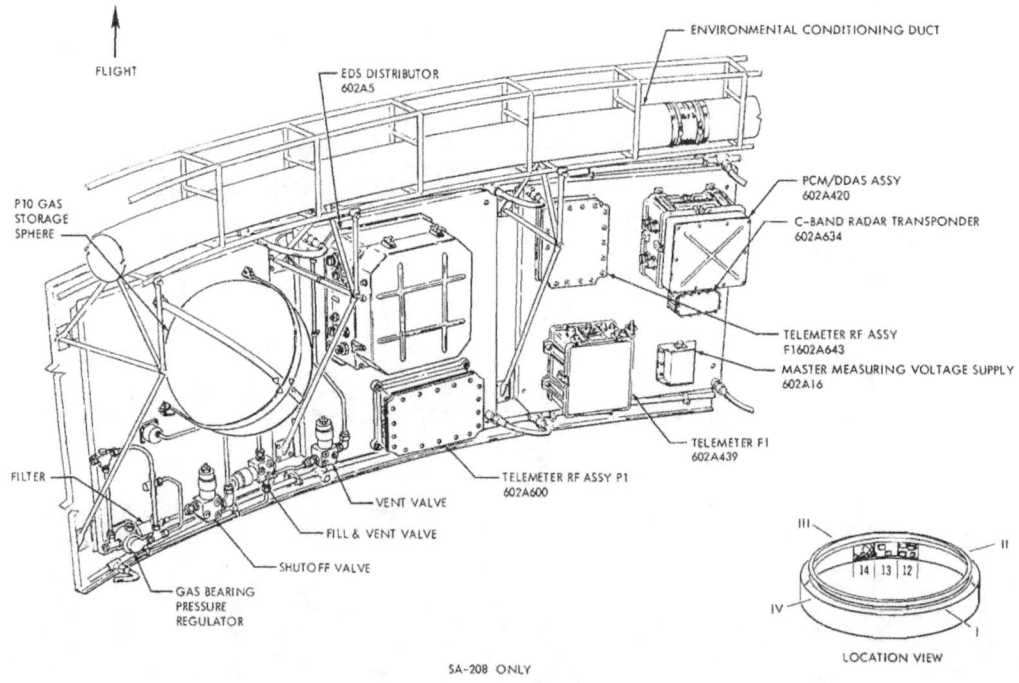

Figure 6-4 (Sheet 3 of 6)

Section VI Instrument Unit

IU COMPONENTS

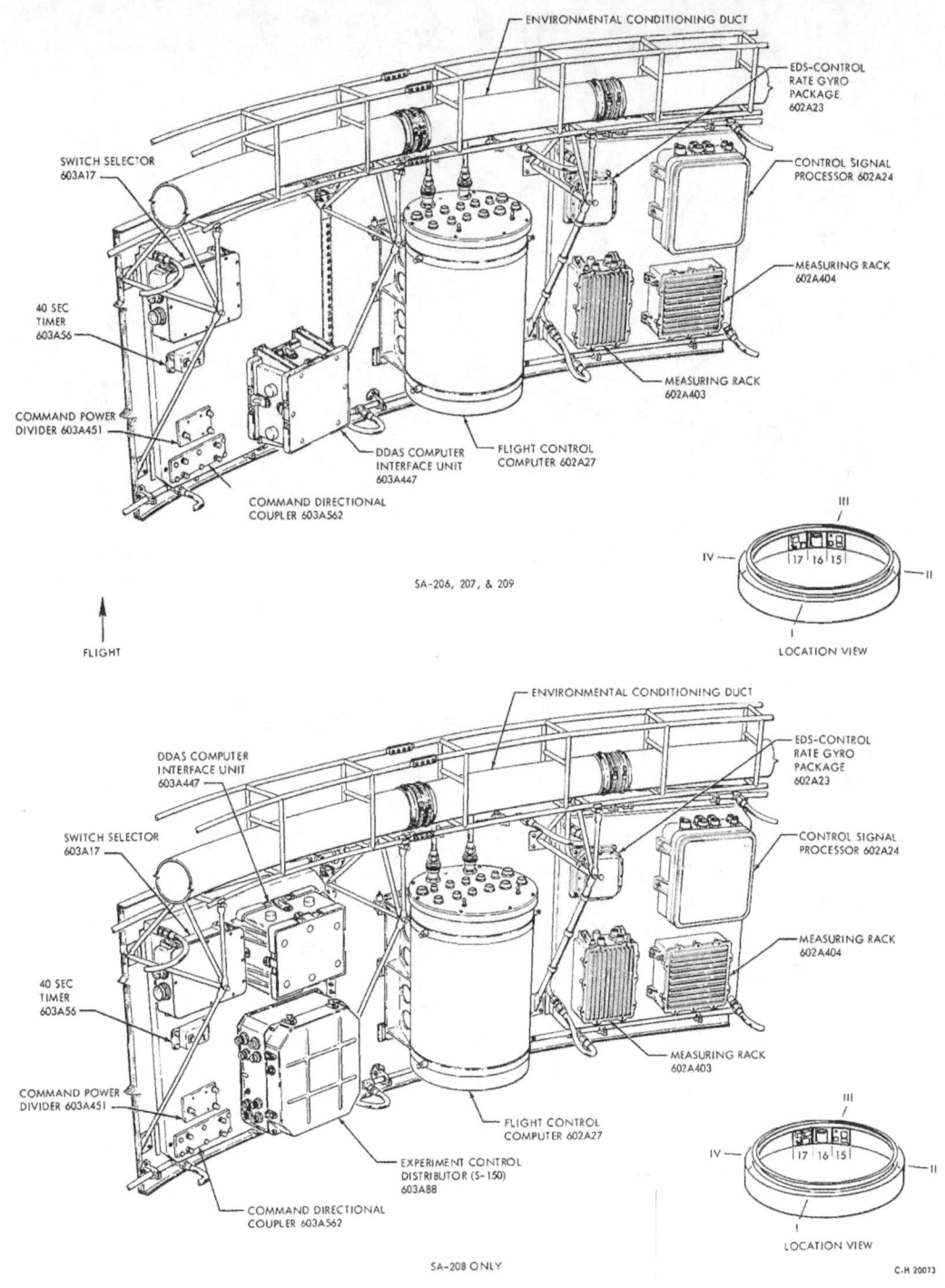

Figure 6-4 (Sheet 4 of 6)

Section VI Instrument Unit

IU COMPONENTS

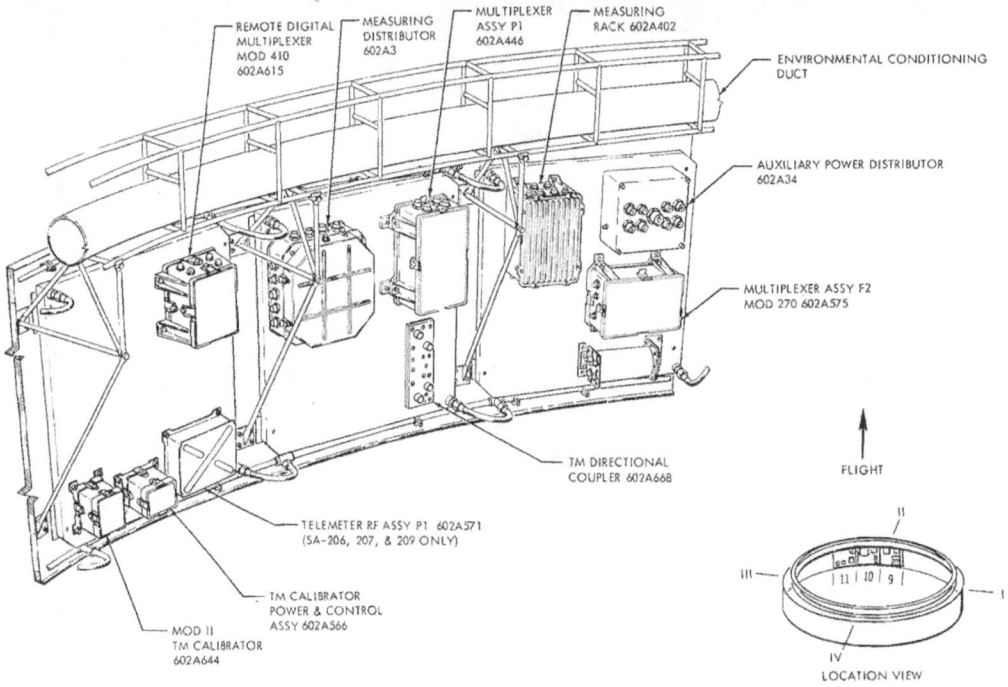

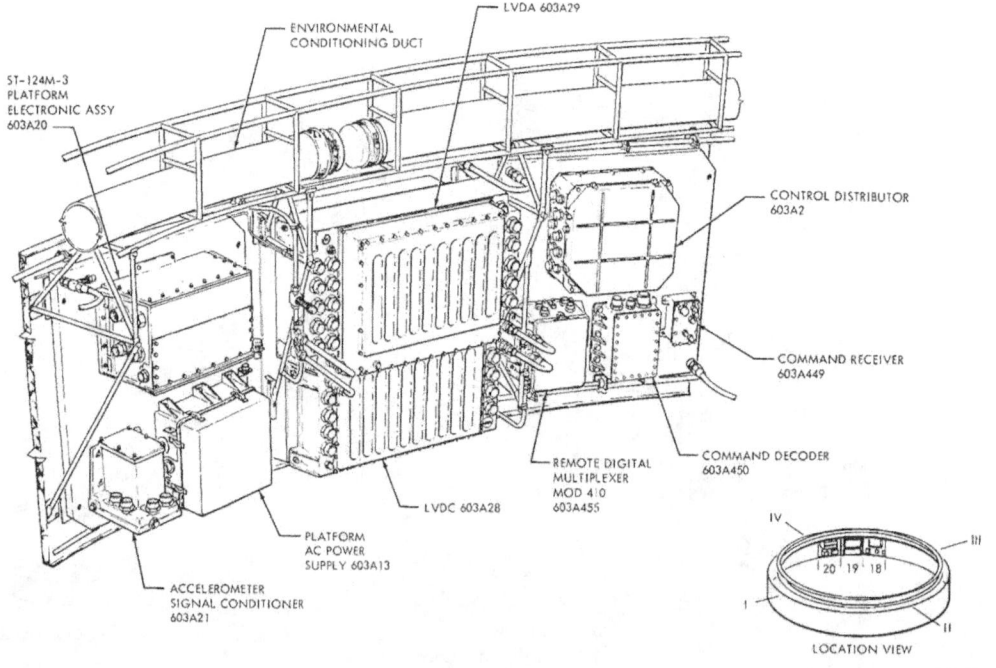

Figure 6-4 (Sheet 5 of 6)

Section VI Instrument Unit

IU COMPONENTS

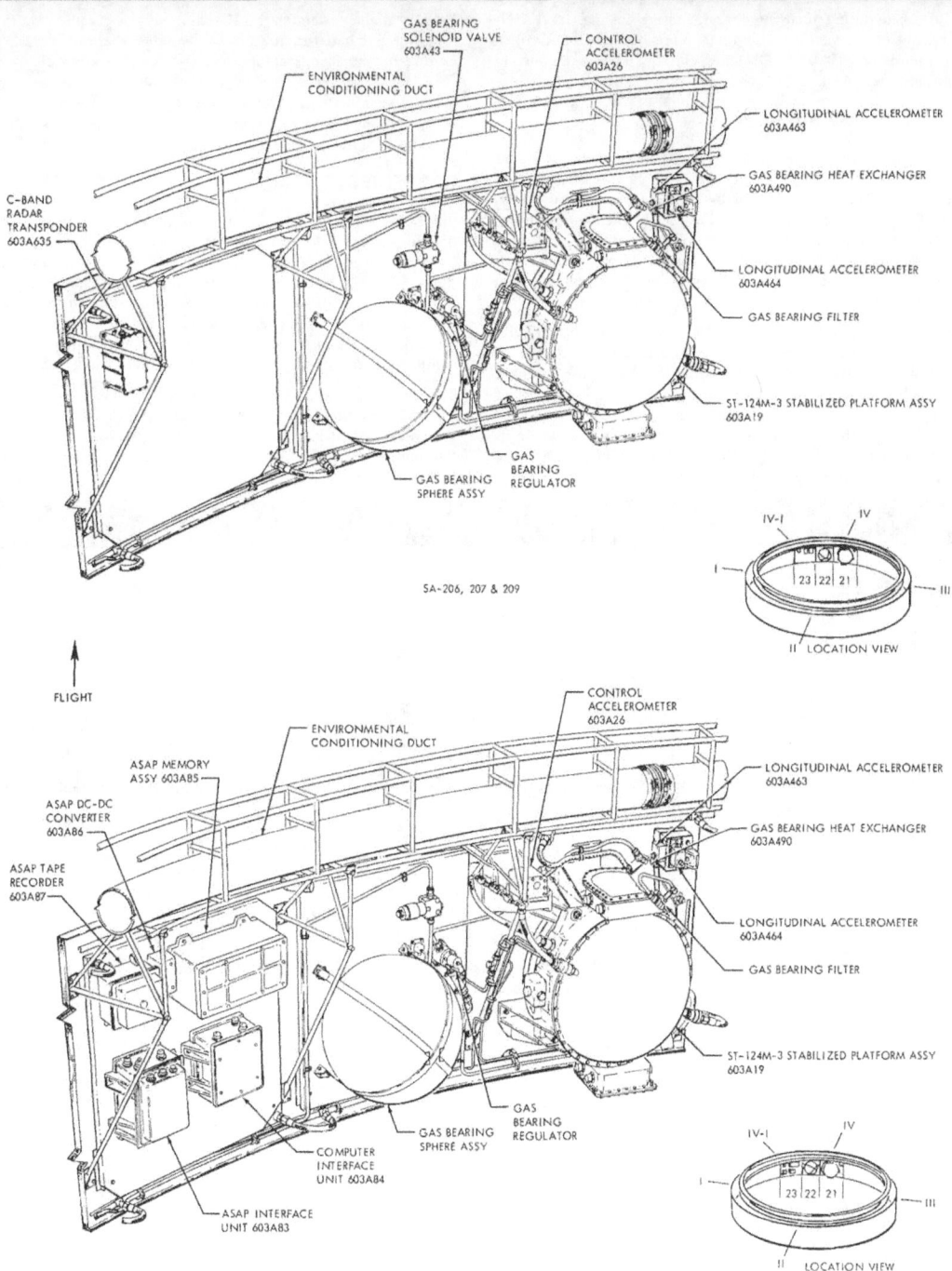

Figure 6-4 (Sheet 6 of 6)

Section VI Instrument Unit

interstage compartment. The air or GN_2 is distributed through a flexible duct system mounted above the payload interface as shown in figure 6-1. Ventilating air for temperature and humidity control is furnished during preflight phases. During fueling, inert GN_2 is furnished to prevent the accumulation of a hazardous and corrosive atmosphere. The air or GN_2 flows through holes in the ducting previously described.

GAS BEARING SUPPLY.

Gaseous nitrogen, for the ST-124M-3 inertial platform, is stored in a two cubic foot sphere in the IU at the pressure of 3,000 psig (figure 6-5). The sphere is filled by applying high pressure GN_2 through the umbilical under control of the IU pneumatic console. A low pressure switch monitors the sphere; and, if the pressure falls below 1,000 psig, the ST-124M-3 stable platform is shut down to preclude damage to the gas bearings during ground checkout operations. This switch is inactive during flight.

Output of the sphere is through a filter and a pressure regulator. The regulator reduces the sphere pressure to a level suitable for gas bearing lubrication. Pressure internal to the platform is sensed and applied as a control pressure to the regulator. This provides for a constant pressure differential across the gas bearings. The gas flows from the regulator through a heat exchanger, where its temperature is stabilized, then through another filter and on to the gas bearings. Spent gas is then vented into the IU compartment.

ELECTRICAL POWER SYSTEMS.

Primary flight power for the IU equipment is supplied by three (3) silver-zinc batteries at a nominal voltage level of $28(\pm 2)$ Vdc. During prelaunch operations, primary power is supplied by the GSE. Where ac power is required within the IU, it is developed by solid state dc-to-ac inverters. Power distribution within the IU is accomplished through power distributors which are, essentially, junction boxes and switching circuits.

BATTERIES.

Silver-zinc primary flight batteries are installed during prelaunch operations at the locations shown in figure 6-4, sheet 2. These batteries are identical, each having the characteristics shown in figure 6-7. Each battery is connected to a separate bus in a power distributor. The D10 and D30 batteries are connected to a common bus through isolation diodes to provide a redundant power source for critical IU platform, switch selector, and control functions. Flight components are connected to the buses in the various distributors.

The silver-zinc batteries are characterized by their high efficiency. Their ampere-hour rating is about four times as great as that of a lead-acid or nickel-cadmium battery of the same weight. The low temperature performance of the silver-zinc batteries is also substantially better than the others.

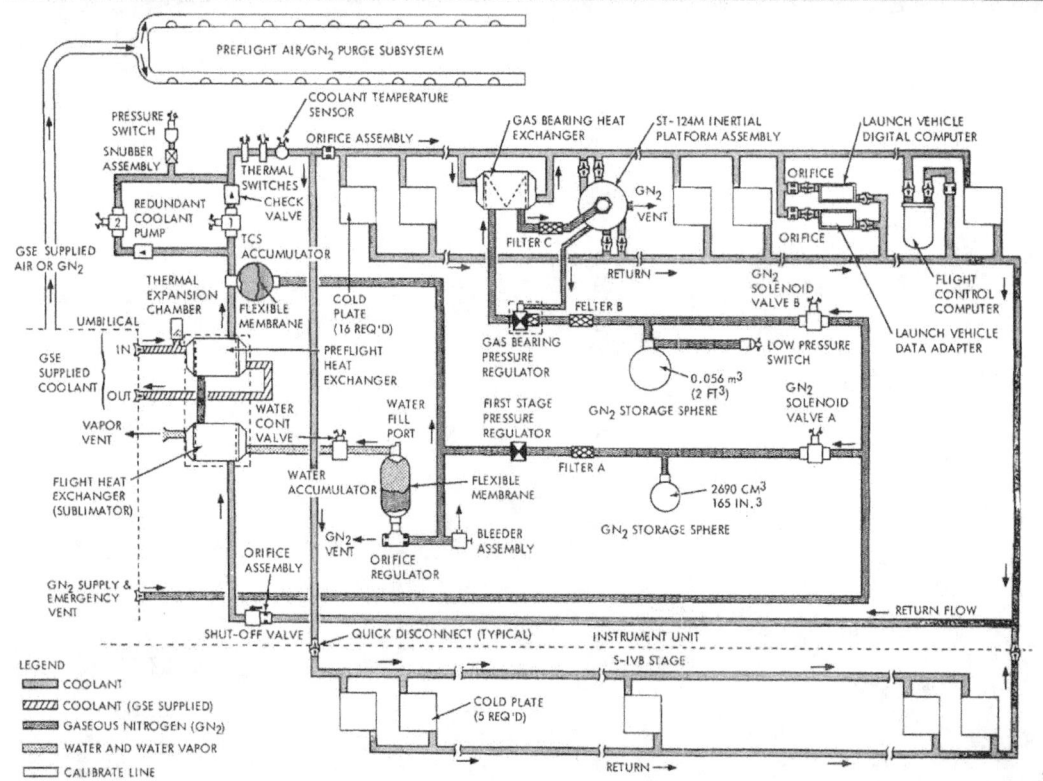

Figure 6-5

SUBLIMATOR DETAILS

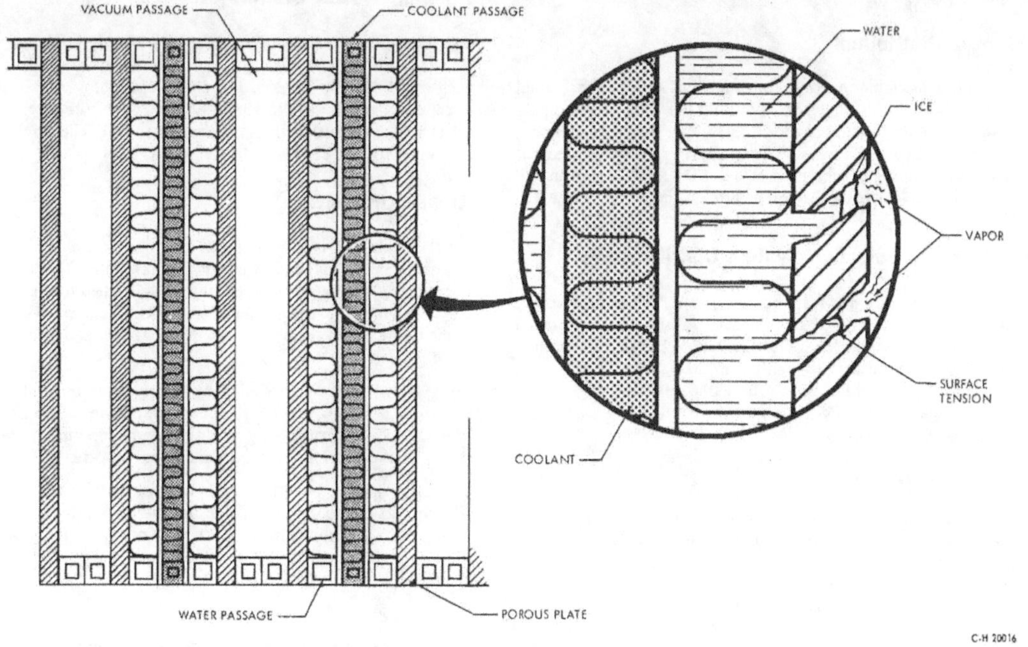

Figure 6-6

POWER CONVERTERS.

The IU electrical power systems contain a 56-V power supply and a 5-V measuring voltage supply.

56-V Power Supply.

The 56-V power supply furnishes the power required by the ST-124M-3 platform electronic assembly and the accelerometer signal conditioner. It is basically a dc-to-dc converter that uses a magnetic amplifier as a control unit. It converts the unregulated 28 Vdc from the batteries to a regulated 56 Vdc. The 56-V power supply is connected to the platform electronic assembly through the power and control distributors.

5-V Measuring Voltage Supply.

The 5-V measuring voltage supply converts unregulated 28 Vdc to a closely regulated 5 (+.005) Vdc for use throughout the IU measuring system. This regulated voltage is used primarily as excitation for measurement sensors (transducers), and as a reference voltage for inflight calibration of certain telemetry channels. Like the 56-V supply, it is basically a dc-to-dc converter.

DISTRIBUTORS.

The distribution system within the IU is comprised of the following:
 1 Measuring distributor (2 used on SA-206 only)
 1 Control distributor
 1 Emergency Detection System (EDS) distributor
 1 Power distributor
 2 Auxiliary power distributors
 1 Experiment distributor (used on SA-208 only)

Measuring Distributors.

The primary function of the measuring distributors is to collect all measurements that are transmitted by the IU telemetry system, and to direct them to their proper telemetry channels. These measurements are obtained from instrumentation transducers, functional components, and various signal and control lines. The measuring distributors also distribute the output of the 5-V measuring voltage supply throughout the measuring system.

Through switching capabilities, the measuring distributors can change the selection of measurements monitored by the telemetry

IU BATTERY CHARACTERISTICS

TYPE	DRY CHARGE
MATERIAL	ALKALINE SILVER-ZINC
CELLS	20 (WITH TAPS FOR SELECTING 18 OR 19 CELLS IF REQUIRED TO REDUCE HIGH VOLTAGE)
NOMINAL VOLTAGE	1.5 PER CELL
ELECTROLYTE	POTASSIUM HYDROXIDE (KOH) IN DEMINERALIZED WATER
OUTPUT VOLTAGE	+28 ± 2 VDC
OUTPUT CURRENT	35 AMPERES FOR A 10 HOUR LOAD PERIOD (IF USED WITHIN 120 HOURS OF ACTIVATION)
GROSS WEIGHT	165 POUNDS EACH

Figure 6-7

Section VI Instrument Unit

system. The switching function transfers certain measurements to channels which had been allotted to expended functions. If it were not for this switching, these channels would be wasted for the remainder of the flight.

Control Distributor.

The control distributor provides distribution of 28-V power to small current loads and distributes 56 Vdc from the 56-V power supply to the ST-124M-3 inertial platform assembly. The control distributor provides power and signal switching during prelaunch checkout for testing various guidance, control, and EDS functions, requested by the launch vehicle data adapter through the switch selector.

Emergency Detection System Distributor.

The EDS distributor provides the only electrical link between the spacecraft and the LV. All EDS signals from the LV are routed to the logic circuits in the EDS distributor. EDS output signals from these logic circuits are then fed to the spacecraft and to the IU telemetry. Also, EDS signals from the spacecraft are routed back through the IU EDS logic circuits before being sent to the S-IVB and S-IB stages.

Power Distributor.

The power distributor provides primary distribution for all 28-V power required by IU components. Inflight 28-V battery power, or prelaunch ESE-supplied 28-V power, is distributed by the power distributor as shown in figure 6-8.

The power distributor also provides paths for command and measurement signals between the ESE and IU components. The power distributor connects the IU component power return and signal return lines to the IU single point ground and to the umbilical supply return bus. These return lines are connected to the common bus in the power distributor, directly or indirectly, through one of the other distributors.

Auxiliary Power Distributors.

Two auxiliary power distributors supply 28-Vdc power to small current loads. Both auxiliary power distributors receive 28 Vdc from each of the battery buses in the power distributor. Relays in the auxiliary power distributors provide power ON/OFF control for IU components during the prelaunch checkout. These relays are controlled by the ESE.

IU GROUNDING.

All IU grounding is referenced to the outer skin of the LV. The power system is grounded by means of wires routed from the power distributor COM bus to a grounding stud attached to the LV skin. All COM buses in the various other distributors are wired back to the COM bus in the power distributor. This provides for a single point ground.

Equipment boxes are grounded by direct metal-to-metal contact with cold plates or other mounting surfaces which are common to the LV skin. Most cabling shields are grounded to a COM bus in one of the distributors or to the equipment case.

During prelaunch operations, the IU and CSE COM buses are referenced to earth ground. To ensure the earth ground reference until after all umbilicals are ejected, two single-wire grounding cables are connected to the IU below the umbilical plates. These are the final conductors to be disconnected from the IU.

EMERGENCY DETECTION SYSTEM.

The EDS is the principle element of several crew safety systems.

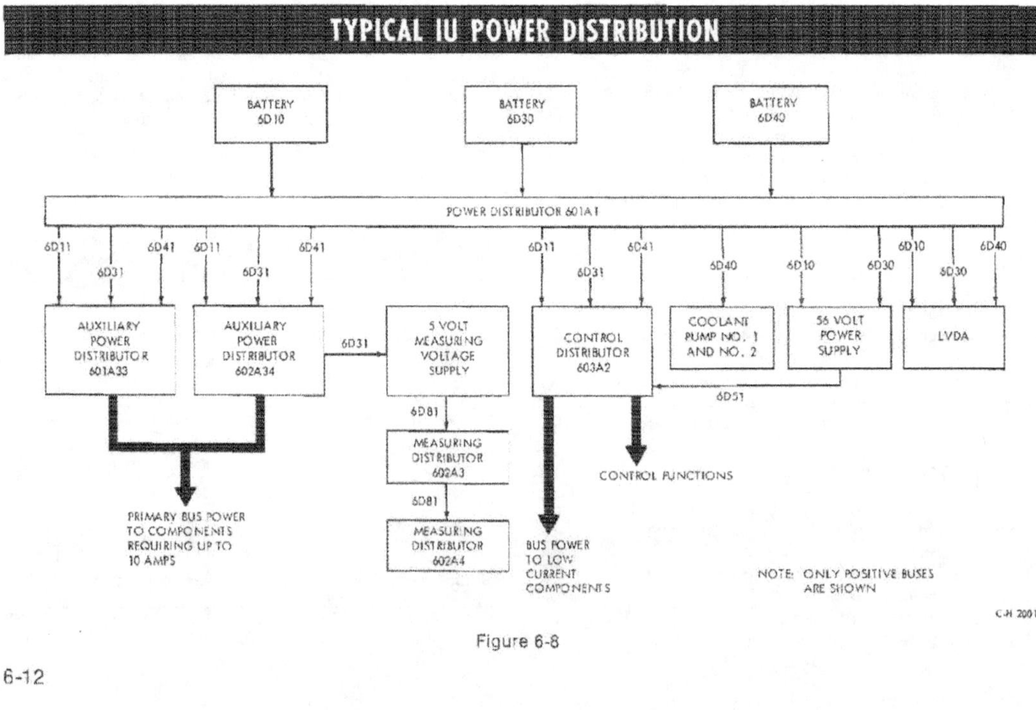

TYPICAL IU POWER DISTRIBUTION

Figure 6-8

Section VI Instrument Unit

EDS design is a coordinated effort of crew safety personnel from several NASA centers.

The EDS senses development of conditions which could ultimately cause vehicle failure. The EDS reacts to these emergency situations in either of two ways. If breakup of the vehicle is imminent, an automatic abort sequence is initiated. If, however, the emergency condition is developing slowly enough, or is of such a nature that the flight crew can evaluate it and take action, only visual indications are provided to the flight crew. Once an abort sequence has been initiated, either automatically or manually, it is irrevocable and runs to completion.

The EDS is comprised of sensing elements, signal processing and switching circuitry, relay and diode logic circuitry, electronic timers and display equipment, all located in various places on the flight vehicle. Only that part of the EDS equipment located in the IU will be discussed here.

There are nine EDS rate gyros installed in the IU. Three gyros monitor each of the three axes (pitch, roll, and yaw) thus providing triple redundancy.

The control signal processor provides power to the nine EDS rate gyros, as well as receiving inputs from them. These inputs are processed and sent to the EDS distributor and to the flight control computer.

The EDS distributor serves as a junction box and switching device to furnish the spacecraft display panels with emergency signals if emergency conditions exist. It also contains relay and diode logic for the automatic abort sequence.

There is an electronic timer, which is activated at liftoff and which produces an output 40 seconds later. This output energizes relays in the EDS distributor which allows multiple engine shutdown, which had been inhibited during the first 40 seconds of launch.

Inhibiting of automatic abort circuitry is also provided by the LV flight sequencing circuits through the IU switch selector. This inhibiting is required prior to normal S-IB engine cutoff and other normal LV sequencing. While the automatic abort capability is inhibited, the flight crew must initiate a manual abort, if an angular-overrate or two-engine-out condition occurs.

See Section III for a more complete discussion of emergency detection and procedures. Section III includes launch vehicle monitoring and control, EDS controls, and abort modes and limits.

LAUNCH VEHICLE NAVIGATION, GUIDANCE, AND CONTROL.

The Saturn IB astrionics system will provide navigation, guidance and control (N, G, & C) of the launch vehicle from liftoff until separation of the Apollo spacecraft from the S-IVB/IU and attitude control until the end of active lifetime. The function of the N, G, & C portion of the astrionics system (figure 6-9) is to steer the launch vehicle along an optimum trajectory (minimum fuel consumption and safe structural loading) into a pre-determined earth orbit. Orbital N, G, & C functions include computations to solve for position and velocity (based on insertion conditions) and vehicle attitude control. Figure 6-10 defines the N, G, & C functions and the equipment or program that will perform the functions. Other functions performed by or interfacing with the N, G, & C hardware are event sequencing and telemetry-data management. Alternate altitude error commands can be provided to the control system from the Apollo spacecraft in the event of a guidance reference failure during any flight phase. During the orbital phase, the spacecraft may provide control system input regardless of the guidance system health.

N, G, AND C SUMMARY.

The ST-124M inertial platform system establishes a space-fixed reference coordinate system and provides instantaneous vehicle attitude (θ) and incremental velocity ($\dot{X}$, $\dot{Y}$, and $\dot{Z}$) with respect to the established coordinate system. The Launch Vehicle Digital Computer/Launch Vehicle Data Adapter (LVDC/LVDA) system uses the velocity output from the platform system during the burn mode of flight to determine the vehicle's present velocity, position, and acceleration. The flight program calculates the desired vehicle pitch and yaw attitude angles as functions of time during S-IB burn. During S-IVB burn, an optimal path to the orbital insertion point is computed using an iterative guidance scheme. During orbit, the vehicle's position and velocity are calculated by using mathematical models of the earth's gravitational field, the atmospheric drag, and the vehicle propellant vent. The orbital guidance mode determines the guidance commands as a function of position, prestored attitude, and time since S-IVB cutoff. Vehicle attitude control is accomplished during all periods by comparing the present vehicle attitude (θ obtained from the platform system) to the desired attitude (X) computed by the flight program. The difference is fed to the flight control computer as an attitude error signal (ψ). The flight control computer sums the vehicle attitude error information with vehicle attitude rate ($\dot{\phi}$) information from the EDS control rate gyros and vehicle lateral acceleration ($\ddot{y}$) information (used during S-IB burn only) from the control accelerometers. The result of this summation is an appropriate steering command output (β_c) that changes the vehicle's thrust vector by either gimballing the S-IB or S-IVB engines or activating the S-IVB auxiliary propulsion system. Figure 6-11 lists the N, G, & C system components. Figure 6-12 is a block diagram of the Saturn IB navigation, guidance, and control subsystem.

NAVIGATION SCHEME.

Powered Flight.

The basic navigation scheme is shown in figure 6-13. Gimbal resolvers supply platform gimbal angles in analog form to the LVDA. An analog-to-digital converter in the LVDA converts the signals to the digital format required by the LVDC.

Platform integrating accelerometers sense acceleration components and mechanically integrate them into velocity. The LVDA processes the analog velocity data for use in the navigation equation. Within the LVDC flight program, initial velocity imparted by the spinning earth, gravitational velocity, and the platform velocities are algebraically summed. This vehicle velocity is then integrated to determine vehicle position.

Orbital Flight.

During orbital coast flight, the navigational program continually computes the vehicle position, velocity, and acceleration from equations of motion initialized at beginning of the orbit mode. In orbit, navigation and guidance information in the LVDC can be updated via the digital command system.

Additional navigational computations are used in maintaining vehicle attitude during orbit. These computations establish a local vertical which is used as a reference for attitude control.

GUIDANCE COMPUTATIONS.

The guidance function of the launch vehicle is accomplished by computing the necessary flight maneuvers to meet the desired end conditions of the flight (e.g., inserting the spacecraft into the desired trajectory). Guidance computations are performed within the LVDC by programmed guidance equations, which use navigation data and mission sequence indicators as their inputs. These compu-

6-13

Section VI Instrument Unit

FLIGHT CONTROL CONFIGURATION

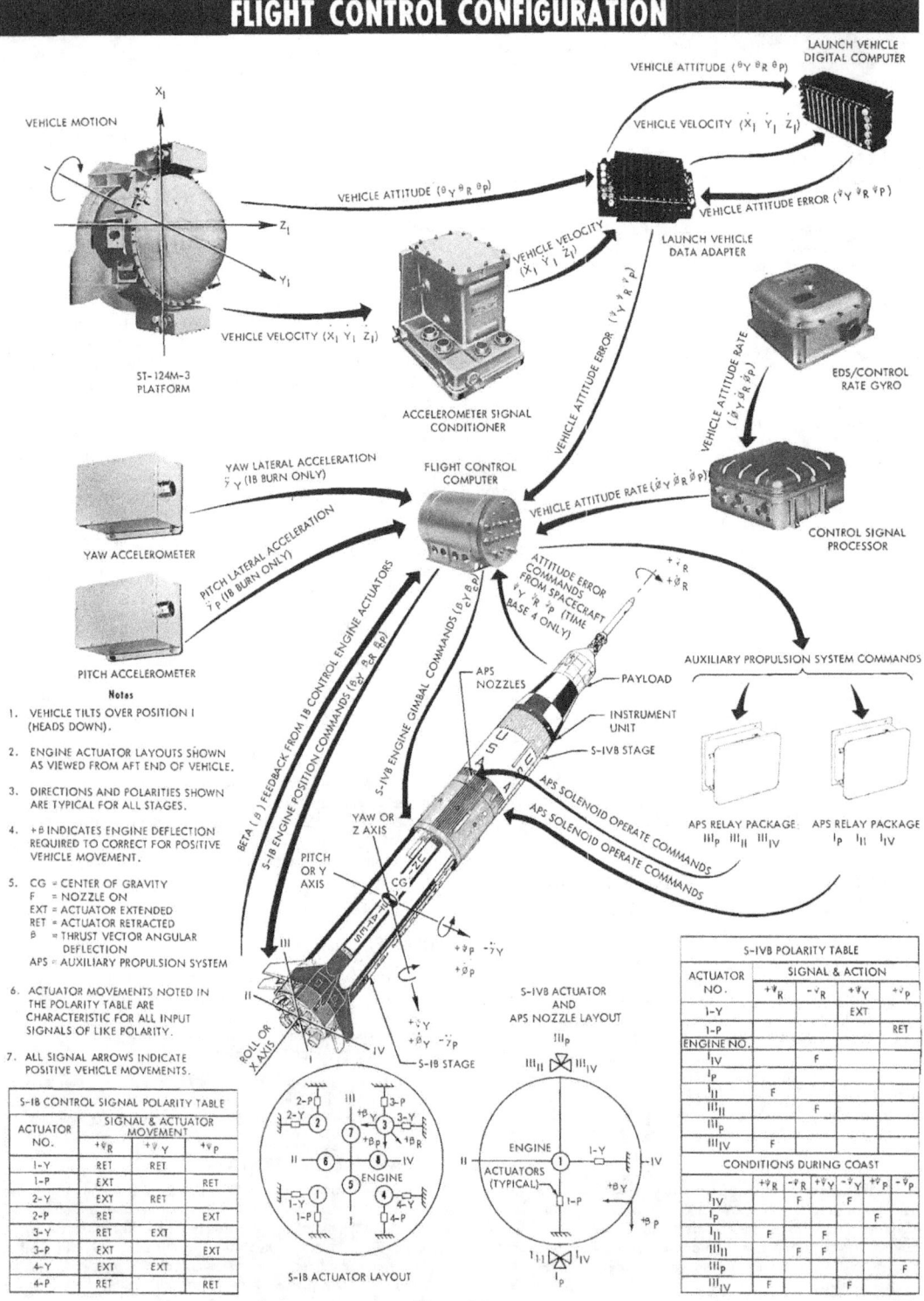

Figure 6-9

Section VI Instrument Unit

NAVIGATION, GUIDANCE AND ATTITUDE CONTROL DEFINITION

FUNCTION	DEFINITION	ASSOCIATED EQUIPMENT
NAVIGATION	DETERMINATION OF VEHICLE POSITION, AND VELOCITY FROM MEASUREMENTS MADE ON BOARD THE VEHICLE.	ACCELEROMETER READINGS FROM THE ST-124M-3 PLATFORM. LVDC/LVDA. LVDC FLIGHT PROGRAM.
GUIDANCE	COMPUTATION OF MANEUVERS NECESSARY TO ACHIEVE THE DESIRED END CONDITIONS OF A TRAJECTORY.	LVDC FLIGHT PROGRAM.
CONTROL	EXECUTION OF NECESSARY MANEUVERS (DETERMINED BY THE GUIDANCE SCHEME) BY CONTROLLING THE PROPER HARDWARE.	GIMBAL ANGLE READING FROM THE ST-124M-3 PLATFORM. LVDC/LVDA. LVDC FLIGHT PROGRAM. FLIGHT CONTROL COMPUTER. EDS/CONTROL RATE GYROS. CONTROL ACCELEROMETERS (IB BURN ONLY). S-IB AND S-IVB ACTUATORS. S-IVB AUXILIARY PROPULSION SYSTEM.

Figure 6-10

NAVIGATION, GUIDANCE AND ATTITUDE CONTROL COMPONENTS

INERTIAL	DIGITAL	ATTITUDE CONTROL
ST-124M-3 INERTIAL PLATFORM ASSEMBLY	LAUNCH VEHICLE DIGITAL COMPUTER (LVDC)	FLIGHT CONTROL COMPUTER (FCC)
PLATFORM ELECTRONICS ASSEMBLY (PEA)	LAUNCH VEHICLE DATA ADAPTER (LVDA)	EDS/CONTROL RATE GYROS (EDS/CRG)
PLATFORM AC POWER SUPPLY		CONTROL SIGNAL PROCESSOR (CSP)
ACCELEROMETER SIGNAL CONDITIONER		CONTROL ACCELEROMETERS (C/A)
56 VDC POWER SUPPLY		S-IB CONTROL ENGINE ACTUATORS
		S-IVB ENGINE ACTUATORS
		AUXILIARY PROPULSION SYSTEM (APS)

Figure 6-11

tations are actually a logical progression with a guidance command as their solution. After the desired attitude has been determined by the "optimal path" program, the guidance command is used in the following control equation: $X - \theta = \psi$ (figure 6-14) where:

X is the desired attitude (guidance command)
θ is the vehicle attitude
ψ is the attitude error command

CONTROL SUBSYSTEM.

The control subsystem (figure 6-15) is designed to control and maintain vehicle attitude by forming the steering commands used to control the engines of the active stage.

Vehicle attitude is achieved by gimbaling the four outboard engines of the S-IB stage or the single engine of the S-IVB stage. These

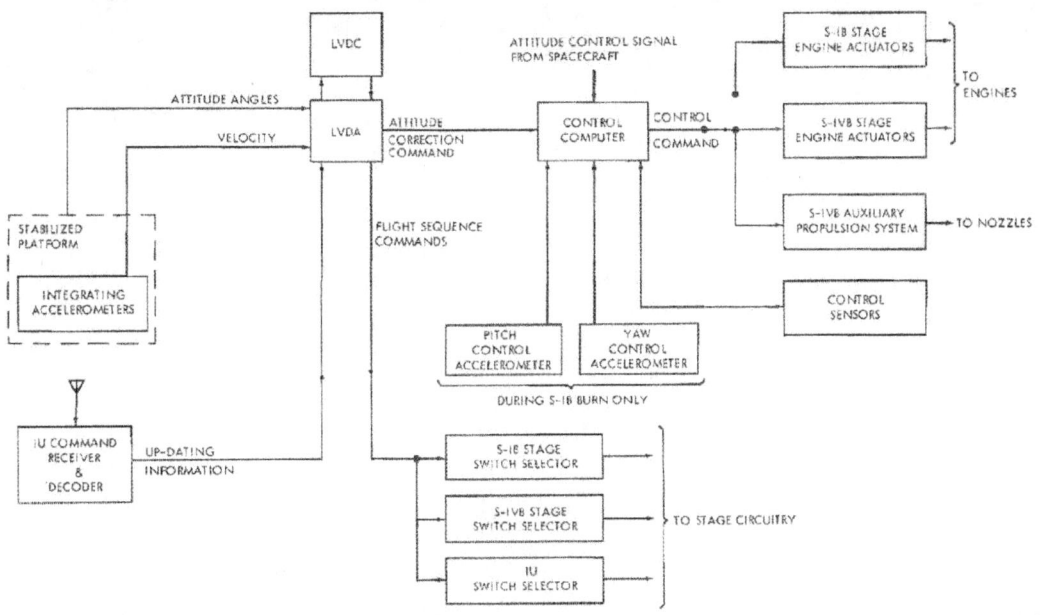

Figure 6-12

6-15

Section VI Instrument Unit

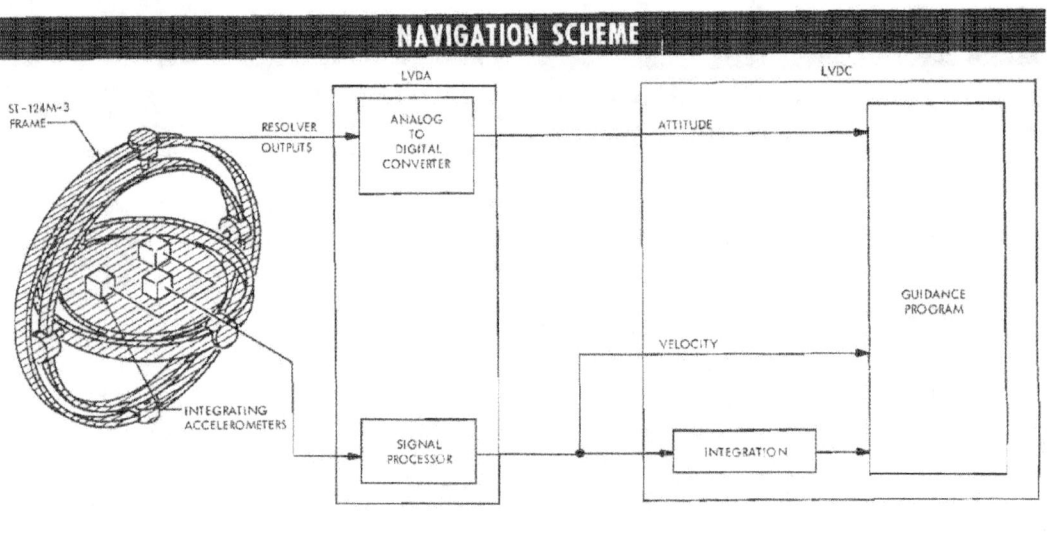

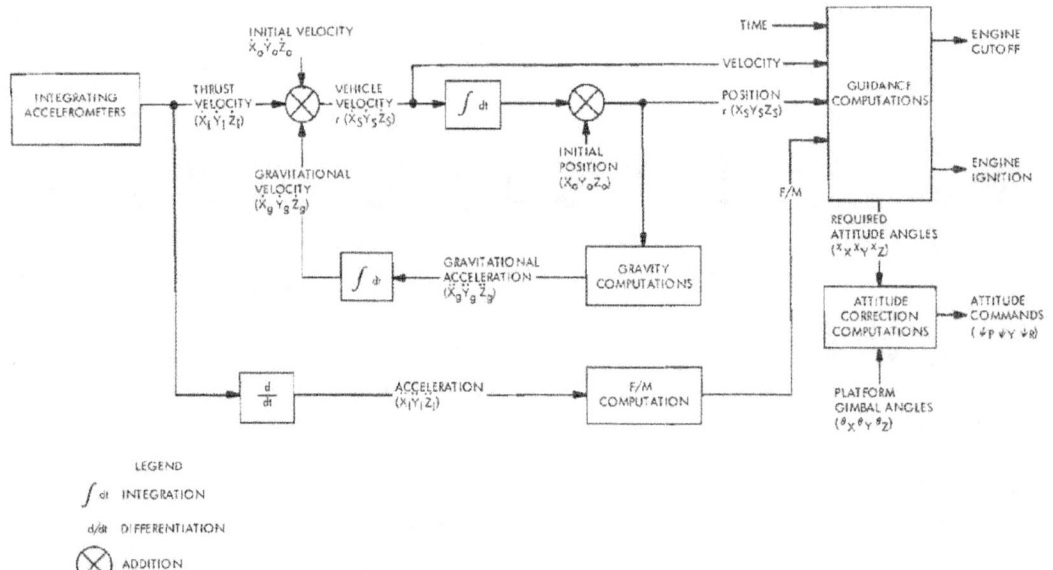

Figure 6-13

engines are gimbaled by hydraulic actuators. Roll attitude control on the S-IVB stage cannot, of course, be controlled with a single engine. Therefore, roll control of the S-IVB stage is accomplished by the APS (figure 6-9). During the coast period of the mission, the S-IVB APS is used to control the vehicle attitude in all three axes.

The control system accepts guidance commands from the guidance system. These commands, which are actually attitude error signals, are then combined with measured data from the various control sensors. The resultant output is the command signal to the various engine actuators and APS nozzles.

The final computations (analog) are performed within the flight control computer. This computer is also the central switching point for command signals. From this point, the signals are routed to their associated active stages and to the appropriate attitude control devices.

NAVIGATION AND GUIDANCE COMPONENTS.

ST-124M-3 Inertial Platform Assembly.

The gimbal configuration of the ST-124M-3 offers unlimited free-

Section VI Instrument Unit

dom about the X & Y axes, but is limited to ±45 degrees about its Z axis (vehicle yaw at launch). See figure 6-16.

The gimbal system allows the inertial gimbal rotational freedom. Three single-degree-of-freedom gyroscopes have their input axes aligned along an orthogonal inertial coordinates system; X_1, Y_1, and Z_1 of the inertial gimbal. A signal generator, which is fixed to the output axis of each gyro, generates electrical signals proportional to torque disturbances. These signals are conditioned by the servo electronics and terminate in the gimbal pivot servotorque motors. The servo loops maintain the inner gimbal rotationally fixed in inertial space.

The inner gimbal has three, pendulous, integrating, gyroscopic accelerometers, oriented along the inertial coordinates X_1, Y_1 and Z_1. Each accelerometer measuring head contains a pendulous, single-degree-of-freedom gyro. The speed of rotation of the measuring head is a measure of acceleration along the input axis of the accelerometer. Since acceleration causes the accelerometer shaft to be displaced as a function of time, the shaft position (with respect to a zero reference) is proportional to velocity, and the accelerometer is referred to as an integrating accelerometer.

Vehicle attitude is measured with respect to the inertial platform, using dual speed (32:1) resolvers located at the gimbal pivot points. The outputs of these angle encoders are converted into a digital count in the LVDA.

During prelaunch, the ST-124M-3 platform is held aligned to the local vertical by a set of gas bearing leveling pendulums. The pendulum output is amplified in the platform, and then transmitted to the ground equipment alignment amplifier. The alignment amplifier provides a signal to the torque drive amplifier, and then to the platform gyro torque generator. The vertical alignment system levels the platform to an accuracy of ±2.5 arc seconds.

The azimuth alignment is accomplished by means of a theodolite on the ground and two prisms on the platform; one fixed and one servo driven. The theodolite maintains the azimuth orientation of the movable prism, and the computer computes a mission azimuth and programs the inner gimbal to that azimuth. The laying system has an accuracy of ±5 arc seconds.

At approximately liftoff minus 17 seconds, the platform is released from an earth reference to maintain an inertial reference initiated at the launch point. At this time, the LVDC begins navigation, using velocity accumulations derived from the ST-124M-3 inertial platform.

Platform Electronic Assembly (PEA).

The PEA contains the following circuitry:

a. Amplifiers, modulators, and stabilization networks for the platform gimbal and accelerometer servo loops

b. Relay logic for signal and power control

c. Amplifiers for the gyro and accelerometer pick-off coil excitation

d. Automatic checkout selection and test circuitry for servo loops

e. Interlocks for the heaters and gas supply circuits.

ST-124M-3 AC Power Supply.

The ST-124-M3 platform ac power supply furnishes the power required to run the gyro rotors and provides excitation for the platform gimbal synchros. It is also the frequency source for the resolver chain references and for gyro and accelerometer servo systems carrier.

The supply produces a three-phase (sine wave) output which is fixed at 26.5 V (rms) line-to-line at a frequency of 400 Hz. Three single-phase, 20-V reference outputs (square wave) of 4.8 kHz, 1.92 kHz, and 1.6 kHz are also provided. With a normal input voltage of 28 Vdc, the supply is capable of producing a continuous 250-VA output.

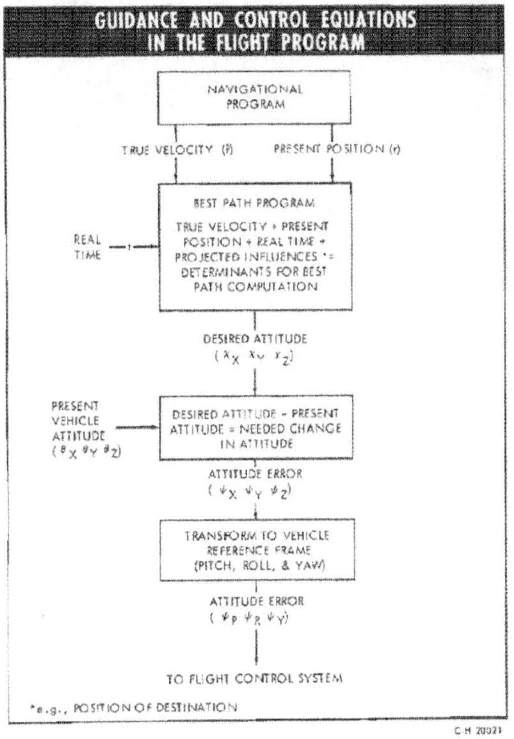

Figure 6-14

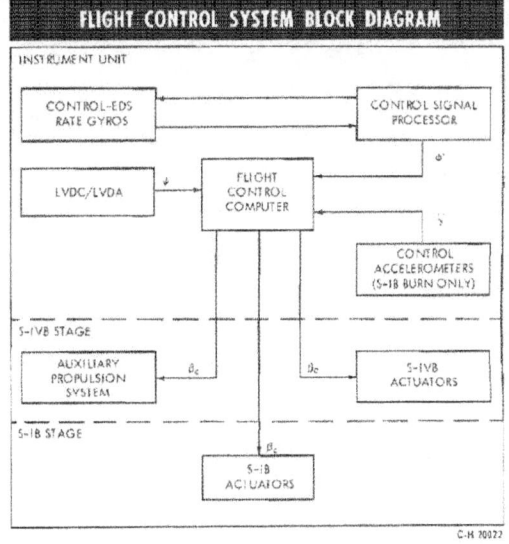

Figure 6-15

Section VI Instrument Unit

PLATFORM GIMBAL CONFIGURATION

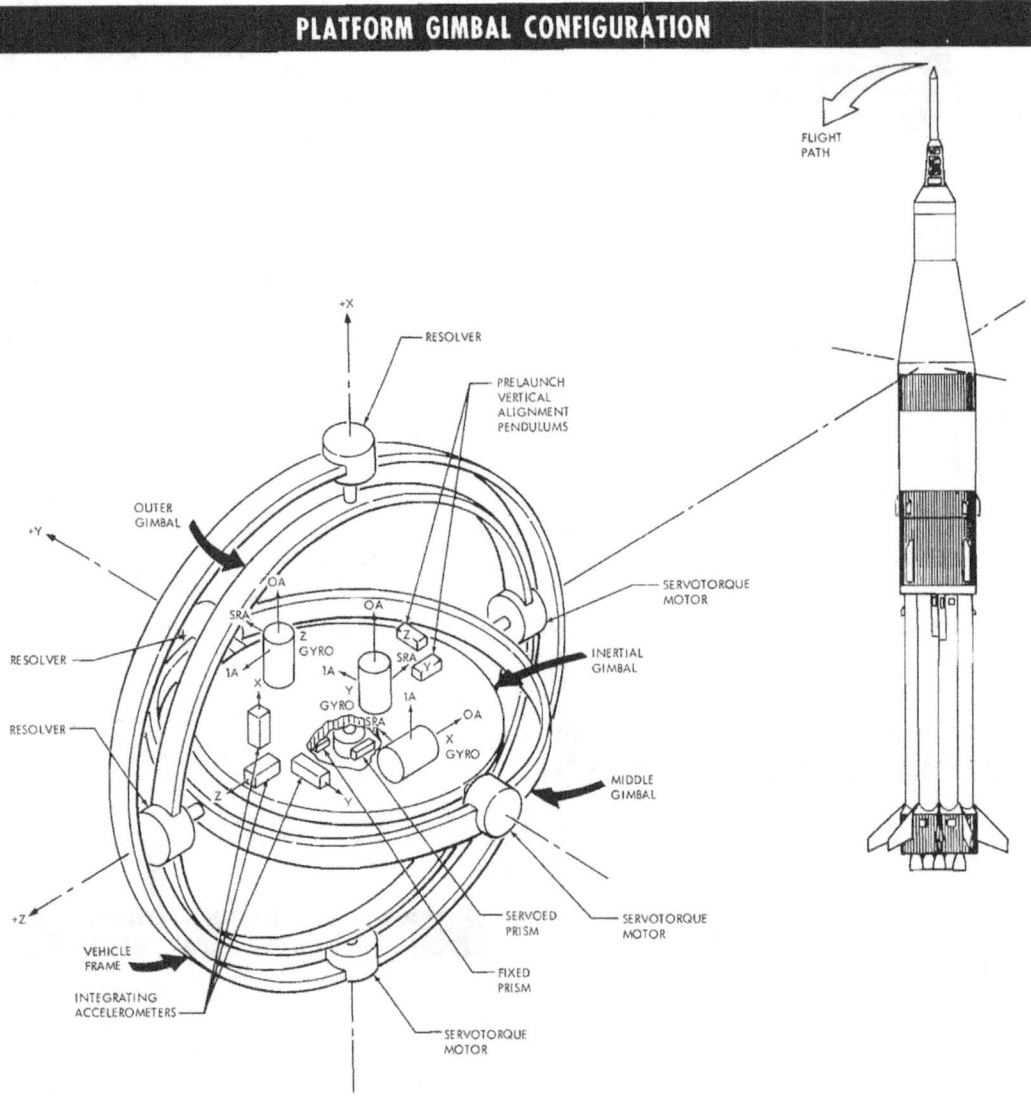

Figure 6-16

Accelerometer Signal Conditioner.

The accelerometer signal conditioner accepts the velocity signals from the accelerometer optical encoders and shapes them before they are passed on to the LVDA/LVDC. Additional outputs are provided for telemetry and ground checkout.

LV Digital Computer and LV Data Adapter.

The LVDC and LVDA form an electronic digital computer system. The LVDC is a relatively high-speed computer with the LVDA serving as its input/output device. Any signal to or from the computer is routed through the LVDA. See figure 6-17 and 6-18 for LVDC and LVDA characteristics.

The LVDA and LVDC are involved in four main operations:

a. Prelaunch checkout

b. Navigation and guidance computations

c. Vehicle sequencing

d. Orbital checkout

The LVDC is a general purpose computer which processes data under control of a stored program. Data is processed serially in two arithmetic functional areas which can, if so programmed, operate concurrently. Addition, subtraction, and logical extractions are performed in one arithmetic functional area while multiplication and division are performed in the other.

Section VI Instrument Unit

The principal storage device is a random access, ferrite-core memory with separate controls for data and instruction addressing. The memory can be operated in either a simplex or duplex mode. In duplex operation, memory modules are operated in pairs with the same data being stored in each module. Readout errors in one module are corrected by using data from its mate to restore the defective location. In simplex operation, each module contains different data, which doubles the capacity of the memory. However, simplex operation decreases the reliability of the LVDC, because the ability to correct readout errors is sacrificed. The memory operation mode is program controlled. Temporary storage is provided by static registers, composed of latches, and by shift registers, composed of delay lines and latches.

Computer reliability is increased within the logic sections by the use of triple modular redundancy. Within this redundancy scheme, three separate logic paths are voted upon to correct any errors which develop.

CONTROL SUBSYSTEM COMPONENTS.

The control subsystem for the Saturn IB vehicles is composed of the control accelerometers in addition to the control/EDS rate gyros, the control signal processor, and the flight control computer.

Control Accelerometers.

The body mounted control accelerometers provide lateral acceleration measurements along the vehicle pitch and yaw axes. This includes sensing the tangential component of rotational acceleration about the pitch and yaw axes to help minimize the angle of attack during the early portion of S-IB Burn.

The S-IB control law for the thrust vector deflection angle (β) is:

$$\beta = A_0 \Psi + A_1 \dot{\phi} + g_2 \ddot{Y}$$

where $\ddot{Y}$ is the lateral acceleration along the pitch or yaw axis as designated, and g2 is the corresponding gain factor. The control accelerometers sensitive axes are perpendicular to the vehicle longitudinal axis.

The lateral acceleration control is used during S-IB stage propulsion to reduce structural loads from aerodynamic forces and to provide minimum-drift control.

Control/EDS Rate Gyros.

The vehicle angular rate error signals ($\dot{\phi}_p, \dot{\phi}_r, \dot{\phi}_s$) are supplied by the control/EDS rate gyro package through the control signal processor. The outputs are provided as redundant signals and are the attitude angular rates of the vehicle about it's pitch, roll, and yaw axes.

The angular rate error signals control the rate of response of the vehicle to the Beta command and eliminate overshoot at the end of the correction (reference the $A_1 \dot{\phi}$ term given in the control formula in the previous paragraph on Control Accelerometers).

Control Signal Processor.

The control signal processor demodulates the ac signals from the control-EDS rate gyros into dc signals, required by the flight control computer. The control signal processor compares the output signals from the triple redundant gyros and selects one each of the pitch, yaw, and roll signals for the flight control computer. The control signal processor supplies the control-EDS rate gyro package with the necessary control and reference voltages. EDS and DDAS rate gyro monitoring signals also originate within the control signal processor, thus accounting for the EDS portion of the control-EDS rate gyro name.

Flight Control Computer.

The flight control computer is an analog computer which converts

LVDC CHARACTERISTICS

ITEM	DESCRIPTION
TYPE	GENERAL PURPOSE, DIGITAL, STORED PROGRAM
MEMORY	RANDOM ACCESS, FERRITE (TORODIAL) CORE, WITH A CAPACITY OF 32,768 WORDS OF 28 BITS EACH.
SPEED	SERIAL PROCESSING AT 512,000 BITS PER SECONDS
WORD MAKE-UP	MEMORY = 28 BITS, INCLUDING 2 PARITY BITS DATA = 26 BITS INSTRUCTION = 13 BITS
PROGRAMMING	18 INSTRUCTION CODES 10 ARITHMETIC 6 PROGRAM CONTROL 1 INPUT/OUTPUT 1 STORE
TIMING	COMPUTER CYCLE = 82.02 μ SEC. BIT TIME = 1.95 μ SEC. CLOCK TIME = 0.49 μ SEC.
INPUT/OUTPUT	EXTERNAL, PROGRAM CONTROLLED

Figure 6-17

LVDA CHARACTERISTICS

ITEM	DESCRIPTION
INPUT/OUTPUT RATE	SERIAL PROCESSING AT 512,000 BITS PER SECOND
SWITCH SELECTOR	8 BIT INPUT 15 BIT OUTPUT
TELEMETRY COMMAND RECEIVER	14 BITS FOR INPUT DATA
DATA TRANSMITTER	38 DATA AND IDENTIFICATION BITS PLUS VALIDITY BIT AND PARITY BIT
COMPUTER INTERFACE UNIT	15 BITS ADDRESS PLUS 1 DATA REQUEST BIT 10 BITS FOR INPUT DATA PLUS 1 BIT FOR DATA READY INTERRUPT
DELAY LINES	3 FOUR-CHANNEL DELAY LINES FOR NORMAL OPERATION 1 FOUR-CHANNEL DELAY LINE FOR TELEMETRY OPERATIONS
OUTPUT TO LAUNCH COMPUTER	41 DATA AND IDENTIFICATION BITS PLUS DISCRETE OUTPUTS
INPUT FROM RCA-110 GCC	14 BITS FOR DATA PLUS INTERRUPT

Figure 6-18

attitude correction commands (ψ), angular change ($\dot{\phi}$), and vehicle lateral acceleration ($\ddot{Y}$) information (during S-IB burn mode only) into APS thruster nozzle and/or engine actuator positioning commands.

Input signals to the flight control computer include:

a. Attitude correction commands (ψ) from the LVDC/LVDA or spacecraft.

b. Angular rates ($\dot{\phi}$) from the control-EDS rate gyro package, via the control signal processor.

Section VI Instrument Unit

c. Vehicle lateral accelerations ($\ddot{y}$) from the IU-mounted control accelerometers (during S-IB burn only).

Output signals from the flight control computer include:

a. Command signals to the engine actuators (βc)

b. Command signals to the APS thruster nozzles (βc)

c. Telemetry outputs which monitor internal operations and functions.

FLIGHT PROGRAM.

The flight program which is structured modularly, is composed of two basic subsystems: the control subsystem and the application subsystem. The control subsystem controls the sequence and order of execution of all programmed functions in the application subsystems.

The control subsystem consists of the control program and a common communications area for inter-module communications, common data and indicators, and mission or vehicle dependent parameters. The control program, composed of sub-programs and tables, controls the execution of application modules, services interrupts, and routes control to the appropriate application module on a priority basis, and provides utility operations.

The application subsystem provides a master pool of all defined LVDC functions which can be used to generate a total flight program configuration for any given mission task. It is composed of a collection of relatively independent, closed program modules. Each application module is designed to perform one or a number of related functions. These functions include navigation, guidance, attitude control, event sequencing, redundancy and data management, ground command processing, and hardware evaluation.

For purposes of discussion, the flight program is divided into five subelements: the powered flight major loop, the orbital flight program, the minor loop, interrupts, and telemetry. There is also an LVDC pre-flight program which supports launch vehicle checkout and performs the functions that prepare the LVDC for entering the flight mode.

Prelaunch and Initialization.

Until just minutes before launch, the LVDC is under control of the ground control computer (GCC). At approximately T-10 minutes, the GCC issues a prepare-to-launch (PTL) command to the LVDC. The PTL routine performs the following functions:

a. Monitors accelerometer inputs, calculates the platform-off-level indicators, and telemeters accelerometer outputs and time

b. Performs reasonableness checks on particular discrete inputs and alerts

c. Interrogates the LVDC error monitor register

d. Keeps all flight control system ladder outputs zeroed, which keeps the engines in a neutral position for launch

e. Processes the GRR interrupt and transfers LVDC control to the flight program

f. Samples platform gimbal angles.

At T-22 seconds, the launch sequencer issues a GRR alert signal to the LVDC and GCC. At T-17 seconds, a GRR interrupt signal is sent to the LVDC and GCC. With the receipt of this signal, the PTL routine transfers control of the LVDC to the flight program.

When the GRR interrupt is received by the LVDC, the following events take place:

a. The LVDC sets time base zero (T_0)

b. Gimbal angles and accelerometer values are sampled and stored for use by flight program routines.

c. Time and accelerometer readings are telemetered

d. All flight variables are initialized

e. The GCC is signaled that the LVDC is under control of the flight program.

During the time period between GRR and liftoff, the LVDC begins to perform navigational calculations and process minor loops. At liftoff time base 1 (T_1) is initiated.

Powered Flight Major Loop.

The major loop contains the navigation and guidance calculations, timekeeping, and other repetitive operations of the flight program that do not occur on an interrupt basis. Its various routines are subdivided by function. Depending upon mode of operation and time of flight, the program will follow the appropriate sequence of routines.

The accelerometer processing routine accomplishes two main objectives: it accumulates velocities as measured by the platform and detects velocity measurement errors through "reasonableness" tests.

The boost navigation routine combines gravitational acceleration with measured platform data to compute position and velocity.

The "pre-iterative" guidance mode, or "time-tilt" guidance program, is that part of the flight program which performs from liftoff until the end of the S-IB burn. The guidance commands issued during the time-tilt phase are functions of time and engines out only. This phase of the program is referred to as open loop guidance, since vehicle dynamics do not affect or influence the guidance commands. When the launch vehicle has cleared the mobile launcher, the time-tilt program then initiates a roll maneuver to align the vehicle with the proper azimuth and a time-tilt pitch maneuver. Time-tilt yaw guidance commands are computed as a tabular function of time. The roll maneuver is completed at approximately T + 38 seconds while the pitch and yaw time-tilts continue until the guidance commands are frozen (tilt arrest) prior to start of Time Base 2 (Low Level Sense) and Inboard Engine cutoff.

Provisions are made for early engine out guidance modifications for the first detected engine failure only. In Time Base 1, this consists of freezing the pitch guidance command for a specified length of time and modification of the time-tilt computations. In addition, certain other parameter adjustments are made. In Time Base 2, the only change is in the value of the sine function in the zero test computations for outboard engine failure. Time base 3 (T_3) commences where the outboard engine thrust delay is sensed following cutoff.

The iterative guidance mode (IGM) routine, or "path adaptive" guidance, commences approximately 35 seconds after S-IVB stage ignition, and continues until the end of the S-IVB burn. Cutoff is commanded when the velocity required for the target orbit has been reached. IGM employs optimizing techniques, based on the calculus of variations, to determine a minimal propellant flight path which satisfies mission requirements. Since the IGM reacts to vehicle dynamics, it is referred to as closed loop guidance.

Orbital Flight Program.

The orbital flight routines consist of an executive routine, telemetry time-sharing routines to be employed while the vehicle is over receiving stations, navigation, guidance, and timekeeping computations.

When in orbital mode, the flight program will process telemetry acquisition and loss determination once per eight seconds, event sequencing and interrupt processing as required, in addition to orbital navigation, guidance, and control. Other functions include minor loop, ten per second; minor loop support and discrete processing, once per second; and gimbal angle read for orbital navigation, once every four seconds (orbital navigation parameters are calculated once per eight seconds).

Minor Loop.

The minor loop contains control system computations. Since the minor loop is used for vehicle control, minor loop computations are executed at the rate of 25 times per second during the powered phase of flight. However, in earth orbit, a rate of only ten executions per second is required for satisfactory vehicle control. Rate limiting of the output commands prevents the flight control systems from maneuvering the LV at rates that exceed safe limits.

The supporting control functions of computing attitude change increments and coefficients for gimbal-to-body transformation required for attitude error command computations are called minor loop support functions. These functions must be performed once per computation cycle during boost and once per second during orbit.

Interrupts.

An interrupt routine permits interruption of the normal program operation to free the LVDC for priority work, and may occur at any time within the program sequence. When an interrupt occurs, the interrupt transfers LVDC control to a special subroutine which identifies the interrupt source, performs the necessary subroutines, and then returns to the point in the program where the interrupt occurred. Figure 6-19 lists the LVDC interrupts.

Telemetry Routine.

A programmed telemetry feature is also provided as a method of monitoring LVDC and LVDA operations. The telemetry routine transmits specified information and data to the ground via IU telemetry equipment. In orbit, telemetry data must be stored at times when the vehicle is not within range of a ground receiving station. This operation is referred to as data compression. The stored data is transmitted on a time-shared basis with real-time telemetry when the LV is within range of a station.

DISCRETE BACKUPS.

Certain events, are particularly important to the flight program since they are time base references. These events are indicated to the flight program by primary and backup signal paths. The primary signals are recognized as program interrupts; the backup signals are treated as periodically serviced discretes.

Because switch selector commands are functions of time (relative to one of the time bases); accurately timed switch selector commands could not be generated if one of the time-base-initiated discrete signals were missed.

The execution time for any given major loop, including minor loop computations and interrupts, is not fixed because the number of modules executed is dependent on the flight mode, the discrete and interrupt processing requirement is not predictable, and the number of minor loops in each major loop is not fixed.

MODE AND SEQUENCE CONTROL.

Mode and sequence control involves most of the electrical/electronic systems in the launch vehicle. However, in this section, the discussion will deal mainly with the switch selectors and associated circuitry.

The LVDC memory contains a predetermined number of sets of instructions which, when initiated, induce portions of the launch vehicle electrical/electronic systems to operate in a particular mode.

LVDC INTERRUPTS

INTERRUPT STORAGE REGISTER BIT	LVDC DATA WORD BIT POSITION	FUNCTION
1	11	RCA-110A INTERRUPT
2	10	* S-IB LOW LEVEL SENSORS DRY "A"
3	9	RCA-110A INTERRUPT
4	8	S-IVB ENGINE OUT "B"
5	7	* S-IB OUTBOARD ENGINES CUTOFF "A"
6	6	* MANUAL INITIATION OF S-IVB ENGINE CUTOFF "A"
7	5	GUIDANCE REFERENCE RELEASE
8b	4	COMMAND DECODER INTERRUPT "B"
8a	4	COMMAND DECODER INTERRUPT "A"
9	3	SIMULTANEOUS MEMORY ERROR
10	2	SPARE
11	1	
12	Sign	INTERNAL TO THE LVDC

*TIMES ARE MISSION DEPENDENT

Figure 6-19

Each mode consists of a predetermined sequence of events. The LVDC also generates appropriate discrete signals such as engine ignition, engine cutoff, and stage separation.

Most selection and initiation can be accomplished by an automatic LVDC internal command, an external command from ground checkout equipment or IU command system, or by the flight crew in the spacecraft.

The flexibility of the mode and sequence control scheme is such that no hardware modification is required for mode and flight sequence changes. The changes are accomplished by changing the instructions and data in the LVDC memory.

Switch Selector.

Many of the sequential operations controlled by the LVDC are performed through a switch selector located in each stage. The switch selector decodes digital flight sequence commands from the LVDA/LVDC and activates the proper relays, either in the units affected or in the stage sequencer.

Each switch selector can activate, one at a time, up to 112 different relays in its stage. The selection of a particular stage switch selector is accomplished through the command code. Coding of flight sequence commands and decoding by the stage switch selectors reduces the number of interface lines between stages and increases the flexibility of the system with respect to timing and sequence. In the launch vehicle, which contains three switch selectors, up to 336 different functions can be controlled, using only 28 lines from the LVDA. Flight sequence commands may be issued at time intervals as short as 100 milliseconds.

To maintain power isolation between vehicle stages, the switch selectors are divided into sections. The input sections (relay circuits) of each switch selector receive their power from the IU. The output sections (decoding circuitry and drivers) receive their power from the stage in which the switch selector is located. The inputs and outputs are coupled together through a diode matrix. This matrix decodes the 8-bit input code, and activates a transistorized output driver, thus producing a switch selector output.

The output signals of the LVDA switch selector register, with the exception of the 8-bit command, are sampled at the control distributor in the IU and sent to IU PCM telemetry. Each switch selector also provides three outputs to the telemetry system within its stage.

The switch selector is designed to execute flight sequence commands given by the 8-bit code or by its complement. This feature increases reliability and permits operation of the system, despite certain failures in the LVDA switch selector register, line drivers, interface cabling, or switch selector relays.

The flight sequence commands are stored in the LVDC memory, and are issued by the flight program. When a programmed input/output instruction is given, the LVDC loads the 15-bit switch selector register with the computer data.

The switch selector register, bits 1 through 8, represents the flight sequence command. Bits 9 through 13 select the switch selector to be activated. Bit 14 resets all the relays in the switch selectors in the event data transfer is incorrect, as indicated by verification fault information received by the LVDA. Bit 15 activates the addressed switch selector for execution of the command. The switch selector register is loaded in two passes by the LVDC: bits 1 through 13 on the first pass, and either bit 14 or bit 15 on the second pass, depending on the feedback code. The LVDA/LVDC receives the complement of the code after the flight sequence command (bits 1 through 8) has been picked up by the input relays of the switch selector. The feedback (verification information) is returned to the LVDA, and compared with the original code in the LVDC. If the feedback agrees, the LVDC/LVDA sends a read command to the switch selector. If the verification is not correct, a reset

SWITCH SELECTOR FUNCTIONAL CONFIGURATION

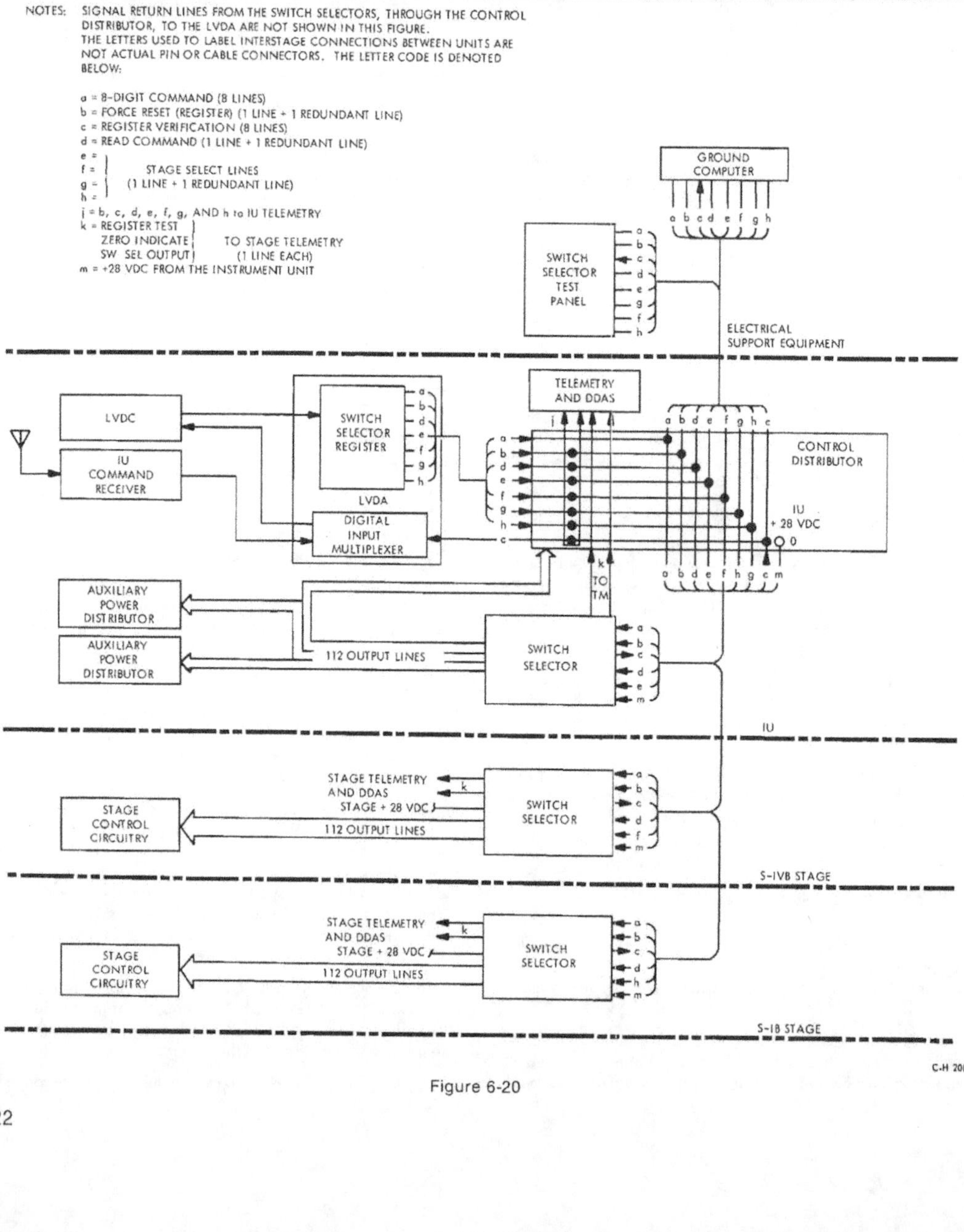
NOTES: SIGNAL RETURN LINES FROM THE SWITCH SELECTORS, THROUGH THE CONTROL DISTRIBUTOR, TO THE LVDA ARE NOT SHOWN IN THIS FIGURE.
THE LETTERS USED TO LABEL INTERSTAGE CONNECTIONS BETWEEN UNITS ARE NOT ACTUAL PIN OR CABLE CONNECTORS. THE LETTER CODE IS DENOTED BELOW:
a = 8-DIGIT COMMAND (8 LINES)
b = FORCE RESET (REGISTER) (1 LINE + 1 REDUNDANT LINE)
c = REGISTER VERIFICATION (8 LINES)
d = READ COMMAND (1 LINE + 1 REDUNDANT LINE)
e = f = g = h = STAGE SELECT LINES (1 LINE + 1 REDUNDANT LINE)
j = b, c, d, e, f, g, AND h to IU TELEMETRY
k = REGISTER TEST, ZERO INDICATE, SW SEL OUTPUT TO STAGE TELEMETRY (1 LINE EACH)
m = +28 VDC FROM THE INSTRUMENT UNIT
GROUND COMPUTER
SWITCH SELECTOR TEST PANEL
ELECTRICAL SUPPORT EQUIPMENT
LVDC
IU COMMAND RECEIVER
SWITCH SELECTOR REGISTER
LVDA
DIGITAL INPUT MULTIPLEXER
TELEMETRY AND DDAS
CONTROL DISTRIBUTOR
IU + 28 VDC
TO TM
AUXILIARY POWER DISTRIBUTOR
AUXILIARY POWER DISTRIBUTOR
112 OUTPUT LINES
SWITCH SELECTOR
IU
STAGE TELEMETRY AND DDAS
STAGE + 28 VDC
STAGE CONTROL CIRCUITRY
112 OUTPUT LINES
SWITCH SELECTOR
S-IVB STAGE
STAGE TELEMETRY AND DDAS
STAGE + 28 VDC
STAGE CONTROL CIRCUITRY
112 OUTPUT LINES
SWITCH SELECTOR
S-IB STAGE
C-H 20027

Figure 6-20

Section VI Instrument Unit

LVDC SWITCH SELECTOR INTERCONNECTION DIAGRAM

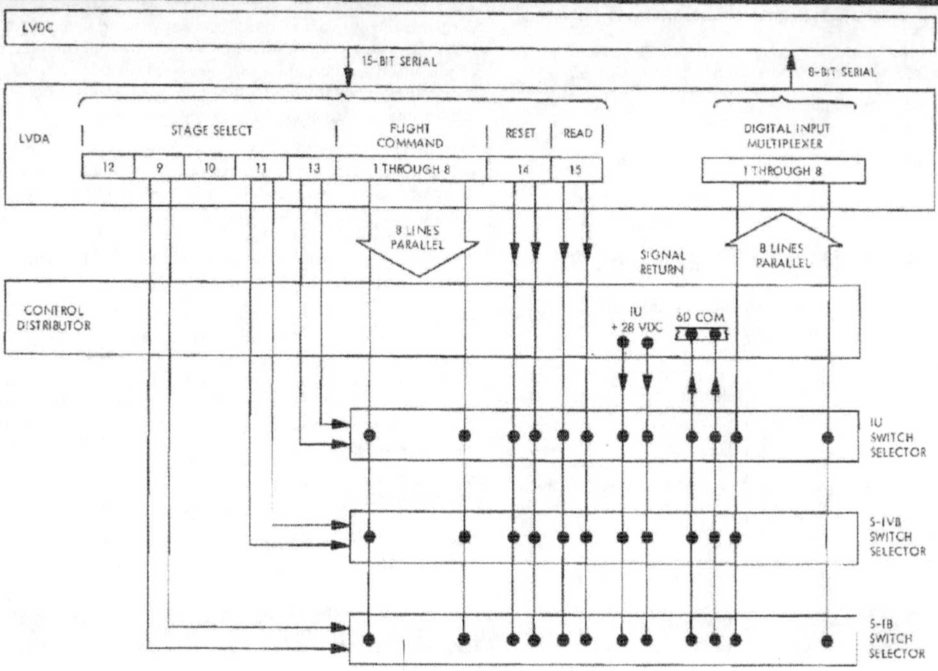

Figure 6-21

command is given (forced reset), and the LVDC/LVDA reissues the 8-bit command in complement form.

Figure 6-20 illustrates the Saturn IB switch selector functional configuration. All switch selector control lines are connected through the control distributor in the IU to the LVDC and the electrical support equipment.

The LVDC switch selector interconnection diagram is shown in figure 6-21. All connections between the LVDA and the switch selectors, with the exception of the stage select inputs, are connected in parallel.

Operation Sequence.

The Saturn IB operation sequence starts during the prelaunch phase at approximately T-14 hours, when the electrical power from the ground support equipment is applied to all stages of the launch vehicle. During this time, the sequencing is controlled from the launch control center/mobile launcher complex, utilizing both manual and automatic control to check out the functions of the entire launch vehicle. After the umbilicals are disconnected, the sequencing is primarily controlled by the flight program within the LVDC.

Since flight sequencing is time-phased, the sequencing operation is divided into four primary time bases. Each time base is related to a particular flight event. These time bases are defined in the following paragraphs.

Time Base No. 1 (T_1). T_1 is initiated by either of the liftoff signals provided by deactuation of the liftoff relays in the IU at umbilical disconnect. Redundancy for this function is provided by three wires through the umbilical supplying ground power to the three disconnect relays. At IU umbilical disconnect, interruption of any two of these three signals will set T_1. If the time since GRR is greater than or equal to 17.4 seconds and less than 150 seconds and either or both liftoff signals are present, T_1 will be set, indicating liftoff has occurred. See figure 6-22 for a logic diagram of these functions.

No "negative backup" (i.e., provisions for the LVDC to return to prelaunch conditions) is provided because, in the event T_1 began by error, the launch vehicle could safely complete T_1 on the pad without catastrophic results.

Time Base No. 2 (T_2). After arming the S-IB propellant level sensors through the S-IB switch selector, the LVDC will initiate time base No. 2 (T_2) upon receiving either of two redundant fuel or LOX level sensor signals, if sufficient downrange velocity exists at that time. However, if Guidance Reference Failure (GRF) has occurred, the LVDC will bypass the velocity test and initiate time base No. 2.

Use of the downrange velocity reading provides a safeguard against starting T_2 on the pad, should T_1 be started without liftoff. Furthermore, if T_2 is not established, no subsequent time bases can be started. This ensures a safe vehicle, requiring at least one additional failure to render the vehicle unsafe on the pad.

Time Base No. 3 (T_3). After arming (Tops grouping) the S-IB LOX Depletion Sensors through the S-IB switch selector, the LVDC shall initiate time base No. 3 upon receiving either of two redundant outboard cutoff signals. The S-IB outboard engines cutoff "A" signal (INT 5) is the primary signal, with the S-IB outboard engines cutoff "B" (DI23) as the backup signal, for starting T_3.

Time Base No. 4 (T_4). Anytime after T_3 + 10 seconds, any of the following four combinations of events will start T_4.

6-23

Section VI Instrument Unit

a. Both S-IVB engine out "A" and "B" indications from the thrust OK pressure switch when the S-IVB engine shuts down.

b. Either engine out indication and the velocity cutoff command.

c. Either engine out indication and a velocity change indication from the LVDC signaling a loss of thrust.

d. The velocity cutoff command and LVDC indication of loss of thrust.

Redundant S-IVB cutoff commands are issued at the start of T_4 to ensure against starting this time base with S-IVB engine thrust present.

MEASUREMENT AND TELEMETRY.

Measurement and telemetry instrumentation within the IU consists of a measuring subsystem, a telemetry subsystem, and an antenna subsystem. This instrumentation monitors certain conditions and events which take place within the IU and transmits monitored signals to ground receiving stations. Telemetry data is used on the ground for the following purposes:

a. Preflight checkout of the launch vehicle.

b. During vehicle flight, for immediate determination of vehicle condition and for verification of commands received by the IU command system.

c. Postflight scientific analysis of the mission.

MEASUREMENTS.

The requirement for measurements of a wide variety has dictated the use of many types of transducers at many different locations. However, a discussion of each transducer type is beyond the scope of this manual. The parameters measured include such things as acceleration, angular velocity, flow rate, position, pressure, temperature, voltage, current, and frequency.

Conditioning of measured signals is accomplished by amplifiers or converters located in measuring racks. Each measuring rack has a capacity of 20 signal conditioning modules. In addition to its conditioning circuitry, most of the signal conditioning modules contain the capability for simulating the transducer inputs. This capability is used for prelaunch calibration of the signal conditioners.

Measurement signals are generally routed to their assigned telemetry channel by the measuring distributors. The measuring distributors contain the capability to connect different measurements to the same telemetry channels during different flight periods. Selected FM measurements are switched to digital data acquisition system (DDAS) channels for ground checkout and then returned to the FM link after checkout.

TELEMETRY.

The function of the IU telemetry (TM) system is to format and transmit measurement signals. See figure 6-23 for a diagram of

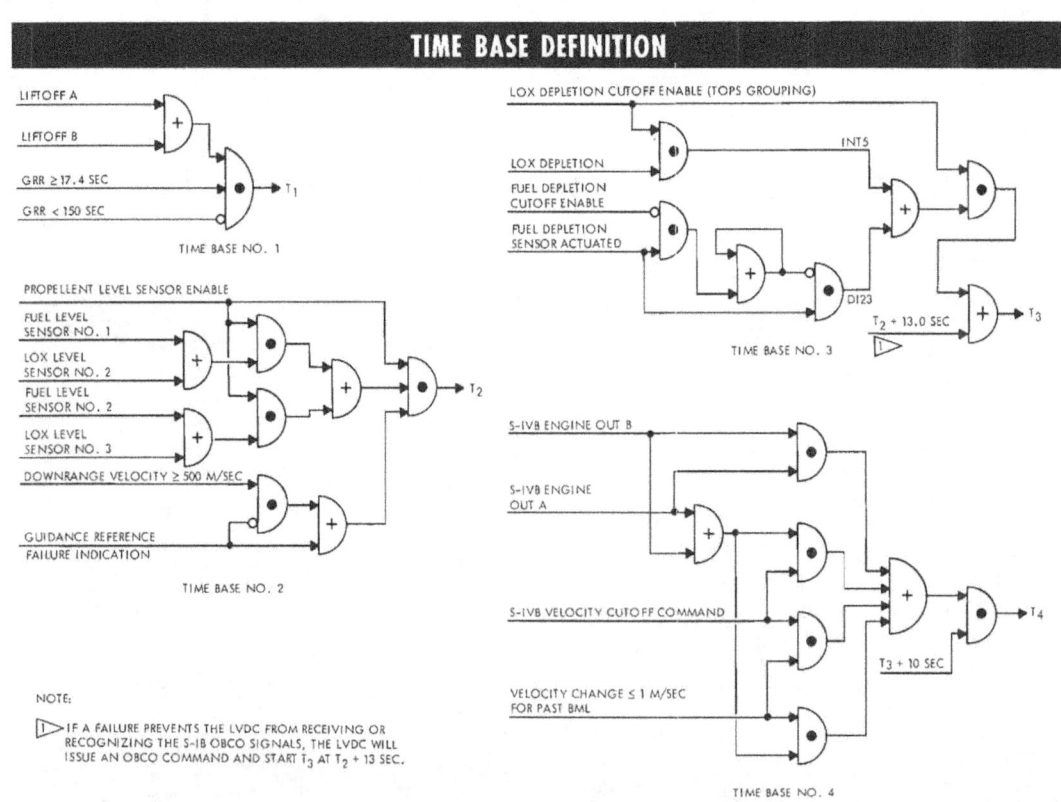

Figure 6-22

the IU TM system. The approximately 225 measurements taken on the IU are transmitted via two TM links as follows:

a. Link P1, Pulse Code Modulation/Frequency Modulation (PCM/FM).

b. Link F1, Frequency Modulation/Frequency Modulation (FM/FM).

Multiplexing.

To enable the two TM links to handle the approximately 225 measurements, both frequency sharing and time sharing multiplexing techniques are used.

Two Model 270 time sharing multiplexers are used in the IU telemetry system. Each one operates as a 30 x 120 (30 primary channels, each sampled 120 times per second) multiplexer with 23 of the primary data channels containing provisions for being submultiplexed. Each sub-multiplexed primary channel forms ten sub-channels, each sampled at 12 times per second. Twenty-seven of the 30 primary channels are used for measurement data, while the remaining three are used for references.

The Model 270 also has an integral calibration generator for inflight calibration capability. Upon command from the TM calibrator, the calibration generator seeks the next available master frame, inhibits the normal measurement data input, and applies a sequence of five calibration voltages to all data channels. Each voltage level is sustained for one master frame, thus requiring five frames or approximately 400 milliseconds for a complete calibration sequence.

Two Model 410 remote digital multiplexers are used in the IU TM system. The Model 410 can accept up to 100 inputs in the form of discretes, digital data, or a combination. This data is temporarily stored as 10-bit digital words and then transferred to the Model 301 programmed format.

Low level conditioned analog signals are fed to subcarrier oscillators (SCO) in the F1 TM oscillator assembly (Model B1). The Model B1 has the capability of handling 27 continuous data channel inputs by utilizing IRIG channels 14 and 17 for FM/FM/FM. Each input signal is applied to a separate SCO, and each SCO produces a different output frequency. The SCO outputs are combined and the composite signal frequency modulates the F1 RF assembly.

The PCM/FM system performs a dual function. During flight, it serves as a TM link; during prelaunch checkout, it serves as an interface with the digital GSE. PCM techniques provide the high degree of accuracy required for telemetering certain signal types. The Model 301 unit accepts and digitalizes analog inputs from the Model 270 and serializes these signals with digital signals from the Model 410 and direct discrete inputs. The Model 301 output is a serial train of digital data, which modulates the RF assembly. During prelaunch checkout, the Model 301 output also feeds through coaxial cable to the GSE.

Both TM RF assemblies are VHF/FM transmitters and use combinations of solid state and vacuum tube electronics. The transmitter outputs couple into a single antenna system containing two omni directional antennas.

The PCM/FM systems of the IU and S-IVB are cross-strapped to provide a redundant transmission path for flight control and other critical measurements. This arrangement routes the data from one IU Model 270 to the S-IVB Model 301, while the data from one of the S-IVB Model 270's is routed to the IU Model 301.

COMMAND SYSTEM.

The IU command system is used to transmit digital information from ground stations to the Launch Vehicle Digital Computer (LVDC) in the IU. Figure 6-24 is a block diagram of the command system for Saturn IB.

The commands and data to be transmitted to the vehicle originate in the Mission Control Center (MCC) in Houston, Texas, and are sent to the remote ground stations of the Spaceflight Tracking and Data Network (STDN). At the ground stations, the command messages (in digital form) are processed and temporarily stored. On command from MCC they are modulated onto a 450 MHz RF carrier by a Digital Command System (DCS) and transmitted.

The signal is received and demodulated by the command receiver. The resultant signal is fed to a command decoder where address verification and final message decoding is accomplished. From the command decoder, the command message (in digital form) is sent through the Launch Vehicle Data Adapter (LVDA) to the LVDC. Verification of the acceptance or rejection of the command message is telemetered to the ground station via the IU telemetry system. If a message is rejected, pertinent data concerning the rejection will also be telemetered.

At the ground station, the verification and acceptance data is recovered from the telemetry message and fed to the DCS through special circuitry. The remainder of the telemetry data is forwarded to the MCC. If an acceptance message is not received by the ground station within a preset length of time, the same message will be retransmitted. Presently, a total of four such attempts will be made before the message is abandoned.

The LVDC is programmed to receive two types of command words from the command decoder: mode words and data words.

The LVDC can be programmed to recognize as many as 26 different mode command words. Many of these command words are common to all flights while others are programmed only for particular missions. Common mode commands include:

a. Time base update

b. Navigational update

c. Execute switch selector routine

d. Telemeter memory contents

e. Terminate command routine.

Data words, as the name implies, contain data to supplement mode commands. The number of data words varies with the mode command involved. For example, a time base update requires only one data word while a navigation update requires more than thirty data words.

SATURN TRACKING INSTRUMENTATION.

Radio tracking determines the vehicle's trajectory providing data for mission control, range safety, and postflight evaluation of vehicle performance.

C-BAND RADAR.

The function of the C-band radar transponder is to increase the range and accuracy of the radar ground stations equipped with AN/FPS-16, and AN/FPQ-6 radar systems. C-band radar stations at the Kennedy Space Center, along the Atlantic Missile Range, and at many other locations around the world, provide global tracking capabilities. Two C-band radar transponders are carried in the IU to provide radar tracking capabilities independent of the vehicle attitude. Two antennas (one for each transponder) are

Section VI Instrument Unit

IU TELEMETRY SYSTEM

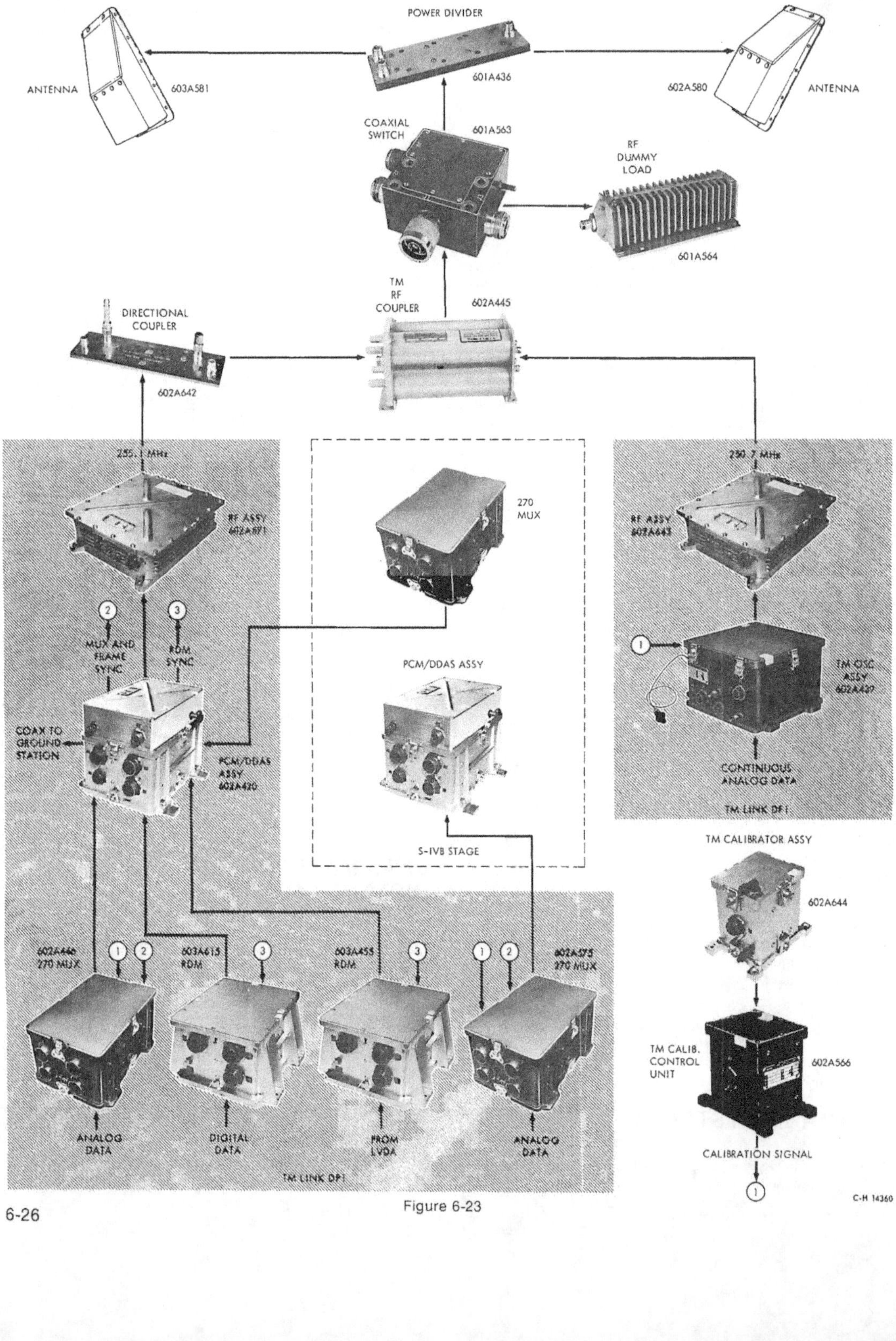

Figure 6-23

6-26

Section VI Instrument Unit

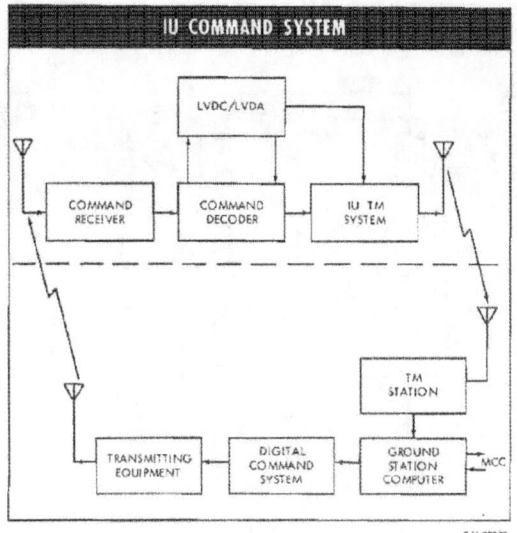

Figure 6-24

located 180° apart on the IU outer surface providing an omni antenna pattern around the vehicle.

The transponder consists of a single compact package. Major elements include an integrated RF head, an IF amplifier, a decoder, over-interrogation protection circuitry, a fast-recovery solid-state modulator, a magnetron, a secondary power supply, and transducers for telemetry channels.

The transponder receives a coded double pulse interrogation from ground stations and transmits a single-pulse reply in the same frequency band.

Two conditioned telemetry outputs are provided to the telemetry system: input signal level and input PRF.

The characteristics of the C-band radar transponder are given in figure 6-25.

GROUND SUPPORT EQUIPMENT.

The IU, because of its complex nature, requires the services of many types of GSE (mechanical, pneudraulic, electrical, electronic) and personnel. This section of the manual is limited to a very brief description of the IU GSE.

There are three primary interfaces between the IU and its GSE. One is the IU access door, used during prelaunch preparations for battery installation, ordnance servicing, servicing IU equipment and S-IVB forward dome. The second interface is the umbilical, through which the IU is furnished with ground power, purging air/GN_2, coolant for environmental control and hardwire links with electrical/electronic checkout equipment. The third interface is the optical window through which the guidance system ST-124M-3 inertial platform is aligned.

IU ACCESS DOOR.

The structure of the IU consists of three 120-degree segments of aluminum honeycomb sandwich, joined to form a cylindrical ring. After assembly of the IU, a door assembly provides access to the electronic equipment inside the structure. This access door has been designed to act as a load supporting part of the structure in flight.

Work platforms, lights, and air-conditioning are used inside the IU to facilitate servicing operations. When the spacecraft is being

C-BAND TRANSPONDER CHARACTERISTICS

RECEIVER CHARACTERISTICS

FREQUENCY (TUNABLE EXTERNALLY)	5690 ± 2 MHz
BANDWIDTH (3 db)	8-12 MHz
OFF-FREQUENCY REJECTION	50 db IMAGE; 80 db MINIMUM, 0.15 TO 10,000 MHz
SENSITIVITY (99% REPLY)	-65 dbm OVER ENTIRE FREQUENCY RANGE AND ALL ENVIROMENTS
MAXIMUM INPUT SIGNAL	+20 dbm
INTERROGATION CODE	DOUBLE PULSE
PULSE WIDTH	0.25 TO 1.0 μSEC (DOUBLE PULSE)
PULSE SPACING	8 ± 0.1 μSEC

TRANSMITTER CHARACTERISTICS

FREQUENCY (TUNABLE EXTERNALLY)	5765 ± 2 MHz
PEAK POWER OUTPUT	400 WATTS MINIMUM, 700 WATTS NOMINAL
PULSE WIDTH	1.0 ± 0.1 μSEC
PULSE JITTER	0.020 μSEC MAXIMUM FOR SIGNALS ABOVE -55 dbm
PULSE RISE TIME (10% TO 90%)	0.1 μSEC MAXIMUM
DUTY CYCLE	0.002 MAXIMUM
VSWR OF LOAD	1.5:1 MAXIMUM
PULSE REPETITION RATE	10 TO 2000 pps; OVERINTERROGATION PROTECTION ALLOWS INTERROGATION AT MUCH HIGHER RATES WITH COUNT-DOWN; REPLIES DURING OVERINTERROGATION MEET ALL REQUIREMENTS

TRANSPONDER CHARACTERISTICS

RECOVER TIME	50 μSEC SINGLE PULSE, 62 μSEC DOUBLE PULSE MAXIMUM FOR INPUT SIGNAL LEVELS DIFFERING BY UP TO 65 db (RECOVERS TO FULL SENSITIVITY WITH NO CHANGE IN TRANSMITTER REPLY POWER OR FREQUENCY WITH MULTIPLE RADARS INTERROGATING SIMULTANEOUSLY)
FIXED DELAY	3.0 ± 0.01 μSEC
DELAY VARIATION WITH SIGNAL LEVEL	50 NANOSECONDS MAXIMUM FROM -65 dbm TO 0 dbm
POWER REQUIREMENTS	24 TO 30 VOLTS
PRIMARY CURRENT DRAIN	0.7 AMPERE STANDBY; 0.9 AMPERE AT 1000 pps

Figure 6-25

Section VI Instrument Unit

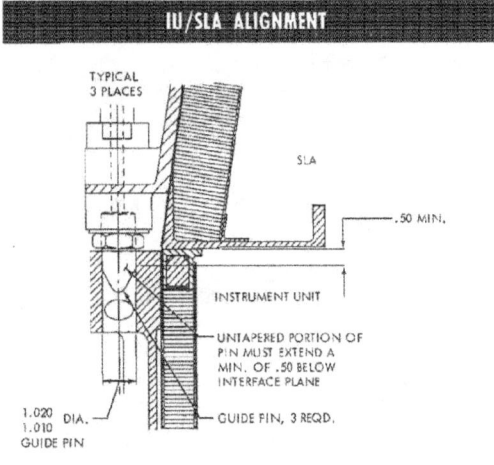

Figure 6-26

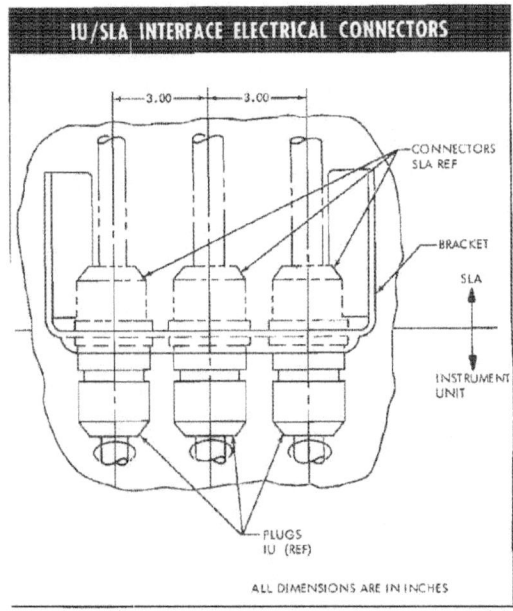

Figure 6-28

IU UMBILICAL.

The physical link between the IU and the GSE is through the umbilical connection, located adjacent to the access door, that mates with the service arm umbilical. The umbilical is made up of numerous electrical connectors, two pneudraulic couplings and an air conditioning duct. The electrical connectors provide ground power and the electrical/electronic signals necessary for prelaunch checkout of the IU equipment. The quick-disconnect couplings provide for circulation of GSE supplied Oronite coolant fluid through the onboard IU/S-IVB ECS cooling system until umbilical disconnect at liftoff. The air conditioning duct provides for compartment cooling air or purging GN_2.

The service arm umbilical is retracted at liftoff and a spring loaded door on the IU closes covering the umbilical plate.

OPTICAL ALIGNMENT.

The IU contains a window through which the ST-124M-3 inertial platform has its alignment checked and corrected by a theodolite located in a hut on the ground and a computer feedback loop. By means of this loop, the launch azimuth can be monitored, updated, and verified to a high degree of accuracy.

IU/SLA INTERFACE.

MECHANICAL INTERFACE.

The IU and spacecraft-LM adapter (SLA) are mechanically aligned with three guide pins and brackets as shown in figure 6-26. These pins facilitate the alignment of the close tolerance interface bolt holes as the two units are joined during vehicle assembly. Six bolts are installed around the circumference of the interface and sequentially torqued using a special MSFC designed wrench assembly.

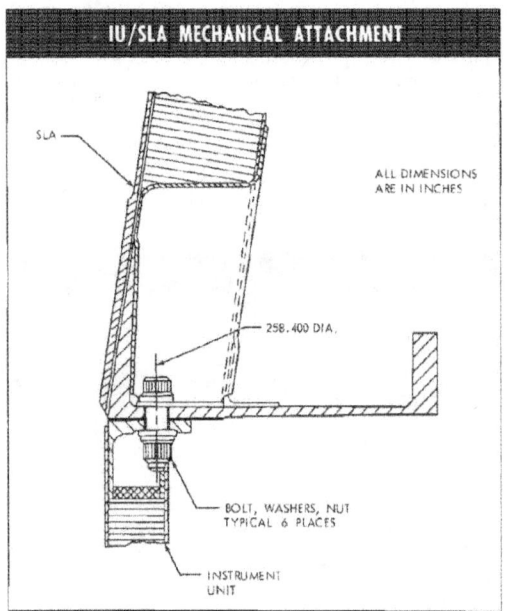

Figure 6-27

fueled through the IU access door, a special protective cover is installed inside the IU to protect components from any possible volatile fuel spillage.

Approximately 20 hours before launch, the IU flight batteries, each weighing 165 pounds, are activated in the battery shop and installed in the IU through the access door.

At approximately L-6 hours, the service equipment is removed and the access door is secured.

These six bolts secure the IU/SLA mechanical interface. (See figure 6-27.)

Electrical Interface.

The electrical interface between the IU and spacecraft consists of three 61-pin connectors. (See figure 6-28.) The definition and function of each connector is presented in the following paragraphs.

IU/Spacecraft Interface Connector J-1.

This connector provides lines for power and for control, indication, and EDS circuitry.

IU/Spacecraft Interface Connector J-2.

This connector provides lines for power, control, and indications for the Q-ball circuitry and the EDS circuitry.

IU/Spacecraft Interface Connector J-3.

This connector provides lines for power and control, indication, and EDS circuitry.

EXPERIMENTS.

M415, THERMAL CONTROL COATINGS.

The objective of Experiment M-415 is to determine the degradation effects of prelaunch, launch, and space environments on the absorptivity/emissivity and stability characteristics of various materials/coatings commonly utilized for passive thermal control. This data will be useful in obtaining correlation of space-environment simulation experiments.

Design Concept.

The experiment objective will be satisfied by subjecting three different coating materials to a series of four exposure conditions on the exterior of the SA-206 IU during pre-launch, launch, and orbital operations. Each of the specimen materials will be mounted on a temperature sensor and will be thermally isolated from surrounding structure. Temperature measurements obtained from the sensors during orbit will allow determination of thermal degradation characteristics. The degradation environments to which the specimen thermal control coating materials will be exposed are categorized as follows:

Pre-Launch Environment. This includes effects of moisture absorption and dust accumulation immediately prior to launch.

Launch Environment. This includes heat and erosion effects during launch ascent as well as effects of retrorocket firing and spacecraft tower jettison.

Space Environment. One set of specimen coatings will be exposed to all environments. A second set will be exposed to S-IB/interstage retrorocket firing, Launch Escape System (LES) jettison and space environment. A third set will be exposed to LES jettison and space environment, while the fourth series will be exposed to space environment only. The test specimens will be located at two positions on the IU to provide two degrees of exposure to retrorocket firing. Aerodynamic fairings will be provided to assure a laminar flow stream for the test specimens during launch ascent.

Equipment Description and Function.

The principal elements of this experiment consist of two sensor panels, each containing 12 thermal sensors arranged in four rows of three. See figure 6-29. Three different thermal control coating specimens are mounted on the sensors in each row, with each column containing the same specimen material. Three of the four sets of test specimens on each panel are protected with covers attached by armament thrusters. The remaining set is protected by a bolt-on cover which will be removed prior to launch. In order to provide a worst-case baseline for data correlation, one specimen on each panel is covered with a black, totally absorbing paint. See figure 6-30.

The two sensor panels will be externally mounted on the IU with one panel directly in line with the center of one of the S-IB/interstage retrorockets for maximum exposure to heat and erosion effects. The second panel is located approximately 51° around the IU from the same retrorocket centerline for peripheral exposure. Aerodynamic fairings are provided at each end of the sensor panels. The experiment will utilize the power, functional command, attitude control and data handling capabilities of the IU.

Operation.

Crew participation is not required for any phase of this experiment. Although the IU telemetry system is operating, the experiment is passive during launch ascent and orbital injection phases of the flight except when the IU digital computer commands the expulsion of the specimen covers by firing the armament thrusters. The experiment functions for the duration of the S-IVB/IU lifetime (9 to 12 hours). Experiment data is obtained after the CSM separates from the IU. The S-IVB/IU is rolled about the X axis so that at local noon the sun-line passes through a point on the IU midway between the two sensor panels. The X axis is maintained along the flight vector, perpendicular to local vertical. This attitude is maintained during data acquisition.

S150, GALACTIC X-RAY MAPPING.

The objectives of experiment S150 are to survey a portion of the sky for X-rays in the 200 to 12,000 electron-volt energy range and determine the intensity, location, and spectrum of each X-ray source. In addition, the experiment will determine if there is a continuous background of X-rays or merely discrete sources. The results of the experiment will help answer questions concerning emission mechanisms in X-ray sources, the distance to the sources, and the nature of the interstellar medium.

This experiment was conceived to bridge the data gap between sounding rockets and long-life satellites in the field of X-ray astronomy.

The experiment will be initiated at the completion of the SA-208 vehicle primary mission and after separation of the command and service module. The IU/S-IVB will be oriented with the experiment assembly (Location 24) nearest the earth. Then the experiment sensor will be deployed (figure 6-31) so that the base plane of the experiment sensor casting makes an angle of 62-1/2° with respect to the IU/S-IVB longitudinal axis, swinging away from its stowed position on a specially designed, hinged bracket assembly. An onboard storage and regulation system will be used to supply P10 gas (10% methane, 90% argon) to the X-ray counters in the experiment assembly. An Auxiliary Storage and Playback (ASAP) package, consisting of five components mounted at Location 23, will record and play back data once per orbit when the vehicle is in communications contact with a ground station. Radioactive sources will make possible calibration of the instruments in flight. Star sensors will help determine the pointing direction of the proportional counters.

IU/S-IVB roll maneuvers will be used to allow viewing a different portion of the sky on each orbit. Guidance requirements dictate that the IU remain active throughout the experiment. Hence, the maximum duration of the experiment is determined by the duration of IU expendables. Other limiting factors might be the amount of P10 gas available or the quantity of thruster fuel, but it is anticipated that the experiment will operate for at least three orbits and possibly for five orbits.

Section VI Instrument Unit

M415 SENSOR PANELS

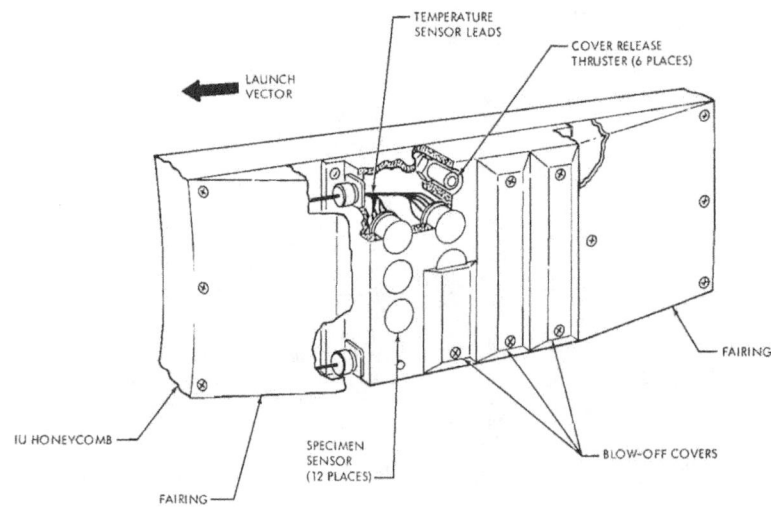

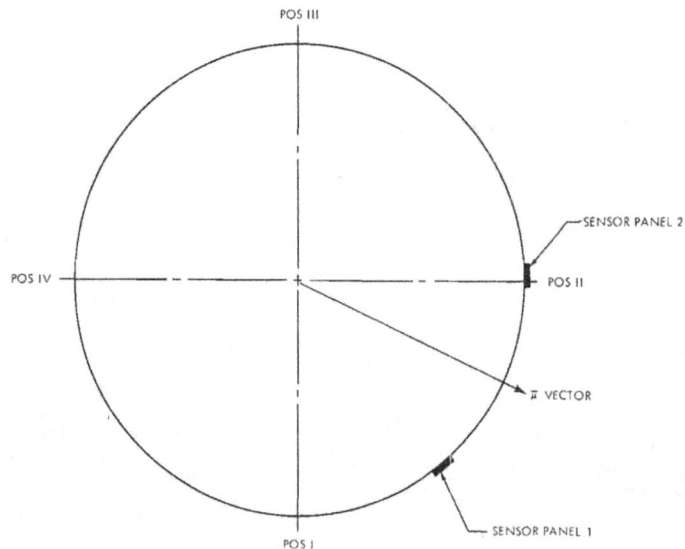

Figure 6-29

Section VI Instrument Unit

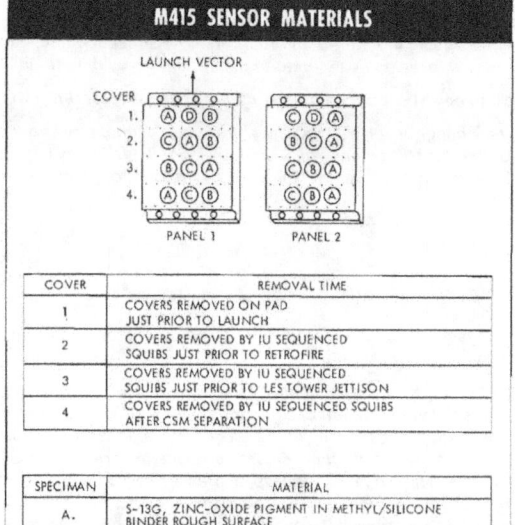

Figure 6-30

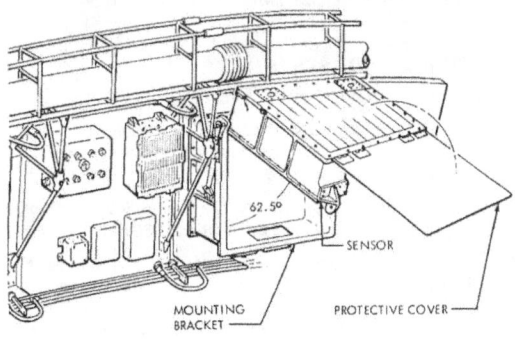

Figure 6-31

MAJOR DIFFERENCES.

The following paragraphs summarize the differences in configuration and function between the Skylab Saturn IB Instrument Units and the Instrument Unit flown on the AS-205 mission.

STRUCTURAL SUBSYSTEM.

Utilize Saturn V ground half umbilical housings to launch Saturn IB vehicles

This configuration will yield the more reliable four-ball-lock mechanism. Implementation of this change allows usage of the Saturn V type housing to launch IB vehicles scheduled to be launched from Saturn V launch complex 39 facilities.

ENVIRONMENTAL CONTROL SUBSYSTEM.

a. Added a 20 micron filter in the inlet port of the gas-bearing regulator

To preclude contamination of the gas-bearing regulator, a 20-micron filter is required.

b. Methanol/water coolant changed to Oronite Flo-Cool 100

Oronite Flo-Cool 100 is less susceptible to corrosive galvanic currents and will reduce the alkaline content which reacts with the cooling system as compared to Methanol/Water.

c. Enlarged volume coolant accumulator

To allow sufficient volume to accommodate Oronite leakage replacement and increased volume due to its expansion, an enlarged volume coolant accumulator is required.

d. Installation of hydraulic snubber assembly (installed between ECS coolant pump outlet and pressure transducer).

Tests have demonstrated that a hydraulic snubber located between the ECS coolant pump outlet and pressure transducer will prevent transducer from responding to dynamic pressure fluctuations from the ECS coolant pump. To extend the life of the pressure transducer by reducing its cycling, a snubber is required.

e. GN_2 storage spheres changed from two 819 cm^3 (50 in^3) to one 2704 cm^3 (165 in^3).

Increased GN_2 volume for longer duration mission. To meet the 6.8-hour mission duration with a safety factor of 1.5, additional GN_2 pressurant is required.

f. Removal of modulating flow control valve, electronic controller assembly, and associated hardware.

Since the deletion of usage of the modulating flow control valve during ground operations and because it is inoperative during flight, the requirement for this hardware is deleted.

g. Delete usage of hazardous gas detection system. With the addition of sample ports to the S-IVB stage for Saturn IB vehicles, the requirement for the IU hazardous gas system is deleted.

h. The IU redundant pump system and related hardware have been incorporated into the design for S-IU-206 through S-IU-212 and S-IU-502 through S-IU-515.

To eliminate the coolant pump as a single point failure, an additional coolant pump is required.

GUIDANCE SUBSYSTEM.

There were no significant changes made directly to the guidance components; however, the following S-IU-206 changes were made that indirectly affect the guidance subsystem.

a. The IU was modified to allow backup guidance commands from the spacecraft platform and computer during first stage burn and manual guidance commands from the spacecraft during second stage burn. In addition, the change imposed the requirement that an IU guidance failure must occur before enabling manual control during boost.

b. The IU was modified to provide dual redundant power inputs to components of the ST-124M-3.

FLIGHT CONTROL SUBSYSTEM.

a. The coolant lines on the S-IU-206 FCC are modified by the addition of an innersleeve.

The innersleeve provides a redundant line in case of a coolant line flare cracking. During QAST of the Saturn V FCC, the flared tubing connecting the front panel coolant ports to the FCC spider cracked in the flared area. Failure analysis revealed that the cracked flares were primarily the result of slow cycle fatigue compounded by high flare stress due to tube misalignment, stress riser due to out-of-tolerance flare radius, or flexing of the flare radius due to normal relative motion between the spider and front panel.

b. Flight control computer spatial filters were added, and also the FCC filters, used in boost flight, were changed to enhance the vehicle dynamic performance.

Analysis of previous flight data, specifically AS-204, indicated that during the S-IVB burn phase a high level ac rate signal was present on the FCC input. The ac signal frequency was approximately 18 Hz and reached an amplitude of 1.75 degrees per second. The overall effect of this condition resulted in a decreased attitude error sensitivity, thus causing the APS engines to fire only with attitude error signals significantly larger than the deadband of the spatial amplifiers.

c. Implement Saturn V backup guidance scheme into Saturn IB FCC.

This change redesigned the FCC inputs so that spacecraft control could be attained during the S-IB burn phase.

The S-IU-205 FCC did not contain the capability for control from the spacecraft during S-IB burn. Additionally, if spacecraft control were energized during S-IVB burn, the error signals were attenuated and limited by the limiter circuits within the FCC.

ECR AAOE-212 established the requirement for manual backup guidance during all modes of flight. In the AS-205 configuration, no backup guidance existed during the S-IB burn mode. This meant an abort condition existed, if the normal IU guidance system had a failure in a non-redundant circuit. The requirement of a gain of 20 during spacecraft control on the roll input DC amplifier is required to make the FCC compatible with spacecraft scale factor for the roll channel.

d. A modification to the FCC inverter detection, to prevent switching during power transfer, was implemented.

The AS-205 inverter detector was sensitive to negative voltage transients on the battery buses feeding the FCC. During the AS-204 and AS-205 countdowns, the inverter detector tripped during the IU power transfer test at T-30 minutes. The trip was caused by the voltage of the battery peaking while unloaded.

The launch mission rules require that the detector be reset at launch. If the detector trips during the last power transfer, a recycle to T-20 minutes is required. If this occurred during the SL-2 countdown, the launch would be slipped to the following day due to the short launch window for rendezvous. Therefore, the detector was redesigned to make it insensitive to voltage transients on the input voltage.

e. Addition of mechanical support to the S-IB FCC center tray

The Saturn IB FCC failed the QAST conducted on November 22, 1971, by IBM. The failure created a loss of 12 electrical signals being monitored on recorders. The subsequent failure analysis revealed five structural discrepancies within the unit.

f. Implement control gains and shaping networks for S-IB FCC

The change involves modifying the gains of these networks for the first 100 seconds of S-IB Stage flight time. The following table shows the "was-is" for each of the control gains:

	Was	Is
a_0	1.55	1.85
a_1	1.70	1.65
g_2	5.0	4.0

The gain change is accomplished by changing resistor values in the networks.

This change was implemented as a result of the AS-206/SL-2 phase II control system design and analysis effort. The report indicated the desirability of increasing the aerodynamic gain margin in the region of maximum aerodynamic pressure (max Q). The gain changes provide a greater aerodynamic gain margin and thereby increase confidence in the ability of the vehicle to accomplish the mission even with severe off-nominal conditions such as an engine out or degradation due to parameter variation.

FLIGHT PROGRAM.

The program has been re-written for the purpose of modularization to make a generalized flight program adaptable to modifications which are mission dependent. Internal functional and mission requirement changes are implemented through changes in processing within the module which performs the function or parameter changes in tables which are processed for the purpose of computation or timing.

The LVDC Equation Defining Document for the Saturn IB Flight Programs should be consulted for information concerning mission dependent requirements as well as general requirements.

ELECTRICAL SUBSYSTEM.

a. S-IB engine cut-off circuitry redesign.

Engine cutoff circuits redesigned to eliminate all single point electrical failures which could cause an inadvertent multiple engine cutoff. Also improved reliability and range safety cutoff.

b. Implement redundancy to enhance probability of completing the prime mission in case of 6D10 or 6D30 battery failure.

The 6D10 battery provided the only source of power for the switch selector and certain control functions. In addition, switch point power for the FCC was supplied only by +6D31 bus. In the event of battery failure, the following unsatisfactory conditions could have developed:

Spacecraft control of Saturn could not be attained

Switch selector commands would not be issued

S-IB outboard engines cutoff, S-IB level sensors dry, and S-IVB engine out "A" and "B" discretes and interrupts into the LVDA would not be issued

The flight control computer switch points were dependent on +6D31 power and loss of the +6D30 battery would result in incorrect FCC channel gains

c. Implement redundant power for ST-124M-3 stabilized platform system.

In order to meet the redundant power requirements to the platform, the method of paralleling the output of the battery buses where required (+6D10 and +6D30) with diodes at the input to the various platform subsystems was chosen rather than attempting to parallel the total output of both batteries with a single set of diodes.

This change provides redundant power to the 56-volt power supply with a separate set of bus isolation diodes, redundant power to the platform AC power supply with another separate set of isolation diodes, and redundant power to the platform electronic assembly with the existing +6D61 bus (redundant power bus).

Battery +6D10 was the single source for the ST-124M-3 prime power.

d. S-IVB passivation will fly in the S-IB vehicles for the first time onboard S-IU-206.

This change will allow the dumping of S-IVB propellants after spacecraft/launch vehicle separation, thereby safing the spent stage. Prior to this change, the S-IVB cutoff signals from the IU locked-out any possibility of dumping the propellants.

e. Addition of telemetry measurement K91-602 allows the monitoring of the redundant EDS cutoff command.

This change converts from an unmanned configuration to a manned configuration.

f. Implement redundant paths for IU/ESE launch critical functions.

The IU has critical functions through the IU/ESE interface connectors which if lost due to an open circuit would cause an engine shutdown after ignition. Other critical functions through the IU/ESE interface connectors if open could prevent a desired engine shutdown after ignition. Due to the vibration from engine ignition until liftoff and the possibility of faulty ESE connectors or of contaminants in the vehicle umbilical connectors, open circuits could occur.

g. Implement redundancy for TB-2 and TB-3 initiate.

The primary indications to the LVDA/LVDC for initiation of time bases 2 and 3 are interstage functions and do not have hardware backups. This change provides redundancy for these critical interstage functions.

h. Deletion of fourth battery (6D20).

Reconfiguration and redesign resulted in no requirement.

INSTRUMENTATION AND COMMUNICATIONS SUBSYSTEM.

a. Implement flight control measurements rechannelization.

There are additional instrumentation and communications subsystem design requirements resulting from the S-IU-206 role in supporting the SL-2 mission. New design requirements are as follows:

Flight control measurements shall appear on the same telemetry channels on S-IU-206, S-IU-207, S-IU-208, and S-IU-513. This required rechannelization of ten measurements on S-IU-206.

b. PCM telemetry design changes (register switch, PSR and digital gate).

Existing PCM telemetry system is highly susceptible to excessive IU and/or vehicle electrical noise environments. Redesign of the Reg. Sw. PSR and digital gate will provide a more nearly optimum noise free PCM telemetry system.

c. Implement thermal control coating experiment (M-415).

The IU I&C subsystem shall be modified to provide signal conditioning and transmission of experiment data. Signal conditioning equipment shall include differential amplifiers and divider networks for a total of 26 resistance thermistors.

Timing shall be provided with all experiment telemetry data to allow correlation with vehicle orbital data and to allow plotting of data as a function of time.

OVERALL RELIABILITY/PRODUCT IMPROVEMENT.

Modify/change hardware and circuitry to make SA-206 thru SA-212 man-rated. (S-IB phase-up).

1. Rework included addition of Q-balls, addition of EDS payload power (+6D91, +6D92 and +6D93), addition of EDS redundant power, change in EDS rate switch circuits, etc.

2. Added battery temperature measurements

3. S-IB Phase-Up, EDS and LV/CSM Electrical Interface

4. Redesign of IU command decoder for solder joint stress relief

5. Implement Changes to RDM Power Supply Assembly

6. Adjust pressure switch setpoint

7. Correct undesirable turnoff of coolant pump number 1

8. Provide line to shut-off coolant pump number 2

9. Desensitize coolant pump pressure switch

10. Implement redesigned water accumulator diaphragm

11. Implement redesigned gas-bearing heat exchanger

12. Improved O-rings for ECS coolant loop

13. Lightning detection devices and associated temporary cabling has been added for monitoring during pre-launch checkout. All cabling will be removed just prior to launch.

SECTION VII
GROUND SUPPORT INTERFACE

TABLE OF CONTENTS

Introduction	7-1
Vehicle Assembly Building	7-1
Launch Control Center	7-1
Mobile Launcher	7-3
Launch Pad	7-10
Mobile Service Structure	7-14
Crawler-Transporter	7-14
Converter-Compressor Facility	7-14
Ordnance Storage Area	7-16
Vehicle Assembly and Checkout	7-16
Hold and Recycle Criteria	7-22
Launch Interlocks	7-22

INTRODUCTION.

Launch Complex 39 (LC-39), Kennedy Space Center, Florida, which originally provided all the facilities necessary to the assembly, checkout, and launch of the Apollo/Saturn space vehicle has been modified to accommodate the Skylab Saturn IB and Saturn V space vehicles. The vehicle assembly building (VAB) provides a controlled environment in which the vehicle is assembled and checked out on a mobile launcher (ML). The space vehicle and the launch structure are then moved as a unit by the crawler-transporter to the launch site, where vehicle launch is accomplished after propellant loading and final checkout. The major elements of the launch complex shown in figure 7-1, are the vehicle assembly building (VAB), the launch control center (LCC), the mobile launcher (ML), the crawler-transporter (C-T), the crawlerway, the mobile service structure (MSS), and the launch pad.

VEHICLE ASSEMBLY BUILDING.

The VAB is located adjacent to Kennedy Parkway, about five miles north of the KSC industrial area. Its purpose is to provide a protected environment for receipt and checkout of the propulsion stages and instrument unit (IU), erection of the vehicle stages and spacecraft in a vertical position on the ML, and integrated checkout of the assembled space vehicle.

The VAB, as shown in figure 7-2 is a totally enclosed structure covering eight acres of ground. It is a structural steel building approximately 525 feet high, 518 feet wide, and 716 feet long. The siding is insulated aluminum except where translucent fiberglass sandwich panels are used in part of the north and south walls.

The principal operational elements of the VAB are the low bay area and high bay area. A 92-foot wide transfer aisle extends through the length of the VAB and divides the low and high bay areas into equal segments (figure 7-3).

LOW BAY AREA.

The low bay area provides the facilities for receiving, uncrating, checkout, and preparation of the S-IVB stage, and the IU (and S-II stage for Saturn V). The low bay area, located in the southern section of the VAB, is approximately 210 feet high, 442 feet wide, and 274 feet long. There are eight stage preparation and checkout cells, four of which are equipped with systems to simulate interface operations between the stages and the IU.

Work platforms, made up of fixed and folded sections, fit about the various sections as required. The platforms are bolted, to permit vertical repositioning, to the low bay structure. Access from fixed floor levels to the work platforms is provided by stairs.

HIGH BAY AREA.

The high bay area provides the facilities for erection and checkout of stages and spacecraft, and integrated checkout of the assembled space vehicle. The high bay area which is located in the northern section of the building, is approximately 525 feet high, 518 feet wide, and 442 feet long. It contains four checkout bays, each capable of accommodating a fully assembled space vehicle.

Access to the vehicle at various levels is provided from air conditioned work platforms that extend from either side of the bay to completely surround the launch vehicle. Each platform is composed of two biparting sections which can be positioned in the vertical plane. The floor and roof of each section conform to and surround the vehicle. Hollow seals on the floor and roof of the section provide an environmental seal between the vehicle and the platform.

Each pair of opposite checkout bays is served by a 250-ton bridge crane with a hook height of 462 feet. The wall framing between the bays and the transfer aisle is open above the 190-foot elevation to permit movement of components from the transfer aisle to their assembly position in the checkout bay.

The high bay doors provide an inverted T-shaped opening 456 feet in height. The lower portion of the opening is closed by doors which move horizontally on tracks. The upper portion of the opening is closed by seven vertically moving doors.

UTILITY ANNEX.

The utility annex, located on the west side of the VAB, supports the VAB, LCC and other facilities in the VAB area. It provides air conditioning, hot water, compressed air, water for fire protection, and emergency electrical power.

HELIUM/NITROGEN STORAGE-VAB AREA.

The gas storage facility at the VAB provides high pressure gaseous helium and nitrogen. It is located east of the VAB and south of the crawlerway. The roof deck of the building is removable to permit installation and removal of pressure vessels through the roof. This facility is serviced from the converter-compressor facility by a 6,000 psig gaseous helium line and a 6,000 psig gaseous nitrogen line.

LAUNCH CONTROL CENTER.

The LCC (figure 7-4) serves as the focal point for overall direction, control, and surveillance of space vehicle checkout and launch. The LCC is located adjacent to the VAB and at a sufficient distance from the launch pad (three miles) to permit the safe viewing of liftoff without requiring site hardening. An enclosed personnel and cabling bridge connects the VAB and LCC at the third floor level.

The LCC is a four-story structure approximately 380 by 180 feet. The ground floor is devoted to service and support functions such

7-1

Section VII Ground Support Interface

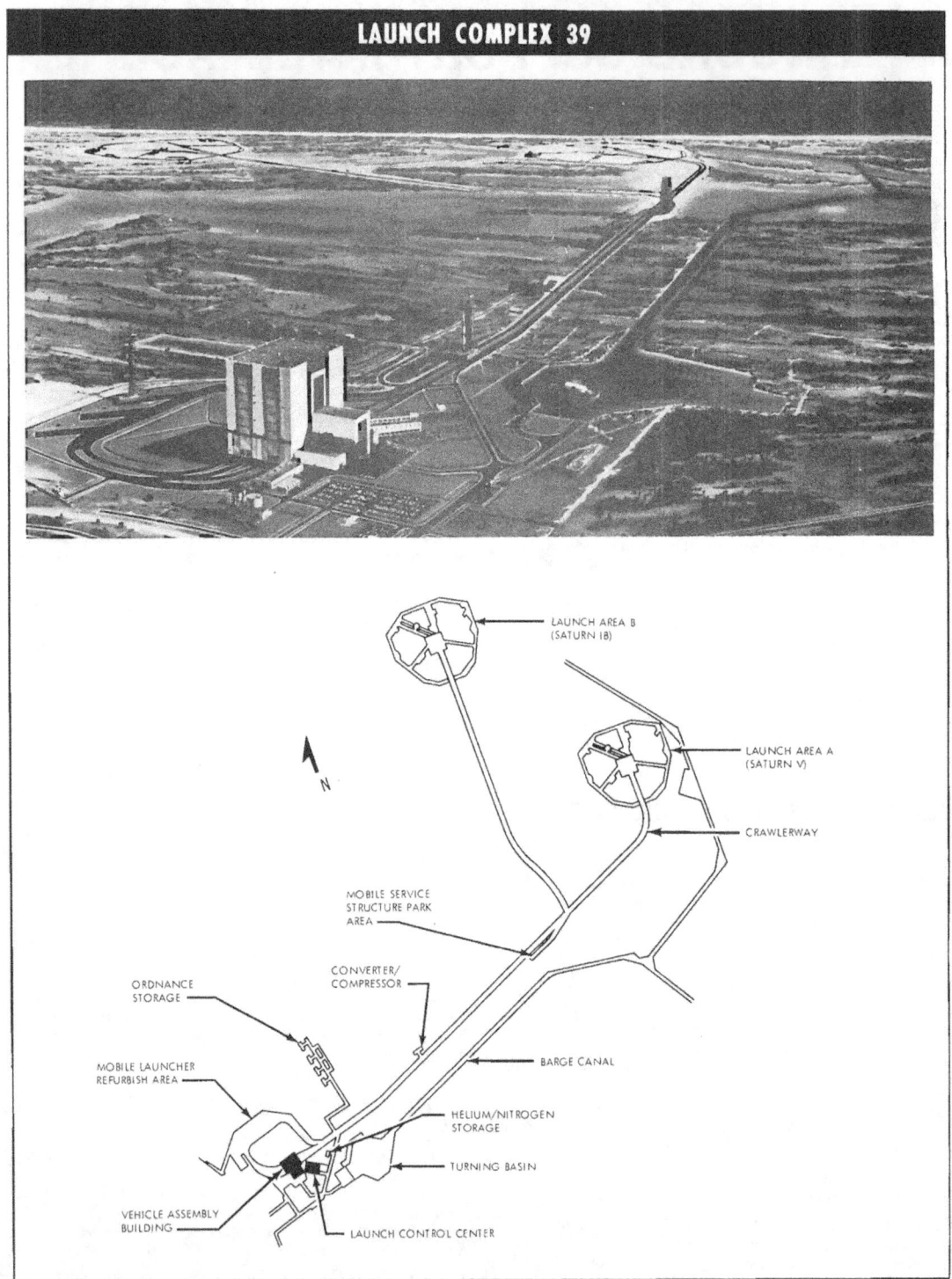

Figure 7-1

as cafeteria, offices, shops, laboratories, the communications control room, and the complex control center. The second floor houses telemetry, RF and tracking equipment, in addition to instrumentation and data reduction facilities.

The third floor is divided into four separate but similar control areas, each containing a firing room, computer room, mission control room, test conductor platform area, visitor gallery, offices and frame rooms. Three of the four firing rooms contain control, monitoring and display equipment for automatic vehicle checkout and launch.

Direct viewing of the firing rooms and the launch area is possible from the mezzanine level through specially designed, laminated, and tinted glass windows. Electrically controlled sun louvers are positioned outside the windows.

The display rooms, offices, launch information exchange facility (LIEF) rooms, and mechanical equipment are located on the fourth floor.

The electronic equipment areas of the second and third floors have raised false floors to accommodate interconnecting cables and air conditioning ducts.

The power demands in this area are large and are supplied by two separate systems, industrial and instrumentation. The industrial power system supplies electric power for lighting, general use receptacles, and industrial units such as air conditioning, elevators, pumps and compressors. The instrumentation power system supplies power to the electronic equipment, computers, and related checkout equipment. This division between power systems is designed to protect the instrumentation power system from the adverse effects of switching transients, large cycling loads, and intermittent motor starting loads. Communication and signal cable provisions have been incorporated into the design of the facility. Cable troughs extend from the LCC via the enclosed bridge to each ML location in the VAB high bay area. The LCC is also connected by buried cableways to the ML refurbishing area and to the pad terminal connection room (PTCR) at the launch pad. Antennas on the roof provide an RF link to the launch pads and other facilities at KSC.

MOBILE LAUNCHER.

The mobile launcher (figure 7-5) is a transportable steel structure which, with the crawler-transporter, provides the capability to move the erected vehicle to the launch pad. The ML is divided into two functional areas, the launcher base and the umbilical tower. The launcher base is the platform on which a Saturn V vehicle is assembled in the vertical position, transported to a launch site, and launched. The umbilical tower, permanently erected on the base, is the means of ready access to all important levels of the vehicle during the assembly, checkout, and servicing periods prior to launch. The equipment used in the servicing, checkout, and launch is installed throughout both the base and tower sections of the ML. The intricate vehicle-to-ground interfaces are established and debugged in the convenient and protected environment of the VAB, and moved undisturbed aboard the ML to the pad.

Mobile launcher 1 is used for Skylab Saturn IB space vehicles, and has been modified by (1) addition of a pedestal, to maintain the spacecraft and upper stages' interfaces with the umbilical tower service arms; and (2) removal or inactivation of S-II and S-IC stage GSE and addition of S-IB stage GSE.

LAUNCHER BASE.

The launcher base is a two story steel structure 25 feet high, 160 feet long, and 135 feet wide. Each of the three levels provides approximately 12,000 square feet of floor space. The upper deck is designated level O. Level A, the upper of the two internal levels, contains 21 compartments and level B has 22 compartments.

A new structural pedestal is installed on the ML deck that raises the Saturn IB launcher platform to the 127 foot level. The basic tower is an open truss welded steel pipe structure. An access bridge is provided from the pedestal deck level to the ML tower. Firing accessories for the S-IB stage were removed from LC-34/37B and are installed on the pedestal deck. There is an opening through the pedestal deck centered over the opening in the ML base for

Figure 7-2

Section VII Ground Support Interface

first stage exhaust. A work platform is provided for the pedestal opening for prelaunch engine servicing. Pneumatics, propellants, water, and other services are routed to the pedestal base, with valve panels installed in an enclosed equipment level below the pedestal deck.

Access to the launcher base interior is provided by personnel/equipment access doors opening into levels A and B and equipment access hatches located on levels O and A.

The base has provisions for attachment to the crawler-transporter, six launcher-to-ground mount mechanisms, and four extensible support columns.

All electrical/mechanical interfaces between vehicle systems and the VAB or the launch site are located through or adjacent to the base structure. A number of permanent pedestals at the launch site provide support for the interface plates and servicing lines.

The base houses such items as the computer systems test sets, propellant loading equipment, hydraulic test sets, propellant and pneumatic lines, air conditioning and ventilating systems, electrical power systems, and water systems. Shock-mounted floors and spring supports are provided so that critical equipment receives less than ±0.5 G mechanically-induced vibrations. Electronic compartments within the ML base are provided with acoustical isolation to reduce the overall rocket engine noise level.

The air conditioning and ventilating system for the base provides environmental protection for the equipment during operations and standby. One packaged air conditioner provides minimal environmental conditioning and humidity control during transit. Fueling operations at the launch area require that the compartments within the structure be pressurized to a pressure of three inches of water above atmospheric pressure and that the air supply originate from a remote area free from contamination.

The primary electrical power supplied to the ML is divided into four separate services: instrumentation, industrial, in-transit and emergency. Instrumentation and industrial power systems are separate and distinct. During transit, power from the crawler-transporter is used for the water/glycol systems, computer air conditioning, threshold lighting, and obstruction lights. Emergency power for

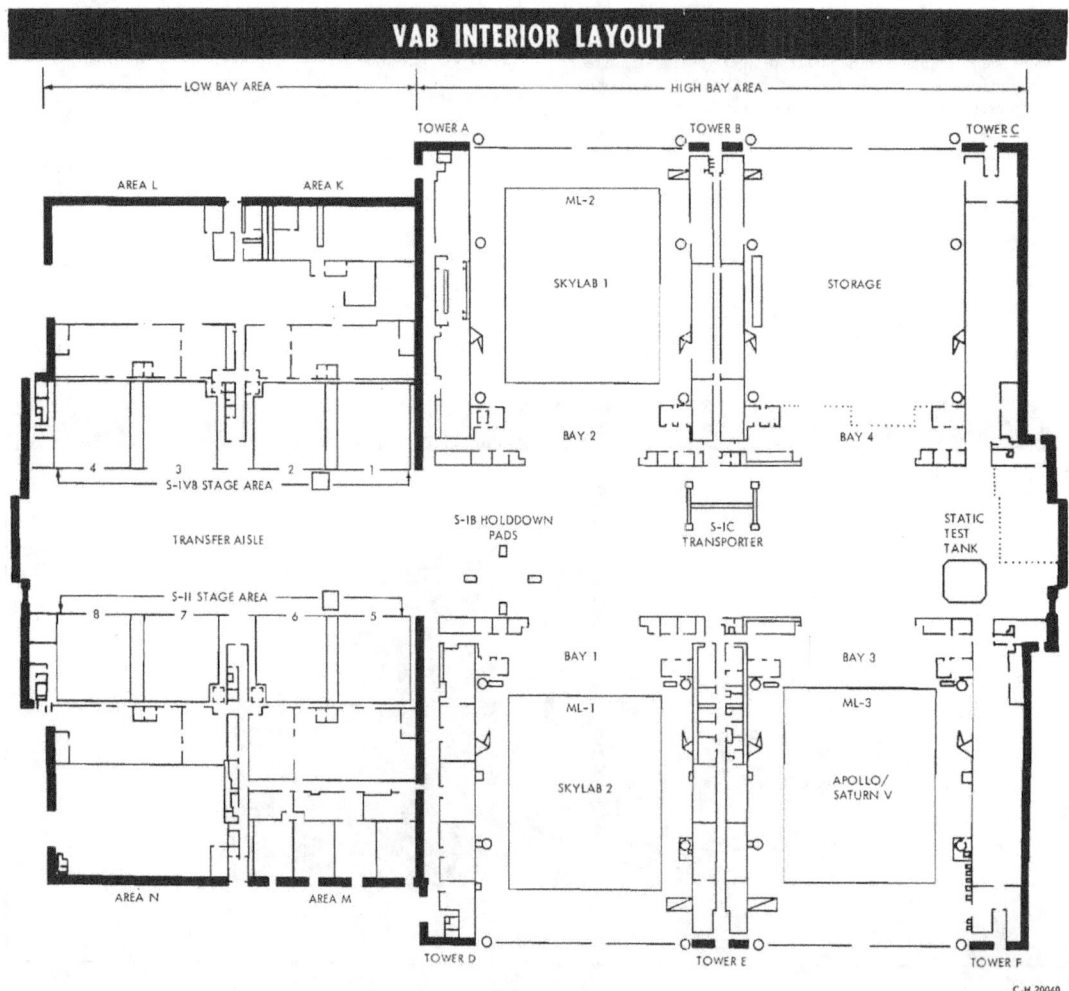

Figure 7-3

Section VII Ground Support Interface

the ML is supplied by a diesel-driven generator located in the ground facilities. It is used for obstruction lights, emergency lighting, and for one tower elevator. Water is supplied to the ML at the VAB and at the pad for fire, industrial and domestic purposes and at the refurbishment area for domestic purposes.

FIRING ACCESSORIES.

The ML firing accessories (figure 7-6) described in the following paragraphs were removed from LC-34/LC-37B and installed on the pedestal deck.

Holddown Arms.

The holddown arms system (figure 7-6) provides support for the vehicle on the launch pedestal and restrains the vehicle from flight until verification of full thrust at launch. The holddown arms system consists basically of eight holddown arms, their holddown release mechanisms, pneumatic separators, and the necessary pneumatic control panel. The pneumatic system is redundant from the inlet supply line on the control panel through to the pneumatic separators. The supply panel, once charged with helium, has a reservoir with the capacity to effect release of the vehicle upon command, even with complete failure of the inlet supply system. The solenoids release gaseous helium at 750 psig which pressurizes the separators through redundant "tuned length" tubing. The tubing is tuned length; that is, the distance and the volume are identical from the solenoid valves to each and all of the pneumatic separators. The criteria here is not total elapsed time but the total time difference between first arm release to last arm release, and the use of helium minimized this time lag. The time from release command to the slowest arm retraction is approximately 140 msec. The activation pressure required to release the arms is 100 psig. An explosive bolt assembly is added to each of the eight holddown arm pneumatic separators. This backup system provides for the activation of the explosive release device if first movement of any or all arms is not received within a specific time period (approximately 190 msec) from the initiation of the release command. The basic holddown arm system has verified its reliability during previous launches. The backup system consists of an explosive bolt assembly that increases the reliability factor.

Boattail Conditioning Lines.

The bottail conditioning and water quench system installation provides the final link connecting the vehicle to the boattail conditioning and water quench system. The water quench system transfers water at a rate of 8000 gpm at 125-psig pressure to the vehicle boattail area when necessary for combating fires prior to vehicle liftoff. In addition, the installation is used in conjunction with the environmental control system and deluge purge system to provide a controlled atmosphere in the boattail area for personnel safety and vehicle protection during various periods while the vehicle is on the launch pedestal. The controlled atmosphere is provided by two purges: air purge and nitrogen (GN_2) deluge purge. The installation automatically disconnects at liftoff after approximately 2-1/4 in. of vehicle travel. The boattail conditioning and water quench system installation is essentially four installations, each in general consisting of a support bracket, disconnect coupling, flexible hose assembly, valve assembly, elbow assembly, and the hardware required for installation. The four installations are installed at the four principal positions and are similar. The flexible hose assembly is covered with a heat-and-blast-protective coating.

Fuel Fill Mast.

The fuel fill mast system is used to provide a means for connecting the fuel transfer line to the launch vehicle and to provide a fast, fail-safe method for disconnecting the line at liftoff. The mast supplies fuel to the vehicle at a pressure of 50 psig and a flowrate of 1500 gpm. The pivotal feature of the fuel fill mast provides for mast erection and retraction, thereby providing adequate liftoff clearance for the vehicle. Mast retraction at vehicle liftoff is an automatic function tied to the firing circuits for proper sequencing; however, manual control for erecting and testing can be performed through use of a pneumatic valve box mounted on the cross member of each mast support stand. The fuel fill mast consists of a retractable coupling assembly, two cylinder assemblies, support bracket assembly, retracting assembly, mast arrestor mounting assembly, valve box lox assembly, support stand, upper and lower pipe weldments, and a hose assembly that provides a flow path for the fuel during propellant tanking operation. The fuel fill mast

Figure 7-4

Section VII Ground Support Interface

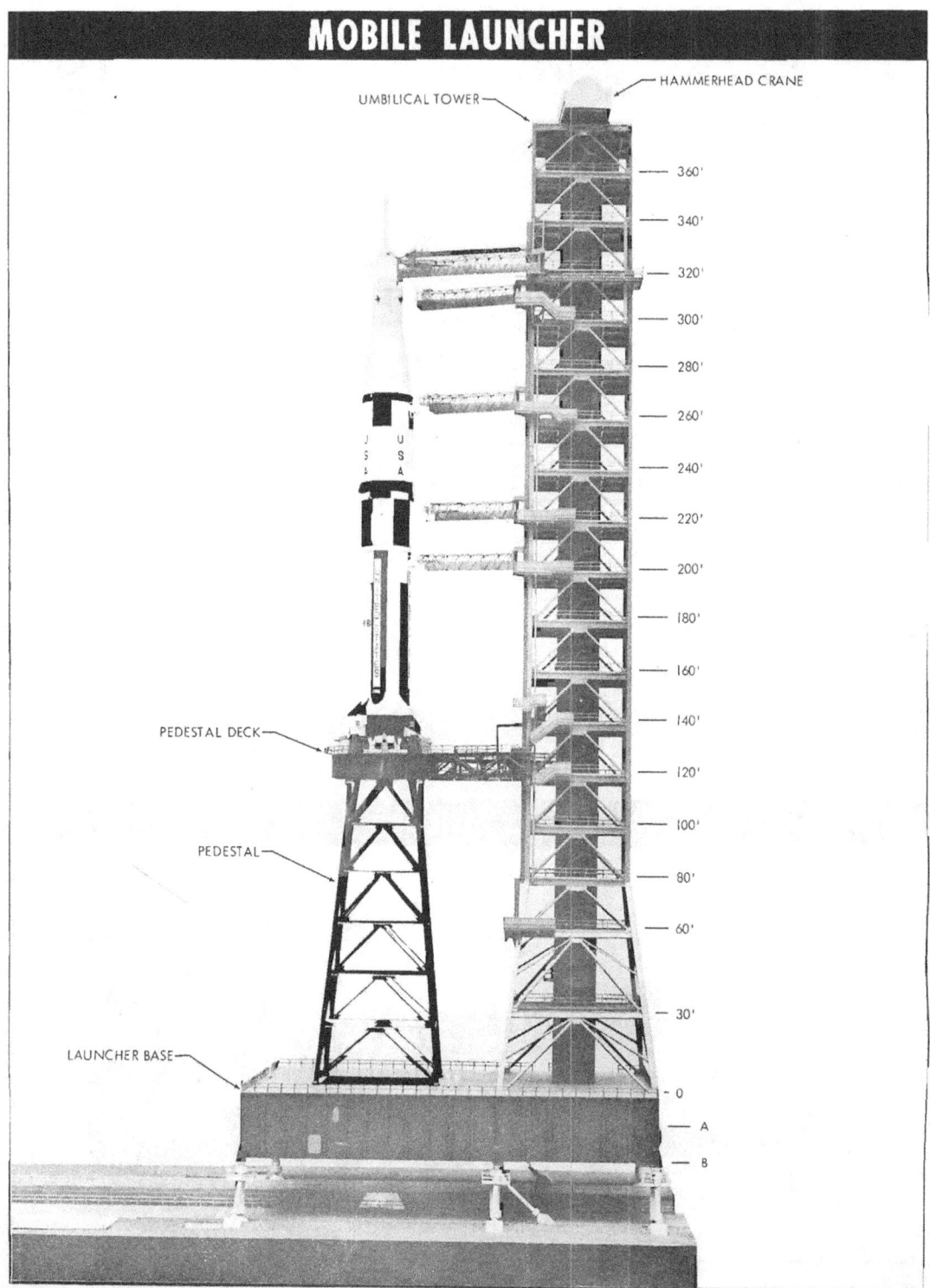

Figure 7-5

Section VII Ground Support Interface

incorporates provisions for vertical, lateral, and retraction adjustments to ensure correct alignment and operation.

Lox Fill Mast.

The lox fill mast system provides a means of connecting the lox transfer line to the launch vehicle and provides a fast, fail-safe method for disconnecting the line at vehicle liftoff. Disconnection of the lox transfer line at liftoff is a function of vehicle motion, as the vehicle simply lifts away from the mast assembly. Mast retraction after vehicle liftoff is a remotely controlled function of the pneumatic circuits. These circuits are controlled by solenoid valves located in the launcher valve box. The lox fill mast system functional requirements are to supply liquid oxygen to the vehicle through a 6-in. coupling at a pressure of 90 psig with a flow rate of 1250 gpm, with inflight disconnect capability. The lox fill mast consists of a retractable coupling assembly, two cylinder assemblies, support bracket assembly, retracting assembly, mast arrestor, mounting assembly, valve box assembly, support stand, upper and lower pipe weldments, hose assembly and pneumatic lines. The retractable coupling assembly, upper and lower pipe weldments, and hose assembly provide a flow path for the lox during propellant tanking operations. The lox fill mast incorporates provisions for vertical, lateral, and retraction adjustments to ensure correct alignment and operation.

Short Cable Masts.

The short cable masts provide a connecting link, structural support, and disconnecting capability for electrical cable and pneumatic

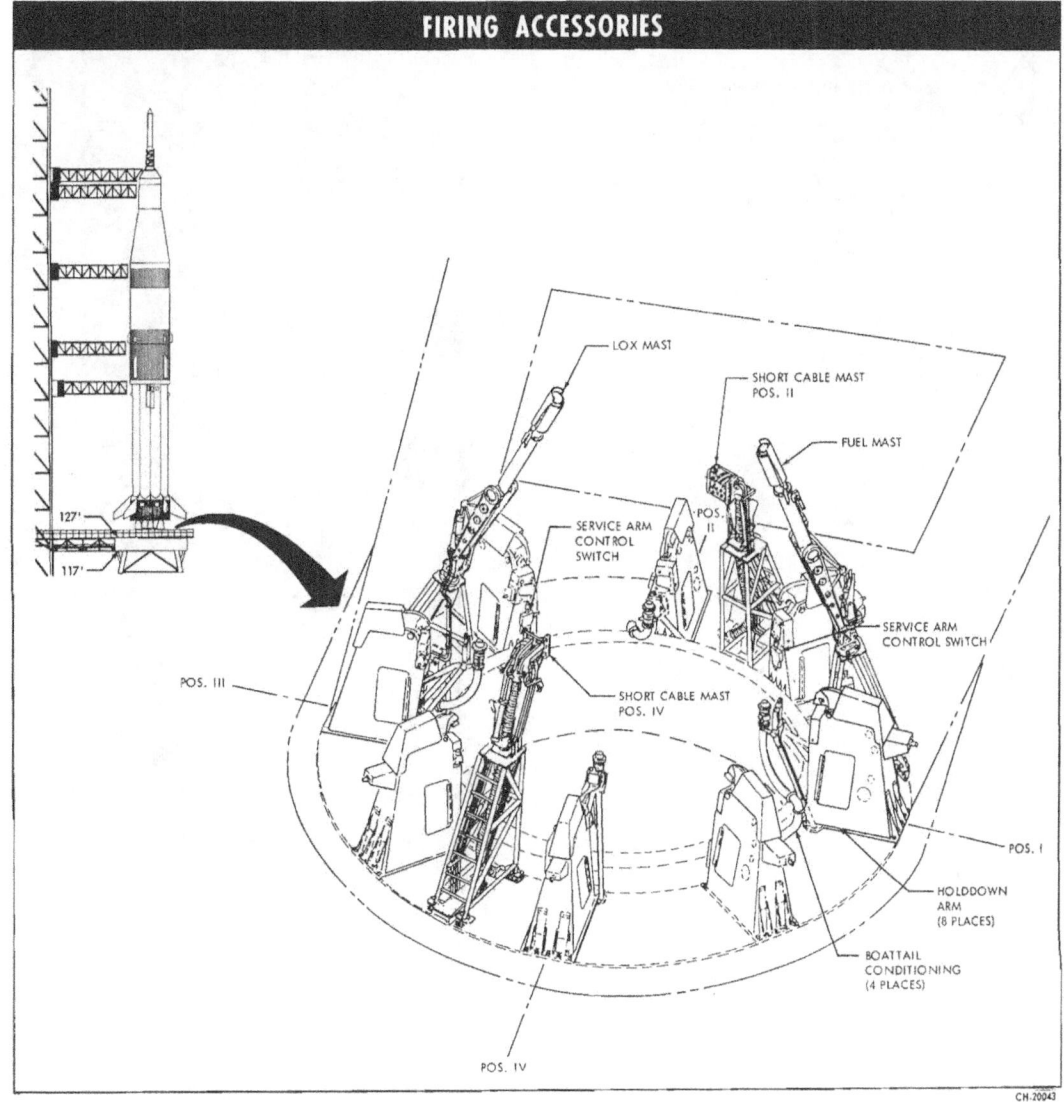

Figure 7-6

Section VII Ground Support Interface

MOBILE LAUNCHER SERVICE ARMS

 S-IB FORWARD (INFLIGHT). PROVIDES PNEUMATIC, ELECTRICAL, AND AIR-CONDITIONING INTERFACES. UMBILICAL WITHDRAWAL BY PNEUMATIC DISCONNECT IN CONJUNCTION WITH PNEUMATICALLY DRIVEN BLOCK AND TACKLE/LANYARD DEVICE. SECONDARY MECHANICAL SYSTEM. RETRACT TIME IS 3.9 SECONDS.

 S-IVB AFT (INFLIGHT). PROVIDES LH_2 AND LOX TRANSFER, ELECTRICAL, PNEUMATIC, AND AIR-CONDITIONING INTERFACES. UMBILICAL WITHDRAWAL SYSTEMS SAME AS S-IVB FORWARD. ALSO EQUIPPED WITH LINE HANDLING DEVICE. RETRACT TIME IS 6.1 SECONDS (MAX).

 S-IVB FORWARD/IU UMBILICAL (INFLIGHT). FOR S-IVB STAGE, PROVIDES FUEL TANK VENT, ELECTRICAL, PNEUMATIC, AIR-CONDITIONING, AND PREFLIGHT CONDITIONING INTERFACES. FOR IU, PROVIDES PNEUMATIC, ELECTRICAL, AND AIR-CONDITIONING INTERFACES. UMBILICAL WITHDRAWAL BY PNEUMATIC DISCONNECT IN CONJUNCTION WITH PNEUMATIC/HYDRAULIC REDUNDANT DUAL CYLINDER SYSTEM. SECONDARY MECHANICAL SYSTEM. ARM ALSO EQUIPPED WITH LINE HANDLING DEVICE TO PROTECT LINES DURING WITHDRAWAL. RETRACT TIME IS 6.1 SECONDS (MAX).

 SERVICE MODULE (INFLIGHT). PROVIDES AIR-CONDITIONING, VENT LINE, COOLANT, ELECTRICAL, AND PNEUMATIC INTERFACES. UMBILICAL WITHDRAWAL BY PNEUMATIC/MECHANICAL LANYARD SYSTEM WITH SECONDARY MECHANICAL SYSTEM. RETRACT TIME IS 6.1 SECONDS (MAX).

 COMMAND MODULE ACCESS ARM (PREFLIGHT). PROVIDES ACCESS TO SPACECRAFT THROUGH ENVIRONMENTAL CHAMBER. ARM MAY BE RETRACTED OR EXTENDED FROM LCC. RETRACTED TO 12° PARK POSITION DURING PERIOD T-43 TO T-5 MINUTES. EXTEND TIME IS 12 SECONDS FROM THIS POSITION.

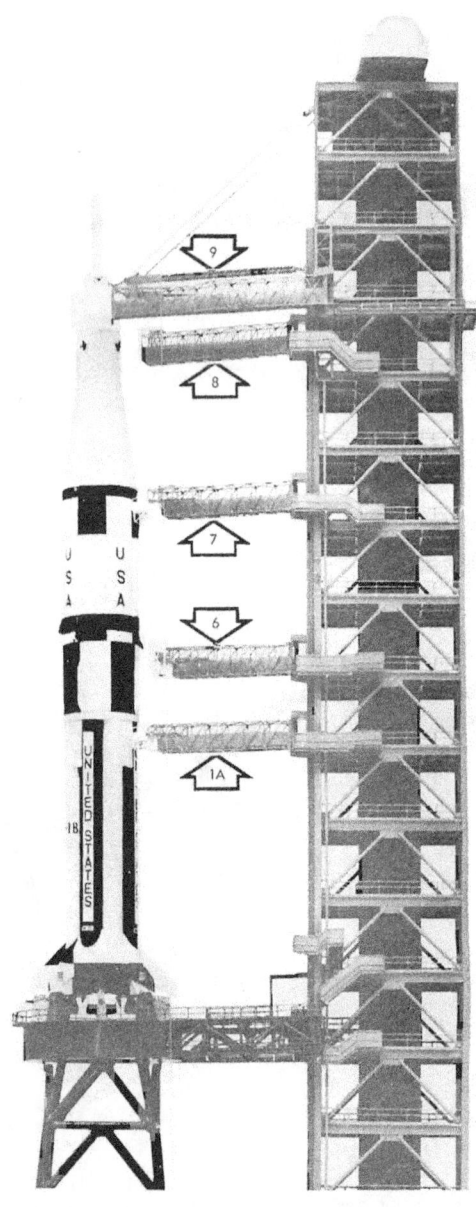

Figure 7-7

Section VII Ground Support Interface

service lines required for checkout and operation of Saturn IB vehicle engine components prior to and during the process of launch. After the vehicle has lifted approximately four inches off the launch pad, the masts disconnect all cables and lines, and retract from the vehicle. Each short cable mast consists of a support platform, a mast weldment, two kickoff cylinders, a retract cylinder, a quick-release housing, a latch-back mechanism, electrical cables, pneumatic service lines, and supporting hardware. The double-action, pneumatic retract cylinder furnishes the primary force for retracting the short cable mast during launch. A 750-psig GN_2 pressure is applied to the top of the cylinder while an opposing constantly venting 50-psig GN_2 pressure is applied at the bottom of the cylinder and acts as a cushioning pressure for the mast at the end of retraction. Electrical cables are routed through an 8-in. flexible shield to the quick-release housing. These cables connect electrical circuitry to the S-IB stage engine compartment from the beginning of checkout until the mast is disconnected after vehicle liftoff. The pneumatic lines are mounted along the left and right sides of the mast weldment. These lines supply pneumatic pressure to the S-IB stage for standby, checkout, and launch.

UMBILICAL TOWER.

The umbilical tower is an open steel structure 380 feet high which provides the support for umbilical service arms, access arm, work and access platforms, distribution equipment for the propellant, pneumatic, electrical and instrumentation subsystems, and other ground support equipment. The distance from the vertical centerline of the tower to the vertical centerline of the vehicle is approximately 80 feet. The distance from the nearest vehicle column of the tower to the vertical centerline of the vehicle is approximately 60 feet. Two high speed elevators service 18 landings, from level A of the base to the 340-foot tower level.

The hammerhead crane is located on top of the umbilical tower. The load capacity of the crane is 25 tons with the hook extended up to 50 feet from the tower centerline. With the hook extended between 50 and 85 feet from the tower centerline, the load capacity is 10 tons. The hook can be raised or lowered at 30 feet per minute for a distance of 468 feet. The trolley speed is 110 feet per minute. The crane can rotate 360 degrees in either direction at one revolution per minute. Remote control of the crane from the ground and from each landing between levels 0 and 360 is provided by portable plug-in type control units.

Tower modifications for Skylab Saturn IB propellant systems consisted of (1) adding an RP-1 supply line from the ML base to an RP-1 skid added at the 100 foot level, thence to the S-IB stage fuel mast; (2) adding lox line to the S-IB stage lox mast, assignment of the 1000 gpm vacuum-jacketed lox replenish system (for Saturn V) to accomplish both fill and replenish functions for Saturn IB, and inactivation of the un-insulated 10,000 gpm lox supply lines and SIC skid used on Saturn V.

SERVICE ARMS.

The service arms provide access to the launch vehicle and support the service lines that are required to sustain the vehicle as described in figure 7-7. The service arms are designated as either preflight or inflight arms. The preflight Command Module Access Arm is retracted to a 12 deg park position during final launch preparation and then is retracted and locked against the umbilical tower prior to liftoff. The inflight arms retract at vehicle liftoff, after receiving a command signal from the service arm control switches located in holddown arms.

The inflight service arm launch retract sequence typically consists

Figure 7-8

Section VII Ground Support Interface

of the following operations: umbilical carrier release, carrier withdrawal, and arm retraction and latchback. When the vehicle rises 3/4-inch, the primary liftoff switches on the holddown arms activate a pneumatic system which unlocks the umbilical carriers and pushes each carrier from the vehicle. If this system fails, the secondary mechanical release mechanism will be actuated 500 milliseconds after T-O by the service arm control switch back-up timer. Upon carrier ejection, a double pole switch activates both the carrier withdrawal and arm retraction systems. If this switch fails, it will be by-passed by a signal from the backup timer. Line handling devices on the S-IVB forward and aft arms are also activated on carrier ejection. Carrier withdrawal and arm retraction is accomplished by pneumatic and/or hydraulic systems.

Service arm modifications for Skylab Saturn IB are:

SA-1A Previously an S-IC forward arm, modification consists of making it an in-flight arm, installing S-IB stage umbilical plates, and adding an extension of 68" to lengthen the arm. Minor modifications are made to the service arm skid. SA-1A is installed in the former position of SA-5.

SA-6 The carrier is modified to interface with the Saturn IB cylindrical S-IVB adapter.

SA-7 The umbilical services for the LEM are deleted.

Changes are not required of SA-8 and SA-9. The service arms for the S-IC and S-II stages are removed (SA-1, SA-2, SA-3, SA-4, SA-5).

LAUNCH PAD.

The launch pad shown in figure 7-8 is typical of launch pads 39-A and 39-B. The following details are applicable to both pads except where differences are specified.

The launch pad provides a stable foundation for the ML during launch and prelaunch operations and an interface to the ML for ML and vehicle systems. The two pads at LC-39 are located approximately three miles from the VAB area. Each launch site is an eight-sided polygon measuring approximately 3,000 feet across. Pad B is used for Skylab Saturn IB space vehicles.

LAUNCH PAD STRUCTURE.

The launch pad is a cellular, reinforced concrete structure with a top elevation of 48 feet above sea level (42 feet above grade elevation). The longitudinal axis of the pad is oriented north-south, with the crawlerway and ramp approach from the south.

Located within the fill under the west side of the structure (figure 7-9) is a two-story concrete building to house environmental control and pad terminal connection equipment. On the east side of the structure, within the fill, is a one-story concrete building to house the high pressure gas storage battery. On the pad surface are elevators, staircases and interface structures to provide service to the ML and the mobile service structure (MSS). A ramp, with a five percent grade, provides access from the crawlerway. This is used by the C-T to position the ML/space vehicle and the MSS on the support pedestals. The azimuth alignment building is located on the approach ramp in the crawlerway median strip. A flame trench 58 feet wide by 450 feet long, bisects the pad. This trench opens to grade at the north end. The 700,000-pound mobile wedge-type flame deflector is mounted on rails in the trench.

An escape chute is provided to connect the ML to an underground, hardened room. This room is located in the fill area west of the

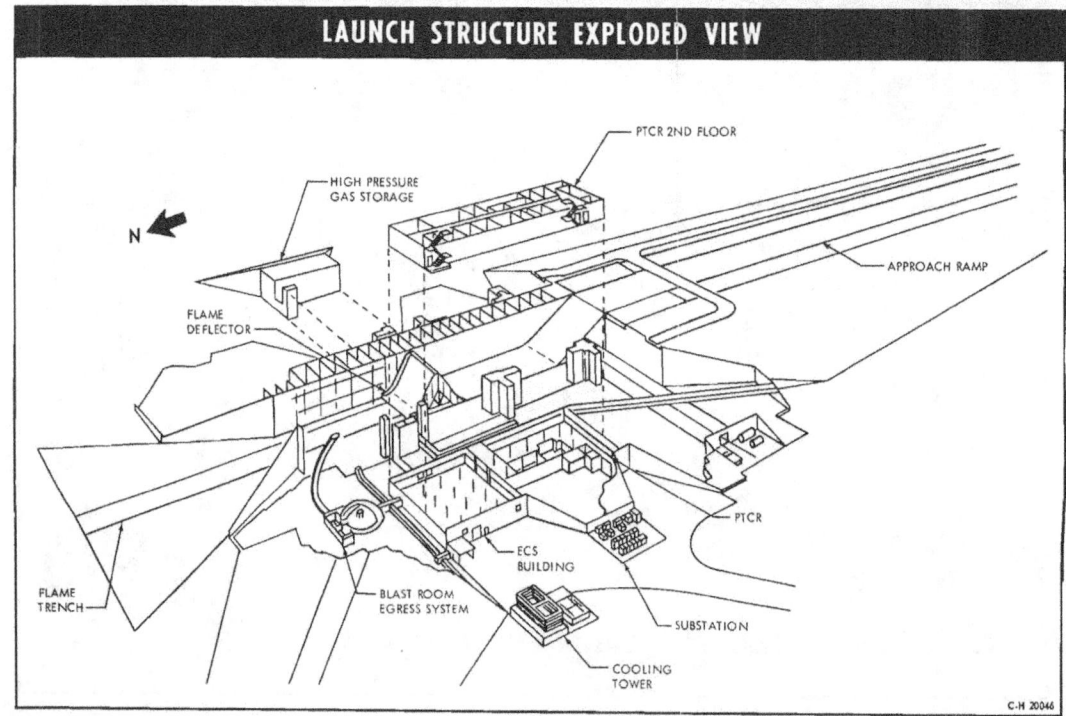

Figure 7-9

Section VII Ground Support Interface

support structure. This is used by astronauts and service crews in the event of a malfunction during the final phase of the countdown.

PAD TERMINAL CONNECTION ROOM.

The pad terminal connection room (PTCR) (figure 7-9) provides the terminals for communication and data link transmission connections between the ML or MSS and the launch area facilities and between the ML or MSS and the LCC. This facility also accommodates the electronic equipment that simulates the vehicle and the functions for checkout of the facilities during the absence of the launcher and vehicle.

The PTCR is a two-story hardened structure within the fill on the west side of the launch support structure. The launch pedestal and the deflector area are located immediately adjacent to this structure. Each of the floors of this structure measures approximately 136 feet by 56 feet. Entry is made from the west side of the launch support structure at ground level into the first floor area. Instrumentation cabling from the PTCR extends to the ML, MSS, high pressure gas storage battery area, lox facility, RP-1 facility, LH_2 facility, and azimuth alignment building. The equipment areas of this building have elevated false floors to accommodate the instrumentation and communication cables used for interconnecting instrumentation racks and terminal distributors.

The air conditioning system, located on the PTCR ground floor, provides a controlled environment for personnel and equipment. The air conditioning system is controlled remotely from the LCC when personnel are evacuated for launch. This system provides chilled water for the air handling units located in the equipment compartments of the ML. A hydraulic elevator serves the two floors and the pad level.

Industrial and instrumentation power is supplied from a nearby substation.

ENVIRONMENT CONTROL SYSTEM.

The ECS room located in the pad fill west of the pad structure and north of the PTCR (figure 7-9) houses the equipment which furnishes temperature and/or humidity controlled air or nitrogen for space vehicle cooling at the pad. The ECS room is 96 feet wide by 112 feet long and houses air and nitrogen handling units, liquid chillers, air compressors, a 3000-gallon water-glycol storage tank and other auxiliary electrical and mechanical equipment.

HIGH PRESSURE GAS SYSTEM.

The high pressure gas storage facility at the pad provides the launch vehicle with high pressure helium and nitrogen. This facility is an integral part of the east portion of the launch support structure. It is entered from ground elevation on the east side of the pad. The high pressure (6,000 psig) facilities at the pad are provided for high pressure storage of 3,000 cubic feet of gaseous nitrogen and 9,000 cubic feet of gaseous helium.

LAUNCH PAD INTERFACE STRUCTURE.

The launch pad interface structure (figure 7-10) provides mounting support pedestals for the ML and MSS, an engine service platform transporter, and support structures for fueling, pneumatic, electric power and environment control interfaces.

The ML at the launch pad (as well as the VAB and refurbish area) is supported by six mount mechanisms which are designed to carry vertical and horizontal loading. Four extensible columns, located near each corner of the launcher base exhaust chamber,

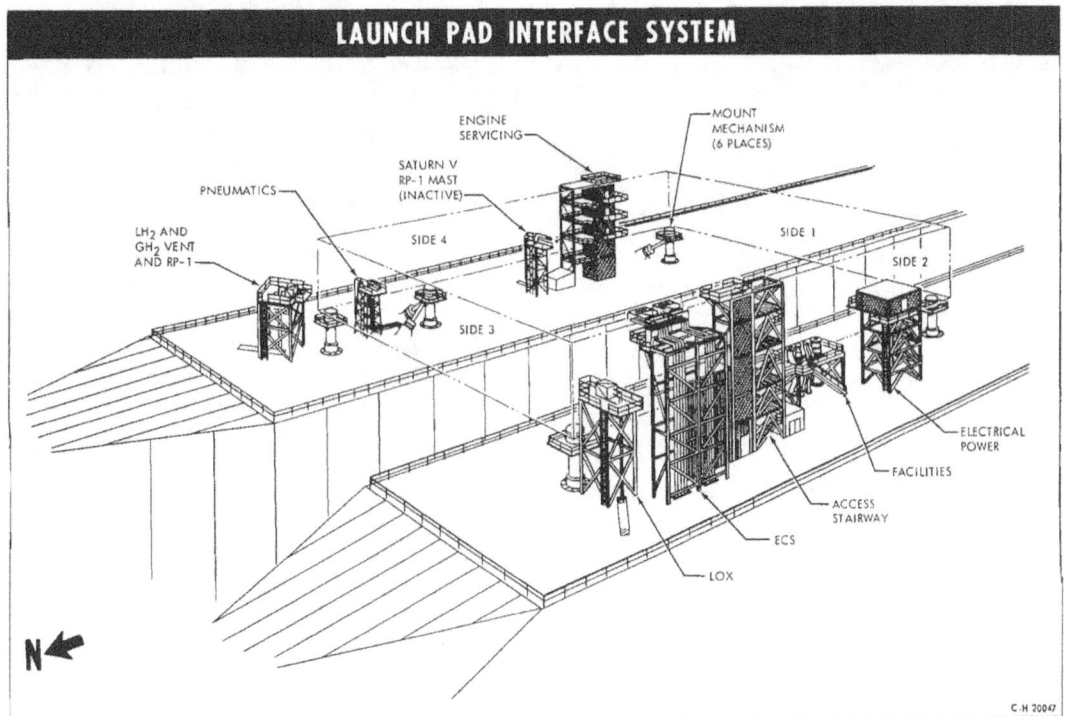

Figure 7-10

Section VII Ground Support Interface

also support the ML at the launch site. These columns are designed to prevent excessive deflections of the launcher base when the vehicle is fueled and from load reversal in case of an abort between engine ignition and vehicle liftoff.

The MSS is supported on the launch pad by four mounting mechanisms similar to those used to support the ML.

The engine servicing platform provides access to the pedestal deck area for servicing of the S-IB engines.

Interface structures are provided on the east and west portions of the pad structure (figure 7-10) for propellant, pneumatic, power, facilities, environmental control, communications, control and instrumentation systems.

Modification for Skylab Saturn IB consists of re-routing the RP-1 supply line interface to the LH_2 mast and deactivating the RP-1 mast used for Saturn V.

EMERGENCY INGRESS/EGRESS AND ESCAPE SYSTEM.

The emergency ingress/egress and escape system (figure 7-11) provides access to and from the Command Module (CM) plus an escape route and safe quarters for the astronauts and service personnel in the event of a serious malfunction prior to launch. Depending upon the time available, the system provides escape by either slide wire or elevator. Both means utilize the CM access arm as a component.

The slide wire egress system (figure 7-11) provides the primary means of escape. A 1 1/8-inch diameter steel cable extends from the 341.72-foot level of the ML to a tail tower approximately 2,200 feet west of the ML. Astronauts and technicians evacuate the white room, cross the access arm, and follow a catwalk along the east and north sides of the ML to the egress platform at the 320-foot level. Here, they board the 9-man cab transporter suspended from the cable and snubbed against the egress platform. The cab is released by levers in the cab. It rides the slide wire down to the landing area where it is decelerated and stopped by an arresting gear assembly.

The secondary escape and normal egress means are the tower high speed elevators. These move between the 340 foot level of the tower and level A at 600 feet per minute. At level A, egressing personnel move through a vestibule to elevator No. 2 which takes them down to the bottom of the pad. Armored personnel carriers are available at this point to remove them from the pad area.

When the state of the emergency allows no time for retreat by motor vehicle, egressing personnel upon reaching level A of the ML slide down the escape tube into the rubber-lined blast room vestibule (figure 7-11). The escape tube consists of a short section which extends from the elevator vestibule at ML level A to side 3 of the ML base where it interfaces with a fixed portion that penetrates the pad at an elevation of 48 feet. At the lower extremity of the illuminated escape tube, a deceleration ramp is provided to reduce exit velocity, permitting safe exit for the user.

Entrance to the blast room is gained through blast-proof doors controllable from either side. The blast room floor is mounted on coil springs to reduce outside acceleration forces to 3 to 5 G's. Twenty people may be accommodated for 24 hours. Communication facilities are provided in the room including an emergency RF link in which the receiving antenna is built into the ceiling. In the event that escape via the blast-proof doors is not possible, a hatch in the top of the blast room is accessible to rescue crews.

An underground air duct from the vicinity of the blast room to the remote air intake facility permits egress from the pad structure to the pad perimeter. Provision is made to decrease air velocity in the duct to allow personnel movement through the duct.

Emergency ingress to the CM utilizes the tower high speed elevators and the CM access arm.

ELECTRICAL POWER.

The electrical power for the launch pad is fed from the 69 kv main substation to switching station No. 1, where it is stepped down to 13.8 kv. The 13.8 kv power is fed to switching station No. 2 from where it is distributed to the various substations in the pad area. The output of each of the substations is 480 volts with the exception of the 4160-volt substations supplying power to the fire protection water booster pump motors and the lox pump motors.

FUEL SYSTEM FACILITIES.

The fuel facilities, located in the northeast quadrant of the pad approximately 1,450 feet from pad center, store RP-1 and liquid hydrogen.

The RP-1 facility consists of three 86,000-gallon (577,000-pound) steel storage tanks, a pump house, a circulating pump, a transfer pump, two filter-separators, an 8-inch stainless steel transfer line, RP-1 foam generating building and necessary valves, piping, and controls. Two concrete RP-1 holding ponds, 150 feet by 250 feet with a water depth of two feet, are located north of the launch pad, one on each side of the north-south axis. The ponds collect spilled RP-1 and water drainage from the pad area. Traps permit outflow of water from the holding ponds, but retain the fuel for skimming and disposal.

The LH_2 facility consists of one 850,000-gallon spherical storage tank, a vaporizer/heat exchanger which is used to pressurize the storage tank to 65 psig, a vacuum-jacketed, 10-inch, Invar transfer line and a burn pond venting system. The internal tank pressure, maintained by circulating LH_2 from the tank through the vaporizer and back into the tank, is sufficient to provide the proper flow of LH_2 from the storage tank to the vehicle without using a transfer pump. Liquid hydrogen boiloff from the storage and ML areas is directed through vent piping to bubblecapped headers submerged in the burn pond. The hydrogen is bubbled to the surface of the 100-foot square, water filled, concrete pond where a hot wire ignition system maintains the burning process.

LOX SYSTEM FACILITY.

The lox facility is located in the northwest quadrant of the pad area, approximately 1,450 feet from the center of the pad. The facility consists of one 900,000-gallon spherical storage tank, a lox vaporizer to pressurize the storage tank, 1000 gpm fill pumps, a vacuum jacketed transfer line, and a 10 in. uninsulated vent line from lut to drain basin.

Pad B lox system modifications for Skylab Saturn IB consist of using the 1000 gpm pumps and vacuum-jacketed line for fill and replenish; and disabling the un-insulated 10,000 gpm main lox fill pumps and line used for Saturn V.

GASEOUS HYDROGEN FACILITY.

This facility is located on the pad perimeter road northwest of the liquid hydrogen facility. The facility provides GH_2 at 6,000 psig to the launch vehicle. The facility consists of four storage tanks having a total capacity of 800 cubic feet, a flatbed trailer on which are mounted liquid hydrogen tanks and a liquid-to-gas converter, a transfer line and necessary valves and piping.

AZIMUTH ALIGNMENT BUILDING.

The azimuth alignment building is located in the approach ramp to the launch structure in the median of the crawlerway about 700 feet from the ML positioning pedestals. The building houses

Section VII Ground Support Interface

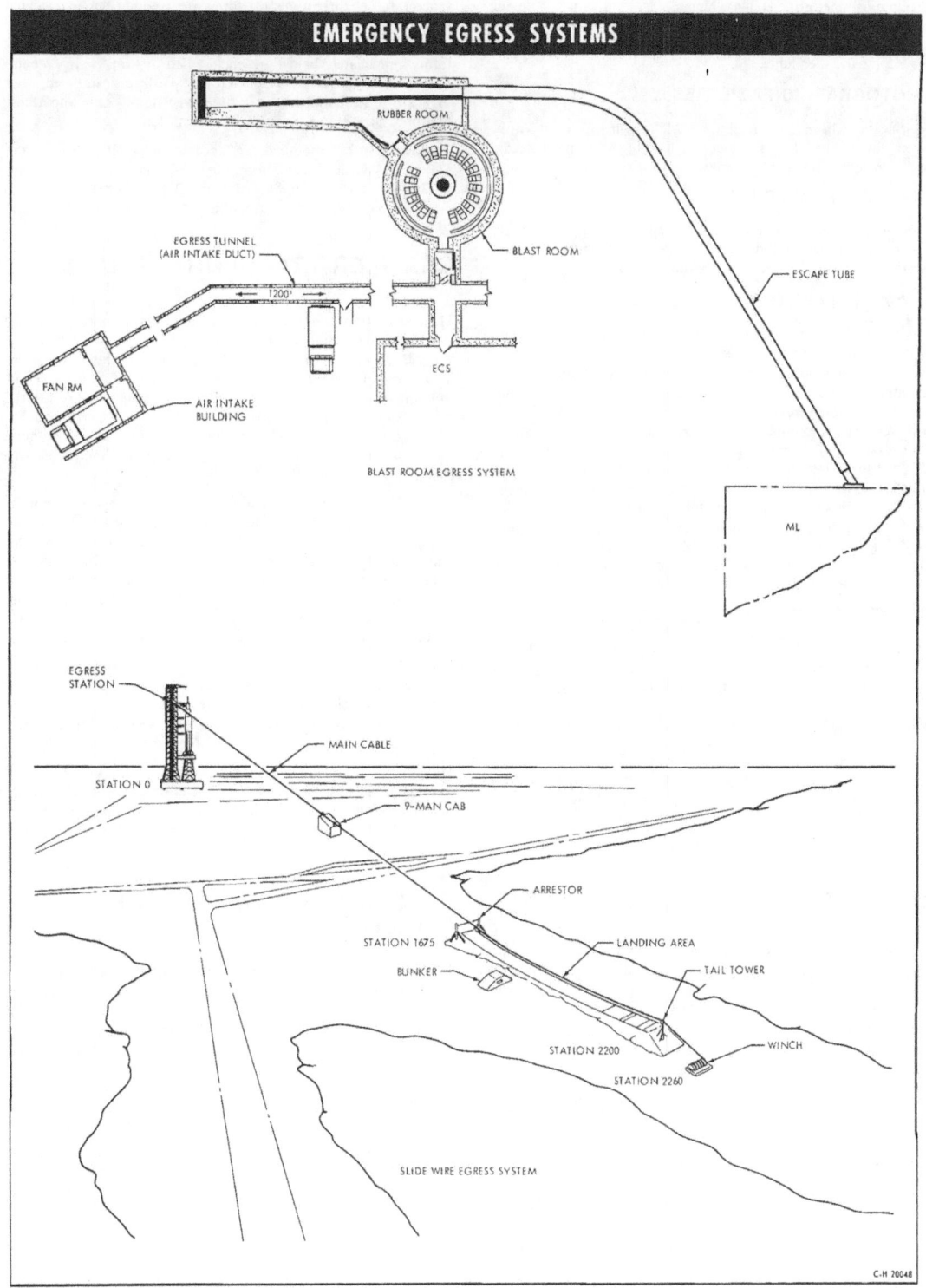

Figure 7-11

the auto-collimator theodolite which senses, by a light source, the rotational output of the stable platform. A short pedestal, with a spread footing isolated from the building, provides the mounting surface for the theodolite.

PHOTOGRAPHIC FACILITIES.

Modifications have been made for Skylab Saturn IB to adjust the water flow to the ML, to reduce the flow to the flame deflector, and to delete excess storage tanks.

AIR INTAKE BUILDING.

This building houses fans and filters for the air supply to the PTCR, pad cellular structure and the ML base. The building is located west of the pad, adjacent to the perimeter road.

FLAME DEFLECTOR.

The flame deflector is positioned under the vehicle in the flame trench and routes the booster exhaust away from the pad complex. These facilities support photographic camera and closed circuit television equipment to provide real-time viewing and photographic documentation coverage. There are six camera sites in the launch pad area, each site containing an access road, five concrete camera pads, a target pole, communication boxes and a power transformer with a distribution panel and power boxes. These sites cover prelaunch activities and launch operations from six different angles at a radial distance of approximately 1,300 feet from the launch vehicle. Each site has four engineering sequential cameras and one fixed, high speed, metric camera (CZR). A target pole for optical alignment of the CZR camera is located approximately 225 feet from the CZR pad and is approximately 86 feet high.

PAD WATER SYSTEM FACILITIES.

The pad water system facilities supply water to the launch pad area for fire protection, cooling, and quenching. Specifically, the system furnishes water for the industrial water system, flame deflector cooling and quench, ML and pedestal deck cooling and quench, ML tower fogging and service arm quench, sewage treatment plant, Firex water system, lox and fuel facilities, ML and MSS fire protection and all fire hydrants in the pad area. The water is supplied from three 6-inch wells, each 275 feet deep. The water is pumped from the wells through a desanding filter and into a 1,000,000-gallon reservoir.

MOBILE SERVICE STRUCTURE (MSS).

The mobile service structure (figure 7-12) provides access to those portions of the space vehicle which cannot be serviced from the ML while at the launch pad. During nonlaunch periods, the MSS is located in a parked position along side of the crawlerway, 7,000 feet from the nearest launch pad. The MSS is transported to the launch site by the C-T. It is removed from the pad a few hours prior to launch and returned to its parking area.

The MSS is approximately 402 feet high, measured from ground level, and weighs 12 million pounds. The tower structure rests on a base 135 feet by 135 feet. The top of the MSS base is 47 feet above grade. At the top, the tower is 87 feet by 113 feet.

The MSS is equipped with systems for air conditioning, electrical power, communications networks, fire protection, nitrogen pressurization, hydraulic pressure, potable water and spacecraft fueling.

The structure contains five work platforms which provide access to the space vehicle. The outboard sections of the platforms are actuated by hydraulic cylinders to open and accept the vehicle and to close around it to provide access to the launch vehicle and spacecraft. The three upper platforms are fixed but can be relocated as a unit to meet changing vehicle configurations. The uppermost platform is open, with a chain-link fence for safety. The two platforms immediately below are enclosed to provide environmental control to the spacecraft. The two lowest platforms can be adjusted vertically to serve different parts of the vehicle. Like the uppermost platform, they are open with a chain-link fence for safety.

Platform No. 2 is modified for the exterior configuration of the Saturn IB vehicle, and Platform No. 1 is modified for the Skylab Saturn V vehicle. After Skylab 1 and 2 launches, Platform No. 1 will be modified to be compatible with the Saturn IB vehicle.

CRAWLER-TRANSPORTER (C-T).

The crawler-transporter (figure 7-13) is used to transport the mobile launcher and the mobile service structure. The ML, with the space vehicle, is transported from the vehicle assembly building to the launch pad. The MSS is transported from its parking area to and from the launch pad. After launch, the ML is taken to the refurbishment area and subsequently back to the VAB. The C-T is capable of lifting, transporting, and lowering the ML or the MSS without the aid of auxiliary equipment. The C-T supplies limited electric power to the ML and the MSS during transit.

The C-T consists of a rectangular chassis which is supported through a suspension system by four dual-tread crawler-trucks. The overall length is 131 feet and the overall width is 114 feet. The unit weighs approximately 6 million pounds. The C-T is powered by self-contained, diesel-electric generator units. Electric motors in the crawler-trucks propel the vehicle. Electric motor-driven pumps provide hydraulic power for steering and suspension control. Air conditioning and ventilation are provided where required.

The C-T can be operated with equal facility in either direction. Control cabs are located at each end and their control function depends on the direction of travel. The leading cab, in the direction of travel, will have complete control of the vehicle. The rear cab will, however, have override controls for the rear trucks only.

Maximum C-T unloaded speed is 2 mph, 1 mph with full load on level grade, and 0.5 mph with full load on a five percent grade. It has a 500-foot minimum turning radius and can position the ML or the MSS on the facility support pedestals within ± two inches.

CONVERTER/COMPRESSOR FACILITY.

The primary gaseous nitrogen source for LC-39 is via a cross country line from an industrial supplier; however, the on-site Converter/Compressor Facility (CCF) operates as a back-up source. For gaseous helium the CCF compresses helium brought in by trucks and railroad cars. The CCF is located in the northside of the crawlerway between the VAB and the MSS park area. The CCF converts liquid nitrogen to low pressure and high pressure gaseous nitrogen and compresses gaseous helium. It then supplies the gases to the storage facilities at the launch pad and at the VAB.

The facility includes a 500,000-gallon storage tank for liquid nitrogen, tank vaporizers, high pressure liquid nitrogen pump and vaporizer units, high pressure helium compressor units, helium and nitrogen gas driver/purifiers, rail and truck transfer facilities and a data link transmission cable tunnel.

After processing, the gaseous nitrogen is piped to the distribution lines supplying the VAB area (6,000 psig) and the pad (150 psig and 6,000 psig). The gaseous helium is stored in tube-bank rail

Section VII Ground Support Interface

cars. The helium passes through the CCF helium compressors which boost its pressure from the tube-bank storage pressure to 6,000 psig after which it is piped to the VAB and pad high pressure storage batteries. Mass flow rates of high pressure helium, high

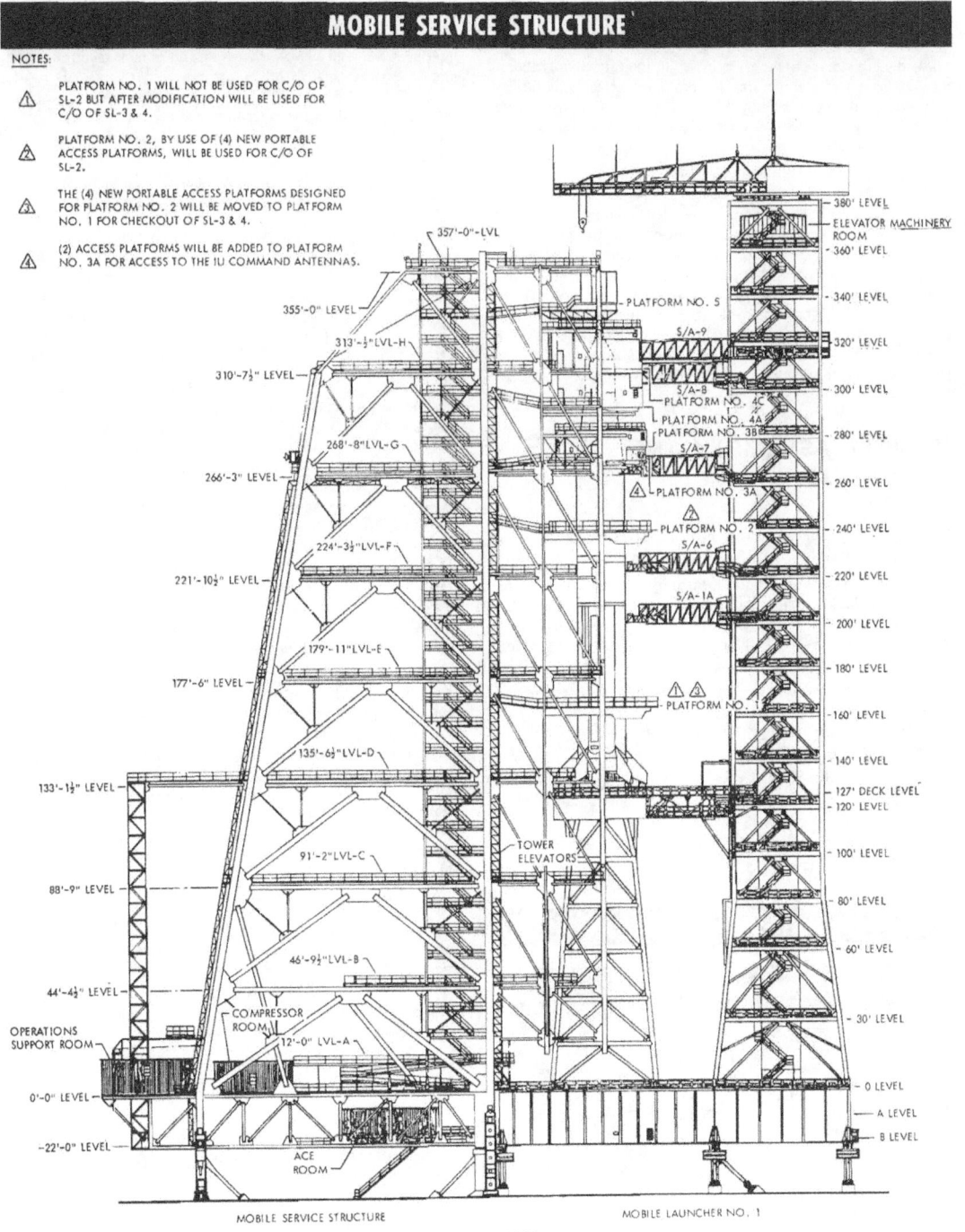

Figure 7-12

Section VII Ground Support Interface

pressure nitrogen, and low pressure nitrogen gases leaving the CCF are monitored on panels located in the CCF.

ORDNANCE STORAGE AREA.

The ordnance storage area serves LC-39 in the capacity of laboratory test area and storage area for ordnance items. This facility is located on the north side of the crawlerway and approximately 2,500 feet northeast of the VAB. This remote site was selected for maximum safety.

The ordnance storage installation, enclosed by a perimeter fence, is comprised of three archtype magazines, two storage buildings, one ready-storage building, an ordnance test building and a guard service building. These buildings, constructed of reinforced concrete and concrete blocks, are over-burdened where required. The facility contains approximately 10,000 square feet of environmentally controlled space. It provides for storage and maintenance of retrorockets, ullage rockets, explosive separation devices, escape rockets and destruct packages. It also includes an area to test the electro-explosive devices that are used to initiate or detonate ordnance items. A service road from this facility connects to Saturn Causeway.

VEHICLE ASSEMBLY AND CHECKOUT.

The vehicle stages and the instrument unit (IU) are, upon arrival at KSC, transported to the VAB by special carriers.

The S-IB stage is erected on hold-down arms installed in the transfer aisle for installation of fins then is lifted onto the pedestal of ML-1 in High Bay 1. The S-IVB stage and the IU are delivered to preparation and checkout cells in the low bay area for inspection, checkout, and pre-erection preparations.

The S-IVB stage and IU are then moved to the transfer isle and over the diaphram into High Bay 1 and assembled vertically on the ML with the S-IB stage. The spacecraft and launch escape system are then assembled.

Following assembly, the space vehicle is connected to the LCC via a high speed data link for integrated checkout and a simulated flight test. When checkout is completed, the crawler-transporter (C-T) picks up the ML, with the assembled space vehicle, and moves it to the launch site over the crawlerway.

Figure 7-13

For the initial Skylab Saturn IB mission, SL-2, a boilerplate spacecraft (BP-30) is installed and the vehicle is moved to the Pad for facilities checkout. The vehicle is then returned to the VAB for installation of the flight spacecraft.

At the launch site, the ML is emplaced and connected to system interfaces for final vehicle checkout and launch monitoring. The mobile service structure (MSS) is transported from its parking area by the C-T and positioned on the side of the vehicle opposite the ML. A flame deflector is moved on its track to its position beneath the blast opening of the ML to deflect the blast from the S-IB stage engines.

For launch of the SL-1, the MSS will be moved briefly to Pad A for inspection of the vehicle, then returned to Pad B.

During the prelaunch checkout, the final system checks are completed, the MSS is removed to the parking area, propellants are loaded, various items of support equipment are removed from the ML and the vehicle is readied for launch. After vehicle launch, the C-T transports the ML to the VAB for refurbishment.

TEST SYSTEM.

The requirement to launch Saturn IB vehicles from the Saturn V launch site mandated a reconfiguration of LC-39-3 (FR-3, ML-1, Pad B, and High Bay 1). The launch vehicle ground support equipment (LVGSE) being used to check out and launch the Skylab Saturn IB vehicles is basically Saturn V equipment located at LC-39-3 supplemented by hardware relocated from LC-34/37B and by some new equipment. The S-IB ESE, for example, utilizes converted S-II ESE, augmented by S-IC ESE and some LC-34/37B hardware. The Saturn V ML-3 Terminal Countdown Sequencer (TCS) will be used for the Saturn IB automatic sequence. MGSE systems will utilize Saturn V MGSE, as required and the S-IB stage calips console relocated from LC-37B. The LC-39-110A ground computer and display system will be utilized for Saturn IB, but new software programs will be prepared.

A computer controlled automatic checkout system is used to accomplish the VAB (high bay) and pad testing. A 110A ground computer and the equipment necessary to service and check out the launch vehicle are installed on the ML. Also a 110A ground computer and the display and control equipment necessary to monitor and control the service and checkout operations are installed in the LCC. The computers operate in tandem through a data link. The computer in the ML receives commands from and transmits data to the computer in the LCC. The physical arrangement of the LCC and the ML are illustrated in figures 7-14 and 7-5 respectively.

Test System Operation.

Test system operation for Saturn IB launch vehicle checkout is conducted from the firing room (see figure 7-15). During prelaunch operations, each stage is checked out utilizing the stage control and display console. Each test signal is processed through the computer complex, and is sent to the vehicle. The response signal is sent from the vehicle, through the computer complex, and the result is monitored on the display console. The basic elements of the test system and their functional relationship are shown in figure 7-16.

A switch on the control console can initiate individual operation of a system component or call up a complete test routine from the computer. The insertion of a plastic coded card key, prior to console operation, is a required precaution against improper program callup. Instructions, interruptions and requests for displays are entered into the system by keying in proper commands at the console keyboards.

A complete test routine is called up by initiating a signal at the control panel. The signal is sent to the patch distributor located

Section VII Ground Support Interface

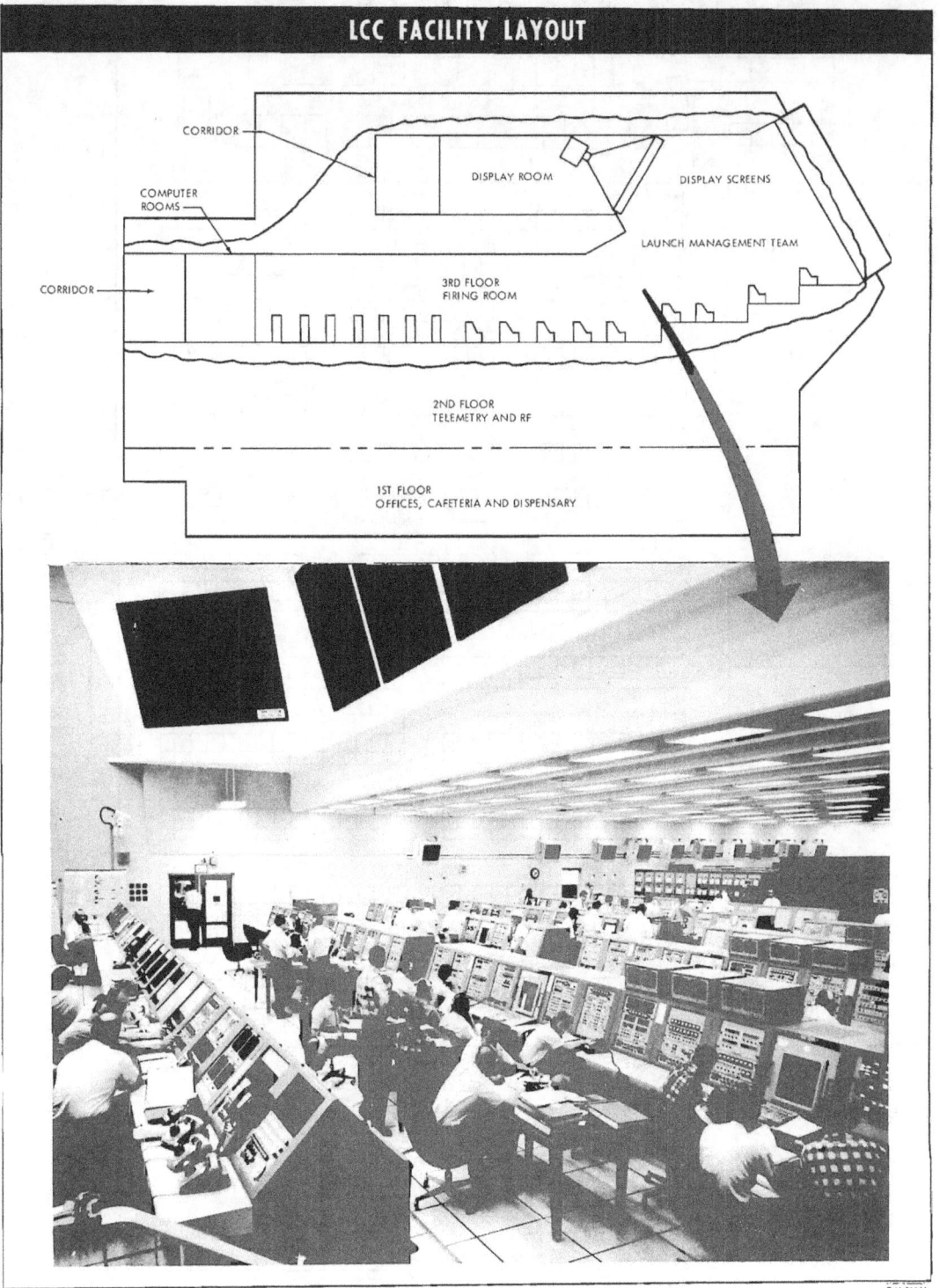

Figure 7-14

7-17

Section VII Ground Support Interface

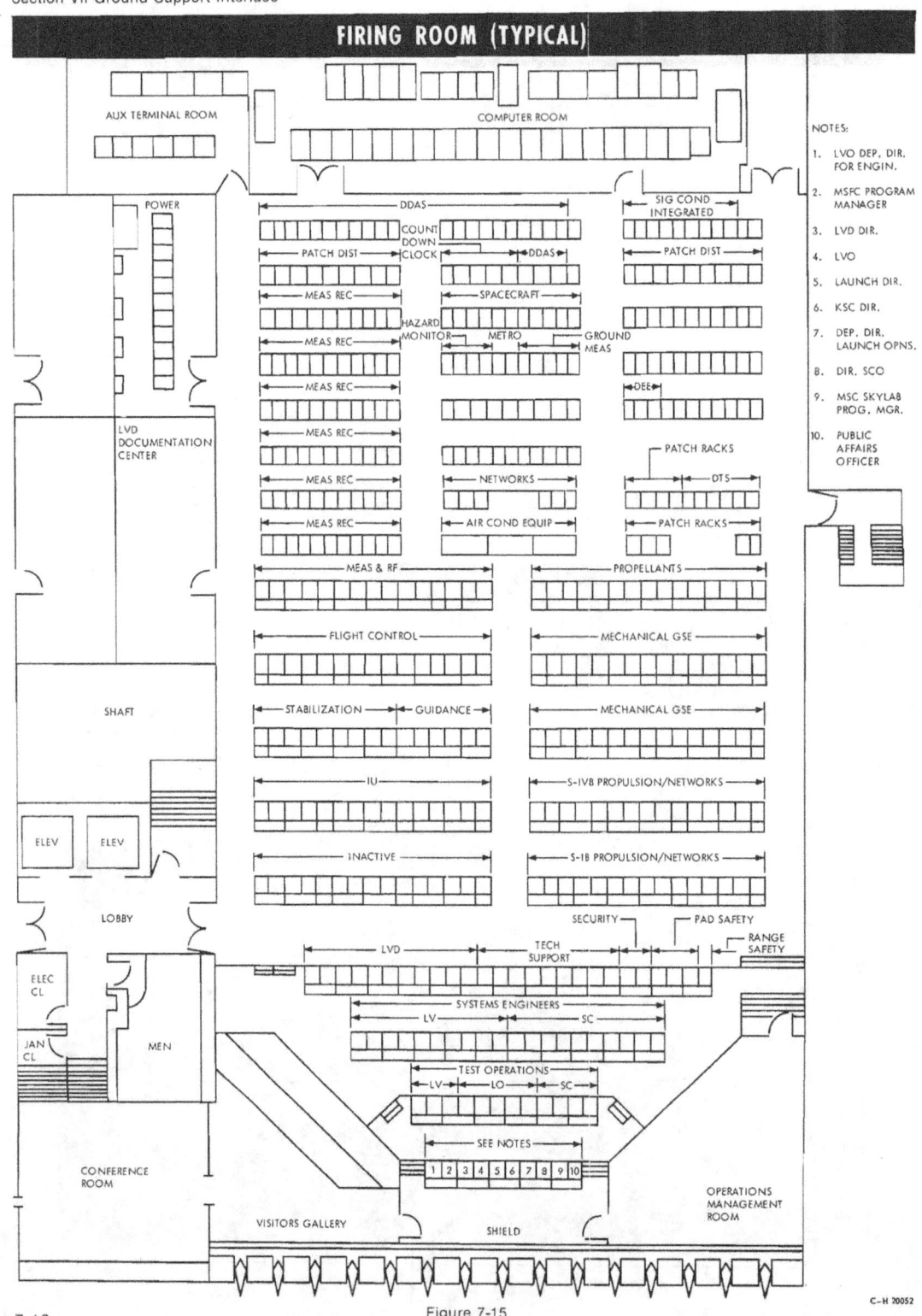

Figure 7-15

Section VII Ground Support Interface

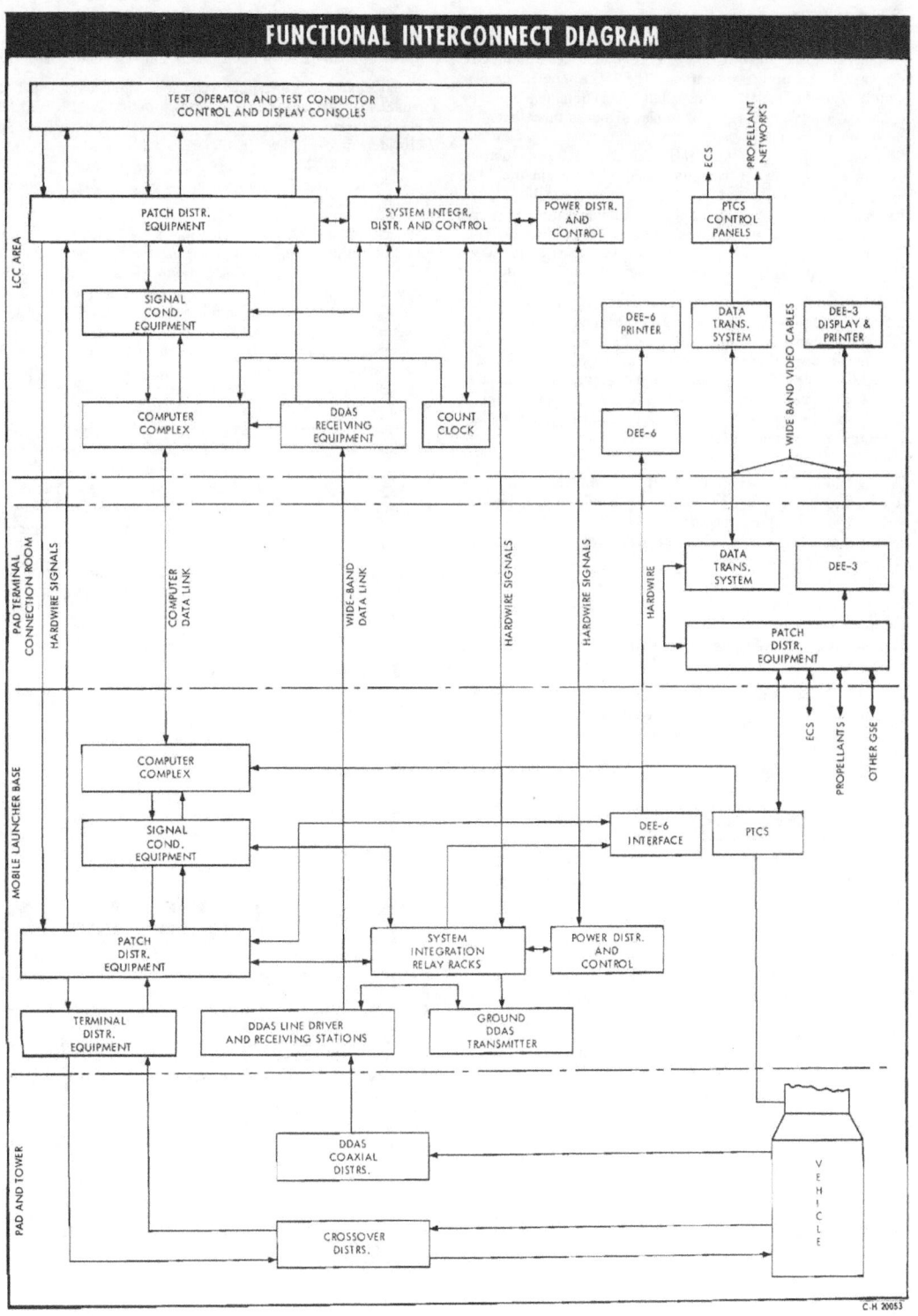

Figure 7-16

Section VII Ground Support Interface

in the LCC and is routed to the appropriate signal conditioning equipment where the signal is prepared for acceptance by the LCC computer complex. The LCC computer communicates with the ML computer to call up the test routine. The ML computer complex sends the signal to the ML signal conditioning equipment and then to the stage relay rack equipment. The signal is then routed to the terminal distribution equipment and through the crossover distributor to interrogate the vehicle sensors. The sensor outputs are sent back to the ML computer complex for evaluation. The result is then sent to the LCC computer complex which routes the result to the stage console for display. Manual control of vehicle functions is provided at the control consoles. This control bypasses the computers and is sent to the vehicle by means of hardwire. The result is also sent back to the display console by hardwire.

The digital data acquisition system (DDAS) collects the vehicle and support equipment responses to test commands, formats the test data for transmission to the ML and LCC, and decommutates the data for display in the LCC. Decommutated test data is fed to the LCC computers for processing and display, and for computer control of vehicle checkout. The DDAS consists of telemetry equipment, data transmission equipment and ground receiving stations to perform data commutation, data transmission and data decommutation.

The digital event evaluators (DEE) are used to monitor the status of input lines and generate a time tagged printout for each detected change in input status. High speed printers in the LCC are connected to each DEE to provide a means for real-time or post-test evaluation of discrete data. Two systems (DEE-3 and DEE-6) are used to monitor discrete events.

Two DEE-3's are located in the PTCR with a printer located in the LCC. One unit monitors 768 inputs and the other monitors 1032 inputs, all associated with propellant loading, environmental control, and water control.

The DEE-6 processor, offline printer and remote control panel are located in the LCC, while interface equipment is located in the ML base. Several remote printers are located strategically within the firing room. The DEE-6 monitors up to 4320 discrete signals from the vehicle stage umbilicals, pad and tower ground support equipment.

The computer complex consists of two 110A general purpose computers and peripheral equipment. This equipment includes a line printer, card reader, card punch, paper tape reader and magnetic tape transports. The peripheral equipment provides additional bulk storage for the computer, acts as an input device for loading test routines into the computer memory and as an output device to record processed data. One computer is located in the ML base and the other in the LCC. The computers are connected by underground hardwire. The LCC computer, the main control for the system accepts control inputs from test personnel at the consoles in the firing room as well as inputs from tape storage and transmits them as test commands to the ML computer. The ML computer has the test routines stored in its memory banks. These routines are called into working memory and sent as discrete signals to the launch vehicle in response to the commands received from the LCC computer. The ML computer reports test routine status, data responses and results of test to the LCC computer. It is through this link that the control equipment and personnel in the firing room are informed of the test progress.

The propellant tanking computer system (PTCS) determines and controls the quantities of fuel and oxidizer on board each stage. Optimum propellant levels are maintained and lox and LH_2 are replenished as boiloff occurs during the countdown. The propellant tanking operation is monitored on the PTCS control panel.

The propellant Data Transmission System (DST) is the digital communication link between the LCC and the PTCR for propellant loading and environmental control command and monitor data. The system operates on a time-shared mode with data transmitted over video pairs. Each input signal is sampled about once a second.

Visual surveillance of launch vehicle checkout is provided to the launch management team and for distribution to MSC and MSFC through the operational television system (OTV). Fifty two cameras provide this capability, 18 of which are located on the ML, in the pad area, 12 on the MSS and 6 in the LCC. Any camera may be requested for viewing on the 10 x 10 foot screens in the firing room.

Vehicle and ground instrumentation measurements may be monitored by firing room personnel using the television data display system (TDDS). Stage DDAS, ground DDAS, and facility and environmental measurements are processed in the CIF telemetry ground station, and prepared for computer entry. The CIF GE-635 scientific computer is used to process these measurements to provide engineering units to be displayed on 10 dual eight-inch Operational Television (OTV) monitors in the firing room. The computer can also check each measurement for out-of-limit conditions, and alert the operator either visually or audibly if pre-selected limits are exceeded. The operator has the option of selecting pre-defined pages or operator defined pages which contain up to sixteen measurements, pre-defined sets of measurements where only those that have out-of-limit conditions are displayed, or a graphical presentation of an individual measurement. The graph can be generated in real time, or from the last 51 minutes of history, or a combination of both.

Certain major events may be observed by members of the launch management team who occupy the first four rows in the firing room. The significant launch vehicle events which are displayed on the 10 x 10 foot screen are shown in figure 7-17.

PRELAUNCH OPERATIONS.

The prelaunch operations (figure 7-18) take place in the Manned Spacecraft Operations (MSO) Building, the VAB and the launch pad.

MSO Building Activities.

After receipt of the spacecraft stages at KSC, inspection, testing, assembly and integrated checkout of the spacecraft take place in the MSO building. The assembled spacecraft is transported to the VAB and mated to the launch vehicle.

		MAJOR EVENTS		
AS-206	LAUNCH SEQUENCE START	S-IB ON INTERNAL POWER	S-IB STAGE CUTOFF	
RANGE SAFE	S-IVB LOX TANK PRESSURIZED		SIB/S-IVB SEP LOGIC ZERO	
	S-IVB LH₂ TANK PRESSURIZED		S-IVB ENGINE START	
	S-IVB PROPELLANTS PRESSURIZED		S-IVB CUTOFF	
LAUNCH SUPPORT PREPS COMPLETE	S-IB FUEL TANKS PRESSURIZED	LSE READY FOR IGNITION		
S-IVB PREP COMPLETE	S-IB LOX TANKS PRESSURIZED	READY FOR S-IB IGNITION		RF SILENCE
IU READY	S-IB PROPELLANTS PRESSURIZED	START S-IB IGNITION		
				TEST HOLDING
EDS READY	AUDIO SEL POSIT NO. 3 INDICATION	COMMIT		TEST COUNTING
S-IB PREP COMPLETE	AUDIO SEL POSIT NO. 4 INDICATION	LIFTOFF		EVENT SYSTEM CALIBRATING

Figure 7-17

Section VII Ground Support Interface

VAB Activities.

The VAB activities are the assembly and checkout activities which are completed in two major areas of the VAB; the high bay and the low bay.

Low Bay Activities. The low bay activities include receipt and inspection of the S-IVB stage and IU and the assembly and checkout of the S-IVB stage.

The S-IVB stage is brought into the low bay area and positioned on the checkout dolly. A fuel tank inspection, J-2 engine leak test, hydraulic system check and propellant level sensor electrical checks are made.

High Bay Activities. High bay activities include S-IB stage checkout, stage mating, stage systems tests, launch vehicle integrated tests, space vehicle overall tests, and a simulated flight test. High bay checkout activities are accomplished using the consoles in the firing room, the computer complex, and display equipment.

The S-IB stage is positioned and secured to the ML and access platforms are installed. The umbilicals are secured to the vehicle plates. Prepower and power-on checks are made to ensure electrical continuity. Pneumatic, fuel, lox and H-1 engine leak checks are made. Instrumentation, and range safety system checks are made.

The S-IVB stage is mated to the S-IB stage and the IU is mated to the S-IVB. The S-IVB and IU umbilicals are secured to the vehicle plates. Pre-power and power-on checks are made to ensure electrical continuity. S-IVB engine hydraulic, pressurization and auxiliary propulsion system leak checks are made. S-IVB propellant, propulsion, pressurization and range safety system checks are made. IU S-band, and C-band and guidance and navigation system checks are made.

Following completion of the stage system tests, launch vehicle integrated checks are accomplished. Vehicle separation, flight control, sequence malfunction and emergency detection system checks are made. The spacecraft is then mated to the launch vehicle.

For the first checkout flow with SA-206, a boilerplate spacecraft (BP-30) is installed and the vehicle is moved to the pad for facilities checkout. The vehicle is then returned to the VAB for installation of the flight spacecraft.

After the spacecraft is mated, two space vehicle tests are made. Test number 1 is performed to verify RF, ordnance, pressurization, propulsion, guidance and control, propellant and emergency detection system operation. Test number 2 is performed to verify proper operation of all systems during an automatic firing sequence and flight sequence. This includes a simulated holddown arm release, electrical umbilical ejection, swing arm retraction and firing of live ordnance in test chambers; and a simulated flight sequence that verifies proper operation of the space vehicle during a normal minus count and an accelerated plus count. A normal mission profile is followed during this time. The simulated flight test ensures that the space vehicle is ready for transfer to the pad. The launch escape system is installed on the command module of the spacecraft. The ML and space vehicle are now ready for transport to the pad.

Vehicle Transfer and Pad Mating Activities.

After completion of the VAB activities, the ML transports the assembled space vehicle to the launch pad. Approximately eight hours are required for this operation. The space vehicle/ML are then interfaced with the launch pad.

Pad Activities.

For the facility checkout flow of the pad, SA-206 and boilerplate spacecraft BP-30 will be moved to the pad. Facility checkout activities will include an LV cryogenic loading sequence; a LUT/PAD water system test; and an RP-1 loading sequence. After facility checkout is completed, the vehicle is returned to the VAB for installation of the flight spacecraft. The space vehicle will then be moved to the pad for prelaunch checkout and servicing.

In general, once the vehicle and ML have been mated to the pad facility, two major operations must be performed. The first is to verify the readiness of the launch vehicle, spacecraft and launch facility to perform the launch sequence and the second is to complete the launch operation.

The Countdown Demonstration Test (CDDT) verifies that the launch vehicle and the ground support equipment are in launch status. The CDDT is performed in two phases, the wet CDDT and the dry CDDT.

The wet CDDT is performed the same as the launch countdown with the following major exceptions:

1. Service arms are pressure tested.

2. Digital range safety command system test code plugs are used instead of flight code plugs.

3. Hypergol cartridges, igniters, initiators, safe and arm devices, and exploding bridgewire detonators are inert.

4. Astronauts do not board the spacecraft.

5. Terminal count sequence is interrupted at time for ignition.

The dry CDDT is performed the same as the last 3 1/2 hours of the launch countdown with the following major exceptions:

1. Launch vehicle cryogenic propellants are not on board.

2. The primary damper is not disconnected.

3. Service Arm No. 9 is reconnected as soon as the system has stabilized in the park position.

4. Hypergol cartridges, igniters, initiators, safe and arm devices, and exploding bridgewire detonators are inert.

5. Terminal count sequence is not initiated.

PRELAUNCH OPERATIONS	
OPERATIONS	COMPLETE * (DAYS BEFORE LAUNCH)
CSM OPERATIONS	
COMBINED SYSTEM TEST	L-228
UNMANNED ALTITUDE RUN	L-130
MANNED ALTITUDE RUN	L-110
CSM/SLA MATE	L-100
ORDNANCE INSTALLATION	L-98
LV VAB HIGH BAY OPERATIONS	
S-IB ERECTION	L-242
LV ERECTION	L-235
LV ELECTRICAL SYSTEM TEST	L-213
LV MALFUNCTION OVERALL TEST	L-173
LV SERVICE ARM OVERALL TEST	L-131
SPACECRAFT ERECTION	L-81
SPACE VEHICLE VAB OPERATIONS	L-71
TRANSFER TO PAD	L-69
PAD OPERATIONS	
LV POWER ON	L-67
SV PLUGS IN TEST	L-41
SV FLIGHT ELECTRICAL MATING	L-40
SV BACKUP GUIDANCE TEST	L-38
SV FLIGHT READINESS TEST	L-34
SV HYPERGOLIC LOADING	L-21
S-IB RP-1 LOADING	L-19
CDDT-WET/DRY	L-10
COUNTDOWN	L-0

C-H 20055

Figure 7-18

Section VII Ground Support Interface

Following the CDDT, preparations for the actual countdown are started. The preparations include items which would either compromise the safety of the vehicle if done later in the countdown or impose additional constraints on pad access during the final phases of the countdown.

Approximately six days before the launch readiness day, the countdown begins and the space vehicle is subjected to the final checkout and servicing operations required for launch.

Concurrently, the Skylab 1 space vehicle is undergoing prelaunch checkout and launch operations on Pad A. The countdown for Skylab 2 is timed such that the countdown clock will be started (at T-22:30 hours) upon lift-off of SL-1.

The final phase of the countdown starts approximately nine hours prior to liftoff. During the final phase, the cryogenics are loaded, conditioned and pressurized. Final checks are made on all subsystems. The propulsion systems are serviced and prepared for launch. All onboard spheres are brought up to flight pressure and the crew mans the Command Module.

By the time spacecraft closeout is complete, most major operations have been completed. Propellants are being replenished as required to supplement cryogens lost due to boiloff. Boiloff will continue until the various stage vent valves are closed for tank prepressurization and some vapor may be noticeable.

With the start of the automatic sequence at T-187 seconds, the final operations required for launch begin. All pneumatic and propellant supply lines are vented and purged to prevent damage to the vehicle at umbilical release. The vehicle is switched to internal power, necessary purges are put in launch mode and some services are retracted.

At T-3 seconds, the S-IB ignition command is given. At T-0 seconds, the launch commit signal is given, causing the eight holddown arms to retract. These arms restrain the launch vehicle until a satisfactory thrust level is achieved after which the controlled release assemblies provide for gradual release of the vehicle during liftoff.

HOLD AND RECYCLE CRITERIA.

Interruption of the countdown due to equipment failure, weather or other causes may occur at any time. When the countdown is interrupted, subsequent feasible actions depend on the function taking place at the time. These actions include holding and/or recycling. Feasible actions also, in some cases, are affected by previous operations conducted on the vehicle, such as the number of pressure cycles the propellant tanks have undergone.

A hold is defined as an interruption of the countdown for unfavorable weather, repair of hardware, or correction of conditions unsatisfactory for launch or flight.

In a recycle, the countdown is stopped and returned to a designated point as specified in the launch mission rules.

For a scrub, the launch attempt must be rescheduled for a later window.

A turnaround comprises the actions required to recycle, hold until the countdown can be resumed for a specific window, and complete the countdown from the re-entry point to T-0.

Decision/repair time is the time available to make decisions and/or conduct repair operations before initiating the count.

LAUNCH CONSTRAINTS.

Various operational, launch vehicle, spacecraft and support equipment factors affect hold/recycle processing actions. Several of these factors are briefly discussed in the following paragraphs.

The length of the launch window on any launch date is a mission peculiar constraint. This constraint determines the maximum hold limit for the countdown period between start of cryogenic loading and commence S-IVB stage start bottle chilldown.

Launch vehicle batteries have a life of 120 hours following activation. The batteries are installed in the vehicle at T-53.5 hours and are activated 33.0 hours prior to their installation. Assuming a countdown programmed hold of 15 hours, the battery life expended in a normal count is 88.0 hours, leaving an available battery life of 32.0 hours.

The Safe and Arm (S&A) devices are remotely controlled ordnance items used to make safe and/or arm the launch vehicle propellant dispersion systems. The devices are certified at T-108 hours. Recertification is required in seven calendar days. Device removal from the vehicle is required to perform recertification. The devices are installed at T-110 hours and connected at T-46.5 hours. In a normal count, the allowable S&A device life is 59 hours.

The S-IVB Auxiliary Propulsion System (APS) modules are serviced with hypergolic propellants. Pressurization of the system is done at T-40 minutes. Depressurization and gas removal must be accomplished in the event of a scrub if access is required in the S-IVB access control area. This task takes two hours and requires use of the Mobile Service Structure (MSS).

The Command and Service Module (CSM) fuel cell cryogenics provide the electrical power for the spacecraft. The water resulting from the reaction in the fuel cell is used for drinking purposes during space missions. The cryogenic tanks are loaded to sufficient capacity to tolerate a 56 hour delay for a normal count. Water generated by fuel cell operation must be drained if a hold will exceed 17 hours. Cryogenic replenishment is normally required if turnaround exceeds 56 hours. The MSS is required for CSM cryogenic servicing.

Capacities of launch support facilities and equipment such as the gaseous hydrogen and helium facilities, the cryogenic storage facilities and the ground hydraulic supply unit affect hold and recycle capability. For example, the gaseous hydrogen system can support four complete cycles of start tank chilldown operations from T-22 minutes to T-4.5 minutes before recharging of the storage battery is required. Recharging is accomplished by mobile units which cannot be moved into position until launch vehicle cryogenics are downloaded.

LAUNCH INTERLOCKS.

The terminal countdown sequencer (TCS) automatically controls the space vehicle countdown beginning at T-187 seconds. Interlocks are provided to prevent initiation of this sequence or to terminate the sequence if conditions do not remain proper for launch. Once the automatic sequence has been initiated, it can only be stopped by a cutoff signal. There are no provisions for holding.

FIRING PREPARATION COMPLETE INTERLOCKS.

These interlocks insure that all vehicle and ground support equipment systems are in proper condition for starting the automatic sequence.

S-IB Stage ESE.

The following interlocks functions are provided by the S-IB stage ESE.

a. Lox leak failure
b. No conax fired
c. Ignition armed

Section VII Ground Support Interface

d. Safety switches armed

e. Flight sequence zero

f. Propellant dispersion system ready

g. Terminal countdown sequencer power supply OK

h. Spacecraft ready for launch

i. Instrument unit ready for launch

j. EDS ready

k. Range safety

l. Launch support equipment preparations complete

m. S-IVB systems ready

n. Prevalves open

o. Purges armed

p. Terminal countdown sequencer ready

q. +1D161 (launch bus) supervision

S-IVB Stage ESE.

The following interlocks functions, which are summed as the "S-IVB systems ready" interlock in the S-IB stage ESE, are provided by the S-IVB stage ESE.

a. LH_2 tank wet

b. Non-programmed engine cutoff off

c. Passivation relays reset

d. S-IVB engine cutoff

e. Engine ready

f. Aux hydraulic pump flight mode relay not reset

g. Aux hydraulic pump power on

h. Ordnance OK

i. Ullage rocket pilot relays reset

j. APS No. 1 or No. 2 engine valve power on

k. Lox and LH_2 prevalves emergency close command off

l. Heat exchanger bypass valve enable relay reset

m. Mixture ratio control valve relays reset

n. Engine start relay reset

o. TM PI transmitter on

Instrument Unit ESE.

The following interlock functions, which are summed as "Instrument Unit ready for launch" in the S-IB stage ESE, are provided by the IU ESE.

a. LVDA firing commit inhibit A and B

b. IU guidance failure A and B

c. 56 volts OK

d. 400 Hz power A and B

e. Control voltage OK

f. Control attenuator timer at zero

g. Control accelerometer on

h. Gyro on

i. ST-124M system ready

j. Power transfer command and T-47

k. IU power transfer complete

l. Switch selector internal power

m. Guidance alert and release

n. LVDA firing commit enable

EDS ESE.

The following interlock functions are provided by the EDS ESE and are summed as the "EDS ready for launch" interlock in the S-IB ESE.

a. EDS not ready

b. Ready circuits enable

READY FOR IGNITION INTERLOCKS.

These interlocks insure that all vehicle and ground support equipment systems are in the proper condition for starting the S-IB engine ignition sequence. If the interlocks are not met at time for ignition, the countdown will terminate automatically. The count will then have to be recycled to the start of automatic sequence. Once the ignition sequence has started, a change of state of these functions will not result in automatic termination of the countdown.

S-IB Stage ESE.

The following interlock functions are provided by the S-IB stage ESE.

a. Lox leak failure

b. No conax fired

c. Ignition armed

d. Safety switches armed

e. Flight sequence zero

f. Propellant dispersion system ready

g. Terminal countdown sequencer power supply OK

h. Spacecraft ready for launch

i. Instrument unit ready for launch

j. EDS ready

k. Range safety

l. S-IVB stage ready for launch

m. Propellants pressurized

n. S-IB power transfer complete

o. Gas generator lox injector purge on

p. Thrust chamber fuel injector purge on

q. Launch bus (+1D161) supervision

S-IVB Stage ESE.

The following interlock functions are provided by the S-IVB stage ESE. These are summed as "S-IVB stage ready for launch" in the S-IB stage ESE.

a. LH_2 tank wet

b. Non-programmed engine cutoff off

c. Passivation relays reset

d. S-IVB engine cutoff

e. Engine ready

f. Aux hydraulic pump flight mode relay not reset

Section VII Ground Support Interface

g. Aux hydraulic pump power on

h. Ordnance OK

i. Ullage rocket pilot relays reset

j. APS No. 1 or No. 2 engine valve power on

k. LH_2 directional vent in flight position

l. Power transfer complete

Instrument Unit ESE.

The following interlock functions are provided by the IU ESE. These are summed as "IU ready for launch" in the S-IB stage ESE.

a. LVDA firing commit inhibit A and B

b. IU guidance failure A and B

c. 56 volts OK

d. 400 Hz power A and B

e. Control voltage OK

f. Control attenuator timer at zero

g. Control accelerometer on

h. Gyro on

i. ST-124M systems ready

j. Power transfer command and T-47

k. IU power transfer complete

l. Switch selector internal power

m. Guidance alert and release

n. LVDA firing commit enable

EDS ESE.

The "EDS ready" interlock in the S-IB stage ESE is the "EDS not ready" interlock in the EDS ESE. This interlock ensures that the EDS control rate gyro rates are not excessive, enable logic is zero, cutoff enable 1 and 2 are not enabled, cutoff commands 1, 2, and 3 from the spacecraft have not been generated, and EDS unsafe A and B have not been generated.

CUTOFF INTERLOCKS.

Automatic Sequence Start Until Time for Ignition.

Cutoff interlocks for automatic sequence termination from the start of automatic sequence until time for ignition are as follows:

a. Manual cutoff (switch function on S-IB networks panel)

b. Emergency manual cutoff (switch function on S-IB firing panel)

c. Premature ignition

d. Premature commit

e. Terminal countdown sequencer failure

f. Voltage failure (S-IB stage)

g. Sequence failure

Ignition Until Commit.

The cutoff interlocks from ignition to commit are as follows:

a. Time for commit

b. All engines running

c. Absence of cutoff

Time for Ignition Until Commit.

The following functions are interlocked to provide cutoff between time for ignition and commit if conditions warrant.

a. Manual cutoff (switch function on S-IB networks panel)

b. Emergency manual cutoff (switch function on the S-IB firing panel)

c. Terminal countdown sequencer failure

d. IU failure

e. EDS failure

f. Lox leak failure

g. Any engine cutoff

h. Thrust failure

Commit Until Plugs Separation.

The following functions are interlocked for cutoff between commit and umbilical connector separation from the vehicle.

a. Manual cutoff (switch function on S-IB networks panel)

b. Emergency manual cutoff (switch function on S-IB firing panel)

c. Launch failure (liftoff has not occurred before launch failure timers have expired)

SECTION VIII
MISSION CONTROL MONITORING

TABLE OF CONTENTS

Introduction	8-1
Mission Control Center	8-1
Launch Control Center	8-3
Abort Ground Rules	8-7
Huntsville Operations Support Center	8-8
Launch Information Exchange Facility	8-10
Spaceflight Tracking & Data Network	8-12

INTRODUCTION.

The basic purpose of mission control monitoring is to provide guidance for launch and flight operations. This is intended to ensure the successful achievement of mission objectives. Successful mission accomplishment requires that launch and flight monitoring occur during real time so that rapid decisions can be made. Data must be analyzed rapidly in order to identify malfunctions and deviations. Effects must be correlated with causes accurately, so that corrective actions to the launch and flight operations can be taken. Mission rules are finalized prior to the final countdown. These rules are designed to minimize the real-time rationalizations required to cope with non-nominal situations.

MISSION CONTROL MANAGEMENT.

The mission director (MD) has overall authority during the mission. During mission operations, the MD is usually located at the mission control center in Houston (MCC), but he may be located at the launch control center in Kennedy Space Center (LCC-KSC). Both the flight and launch directors are responsible to the MD. The flight director at MCC exercises control over flight operations from liftoff of the space vehicle through spacecraft recovery. The launch director at LCC-KSC exercises prime control over launch operations from the beginning of the final launch countdown until the space vehicle has cleared the umbilical tower. Centralized mission control is accomplished at MCC and is supported by the network remote stations of the spaceflight tracking and data network (STDN), the Huntsville operations support center (HOSC) at Marshall Space Flight Center (MSFC), and the LCC-KSC. Specialists in the areas of navigation, electrical networks, instrumentation, and propulsion are located at the LCC and HOSC to provide MCC timely information as required.

MISSION CONTROL CENTER.

The flight operations director (FOD) is responsible for all operational aspects of Skylab spaceflight missions and provides the management interface between the flight director and program management. Direct responsibility for mission control, however, will be vested in the flight director at the mission control center (MCC) throughout the mission. Flight control functions will be effected by flight controllers at the MCC utilizing the remote sites of the STDN. Mission control will be accomplished by providing in-flight analysis of the mission (mission trajectory, vehicle systems, experiment systems, scientific data, and flight plan) and by controlling the progress of the in-flight phase of the mission through the utilization of voice, telemetry, tracking, and update capabilities remoted from the MCC and STDN facilities.

The general mission control functions to be accomplished by flight controllers are as follows:

a. Monitor and evaluate, in real and delayed time, the vehicle systems, experiment systems, scientific data, and trajectory data. Based upon these data, decisions will be made concerning the progress of the mission toward satisfying primary mission objectives and mandatory, principal, and secondary detailed test objectives (DTO's) and the need for proceeding to alternate flight plans, contingency plans, or mission aborts.

b. Monitor and evaluate the condition of the flight crew. Based upon these data, decisions affecting crew health and safety will be made.

c. Perform ephemeris and maneuver updating in real time. Updated information will be passed to the spacecraft via the up-data link and voice from the MCC.

d. Monitor, evaluate, and update flight plan activity, including experimental tasks, work/rest cycles, and equipment checks. Decisions to alter flight plan activity will be based upon such factors as crew status, spacecraft status, experiment status, STDN status, and weather.

e. Advise the flight crew of updated mission instructions, anomalies in spacecraft systems found during ground monitoring, ground evaluation and recommendations to solve or circumvent any spacecraft anomalies, and recovery-area weather conditions.

MISSION OPERATIONS CONTROL ROOM (MOCR).

MCC operations are conducted from a central MOCR where flight controllers monitor and analyze mission status in order to make decisions and take corrective actions consistent with flight plan objectives and mission rules. (Figure 8-1 shows the MOCR console layout.) Here the flight control team has rapid access to information concerning mission progress, but are not distracted by support function activities or by routine evaluation tasks. Operations in the MOCR are assisted by several staff support rooms (SSR), science rooms, and other supporting areas similar to those in Apollo operations. The MOCR staff is divided into three groups (figure 8-2): vehicle systems, experiments, and biomed; mission command and control; and flight dynamics/booster/EREP/EVA. Launch vehicle responsibilities lie with three booster systems engineers. Booster systems engineer No. 1 is responsible to the flight director for integrating all launch vehicle activities, executing command action as required and monitoring stage functions and propulsion performance. Booster systems engineer No. 2 is responsible for integrating all stage systems. Booster systems engineer No. 3 is responsible for integrating all IU, electrical, and instrumentation systems monitoring and troubleshooting.

FLIGHT OPERATIONS MANAGEMENT SUPPORT.

In the Apollo program, the spacecraft analysis (SPAN) activity served as an interface between the flight operations organization and the Apollo spacecraft program office (ASPO). This interface was required by the FOD for detailed technical support from the design, checkout and testing organizations through the ASPO representative. This support consisted primarily of engineering judgement of the design personnel on the operation of the vehicle

Section VIII Mission Control Monitoring

MOCR CONSOLE LAYOUT

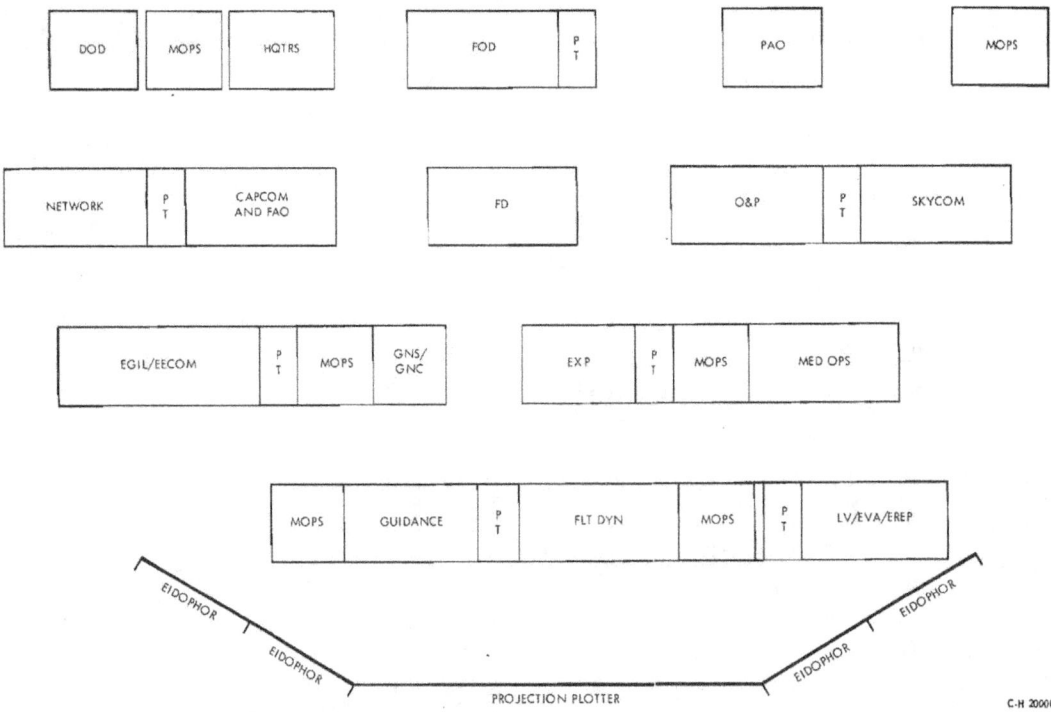

Figure 8-1

systems in off-nominal situations. The SPAN interface also provided the program office and other program elements with a system for making recommendations to the flight operation and served as a channel for receiving and evaluating certain near-real-time information and summary data as the operation progressed. For the Skylab program, the MSC Skylab program office will provide a similar interface capability for the MSC spacecraft and experiment hardware and for MSC flight software. Similarly, the Marshall Space Flight Center will provide an interface for the MSFC-designed or MSFC-contracted flight systems software, and experiments. For Skylab, the designation of this function will be changed to flight operations management support, and the designation for the present SPAN room will be changed to the flight operations management room (FOMR).

An additional role of the FOMR will be to accommodate the complexities of specifying mission activity priorities (on request from the flight operations team) where no clear-cut guidelines have been established or where mission problems necessitate tradeoffs. Thus, the FOMR operation will accomplish the following major functions. First, it will provide detailed technical support to the flight operations team for all flight systems hardware and software. Second, it will have a modified role in providing required policy level adjustments to the experiment priorities and planned mission activities as the flight progresses. Finally, the FOMR operation will provide support to the Skylab flight management team. The flight operations management support organization is illustrated in figure 8-3.

MISSION CONTROL PHASES.

Mission control during the Skylab mission phases involving the Saturn IB vehicle are discussed below.

Prelaunch Tests.

For SL-2, two MCC command interface tests will be conducted with each vehicle (CSM and LV). As with the SWS, this will be the first test in which the MCC will directly interface with each SL-2 vehicle for MCC command and telemetry verification. A second test will be repeated as close to launch as possible to serve as a final interface verification and will include all late telemetry and command changes to both flight and ground hardware and software. The MCC interface tests are required only for the first CSM mission (SL-2) unless telemetry or command changes to either onboard or ground equipment invalidate the test.

For each CSM flight the MCC may support an FRT and will support a CDDT test.

Countdown.

The launch countdown may be monitored (voice and telemetry) on a limited basis as early as T-24 hours. However, full MCC support will not be provided until approximately T-6 hours.

Spacecraft systems checkout will be accomplished by KSC using the automatic checkout equipment (ACE), and the status and gross system appraisal will be relayed to the MSFC and the MCC. This information, coupled with MCC's spacecraft checkout information and network evaluation, will provide the basis for mission GO/NO-GO decision. Specific flight control procedures during the prelaunch phases are as follows:

a. Monitor spacecraft checkout—KSC and MCC.

b. Monitor launch vehicle systems status—launch control center (LCC), MSFC, and MCC.

c. Verify network status—MCC and Goddard Space Flight Center (GSFC).

Section VIII Mission Control Monitoring

MOCR ORGANIZATIONAL STRUCTURE

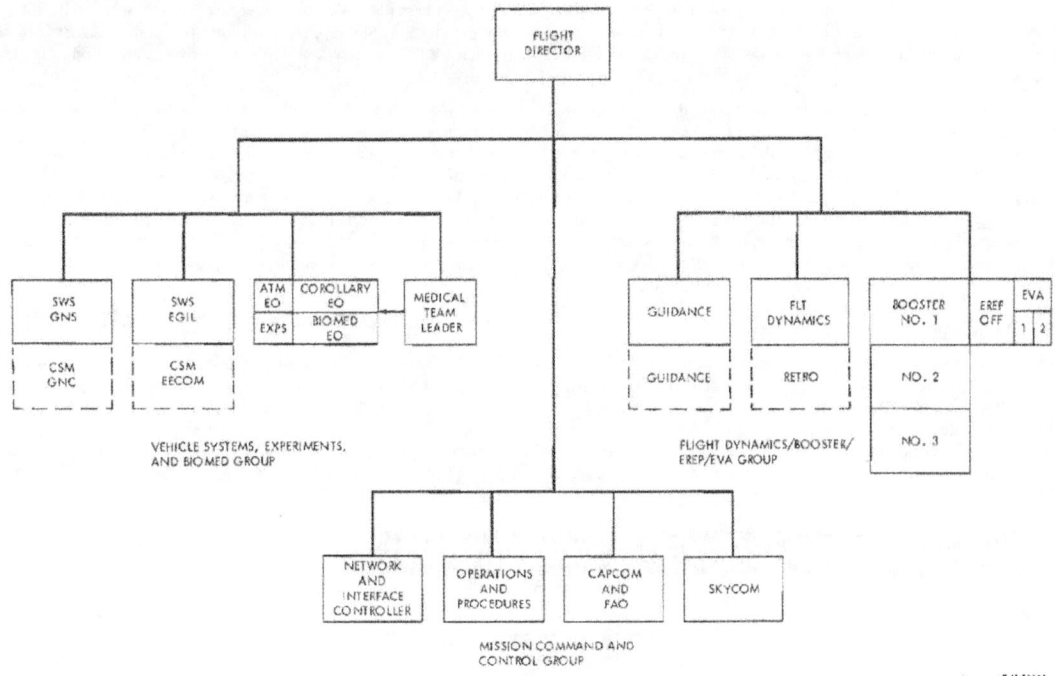

Figure 8-2

d. LCC and MCC control-display status.

e. Recovery-force status and weather decision.

f. Transmit SL-1 targeting parameters to the SL-2 launch vehicle digital computer in the IU. Command loads will be generated at the MCC based on latest tracking data and will be transmitted to KSC at approximately T-8 hours for loading into the IU computer via the DCS. This load is also sent to the Huntsville operations support center (HOSC) for verification. A final update on the target load and lift-off time will made between T-37 and T-15 minutes. This update will again be based on MCC computation of the targeting parameters from the latest available remote site tracking data.

Both a prime and backup mode will be available for entering and updating the targeting parameters. The prime mode is the DCS, and the backup will be hardline via the RCA 110A computer at KSC using a teletype (TTY) link from the MCC.

Launch.

Flight control procedures for the manned launch phase will consist of the evaluation of spacecraft systems, launch vehicle systems, flight crew condition, and space vehicle dynamics and trajectory. Mission abort capability will exist for cases of adverse launch vehicle and spacecraft system or trajectory conditions. Abort can be initiated automatically by the emergency detection system (from lift-off to T+171 seconds) or manually by the crew. Abort request can be initiated by the flight director on recommendations from the CSM systems engineer, the booster systems engineer, and the flight dynamics officer. Also, abort request can be initiated by the flight dynamics officer and the booster systems engineer when specific mission rules are violated. Abort request is relayed to the flight crew by voice and/or by digital uplink command. (The digital uplink command lights the ABORT light on the crew display panel.) Specific flight control procedures which are accomplished between lift-off and orbital insertion are as follows:

a. Monitor and evaluate launch vehicle systems.

b. Monitor and evaluate spacecraft systems.

c. Monitor and evaluate the condition of the flight crew.

d. Monitor and evaluate space vehicle dynamics and trajectory.

e. Call marks on abort modes.

f. Send abort request if mission rules are violated.

g. Provide contingency maneuvers.

h. Monitor the flight plan and recommend alternate procedures to the flight crew in contingency situations.

i. Advise the flight crew of launch vehicle, spacecraft, and trajectory status.

LAUNCH CONTROL CENTER.

Master control of launch operations for the Skylab program is provided by the launch complex 39 launch control center (LCC). The LCC supported by the central instrumentation facility (CIF) at KSC and by the Huntsville operations support center at MSFC provides management control over the prelaunch checkout and launch operations. After the vehicle clears the umbilical tower, the LCC supports the MCC which then assumes control of the mission. For a description of the launch facility, see Section VII.

LAUNCH TEAM.

The Skylab launch team (or test team) is a group of NASA and contractor personnel responsible to the KSC center director for performing all prelaunch, launch, and postlaunch activities for

Section VIII Mission Control Monitoring

an assigned mission. This includes, but is not limited to, the receipt, preparation, buildup, modification, maintenance, operation, rework, test, and checkout of individual and integrated space vehicle systems and their associated ground support equipment, and also the preparation of required documentation. The purpose of the launch team is to draw together all of the KSC resources required to perform the above.

The formal line/staff organization is the major functional organization from which the launch team draws personnel required to accomplish a specific mission. The directorate of launch operations, including the launch vehicle operations and spacecraft operations directorates, the directorate of technical support, including the information systems and support operations directorates, and supported by the directorate of installation support are the formal organizational elements which provide the launch team personnel. The launch team is a semi-formal organization that has responsibility for accomplishing the operations, tests, and countdown required for a specific mission.

The launch team operates in two modes, on-station and off-station operations. In the on-station operational mode, the launch team (figure 8-4) mans the test/launch consoles for the purpose of checking out individual or integrated space vehicle systems and associated ground support equipment or conducting integrated test and checkout procedures, such as the launch countdown which requires LCC firing room and/or ACE control room support. In the off-station operational mode, the launch team operates during the period of time when the flight hardware and related support equipment are not undergoing integrated testing, specifically during the preparation, buildup, modification, troubleshooting, rework of vehicle stages, spacecraft, other flight hardware, and support equipment.

Launch operations management is divided into two phases, test operations planning and test operations execution. The test operations planning phase includes the responsibility of developing, integrating, and coordinating the planning, scheduling, and technical documentation required to prepare and launch space vehicles. The test operations execution phase includes performing the test, checkout, and launch of the space vehicle in accordance with test and checkout plans and procedures prepared in the planning phase.

CENTRAL INSTRUMENTATION FACILITY (CIF).

The CIF houses the following: centralized KSC data systems for telemetry and launch area data reception, processing, display, distribution, and transmission; KSC central timing station; instrumentation and standards laboratories; automatic data processing (ADP) (administrative computing) systems; and the NASA/KSC data office. The functions performed at the CIF are a part of the KSC

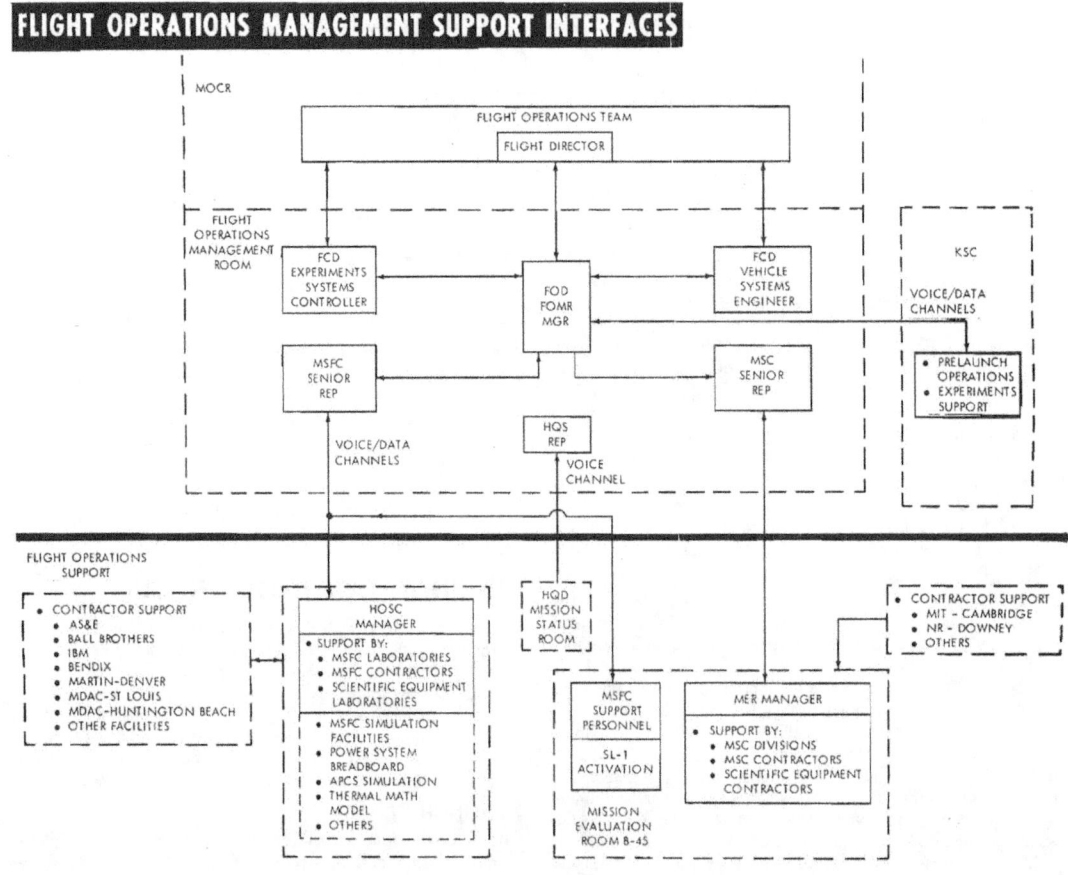

Figure 8-3

Section VIII Mission Control Monitoring

SL-1 & SL-2 ON-STATION TEST TEAM

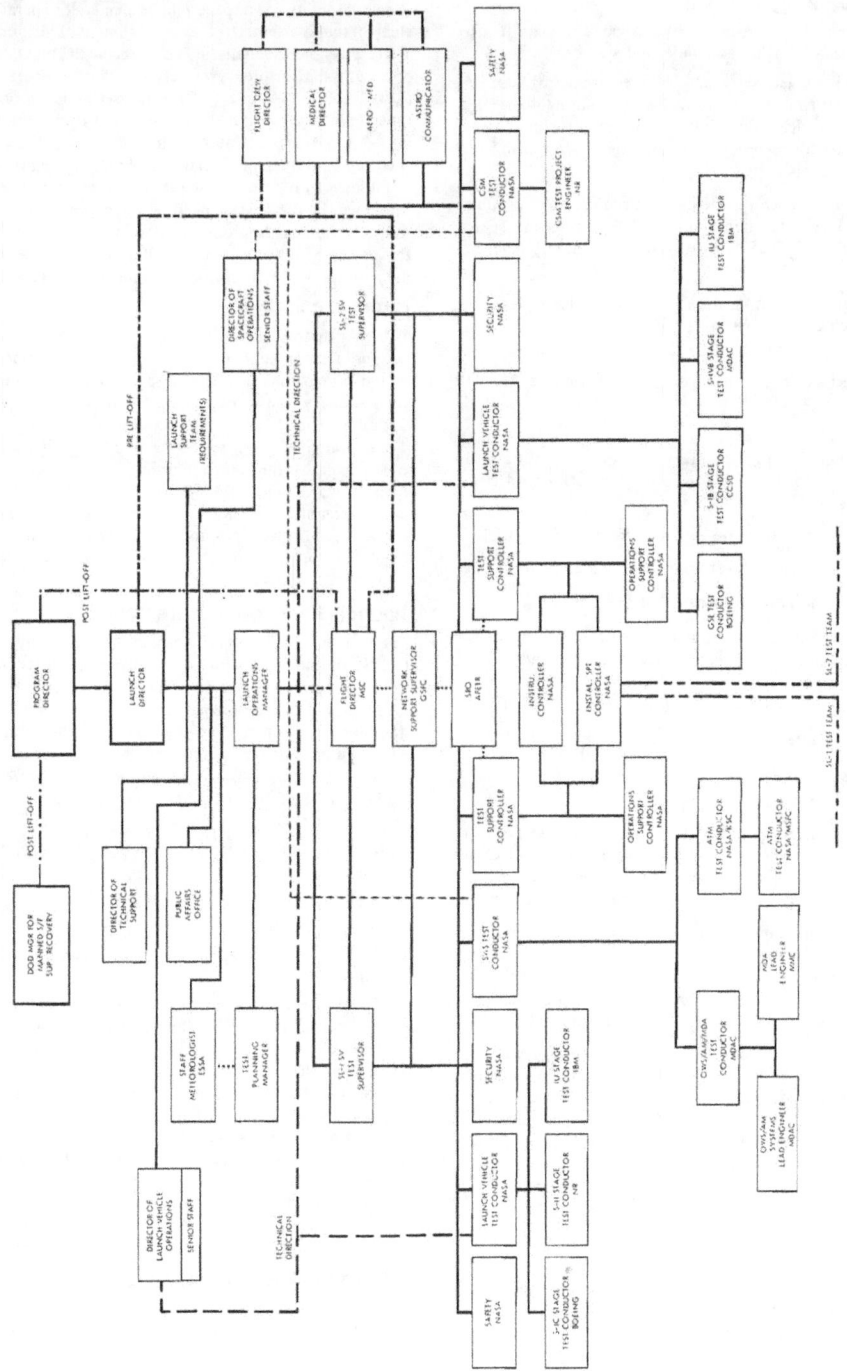

Figure 8-4

Section VIII Mission Control Monitoring

support for space vehicle testing, countdown, launch, and flight. Prime critical power for selected electronic equipment is supplied by the on-site generating plant.

Real Time Data Input to CIF.

The following real time data inputs are provided to the CIF telemetry station via KSC cable and wideband systems:

a. VHF and S-band telemetry are provided from two receiving sites. The CIF antenna site is located a mile north of the KSC industrial area in an RF quiet zone. This site has VHF and S-band antennas, predetection recorders, receivers, and video transmission of the received signals.

b. RF and hardline launch vehicle data are acquired by the launch vehicle digital data acquisition system (DDAS) when the vehicle is on the ground. The DDAS data are provided to the CIF telemetry station until T-0.

c. Facility measurements from the information systems measuring system at launch complex 39 are provided to the CIF telemetry station.

d. Launch trajectory data from the AFETR are routed to the CIF telemetry station. The AFETR provides vehicle position and velocity data.

e. The flight crew training building provides inputs from the mission simulators for transmission to the mission control center (MCC) in Houston.

f. Space vehicle orbital data are provided from MSFC via LIEF.

g. Other inputs as required to support Skylab data requirements.

Skylab Launch Data System (SLDS).

The SLDS is the information link that interconnects KSC with MCC and other centers. The four subsystems comprising the SLDS are:

a. Countdown and status transmitting subsystem (CASTS).

b. Television subsystem.

c. Launch trajectory subsystem.

d. Command subsystem.

The CASTS transmits up to 120 selected discrete functions and three independent countdown time words from KSC to MCC with an output bit rate of 2.4 kbps. These data are transmitted to MCC independently of all other operational equipment and data circuits.

The television subsystem contains video control consoles in the O&C building that receive 12 television channels from the launch complex 39 operational television system. At the consoles, five channels are selected for transmission to the MCC. SLDS transmits one of these channels to MCC and receives one channel from MCC.

The launch trajectory subsystem collects launch trajectory data from AFETR and transmits it to the real time computer complex. This subsystem provides real-time, smoothed and raw, radar trajectory data from liftoff through the sub-orbital flight phase of the vehicle. It also provides the liftoff event in the computer ID word from the impact prediction computers at AFETR.

The command subsystem processes and transmits commands from MCC to KSC.

LAUNCH PARAMETERS.

The criticality of the parameters monitored by the LCC at KSC fall into two major groups, mandatory and highly desirable. When time permits, the failure of a mandatory or highly desirable item is reported to the mission director by the launch director or the flight director. The initial report includes the position or facility that detected the malfunction. Subsequently, the mission director is informed of the estimated time to repair and the recommended proceed, hold, recycle, or scrub action as it develops.

Mandatory.

A mandatory item is a space vehicle element or operational support element that is essential for accomplishment of the primary mission. These essential elements include prelaunch, flight, and recovery operations that ensure crew safety and effective operational control, as well as the attainment of the primary mission objectives. The time applicable for all mandatory items is from start of countdown to T-5 sec. If a mandatory item fails during the countdown, it is corrected during the prelaunch phase, holding or recycling the countdown as required. If the item cannot be corrected to permit liftoff within the launch window, then the mission director will scrub the launch. Appropriate coordination with the launch director and flight director, and the DOD manager for manned space flight support operations, occurs prior to scrubbing the launch.

Highly Desirable.

A highly desirable item is a space vehicle element or operational support element that supports and enhances the accomplishment of the primary mission or is essential for the accomplishment of the secondary mission objectives. The time applicable for all highly desirable items is start of countdown to start of automatic sequence. Consideration is given to the repair of any highly desirable item but in no case is the launch scrubbed for any single highly desirable item. If two or more highly desirable items fail and if other aggravating circumstances occur, the mission director may scrub the mission, following appropriate coordination with the launch and flight directors and the DOD manager for manned space flight support operations.

Launch Vehicle Measurements.

The specific launch vehicle measurements that must be monitored by KSC during the applicable time periods are listed in the "Launch Vehicle Operations" section of the Apollo/Saturn IB Launch Mission Rules. The measurements are marked mandatory or highly desirable to denote the type of action taken when the measurement values go outside limits. Some inflight measurements are included in the measurement list in addition to the hardwire redline measurements. The following notes apply to the subject measurement list:

a. Instrumentation failure is a probable cause of failure for all parameters.

b. Minimum values are acceptable only when vehicle leakage is within allowable limits.

c. Minimum and maximum values represent limits of acceptable operation.

d. The "time applicable" is the suggested time for checking a particular redline value. Time applicable and parameter values are chosen to simultaneously minimize countdown impact and provide confidence that the parameters will remain acceptable to liftoff. It is assumed that observation is continued as required.

The launch site and MCC verify, whenever possible, telemetry readout discrepancies occurring prior to liftoff. If the MCC loses a parameter but the launch site has a valid readout, the MCC will continue monitoring on the basis of launch site readout. This is true except for those mandatory parameters upon which command action is taken. The measurements transmitted to MCC are marked with an asterisk in the subject measurement list.

GSE and ESE Parameters.

A list of GSE and ESE parameters are also shown in the Apollo/Saturn IB Launch Mission Rules. A description of the malfunction/condition on parameter value limits is shown in conjunction with its relative times, and action to be taken is shown for each item.

Section VIII Mission Control Monitoring

PRE-INSERTION LAUNCH ABORT AREAS

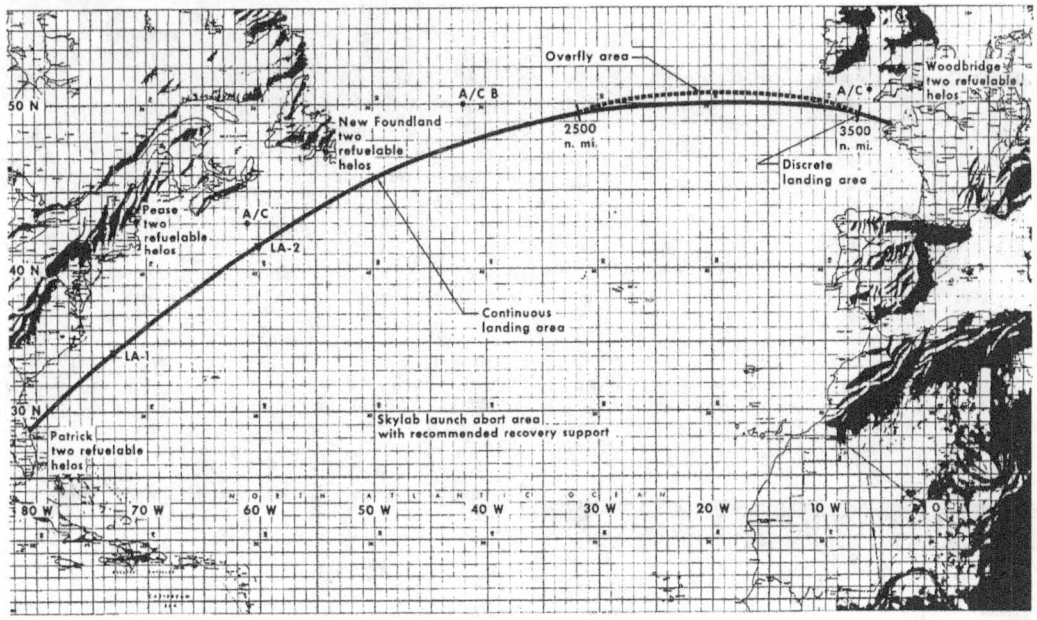

Figure 8-5

The ground support equipment (GSE) and the electrical support equipment (ESE) included in the list are:

a. Fuel, Lox, LH₂ systems

b. Spacecraft piping

c. Pneumatic systems

d. Apollo access arm

e. Electrical power system

f. Environmental control system

g. Firing accessories

h. Umbilical tower elevator

ABORT GROUND RULES.

INTRODUCTION.

Flight crew safety always takes precedence over accomplishment of mission objectives. Abort is defined as mission termination prior to orbital insertion as a result of a spacecraft or launch vehicle malfunction. Unscheduled mission termination at or after orbital insertion is referred to as early mission termination. These abort ground rules concern primarily decisions about manual abort execution rather than automatic abort. The abort request command is defined as a transmission from the STDN or LCC that illuminates the red ABORT light on the command pilot's panel. This light is one of two cues necessary for the crew to abort. A voice report from ground control can provide the second cue. If possible, abort request commands are based on two independent indications of the failure. Failures are categorized according to the following priorities:

a. Priority I failure causes destruction of the space vehicle, abnormal damage to the launch complex, loss of human life, or a hazard to the astronauts.

b. Priority II failure results in damage to the space vehicle, launch complex, or both; or requires rescheduling of the launch date.

c. Priority III failure causes a launch delay of less than 24 hours.

d. Priority IV failure has no significant effect on the launch operation.

If time permits, aborts and early mission terminations are timed for a water landing. The launch abort area extends from the launch site to 3500 NM down range. A continuous recovery area and a discrete recovery area are provided across the Atlantic Ocean for recovery. Figure 8-5 shows the pre-insertion launch abort area.

AUTHORITY.

Prior to liftoff, the launch director is responsible for all actions in the event of launch site emergencies including an abort request. The transfer of control from the launch director to the flight director occurs when the vehicle reaches sufficient altitude to clear the top of the umbilical tower. Within their respective areas of responsibility, the launch director, flight director, DOD manager for manned space flight support operations, and mission director may choose to take any action required for the optimum conduct of the mission. The command pilot may initiate such action as he deems essential for crew safety.

Hold Authority.

The command pilot, the spacecraft test conductor, the launch vehicle test conductor, the launch operations manager, the launch director, the flight director, the DOD manager for manned space flight support operations, or the mission director may call a hold for conditions within their respective areas of responsibility. Only the mission director can scrub a mission.

Abort Request Authority.

At the LCC, the launch operations manager may send an abort request from the time the launch escape system is armed until the vehicle clears the umbilical tower. After tower clearance, control

Section VIII Mission Control Monitoring

shifts to MCC where the flight director, the flight dynamics officer, or the booster systems engineer may send an abort request.

CRITERIA.

Launch Control Center.

Prior to transfer of control to the flight director, the launch operations manager will initiate an abort request as a result of the following events:

a. Major structural failure or explosion
b. Loss of positive vertical motion
c. Uncontrolled vehicle tilting
d. Tower collision resulting in damage requiring immediate abort

If a vehicle collision with the umbilical tower does not require immediate action, the launch operations manager will continue to evaluate the damage and provide information to the flight director. The launch operations manager will inform the flight director when the vehicle has cleared the tower (not including the lightning rod) by stating "clear tower" over the flight directors primary loop. Before the launch operations manager initiates an abort request, the following conditions must exist:

a. The safety of the flight crew must be endangered.
b. An impending catastrophic condition must be observed and reported by a periscope observer and must be confirmed by another periscope observer or by the launch operations manager via television.
c. All launch team personnel have evacuated the pad area.
d. The space vehicle has not cleared the umbilical tower.

After the abort request has been initiated, the launch operations manager will verbally confirm to the flight crew that an abort is recommended and will advise launch site recovery forces that abort has been recommended.

Mission Control Center.

The flight dynamics officer will initiate the abort request for spacecraft systems malfunctions and, when time permits, for trajectory deviations and launch vehicle malfunctions. The flight dynamics officer will initiate the abort request during the flight phase if the vehicle exceeds the flight dynamics envelope, or if booster cutoff conditions preclude a contingency insertion with the SPS and time will not permit forwarding a verbal request for initiation of this command to the flight director. The criterion for defining the flight dynamics envelope is safe recovery of the spacecraft. The booster systems engineer will initiate the abort request if launch vehicle malfunctions will not allow a safe insertion and time will not permit forwarding a verbal request for initiation of this command to the flight director.

RANGE SAFETY.

The range safety officer (RSO) can shut down the launch vehicle engines by transmitting the manual cutoff command, which also illuminates the ABORT light in the spacecraft. The manual cutoff will initiate an automatic abort if transmitted prior to EDS disable. The manual cutoff command starts a 4-sec timer on the ground that enables the destruct command capability. See Section I for details about this system. The RSO will always safe the S-IVB secure range safety command system upon verification of cutoff (following a manual cutoff command) if the destruct command is not to be transmitted.

HUNTSVILLE OPERATIONS SUPPORT CENTER (HOSC).

During pre-mission simulations, prelaunch tests, launch countdown, and flight of the Saturn IB vehicle, MSFC provides real-time support through the HOSC (figure 8-6). The HOSC is arranged on two floors in the MSFC computation laboratory as shown in figure 8-7.

Consoles in the operations support room (figure 8-8) provide real-time vehicle status to systems engineers by means of discrete indicators, analog meters, strip chart recorders, and TV displays. In addition, all areas of the HOSC are served by voice communication and timing systems. Consoles are permanently assigned to Skylab orbital assembly personnel during all mission phases. Console areas assigned to launch vehicle personnel during prelaunch tests and launch activation will be assigned as required to other personnel (e.g. experiment monitoring) during other mission phases. The main conference room will house the main body of support engineers during launch.

GROUND RULES FOR HOSC DATA DISPLAYS.

Saturn launch vehicle measurements specified to be displayed by HOSC determine the content of the real-time data stream. In selecting these measurements, the following ground rules are used:
a. All redline measurements in the Launch Mission Rules Document will be displayed on the HOSC engineering console discrete indica-

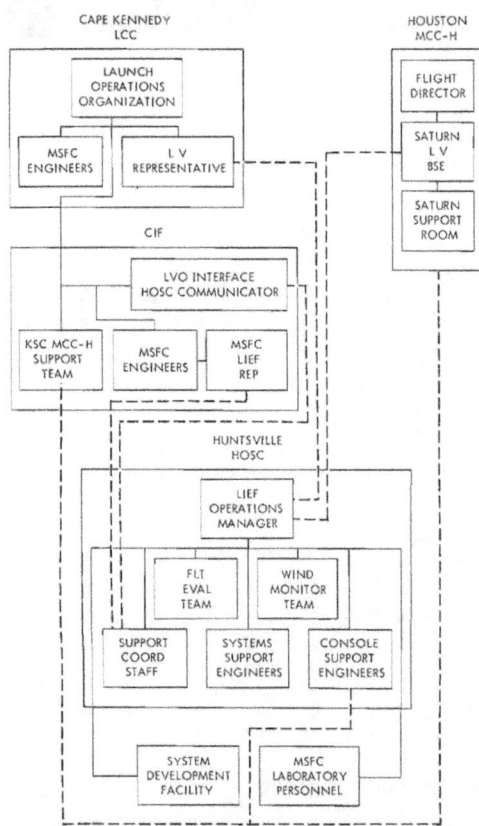

Figure 8-6

Section VIII Mission Control Monitoring

HUNTSVILLE OPERATIONS SUPPORT CENTER

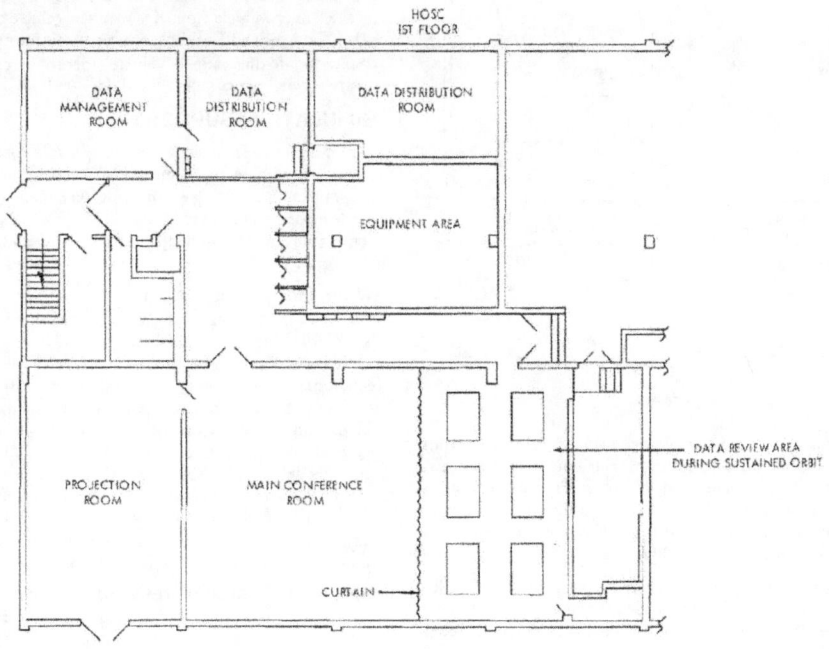

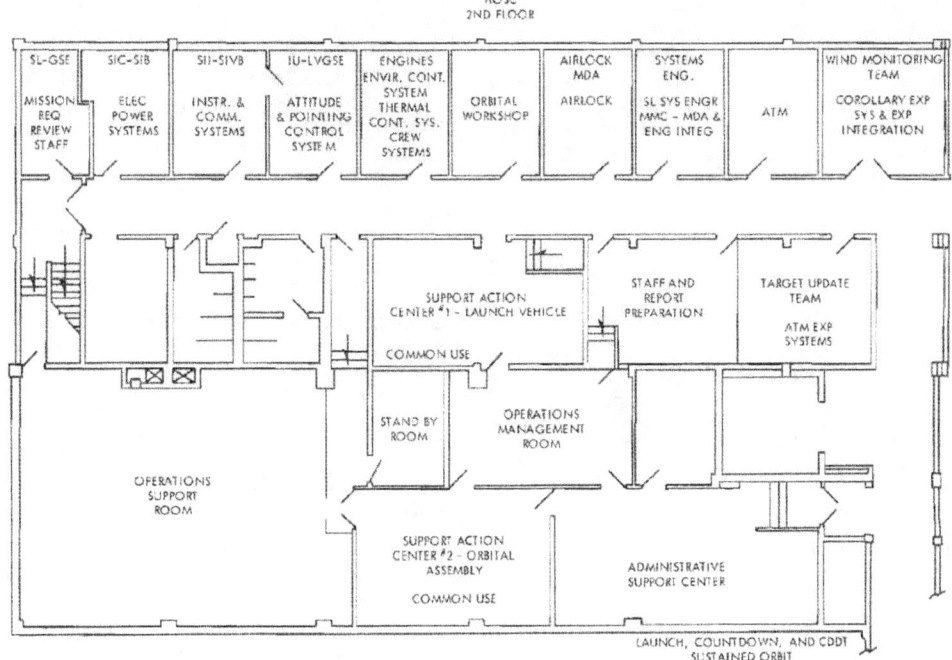

Figure 8-7

Section VIII Mission Control Monitoring

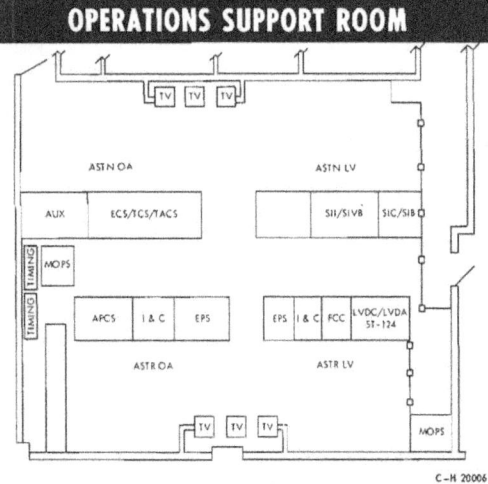

Figure 8-8

tors continuously from the start of the terminal countdown until launch. Each redline discrete indicator will illuminate if the real-time telemetry data indicates that the measurement has violated the minimum or maximum value specified in the Launch Mission Rules Document. Redline indicator displays are identified by a label with a solid red colored background.

b. All launch vehicle measurements identified as MANDATORY or HIGHLY DESIRABLE inflight measurements in the Launch Mission Rules Document will be displayed on the HOSC engineering console display devices in order to provide real-time support to KSC and MCC.

c. Flight control measurements will be displayed on the HOSC engineering console display devices (discrete indicators, analog meters, strip-chart recorders, and/or digital/TV formats).

d. Measurements that are required to support the Saturn launch vehicle prelaunch testing, countdown, and flight operations, and measurements that provide mission sequence of events (commands, events, and system response) and/or provide a basis for real-time launch vehicle systems performance evaluation will be displayed in the HOSC.

e. Measurement displays are limited by: (1) the availability of console display devices and (2) the capacity of the real-time data frame (1023 10-bit words).

CONSOLE SUPPORT ENGINEERS.

Support is provided as required to launch and flight operations from HOSC engineering data consoles. The consoles are manned by systems engineers identified by MSFC Laboratories, who perform detailed system data monitoring and analysis and coordinate support to the Saturn launch vehicle staff support room in mission control center and engineer-to-engineer support to KSC. An observer from each launch vehicle stage contractor is also located at the consoles.

SYSTEMS SUPPORT ENGINEERS.

Systems support engineers are organized into preselected subsystem problem groups including technical and management personnel who provide support in launch vehicle areas that may be the subject of a KSC or MSC request for analysis during terminal launch countdown and flight. The Systems support engineers are composed of both MSFC and stage contractor personnel.

HOSC COMMUNICATOR.

The HOSC communicator is a KSC launch vehicle operations member who mans the LIEF console in the CIF during the terminal countdown and coordinates with the HOSC all requests for support on the launch vehicle from KSC launch vehicle operations personnel. He also provides launch vehicle status information during the countdown to the launch vehicle operations communicator and controls switching of operational TV to HOSC.

SIMULATION SUPPORT.

A number of support activities will provide special systems simulation functions to supplement MCC capabilities. These activities generally utilize existing engineering development hardware and/or math models not provided at the MCC and provide support on contingency and/or scheduled basis. Those activities related to support of the Saturn IB vehicle are as follows.

Wind Monitoring Team.

The wind monitoring team monitors measured ground and flight wind observations at KSC and provides GO/NO-GO recommendations and advisory support to KSC in the event of marginal conditions for launch and certain other pad activities. Vehicle structural bending moment effects are calculated from ground winds, and compared against vehicle capabilities. The actual flight wind profiles, measured prelaunch, are used in a flight dynamics math model to predict the acceptability of structural and control effects. Predicted vehicle in-flight dynamic responses are provided to the MCC flight controllers and the crew. Coordination with a similar spacecraft team at MSC will be maintained for the manned launches and joint recommendations issued.

Saturn IB Systems Development Facility.

The Saturn IB system development facility (breadboard) will be held in a standby mode during major prelaunch test and countdown activities to assist in troubleshooting and testing in the event of problems at KSC. This facility is used pre-mission to develop and verify mission software, and include actual or breadboard launch vehicles and ground computer hardware, software, and related systems in the configuration similar to that utilized at KSC.

Target Update Team.

The target update team, working closely with the MCC flight dynamics personnel, will verify the performance capability of the Saturn IB vehicle to achieve the updated rendezvous target conditions generated by the MCC from tracking of the orbital assembly. The MCC will generate updated Saturn IB guidance targeting parameters after launch of SL-1 and transmit these to the SL-2 and subsequent vehicles at KSC shortly before their launch. The target update team will verify the performance acceptability of the update conditions for MCC and KSC.

OPERATIONS SUPPORT PHASES.

The level of support activities will vary significantly as the total mission activities change. The mission can be divided into five characteristic support phases, some of which are repeated over the course of the mission. These phases and their occurrence during the mission are summarized in figure 8-9.

LAUNCH INFORMATION EXCHANGE FACILITY (LIEF)

The LIEF is a network of communications resources providing close day-to-day exchange between KSC and MSFC of information relating to launch vehicle integration, checkout, and successful mission accomplishment. Direct engineering support is provided in areas of propulsion, navigation, and electrical networks. The joint MSC/MSFC wind monitoring team of the LIEF organization advises the launch director of the acceptability for launch of the actual wind environment.

Section VIII Mission Control Monitoring

OPERATIONS SUPPORT PHASES

SUPPORT PHASES	VEHICLE ACTIVITIES	ACTIVE OPERATIONS SUPPORT INTERFACES				DURATION
		KSC		BSE	FOMR	
		SATURN	SKYLAB			
I PREMISSION PHASE	SL-1 PRELAUNCH SL-2 PRELAUNCH	●	●	**	**	6 MONTHS
II SL-1 LAUNCH PHASE	SL-1 LAUNCH * SL-2 PRELAUNCH	●		●	●	1 DAY
III SATURN IB LAUNCH PHASE	SL-2, SL-3 AND SL-4 LAUNCHES * OA UNMANNED			●	●	12-30 HOURS
IV MANNED OPERATIONS PHASE	OA MANNED				●	SL-2: 27 DAYS SL-3: 55 DAYS SL-4: 55 DAYS
V ORBITAL STORAGE PHASE	OA UNMANNED SL-3 AND SL-4 PRELAUNCH	●		**	●	42-47 DAYS

* LAUNCH PERIODS DEFINED FROM LIFTOFF THROUGH IU ACTIVE LIFETIME
** SIMULATIONS, MCC INTERFACE TESTS

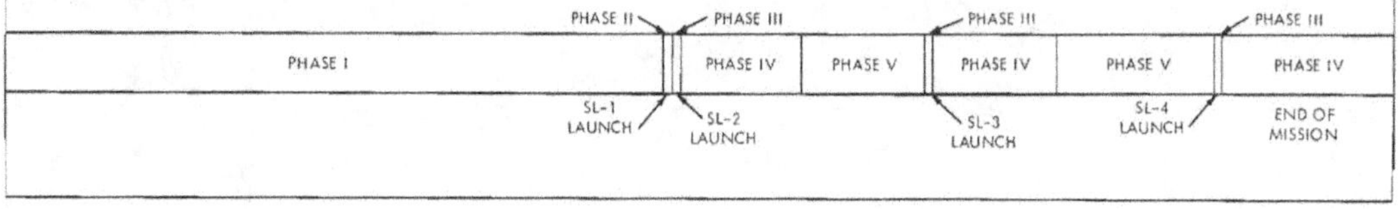

Timeline: PHASE I | PHASE II | PHASE IV | PHASE III | PHASE V | PHASE III | PHASE IV | PHASE V | PHASE III | PHASE IV
Events: SL-1 LAUNCH → SL-2 LAUNCH → SL-3 LAUNCH → SL-4 LAUNCH → END OF MISSION

Figure 8-9

ADVISORY SUPPORT.

An advisory group is convened at the MSFC Huntsville operations support center (HOSC) conference room during countdown and major tests to provide timely response to KSC engineering support requests. Primary communications are by a voice line between the HOSC operations room and the LIEF console at launch complex 39.

REAL-TIME DIGITAL DATA.

MSFC is supplied real-time telemetered data from either of two data cores in the CIF by means of the LIEF wideband circuit (40.8 kbs). MSFC controls parameter selection via the LIEF real-time data request, a serial PCM message transceived at 2.4 kbs via 201B MODEM link.

The real-time data output frame contains 1023 data words of 10 bits each, plus a 30-bit sync word. A complete data request instruction assigns a parameter in each of these 1023-word slots by designating the storage memory address for each. The complete instruction is stored in the data request memory and executed automatically at the 40.8 kbs rate, which is roughly four frames per second. MSFC updating can at any time change one word, a dozen words, or all 1023 words in the data request memory.

MSFC will provide raw space vehicle orbital data from the MSFN to KSC over existing data transmission circuits (MSFC to KSC) at 50 kbs via 303G MODEM link. The STDN data will be displayed at KSC in real-time.

TAPE-TO-TAPE INFORMATION EXCHANGE.

The KSC branch of the information exchange system consists of two Univac 1005 computers at the CIF, interconnected by the 301B MODEM links. Similar installations at MSFC interconnect with the KSC system to form a versatile high-speed (40.8 kbs) digital data information exchange system.

IBM 066-068 CARD TRANSCEIVER NETWORK.

The card-to-card transceiver transmits over a WATS line (4kHz) to any station equipped with a matching unit. The KSC transceiver is in the CIF. The data rate is 10 to 12 cards per minute.

FACSIMILE NETWORKS.

NASA leases alternate-voice facsimile equipment of two types for transmission and reception of documents, drawings, and continuous tone images over standard 4 kHz voice circuits. Magnafax Model 850 stations, each capable of transmitting or receiving approximately 10 pages (8-1/2" x 11") per hour, make up one network. A station directory is published by the NASA network contractor, and calls are made in much the same manner as placing a telephone call.

A high-speed LDX facsimile network links NASA headquarters, KSC, MSC, MSFC, and various NASA contractors through switching facilities located at MSFC. Duplex terminal equipment at each location permits simultaneous transmission and reception of copy at the rate of approximately 60 pages (8-1/2" x 11") each way per hour.

COUNTDOWN TIMING AND LIFTOFF.

KSC countdown timing from the LCC timing racks is transmitted to countdown clocks in the MSFC HOSC via the LIEF console. The LIEF voice circuit is allocated to this 1700 Hz, serial-PDM, encoded signal. Another circuit provides MSFC with a liftoff signal.

CLASSIFIED TELETYPE.

An on-line teletype facility interconnects the KSC communications center in the KSC headquarters building with NASA headquarters and other NASA centers. This is the only facility for exchange

8-11

Section VIII Mission Control Monitoring

of classified launch operations information between NASA elements.

CLOSED CIRCUIT TV.
The LIEF console operator selects and monitors the TV image requested for countdown and launch coverage by the HOSC communicator.

LIEF CIRCUIT RECORDING.
Certain voice circuits are recorded during the countdown for subsequent analysis of launch operations. The LIEF video circuit is also recorded during launch.

SPACEFLIGHT TRACKING & DATA NETWORK (STDN).

The STDN, managed by Goddard Space Flight Center, will be utilized during the Skylab missions. This network consists of 13 land sites (figure 8-10) and the ship Vanguard. These sites are situated arounc the earth to provide as nearly continuous communication with the vehicles as possible.

The basic types of support required from the network are tracking, telemetry record, telemetry real-time display, command communication, voice communication, and television. Figure 8-11 provides a matrix showing station capabilities.

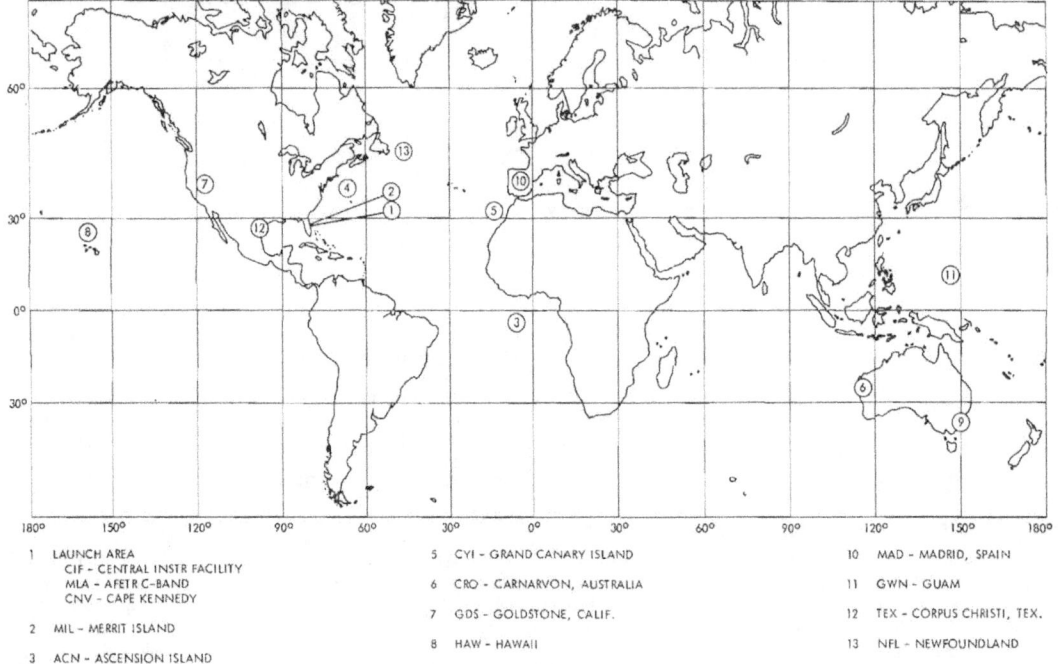

1. LAUNCH AREA
   CIF - CENTRAL INSTR FACILITY
   MLA - AFETR C-BAND
   CNV - CAPE KENNEDY
2. MIL - MERRIT ISLAND
3. ACN - ASCENSION ISLAND
4. BDA - BERMUDA
5. CYI - GRAND CANARY ISLAND
6. CRO - CARNARVON, AUSTRALIA
7. GDS - GOLDSTONE, CALIF.
8. HAW - HAWAII
9. HSK - HONEYSUCKLE, AUSTRALIA
10. MAD - MADRID, SPAIN
11. GWN - GUAM
12. TEX - CORPUS CHRISTI, TEX.
13. NFL - NEWFOUNDLAND

Figure 8-10

Section VIII Mission Control Monitoring

STDN STATION CAPABILITIES

	CNV	CIF	MIL	MLA	BDA	CYI	ACN	MAD	CRO	GWM	HSK	HAW	GDS	TEX	VAN	NFL
TRACKING																
C-BAND RADAR			●		●				●							
USB/CCS			●		●	●	●	●	●	●	●	●	●			
TELEMETRY																
VHF		●	●		●	●	●	●	●	●	●	●	●	●	●	●
USB/CCS			●		●	●	●	●	●	●	●	●	●	●	●	
DATA PROCESSOR		●	●		●	●	●	●	●	●	●	●	●	●	●	●
DATA REMOTING		●	●		●	●	●	●	●	●	●	●	●	●	●	●
BIOMED REMOTING			●		●	●	●	●	●	●	●	●	●	●		
DATA RECORDING		●	●		●	●	●	●	●	●	●	●	●	●	●	●
COMMAND																
UHF	●		●		●	●	●	●	●	●	●	●	●	●	●	●
USB/CCS UPDATA			●		●	●	●	●	●	●	●	●	●	●	●	
CMD PROCESSOR			●		●	●	●	●	●	●	●	●	●	●	●	●
CMD REMOTING	●		●		●	●	●	●	●	●	●	●	●	●	●	●
A/G VOICE																
VHF			●		●	●	●		●	●		●	●	●	●	●
USB			●		●	●	●	●	●	●	●	●	●	●	●	
TV																
S-BAND TV RECORD			●		●	●	●	●	●	●	●	●	●	●	●	
R/T TRANSMISSION			●										●			

Figure 8-11

SECTION IX
MISSION VARIABLES & CONSTRAINTS

TABLE OF CONTENTS

Flight Profile	9-1
Hold & Recycle Rulings	9-1
Launch Window	9-1
Environment	9-1
Flight Mission Rules	9-4
EDS Limits Derivation	9-4
S-IVB/IU Orbital Decay	9-5
S-IU Active Lifetime	9-5
S-IVB Active Lifetime	9-5
Orbital Configuration	9-5

FLIGHT PROFILE.

Several constraining factors govern the Saturn IB launch vehicle flight profile to achieve the desired orbit insertion conditions such as vehicle performance, targeting conditions, range support, and launch complex design. A summary of these factors follows:

a. The SL-2 launch will occur from launch complex 39B into a 50-deg inclination orbit. Since Pad 39B is built with the launch azimuth fixed at 90 deg, a roll command will be required to provide the flight azimuth required (see launch window discussion).

b. To provide proper initial conditions for phasing orbit, the S-IVB cut-off conditions will be targeted for an 81 x 120 NM orbit insertion. The nominal insertion orbit must have a perigee greater than 70 NM.

c. The launch vehicle or spacecraft will not be allowed to enter a flight regime, as defined by the Flight Mission Rules, from which a safe abort cannot be accomplished.

d. The nominal ascent trajectory will be shaped so that aerodynamic loads during a full-lift, free-fall abort from any point along the trajectory shall not exceed 16 G's.

e. The trajectory will be shaped to maintain the aerodynamic heating indicator below an integrated value of 63.6×10^7 $N-m/m^2$ (43.6×10^6 $lbf-ft/ft^2$).

f. For launch aborts, the duration of free fall above the entry interface of 300,00 ft will be at least 100 sec to provide adequate time to perform functions necessary for safe spacecraft entry.

g. To provide compatibility between guidance commands and vehicle reaction, the guidance command angles will be rate limited to 1 deg/sec during powered flight.

h. Continuous range safety command destruct and safing capability is required from launch minus 30 min through insertion plus 60 sec.

i. Continuous STDN coverage of the launch vehicle is desired from start of automatic countdown until 60 sec after insertion.

j. One 3-min pass for telemetry, command, and tracking coverage is desired for the S-IVB stage of the Saturn IB during each revolution through IU powered operation.

k. C-Band track of the S-IVB/IU will be required every TBD revolutions from end of IU life until reentry. This support will be supplied on a non-interference basis with SWS skin tracking support.

l. The S-IVB auxiliary propulsion system will provide attitude control capability for 7 hr 30 min for SL-2 and SL-4 and 4 hr 30 min for SL-3.

HOLD & RECYCLE RULINGS.

Unscheduled hold and recycle rulings depend not only on the type of malfunction, but also on the time period in which the malfunction occurs.

The terminal launch countdown will be held or scrubbed to repair any loss of redundancy if an inflight failure of the affected system, subsystem or function could:

(a) Compromise crew safety, or

(b) Require early termination of the mission, or

(c) Require cancellation or early termination of subsequent missions.

Figure 9-1 lists the actions taken for particular malfunctions occurring at various times from start of space vehicle countdown until liftoff.

LAUNCH WINDOW.

The Saturn IB launch windows are constrained by the phase angle between the SWS and the CSM at insertion and the amount of Saturn IB payload capability sacrificed for the propellant used for yaw steering. Length of the SL-2 launch window is based upon the SL-1 descending node being 153.25 deg, a 5 to 7 orbit rendezvous scheme, and a yaw steering allowance of 700 lbm of propellant. The 700 lbm of yaw steering propellants represent approximately 16 min of launch window capability. The flight azimuth range for this launch window is between 51.82 deg at the opening of the window and 37.68 deg at the close of the window. The launch window times shown in figure 9-2 are current approximations.

ENVIRONMENT.

The accomplishment of mission objectives are contingent upon appropriate environmental conditions. Surface wind, upper air, and weather contingencies are considered below.

SURFACE WIND RESTRICTIONS.

Preliminary results from wind tunnel tests conducted on a 5.5% scale aeroelastic model of the SL-2 vehicle and pedestal in the

Section IX Mission Variables and Constraints

HOLD AND RECYCLE RULINGS

TIME PERIOD	FUNCTION/CONDITION MALFUNCTION	ACTION/COMMENT
1. T-26 HR (START OF S/V COUNT) TO T-12 HR 15 MIN (START OF LV COUNT).	MALFUNCTION OF ANY REPAIRABLE SPACECRAFT SYSTEM.	PROCEED. CORRECT MALFUNCTION IN PARALLEL WITH OTHER OPERATIONS. HOLD AT T-12 HR 15 MIN FOR MANDATORY OR HIGHLY DESIRABLE ITEMS IF ESTIMATES FOR COMPLETION INDICATE ALL WORK (REPAIR AND CLOSEOUT) CANNOT BE ACCOMPLISHED PRIOR TO T-6 HR 30 MIN.
2. T-12 HR 15 MIN (START OF LV COUNT) TO T-6 HR 30 MIN.	MALFUNCTION OF ANY REPAIRABLE SPACE VEHICLE SYSTEM.	PROCEED. CORRECT MALFUNCTION IN PARALLEL WITH OTHER OPERATIONS. HOLD AT T-6 HR 30 MIN FOR HIGHLY DESIRABLE OR MANDATORY ITEMS IF SERVICE STRUCTURE IS REQUIRED FOR REPAIR. HOLD AT T-5 HR 45 MIN IF ONLY THE ACCESS ARM IS REQUIRED FOR REPAIR.
3. T-6 HR 30 MIN (START S-IB LOX LOADING) TO T-4 HR 15 MIN (START S-IVB LH$_2$ LOADING).	MALFUNCTION OF ANY REPAIRABLE SPACE VEHICLE SYSTEM.	PROCEED OR HOLD. PROCEED IF CORRECTION CAN BE ACCOMPLISHED IN PARALLEL WITH NORMAL FUNCTIONS; OTHERWISE HOLD FOR MANDATORY AND HIGHLY DESIRABLE ITEMS. DURING THE HOLD REVIEW CRITICALITY WITH REFERENCE TO PERFORMANCE DEGRADATION. EVALUATE REPAIR TIME WITH RESPECT TO LAUNCH WINDOW TO DETERMINE NECESSITY FOR SCRUB (1 HR TO RETURN SERVICE STRUCTURE).
4. T-4 HR 15 MIN (START OF LH$_2$ LOADING) TO T-3 HR 15 MIN (SC, START OF CABIN CLOSEOUT); TO T-30 MIN (LV).	MALFUNCTION OF ANY REPAIRABLE SPACE VEHICLE SYSTEM.	PROCEED OR HOLD. PROCEED IF CORRECTION CAN BE ACCOMPLISHED IN PARALLEL WITH NORMAL FUNCTIONS; OTHERWISE HOLD FOR MANDATORY AND HIGHLY DESIRABLE ITEMS. REPAIR IF POSSIBLE WITHOUT THE USE OF SERVICE STRUCTURE. IF REPAIR IS NOT POSSIBLE WITHOUT SERVICE STRUCTURE, REVIEW CRITICALITY; MAKE DECISION AS TO PERFORMANCE DEGRADATION, SCRUB IF MANDATORY. HOLD AT T-3 HR 15 MIN FOR COMPLETION OF INTERNAL CM WORK IF REQUIRED.
5. T-3 HR 15 MIN (START OF SC CABIN CLOSEOUT) TO T-30 MIN (CLEAR ACCESS ARM).	PROBLEM IN SPACECRAFT CABIN CLOSEOUT.	HOLD AT T-30 MIN FOR COMPLETION OF CABIN CLOSEOUT.
6. T-30 MIN (CLEAR ACCESS ARM) TO T-10 MIN (S-IVB THRUST CHAMBER CHILLDOWN).	MALFUNCTION OF ANY REPAIRABLE SPACE VEHICLE SYSTEM.	PROCEED OR HOLD. PROCEED IF CORRECTION CAN BE ACCOMPLISHED IN PARALLEL WITH NORMAL FUNCTIONS AND ACCESS TO THE PAD IS NOT REQUIRED; OTHERWISE HOLD AT T-10 MIN FOR MANDATORY AND HIGHLY DESIRABLE ITEMS. REPAIR IF POSSIBLE WITHOUT THE USE OF SERVICE STRUCTURE. IF REPAIR IS NOT POSSIBLE, REVIEW CRITICALITY; MAKE DECISION AS TO PERFORMANCE DEGRADATION, SCRUB IF MANDATORY.
7. T-10 MIN (S-IVB) THRUST CHAMBER CHILLDOWN TO T-3 SEC (IGNITION).	ANY MALFUNCTION OF THE S/V GUIDANCE AND CONTROL SYSTEM. ANY MALFUNCTION OF THE EDS. EXCEEDING OR LOSS OF ANY SPACE VEHICLE REDLINE VALUE APPLICABLE TO THIS TIME PERIOD. A. LOSS OF ANY COMPLETE S/V TELEMETRY LINK TO T-2 MIN 43 SEC. B. LOSS OF ONE OR MORE OF THE FOLLOWING TELEMETRY LINKS TO T-3 SEC, S-IB LINK GP-1, S-IVB LINK CP-1, IU LINK DP-1, ANY COMPLETE S/C LINK. LOSS OF ANY MANDATORY MEASUREMENT. MALFUNCTION OF VEHICLE FIRE DETECTION SYSTEM. (2 OF 4M, 3 OF 4HD). MALFUNCTION OF CIF TM STATION SUCH THAT THE GO/NO-GO STATUS OF THE LAUNCH VEHICLE TM SYSTEMS CANNOT BE DETERMINED. NON-OPERATION OR MALFUNCTION OF THE DIGITAL DATA ACQUISITION SYSTEM WHICH COULD RESULT IN THE PRESENTATION OF ERRONEOUS DATA TO THE OPERATIONAL COMPUTERS AND/OR ESE DISPLAY DEVICES.	T-10 MIN TO T-3 MIN 7 SEC (LAUNCH SEQUENCE START): HOLD. A HOLD OF 5 MIN MAXIMUM CAN BE TOLERATED WITHOUT RECYCLING TO T-10 MIN PROVIDED S-IVB CHILLDOWN OPERATIONS CONTINUE. THIS LIMITATION IS DICTATED BY S-IVB THRUST CHAMBER CHILLDOWN. IF THE HOLD EXCEEDS 5 MIN RECYCLE TO T-10 MIN. A WARMUP PERIOD OF 10 MIN IS REQUIRED PRIOR TO INITIATING CHILLDOWN SEQUENCE AFTER RECYCLING TO T-10 MIN. T-3 MIN 7 SEC TO T-3 SEC: CUTOFF. RECYCLE TO T-10 MIN. MAKE THE DECISION TO HOLD AND REPAIR OR TO SCRUB. A WARMUP PERIOD OF 10 MIN IS REQUIRED BEFORE INITIATING S-IVB CHILLDOWN SEQUENCE AFTER RECYCLING TO T-10 MIN. NOTE: FROM A TECHNICAL STANDPOINT THERE IS NO LIMIT TO THE NUMBER OF RECYCLES TO T-10 MINUTES THAT THE LAUNCH VEHICLE CAN WITHSTAND IN THE TIME PERIOD T-10 MINUTES TO T-3 SEC PROVIDING RED-LINE VALUES REMAIN WITHIN LIMITS.

Figure 9-1 (Sheet 1 of 2)

Section IX Mission Variables and Constraints

presence of a scale model of the LUT indicate that the maximum SL-2 vehicle response should be of no greater magnitude than that of SA-205 for comparable wind speeds. By launching SL-2 from LC-39, the effect of umbilical tower influence on the vehicle response was changed from that of LC-34 (SA-205 launch), and the LC-39 wind damper can be connected to the SL-2 vehicle for operations prior to launch. Maximum vehicle response loads occur on SL-2 with the empty vehicle damper-off configuration, which is comparable to the empty SA-205 configuration. Based on the vehicle response to ground winds and the SA-206 structural characteristics, the preliminary prelaunch and launch wind limits, compared to SA-205, are shown in figure 9-3.

UPPER AIR RESTRICTIONS.

Preflight simulations of the space vehicle response to upper air winds at the time of launch are performed at MSFC by an MSC-MSFC Wind Evaluation Team using wind data provided by KSC. The Cape FPS-16, radar, and the 1.16 (or a suitable replacement in the event 1.16 is not operational) are used to track jimsphere balloons. Jimsphere balloons are released from the launch pad to obtain launch wind information from $T-50$ hr to $T+10$ min on a schedule agreed to by KSC, MSFC, and MSC. Results of the MSFC-MSC wind simulations are provided through the LIEF coordinator to the launch operations manager, or in his absence the test supervisor, in the launch control center. The results may be forwarded periodically as considered appropriate until the release at $T-10$ hr. A report is provided for each balloon released from $T-10$ hr to $T-5$ hr 30 min. Results of the wind simulations are summarized in writing in the HOSC and transmitted by datafax to the LIEF coordinator in the LCC.

If simulations indicate that wind conditions are marginal for a

HOLD AND RECYCLE RULINGS

TIME PERIOD	FUNCTION/CONDITION MALFUNCTION	ACTION/COMMENT
8. T-10 MIN (S-IVB THRUST CHAMBER CHILLDOWN) TO T-3 SEC (IGNITION)	ANY MALFUNCTION OF THE S/C REACTION CONTROL SUBSYSTEM. ANY MALFUNCTION OF THE S/C SERVICE PROPULSION SUBSYSTEM. ANY MALFUNCTION OF THE S/C ELECTRICAL POWER SUBSYSTEM. ANY MALFUNCTION OF THE S/C ENVIRONMENTAL CONTROL SYSTEM. MALFUNCTION OF S/C C-BAND BEACON. MALFUNCTION OF L/V C-BAND SYSTEM OR GLOTRAC BEACON TO T-2 MIN 43 SEC. MALFUNCTION OF MILA USB STATION. MALFUNCTION OF S/C S-BAND SUBSYSTEM. MALFUNCTION OF LAUNCH VEHICLE RANGE SAFETY COMMAND SYSTEM (ALL FOUR CDRS ARE REQUIRED). LOSS OF ETR GROUND SUPPORT INSTRUMENTATION FOR THE SPACE VEHICLE COMMAND SYSTEMS (642B COMPUTER, DRUF, FRW-2A, TRANSMISSION LINES). LOSS OF COUNTDOWN CLOCK (GMT) OUTPUT. MALFUNCTION OF FLYWHEEL GENERATOR.	T-10 MIN TO T-3 MIN 7 SEC (LAUNCH SEQUENCE START). HOLD. A HOLD OF 5 MIN MAXIMUM CAN BE TOLERATED WITHOUT RECYCLING TO T-10 MIN PROVIDED S-IVB CHILLDOWN OPERATIONS CONTINUE. THIS LIMITATION IS DICTATED BY S-IVB THRUST CHAMBER CHILLDOWN. IF THE HOLD EXCEEDS 5 MIN, A WARMUP PERIOD OF 10 MIN IS REQUIRED PRIOR TO INITIATING CHILLDOWN SEQUENCE AFTER RECYCLING TO T-10 MIN. T-3 MIN 7 SEC TO T-3 SEC. CUTOFF. RECYCLE TO T-10 MIN. MAKE THE DECISION TO HOLD AND REPAIR OR TO SCRUB. A WARMUP PERIOD OF 10 MIN IS REQUIRED BEFORE INITIATING S-IVB CHILLDOWN SEQUENCE AFTER RECYCLING TO T-10 MIN. NOTE: FROM A TECHNICAL STANDPOINT THERE IS NO LIMIT TO THE NUMBER OF RECYCLES TO T-10 MIN THAT THE LAUNCH VEHICLE CAN WITHSTAND IN THE TIME PERIOD T-10 MIN TO T-3 SEC PROVIDING RED-LINE VALUES REMAIN WITHIN LIMITS.
9. T-10 MIN (S-IVB THRUST CHAMBER CHILLDOWN) TO T-3 SEC (IGNITION).	A. LOSS OF GROUND COMPUTER SYSTEM SERVICE AND DEE-6 SERVICE TO T-20 SEC. B. LOSS OF GROUND COMPUTER SYSTEM SERVICE OR DEE-6 SERVICE TO T-3 SEC. C. LOSS OF AGCS COMPUTER DISCRETE OUTPUT POWER TO T-3 SEC. MALFUNCTION OF THE AALT TO T-5 SEC. EXCEEDING ANY WEATHER RESTRICTION APPLICABLE TO THIS TIME PERIOD. LOSS OF ANY MANDATORY COMMUNICATION LINK APPLICABLE TO THIS TIME PERIOD.	T-10 MIN TO T-3 MIN 7 SEC (LAUNCH SEQUENCE START). HOLD. A HOLD OF 5 MIN MAXIMUM CAN BE TOLERATED WITHOUT RECYCLING TO T-10 MIN PROVIDED S-IVB CHILLDOWN OPERATIONS CONTINUE. THIS LIMITATION IS DICTATED BY S-IVB THRUST CHAMBER CHILLDOWN. IF THE HOLD EXCEEDS 5 MIN, A WARMUP PERIOD OF 10 MIN IS REQUIRED PRIOR TO INITIATING CHILLDOWN SEQUENCE AFTER RECYCLING TO T-10 MIN. T-3 MIN 7 SEC TO T-3 SEC CUTOFF. RECYCLE TO T-10 MIN. MAKE THE DECISION TO HOLD AND REPAIR OR TO SCRUB. A WARMUP PERIOD OF 10 MIN IS REQUIRED BEFORE INITIATING S-IVB CHILLDOWN SEQUENCE AFTER RECYCLING TO T-10 MIN. NOTE: FROM A TECHNICAL STANDPOINT THERE IS NO LIMIT TO THE NUMBER OF RECYCLES TO T-10 MIN THAT THE LAUNCH VEHICLE CAN WITHSTAND IN THE TIME PERIOD T-10 MIN TO T-3 SEC PROVIDING RED-LINE VALUES REMAIN WITHIN LIMITS.
10. T-3 SEC (IGNITION) TO T-0 (LIFTOFF)	NOT APPLICABLE	NONE. NO HOLDS WILL BE CALLED. NO MANUAL CUTOFF WILL BE GIVEN. AN AUTOMATIC CUTOFF WILL RESULT IN A SCRUB.

Figure 9-1 (Sheet 2 of 2)

Section IX Mission Variables and Constraints

LAUNCH WINDOW				
OPPORTUNITY	DAYS AFTER SL-1 LAUNCH	DATE	FROM (HR:MIN:SEC)	TO (HR:MIN:SEC)
PRIME	1	1 MAY 73	12 NOON	12:09:00 PM
ALTERNATE	2	1 MAY 73	11:28:30 AM	11:34:30 AM
ALTERNATE	6	6 MAY 73	10:00 AM	10:09 AM
ALTERNATE	7	7 MAY 73	9:29 AM	9:35 AM

NOTE: TIME REFERENCED TO EASTERN STANDARD TIME (EST).

Figure 9-2

safe launch, the MSC-MSFC team includes in their report that "Launch winds are marginal for launch" prior to T −2 hr 15 min. Upon receipt of this message, the launch operations manager places a contingency plan into effect that will provide for a new balloon release every hour. The contingency plan remains in effect until liftoff has occurred, until the launch has been scrubbed, or until a subsequent MSC-MSFC wind report states "Launch winds are no longer marginal for launch". The MSC-MSFC team provides a report to the launch operations manager for each release under the contingency plan. Balloon release information is updated prior to the Flight Readiness Report (FRR). Figure 9-4 shows the typical wind speed limits for Skylab Saturn IB/CSM Flights.

WEATHER RESTRICTIONS.

It is highly desirable that launch site area visibility be sufficient for at least one of the long range ground tracking cameras to have an unobstructed view of the space vehicle through the period of maximum dynamic pressure. Failure of optical tracking systems or weather conditions that would result in failure to meet this requirement are considered basis for a hold. It is mandatory that visibility be sufficient for the forward observers to monitor space vehicle conditions. A minimum visibility of 3 NM and a ceiling of 500 ft is considered mandatory.

The space vehicle will not be launched if the nominal flight path will carry the vehicle:

a. Within five statute miles of a cumulonimbus (thunderstorm) cloud or within three statute miles of an associated anvil.

b. Through cold-front or squall-line clouds that extend above 10,000 ft.

c. Through middle cloud layers 6,000 ft or greater in depth where the freeze level is in the clouds.

d. Through cumulus clouds with tops at 10,000 ft or higher.

FLIGHT MISSION RULES.

Flight Mission Rules are procedural statements which provide flight control personnel with guidelines to expedite the decision-making process. The rules are based on an analysis of mission equipment configuration, systems operations and constraints, flight crew procedures, and mission objectives. The rules can be categorized as general and specific. The general mission rules contain the basic philosophies used in the development of rules while the specific mission rules provide the basic criteria from which real-time decisions are made. The Cues/Notes/Comments column of specific mission rules and the background data and flight control procedures provide the flight controller with additional information concerning the condition/malfunction and ruling.

EDS LIMITS DERIVATION.

The EDS limits are established to assure safe separation of the CSM from the launch vehicle in an emergency situation. The Aero-Astrodynamics Laboratory at MSFC determines the EDS limits for the Saturn IB launch vehicle by math model simulations of various failure modes. The simulations determine the vehicle breakup rates; then, the EDS limits are selected between the three-sigma nominal and the breakup rates. See Section III for EDS settings and additional EDS information.

SL-2 SURFACE WIND RESTRICTIONS			
	F.S.	UNPRESSURIZED	PRESSURIZED
F.S.F.	1.4 1.25	51.2 [64] (53) 51.8 [72] (54)	51.5 [64] 73.8 (56)
F.S.E.	1.4 1.25	33.0 [57] (30) 33.2 [63] (33)	72.5 77.4
LAUNCH			29.0 (27)*

VELOCITIES EXPRESSED IN KNOTS AT 530-FT LEVEL

[ ] = DAMPER ENGAGED
( ) = AS-205 LIMITS CORRECTED TO 530-FT REF LEVEL
 * = AS-205 LAUNCH LIMIT BASED ON SPACECRAFT NOT STRUCTURAL CRITERIA
F.S.F. = FREE STANDING FULL
F.S.E. = FREE STANDING EMPTY

Figure 9-3

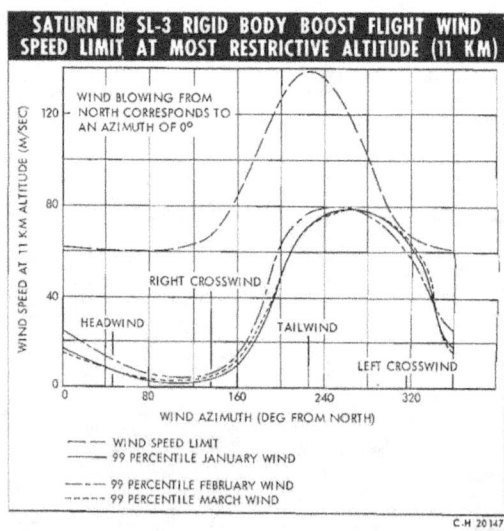

Figure 9-4

Section IX Mission Variables and Constraints

S-IVB/IU ORBITAL DECAY.

The nominal lifetime for the SL-2 S-IVB stage launched into an 81 by 120 NM orbit is 23.1 hr (0.96 days) or approximately 15 revolutions. The plus two-sigma and minus two-sigma densities give lifetimes of 21.6 hr (0.90 days) and 24.8 hr (1.03 days), respectively. The altitude decay history for the SL-2 S-IVB stage launched May 1, 1973, is shown in figure 9-5.

IU ACTIVE LIFETIME.

The IU has a design-goal lifetime of 6 hr 42 min based on a 0.992 reliability figure. However, analyses of individual component and subsystem life expectancies indicate that the pacing item is the 6D40 battery. With the expected electrical load profile, this battery should last approximately 11 hr. Another limiting factor is the GN_2 usage rate, which should deplete the stored supply in approximately 14 hr.

S-IVB ACTIVE LIFETIME.

The S-IVB stage as initially designed has an orbital coast capability of 4.5 hr. This capability has been extended for the SL-2 and SL-4 missions through a number of changes to improve the temperature control of certain components and assemblies. To accommodate the Skylab experiments, these two stages now have a lifetime of approximately 7.5 hr.

APS propellant depletion occurred on SA-205 between station contacts at 15 hr 30 min and 16 hr 20 min. Data indicated that attitude control was normal prior to APS propellant depletion.

ORBITAL CONFIGURATION.

ORBITAL INSERTION.

At orbital insertion 9 min 51.9 sec into flight (predicted to be 10 sec after J-2 engine cutoff) the vehicle consists of the SIVB stage, IU, SLA, and CSM. The S-IB stage separated, the ullage rockets jettisoned, and the launch escape tower jettisoned during the boost phase of flight. See figure 9-6 for vehicle configuration and mission time lines.

SPACECRAFT SEPARATION.

The CSM separates from the SIVB/IU/SLA at vehicle station 2033.799 (reference Section I, Vehicle Profile) approximately 6 min after orbit insertion. Explosive fuse assemblies sever the CSM from the SLA, and sever the adapter longitudinally into four panel sections.

The panels then rotate about hinges attached to the payload adapter with each panel deploying to a 45-deg open position, thus completing the spacecraft separation. (This is true of one Skylab Saturn IB launch vehicle; the others have jettisonable SLA panels same as the Apollo Saturn V launch vehicles.) After separation, the S-IVB/IU/SLA executes maneuvers required to maintain the vehicle in the proper attitude for performing the Thermal Control Coating Experiment (M-415).

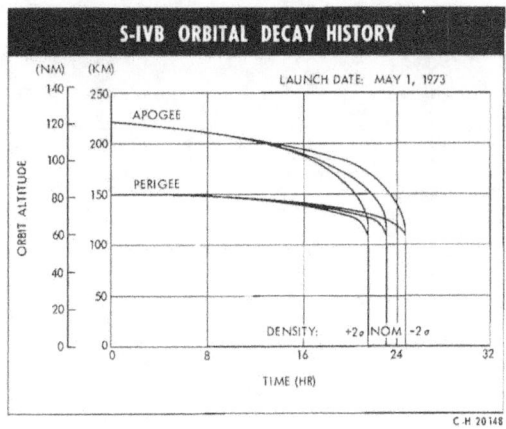

Figure 9-5

Section IX Mission Variables and Constraints

LAUNCH VEHICLE CONFIGURATION AND MISSION TIMELINES

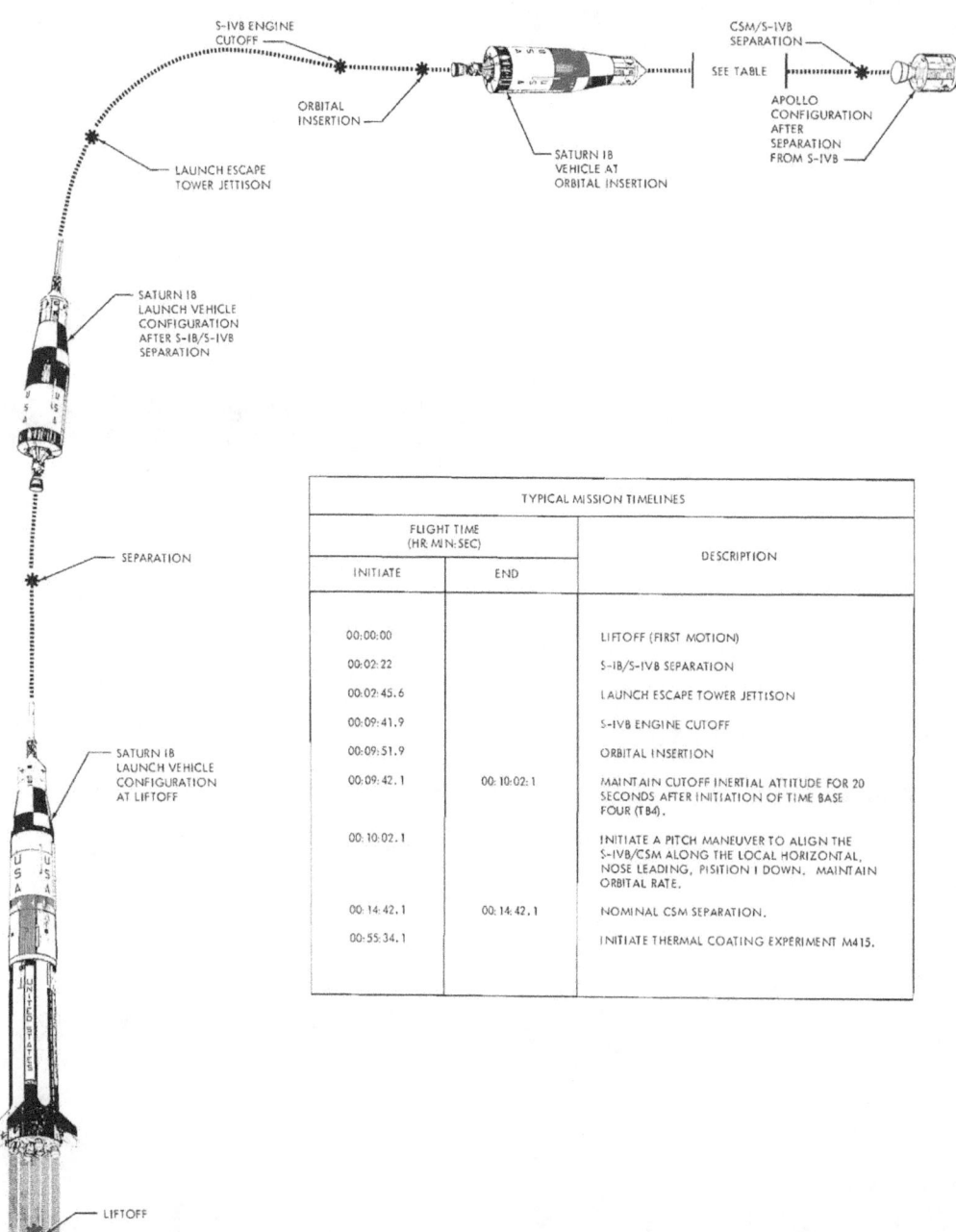

TYPICAL MISSION TIMELINES		
FLIGHT TIME (HR:MIN:SEC)		DESCRIPTION
INITIATE	END	
00:00:00		LIFTOFF (FIRST MOTION)
00:02:22		S-IB/S-IVB SEPARATION
00:02:45.6		LAUNCH ESCAPE TOWER JETTISON
00:09:41.9		S-IVB ENGINE CUTOFF
00:09:51.9		ORBITAL INSERTION
00:09:42.1	00:10:02.1	MAINTAIN CUTOFF INERTIAL ATTITUDE FOR 20 SECONDS AFTER INITIATION OF TIME BASE FOUR (TB4).
00:10:02.1		INITIATE A PITCH MANEUVER TO ALIGN THE S-IVB/CSM ALONG THE LOCAL HORIZONTAL, NOSE LEADING, POSITION I DOWN, MAINTAIN ORBITAL RATE.
00:14:42.1	00:14:42.1	NOMINAL CSM SEPARATION.
00:55:34.1		INITIATE THERMAL COATING EXPERIMENT M415.

Figure 9-6

APPENDIX A
ABBREVIATIONS, SIGNS, AND SYMBOLS

ABBREVIATIONS.

A

A	ampere
ABMA	Army Ballistic Missile Agency
ac	alternating current
accel	accelerometer
ACE	automatic checkout equipment
ACM	actuation control module
AFETR	Air Force Eastern Test Range
alt	altitude
AM	Airlock Module
amb	ambient
AN	ascending node
AOA	angle of attack
approx	approximately
APS	auxiliary propulsion system
ARPA	Advance Research Projects Agency
ASAP	auxiliary storage and playback
ASC	accelerometer signal conditioner
ASD	abort summary document
ASI	augmented spark igniter
assy	assembly
ASTM	American Society for Testing Materials
ATM	Apollo Telescope Mount
atm	atmosphere
att	attenuator or attitude
auto	automatic
aux	auxiliary
avg	average
AVP	address verification pulse

B

BEF	blunt end forward
bhp	brake horsepower
BMAG	body-mounted attitude gyro
bps	bits per second
BSE	booster systems engineer
BTU	British thermal unit

C

calib	calibration
calips	calibrational pressure switch
CAPCOM	spacecraft communicator
CASTS	Countdown and Status Transmission Systems
CAT	control attenuating timer
C/B	circuit breaker
CCW	counterclockwise
C/D	collect/disperse (computer)
CDC	countdown clock
CDDT	countdown demonstration test
CDF	confined detonating fuse
CDSC	Communication Distribution and Switching Center
CDU	coupling data unit
CEI	contract end item
CG	center of gravity
chan	channel
char	characteristics
CIF	Central Instrumentation Facility
CIU	computer interface unit
CKAFS	Cape Kennedy Air Force Station
cm	centimeter
CM	command module
CMC	command module computer
cmd	command
CMR	command module receiver
COD	cross-over detectors
cont	control
convtr	converter
CRES	corrosion resistant (steel)
CRP	computer reset pulse
CSM	command service module
CSP	control signal processor
CST	coast
CT	components test
C-T	crawler-transporter
CW	clockwise
cx/ct	control transmitter/control transformer

D

db	decibel
DB	disagreement bit
dc	direct current
DCS	digital command system
dev	deviation
DDAS	digital data acquisition system
decr	decreasing
DEE	digital events evaluator
deg	degree, angular
dest	destruct
DI	discrete input
dia	diameter
dir	directional
distr	distributor
DN	descending node
DO	discrete output
DOD	Department of Defense
DPF	dynamic pressure feedback
DRS	data receiving station
dyn	dynamic

E

EBW	exploding bridgewire
ECS	environmental control system
EDS	emergency detection system
elec	electrical
ELS	earth landing system
EMR	engine mixture ratio
EMS	entry monitor system
eng	engine
equip	equipment

A-1

Appendix A

err	error		IFV	igniter fuel valve
ESE	electrical support equipment		IGM	iterative guidance mode
ETD	end thrust decay		IMCC	Integrated Mission Control Center
ETR	Eastern Test Range		imp	impulse
			IMU	inertial measurement unit
			IMV	ignition monitor valve
			in.	inch

F

FABU	fuel additive blender unit		inbd	inboard
FCC	flight control computer		incr	increasing
FCSM	flight combustion stability monitor		ind	indication
FCVB	flow control valve box		instl	installation
FDAI	flight director attitude indicator		instr	instrumentation
FIDO	flight dynamics officer		int	internal
FLSC	flexible linear-shaped charge		invtr	inverter
flt	flight		IOA	input/output address
FM	frequency modulation		IODC	input/output data channel
F/M	thrust acceleration (force/mass)		IOR	input/output register (buffer)
FOD	Flight Operations Director		IOS	input/output sense
FOMR	Flight Operations Management Room		IP	impact prediction
FPR	flight performance reserve		IU	Instrument Unit
FRR	flight readiness report			
FRT	flight readiness test			
ft	feet or foot			
fwd	forward			

J

J-2	Saturn IB second stage (S-IVB) engine
jett	jettison

G

g	gram			
G	gravitational constant			
gal	gallon		k	kilo (prefix)
GCS	guidance cutoff signal		KSC	Kennedy Space Center
GDC	gyro display coupler			
gen	generator			

K

L

GG	gas generator		lb	pound
GMTC	Greenwich mean time clock		lbf	pound (force)
G&N	guidance and navigation		lbm	pound (mass)
gnd	ground		LCC	launch control center
GOX	gaseous oxygen		LE	launch escape
gpf	grains per foot		LES	launch escape system
gpm	gallons per minute		LET	launch escape tower
gr	grain		LIEF	Launch Information Exchange Facility
GRR	guidance reference release			
GSCU	ground support cooling unit		LLS	liquid level sensor
GSE	ground support equipment		LM	lunar module
GSFC	Goddard Space Flight Center		lox	liquid oxygen
guid	guidance		L.P.	low pressure
			LPGG	liquid propellant gas generator
			LSC	linear shaped charge

H

HEP	hardware evaluation program		LUT	launcher umbilical tower
HGA	hazardous gas analyzer		LV, L/V	launch vehicle
HOSC	Huntsville Operations Support Center		LVDA	launch vehicle data adapter
			LVDC	launch vehicle digital computer
H.P.	high pressure		LVO	launch vehicle operations
hp	horsepower			
hr	hour			

M

H/W	hardwire		m	milli (prefix) or meter
hyd	hydraulic		M	mega (prefix)
Hz	hertz		man	manual
H-1	Saturn IB first stage (S-IB) engine		max	maximum
H-1C	S-IB stage inboard engine		MCC	Mission Control Center
H-1D	S-IB stage outboard engine		MD	mission director
			MDA	Multiple Docking Adapter
			MDC	main display console

I

IA	input axis		MDF	mild detonating fuse
ICC	Interstate Commerce Commission		meas	measurement
ID	identification or inside diameter		med	medium
IECO	inboard engine cutoff		MESC	master events sequence controller
IF	intermediate frequency		MFCO	manual fuel cutoff

Appendix A

MFCV	modulating flow control valve	POS	position
MFV	main fuel valve	POST	priority of systems tests
MGSE	mechanical ground support equip	ppm	parts per million
mi	mile	prep	preparation
MILA	Merritt Island Launch Activity	press	pressure
min	minute or minimum	prop	propellant
ml	milliliter	prplnt	propellant
ML	mobile launcher	psi	pounds per square inch
MLV	main lox valve	psia	pounds per square inch (atmospheric)
MMH	monomethylhydrazine		
mo	month	psid	pounds per square inch (differential)
MOCR	Mission Operations Control Room		
mod	module or model	psig	pounds per square inch (gage)
MOV	main oxidizer value		
MRCV	mixture ratio control valve	PTCS	propellant tanking computer system
ms	millisecond		
MSC	Manned Spacecraft Center	PU	propellant utilization
MSFC	Marshall Space Flight Center	pwr	power
MSO	Mission Support Operations		
MSS	mobile service structure		

Q

MTVC	manual thrust vector control	Q	dynamic pressure
MUX	multiplexer	Qa	angle-of-attack/dynamic pressure product
		Q-D	quick disconnect

N

R

N	Newton		
N/A	not applicable or not available	RACS	remote automatic calibration system
NASA	National Aeronautics and Space Administration		
		RAD (rad)	a circle segment equal to 180 deg/π
NC	normally closed		
neg	negative	RASM	remote analog submultiplexer
N.G. & C	navigation, guidance, & control		
NM	nautical mile	RC	resistor capacitor
No.	number	R. CAL	radiation calorimeter
NO	normally open	RCC	Recovery Command and Control Center
NP	north pole		
NPSH	net pressure suction head	RCS	reaction control system
NPV	non-propulsive vent	RCVR	receiver
		R&D	research and development
		RDM	remote digital multiplexer

O

OA	output axis	RDSM	remote digital submultiplexer
OAT	overall test		
OECO	outboard engine cutoff	RDX	cyclotrimethylene trinitramine
OETD	outboard engine thrust decay		
OIS	Operations Intercommunications System	ref	reference
		reg	regulator
ord	ordnance	rev	revolution
osc	oscillator	RF	radio frequency
OSR	Operations Support Room	RFI	radio frequency interference
OWS	Orbital Workshop		
oxid	oxidizer	RNG	ranging
oz	ounce (torque)	rpm	revolutions per minute
		RP-1	rocket propellant grade 1 (fuel)

P

		RS	range safety
PAM	pulse amplitude modulation	RSCR	range safety command receiver
PC	pitch control		
PCD	pneumatic control distributor	RSO	range safety officer
PCM	pulse code modulation	rss	root-sum-square
PDS	propellant dispersion system	RTCC	Real-Time Computer Complex
PEA	platform electronics assembly		

S

PETN	pentaerythritol tetranitrate	SA	Saturn or service arm
PIF	Public Information Facility	S&A	safing and arming
pkg	package	S/C	spacecraft
pneu	pneumatics	scfm	standard cubic feet per minute
pos	positive		

A-3

Appendix A

scim	standard cubic inches per minute	TCS	thermal conditioning system or terminal countdown sequencer
SCT	scanning telescope		
sec	second	temp	temperature
SECO	S-IVB engine cutoff	TER	telemetry executive routine
sel	selector		
sep	separation	TLC	simultaneous memory error
seq	sequence		
sf	scaling factor	TLI	translunar injection
S-IB	first stage of Saturn IB launch vehicle	TM	telemetry
		TMR	triple modular redundancy
sig	signal	TOPS	thrust ok pressure switch
SIR	systems integration rack		
S-IVB	second stage of Saturn IB launch vehicle	turb	turbine
		TV	television
SL	Skylab	TVC	thrust vector control
SLA	spacecraft lunar (module) adapter	twr	tower
		typ	typical
SLDS	Skylab Launch Data System		
SM	service module		
SMC	steering misalignment corrections	UDMH	unsymmetrical dimethylhydrazine
SOCR	Sustained Operations Control Room	UHF	ultra-high frequency
		umb	umbilical
sol	solenoid	U/R	ullage rocket
sp	specific	USB	unified S-band
SPAN	spacecraft analysis		
SPGG	solid propellant gas generator		V
		V	volt or velocity
SPS	service propulsion system	VAB	vehicle assembly building
SRA	spin reference axis	Vac	volts alternating current
SSR	Staff Support Room	VCS	Voice Communications System
sta	station		
stby	standby	Vdc	volts direct current
std	standard	VHF	very-high frequency
STDN	Spacecraft Tracking and Data Network	Vpp	volts peak-to-peak
STDV	start tank discharge valve	VSWR	voltage standing wave ratio
STP	standard pressure		W
sw	switch		
SWS	Saturn Workshop	W	watt
sys	system	wt	weight
	T		X
TB	time base	XFER	transfer
TC	thrust chamber	XMTR	transmitter

Appendix A

SIGNS AND SYMBOLS.

@	at
°API	degrees (scale) American Petroleum Institute
°C	degrees Celcius
°F	degrees Fahrenheit
°R	degrees Rankin
℄	centerline
GHe	gaseous helium
GH_2	gaseous hydrogen
GN_2	gaseous nitrogen
He	helium
H_2O	water
KOH	potassium hydroxide
LH_2	liquid hydrogen
N_2O_4	nitrogen tetroxide
q	aerodynamic pressure
∠	angle
Δ	change
≥	greater than or equal to
≤	less than or equal to
>	greater than
<	less than
~	difference
≈	nearly equal to
≡	defined as
Ω	ohm
α	alpha, angle of attack
β	beta, feedback signals from H-1 engine actuators
$β_c$	steering command output
$\ddot{γ}$	gamma, lateral acceleration
θ	theta, platform gimbal angle
$θ_x θ_y θ_z$	platform gimbal angle (vehicle attitude)
λ	lambda, longitude
μ	mu, micro (prefix)
σ	sigma, an occurrence probability symbol expressing a percentage of all possible values in a given parameter.
$\ddot{τ}$	tau, control accelerometer signals
φ	phi, attitude rate
L	launch site latitude
Χ	chi, desired attitude change
$X_r X_p X_y$	guidance command angles (desired vehicle attitude)
$X_x X_y X_z$	guidance command angles (Euler angles)
$\tilde{X}$	thrust direction definition
DX	average change for each minor loop guidance computation
ψ	psi, attitude error
$ψ_r ψ_p ψ_y$	attitude error signals (steering commands)
ψT	total range angle
A_z	azimuth
i	inclination of orbit
P R Y	pitch, roll, and yaw axes
U V W	gravitational coordinate system
$\dot{X}_i$	inertial velocity
$\dot{x} \dot{y} \dot{z}$	incremental velocity
X Y Z	platform gimbal pivot axes
$X_I Y_I Z_I$	measurement coordinate system
$X_s Y_s Z_s$	inertial coordinate system
$X_i Y_i Z_i$	injection coordinate system
$X_4 Y_4 Z_4$ system	

Logic Symbols.

- AND gate
- OR gate
- time delay in sec
- signal, presence of
- signal, absence of

©2013 Periscope Film LLC
All Rights Reserved
ISBN#978-1-937684-20-4
www.PeriscopeFilm.com

www.ingramcontent.com/pod-product-compliance
Lightning Source LLC
Chambersburg PA
CBHW082113230426
43671CB00015B/2684